T0142331

Lecture Notes on Data Engineering and Communications Technologies

Volume 118

Series Editor

Fatos Xhafa, Technical University of Catalonia, Barcelona, Spain

The aim of the book series is to present cutting edge engineering approaches to data technologies and communications. It will publish latest advances on the engineering task of building and deploying distributed, scalable and reliable data infrastructures and communication systems.

The series will have a prominent applied focus on data technologies and communications with aim to promote the bridging from fundamental research on data science and networking to data engineering and communications that lead to industry products, business knowledge and standardisation.

Indexed by SCOPUS, INSPEC, EI Compendex.

All books published in the series are submitted for consideration in Web of Science.

More information about this series at https://link.springer.com/bookseries/15362

Leonard Barolli · Elis Kulla · Makoto Ikeda
Editors

Advances in Internet, Data & Web Technologies

The 10th International Conference on Emerging Internet, Data and Web Technologies (EIDWT-2022)

 Springer

Editors
Leonard Barolli
Department of Information
and Communication Engineering
Fukuoka Institute of Technology
Fukuoka, Japan

Elis Kulla
Department of Information
and Computer Engineering
Okayama University of Science
Okayama, Japan

Makoto Ikeda
Department of Information
and Communication Engineering
Fukuoka Institute of Technology
Fukuoka, Japan

ISSN 2367-4512   ISSN 2367-4520   (electronic)
Lecture Notes on Data Engineering and Communications Technologies
ISBN 978-3-030-95902-9   ISBN 978-3-030-95903-6   (eBook)
https://doi.org/10.1007/978-3-030-95903-6

This Springer imprint is published by the registered company Springer Nature Switzerland AG
The registered company address is: Gewerbestrasse 11, 6330 Cham, Switzerland

Welcome Message of EIDWT-2022 International Conference Organizers

Welcome to the 10th International Conference on Emerging Internet, Data and Web Technologies (EIDWT-2022), which will be held from March 2 to March 4, 2022, at Okayama University of Science, Okayama, Japan.

The EIDWT is dedicated to the dissemination of original contributions that are related to the theories, practices and concepts of emerging Internet and data technologies yet most importantly of their applicability in business and academia toward a collective intelligence approach.

In EIDWT-2022 will be discussed topics related to Information Networking, Data Centers, Data Grids, Clouds, Crowds, Mashups, Social Networks, Security Issues and other Web 2.0 implementations toward a collaborative and collective intelligence approach leading to advancements of virtual organizations and their user communities. This is because, current and future Web and Web 2.0 implementations will store and continuously produce a vast amount of data, which if combined and analyzed through a collective intelligence manner will make a difference in the organizational settings and their user communities. Thus, the scope of EIDWT-2022 includes methods and practices which bring various emerging Internet and data technologies together to capture, integrate, analyze, mine, annotate and visualize data in a meaningful and collaborative manner. Finally, EIDWT-2022 aims to provide a forum for original discussion and prompt future directions in the area.

An international conference requires the support and help of many people. A lot of people have helped and worked hard for a successful EIDWT-2022 technical program and conference proceedings. First, we would like to thank all authors for submitting their papers. We are indebted to program area chairs, program committee members and reviewers who carried out the most difficult work of carefully evaluating the submitted papers. We would like to give our special thanks to Honorary Chair of EIDWT-2022 Prof. Makoto Takizawa, Hosei University,

Japan, for his guidance and support. We would like to express our appreciation to our keynote speakers for accepting our invitation and delivering very interesting keynotes at the conference.

EIDWT-2022 Organizing Committee

Honorary Chair

Makoto Takizawa   Hosei University, Japan

General Co-chairs

Kengo Katayama   Okayama Univ. of Science, Japan
Juggapong Natwichai   Chiang Mai University, Thailand

Program Co-chairs

Elis Kulla   Okayama University of Science, Japan
Omar Hussain   Univ. of New South Wales, Australia

International Advisory Committee

Janusz Kacprzyk   Polish Academy of Sciences, Poland
Arjan Durresi   IUPUI, USA
Wenny Rahayu   La Trobe University, Australia
Fang-Yie Leu   Tunghai University, Taiwan
Yoshihiro Okada   Kyushu University, Japan

Publicity Co-chairs

Tomoya Enokido   Rissho University, Japan
Kin Fun Li   University of Victoria, Canada
Keita Matsuo   Fukuoka Institute of Technology, Japan
Pruet Boonma   Chiang Mai University, Thailand
Flora Amato   Naples University "Frederico II," Italy

International Liaison Co-chairs

David Taniar	Monash University, Australia
Admir Barolli	Alexander Moisiu University, Albania
Santi Caballé	Open University of Catalonia, Spain
Farookh Hussain	Univ. Technology Sydney, Australia
Nadeem Javaid	COMSATS University Islamabad, Pakistan

Local Organizing Committee Co-chairs

Akira Uejima	Okayama University of Science, Japan
Tetsuya Oda	Okayama University of Science, Japan
Masaharu Hirota	Okayama University of Science, Japan

Web Administrators

Kevin Bylykbashi	Fukuoka Institute of Technology, Japan
Ermioni Qafzezi	Fukuoka Institute of Technology, Japan
Phudit Ampririt	Fukuoka Institute of Technology, Japan

Finance Chair

Makoto Ikeda	Fukuoka Institute of Technology, Japan

Steering Committee Chair

Leonard Barolli	Fukuoka Institute of Technology, Japan

PC Members

Akimitsu Kanzaki	Shimane University, Japan
Akio Koyama	Yamagata University, Japan
Akira Uejima	Okayama University of Science, Japan
Alba Amato	National Research Council (CNR)-Institute for High-Performance Computing and Networking (ICAR), Italy
Alberto Scionti	LINKS, Turin, Italy
Antonella Di Stefano	University of Catania, Italy
Arcangelo Castiglione	University of Salerno, Italy
Beniamino Di Martino	Università della Campania "Luigi Vanvitelli," Italy
Bhed Bista	Iwate Prefectural University, Japan
Carmen de Maio	University of Salerno, Italy
Chotipat Pornavalai	King Mongkut's Institute of Technology Ladkrabang, Thailand

Pruet Boonma   Chiang Mai University, Thailand
Raffaele Pizzolante   University of Salerno, Italy
Sajal Mukhopadhyay   National Institute of Technology, Durgapur, India
Salvatore Ventiqincue   University of Campania Luigi Vanvitelli, Italy
Sazia Parvin   Deakin University, Australia
Shigetomo Kimura   University of Tsukuba, Japan
Shinji Sugawara   Chiba Institute of Technology, Japan
Shinji Sakamoto   Kanazawa Institute of Technology, Japan
Sotirios Kontogiannis   University of Ioannina, Greece
Teodor Florin Fortis   West University of Timisoara, Romania
Tomoki Yoshihisa   Osaka University, Japan
Tomoya Enokido   Rissho University, Japan
Tomoya Kawakami   NAIST, Japan
Toshihiro Yamauchi   Okayama University, Japan
Toshiya Takami   Oita University, Japan
Xu An Wang   Engineering University of CAPF, China
Yoshihiro Okada   Kyushu University, Japan

EIDWT-2022 Reviewers

Amato Flora   Kulla Elis
Amato Alba   Leu Fang-Yie
Barolli Admir   Matsuo Keita
Barolli Leonard   Koyama Akio
Bista Bhed   Ogiela Lidia
Chellappan Sriram   Ogiela Marek
Chen Hsing-Chung   Okada Yoshihiro
Cui Baojiang   Palmieri Francesco
Di Martino Beniamino   Paruchuri Vamsi Krishna
Enokido Tomoya   Rahayu Wenny
Fun Li Kin   Spaho Evjola
Gotoh Yusuke   Sugawara Shinji
Hussain Farookh   Takizawa Makoto
Hussain Omar   Taniar David
Javaid Nadeem   Terzo Olivier
Ikeda Makoto   Uehara Minoru
Ishida Tomoyuki   Venticinque Salvatore
Kikuchi Hiroaki   Wang Xu An
Kolici Vladi   Woungang Isaac
Koyama Akio   Xhafa Fatos

EIDWT-2022 Keynote Talks

Mining of Cohesive Groups in Massive Social and Web Graphs

Alex Thomo

University of Victoria, British Columbia, Canada

Abstract. Mining dense subgraphs and discovering hierarchical relations between them is a fundamental problem in graph analysis tasks. For instance, it can be used in visualizing complex networks, finding correlated genes and motifs in biological networks, detecting communities in social and Web graphs, summarizing text and revealing new research subjects in citation networks. Core, truss and nucleus decompositions are popular tools for finding dense subgraphs. A k-core is a maximal subgraph in which each vertex has at least k-neighbors, and a k-truss is a maximal subgraph whose edges are contained in at least k-triangles. Core and truss decompositions have been extensively studied in both deterministic as well as probabilistic graphs. A more recent notion of dense subgraphs is nucleus decomposition which is a generalization of core and truss decompositions that uses higher-order structures to detect dense regions in the graph. In this talk, I will first motivate and illustrate core, truss, and nucleus decompositions for mining dense hierarchical regions in large graphs. Next, I will describe algorithms for computing these decompositions and outline avenues for further research.

Human Centered Approaches in Transformative Computing Applications

Lidia Dominika Ogiela

AGH University of Science and Technology, Krakow, Poland

Abstract. Human centered systems are now recognized as one of the most important solutions in artificial intelligence. They have advantage over other systems from the fact that they still adapt their operation to the changing and unpredictable tasks and functions. The variability of the human analysis process, which is the basis for the operation of such systems, means that the developed IT solutions are constantly evolving, and their development is a determinant of various external factors independent of humans and those that depend on them. Human centered systems allow for the implementation of deep tasks, meaningful analysis and interpretation of various data sets. Their special advantage is the possibility of incorporating characteristic of the human perception processes of automatic data prediction. In human centered systems, transformative computing processes are also carried out, giving the possibility of implementing analysis steps at various levels of inference. The differentiation of the levels at which the interpretation and inference processes are carried out is a characteristic of complex data management structure.

Contents

Data Service Platform for Social and Community to Drive the Royal Project Foundation 1
Suphatchaya Autarrom, Kittayaporn Chantaranimi,
Anchan Chompupoung, Pichan Jinapook, Waranya Mahanan,
Pathathai Na Lumpoon, Juggapong Natwichai, Prompong Sugunsil,
Sumalee Sangamuang, Titipat Sukhvibul, and Pree Thiengburanathum

Implementation of a Local-Community Issues Visualization System Using Open Data and Future Population Projection 11
Tomoyuki Ishida and Mutsuki Kojima

SAE+Bi-GRU Based Security Situation Prediction for Smart Grid 21
Lei Chen, Mengyao Zheng, Zhaohua Liu, Fadong Chen, Kui Zhou,
and Bin Liu

Design of Identity Authentication Scheme for Dynamic Service Command System Based on SM2 Algorithm and Blockchain Technology 31
Jie Deng, Lili Jiao, Lili Zhang, Yongjin Ren, and Wengang Yin

Visual Authentication Codes Generated Using Predictive Intelligence 38
Urszula Ogiela, Makoto Takizawa, and Marek R. Ogiela

Reliable Network Design Problem by Improving Node Reliability 42
Hiroki Yano, Sumihiro Yoneyama, and Hiroyoshi Miwa

Toward Secure K-means Clustering Based on Homomorphic Encryption in Cloud 52
Zheng Tu, Xu An Wang, Yunxuan Su, Ying Li, and Jiasen Liu

On the Insecurity of a Certificateless Public Verification Protocol for the Outsourced Data Integrity in Cloud Storage 63
Xu An Wang, Xiaozhong Pan, Lixian Wei, and Yize Zhao

An Improved Density Peaks-Based Graph Clustering Algorithm 68
Lei Chen, Heding Zheng, Zhaohua Liu, Qing Li, Lian Guo,
and Guangsheng Liang

**Community Division Algorithm Based on Node Similarity
and Multi-attribute Fusion** 81
Du Tiansi, Deng Na, and Chen Weijie

**Research on TCM Patent Annotation to Support Medicine R&D
and Patent Acquisition Decision-Making** 91
Du Tiansi, Deng Na, and Chen Weijie

**An Algorithm for GPS Trajectory Compression Preserving Stay
Points** 102
Shota Iiyama, Tetsuya Oda, and Masaharu Hirota

**Blockchain for Islamic HRM: Potentials and Challenges on
Psychological Work Contract** 114
Olivia Fachrunnisa and Fannisa Assyilah

**Human-Value Orientation as Center for Business Transformation
Model in Digital Era** 123
Ardian Adhiatma and Nurhidayati

**An Energy-Efficient Algorithm to Make Virtual Machines Migrate
in a Server Cluster** 130
Dilawaer Duolikun, Tomoya Enokido, Leonard Barolli,
and Makoto Takizawa

**Energy Consumption Model of a Device Supporting Information Flow
Control in the IoT** 142
Shigenari Nakamura, Tomoya Enokido, and Makoto Takizawa

**A Fuzzy-Based System for Assessment of QoS of V2V Communication
Links in SDN-VANETs** 153
Ermioni Qafzezi, Kevin Bylykbashi, Phudit Ampririt, Makoto Ikeda,
Keita Matsuo, and Leonard Barolli

**Reliable and Low-Cost Digital Transformation Technology Using
Progressive Web Apps in Fog Computing Architecture for Small
and Medium Industries in Indonesia** 163
Zulkifli Tahir, Amil Ahmad Ilham, Ais Prayogi Alimuddin,
Muhammad Zulfadly A. Suyuti, and Charina

**A Low-Cost Solution for Smart-City Based on Public Bus
Transportation System Using Opportunistic IoT** 175
Evjola Spaho and Andrea Koroveshi

A ML-Based System for Predicting Flight Coordinates Considering
ADS-B GPS Data: Problems and System Improvement 183
Kazuma Matsuo, Makoto Ikeda, and Leonard Barolli

Fault Detection from Bend Test Images of Welding Using Faster
R-CNN 190
Shigeru Kato, Takanori Hino, Hironori Kumeno, Shunsaku Kume,
Tomomichi Kagawa, and Hajime Nobuhara

An Efficient Local Search for the Maximum Clique Problem
on Massive Graphs. 201
Kazuho Kanahara, Tetsuya Oda, Elis Kulla, Akira Uejima,
and Kengo Katayama

A Method for Reducing Number of Parameters of Octave
Convolution in Convolutional Neural Networks 212
Yusuke Gotoh and Yu Inoue

A Comparison Study of RIWM with RDVM and CM Router
Replacement Methods for WMNs Considering Boulevard Distribution
of Mesh Clients 223
Admir Barolli, Phudit Ampririt, Shinji Sakamoto, Elis Kulla,
and Leonard Barolli

A Fuzzy-Based System for Safe Driving in VANETs Considering
Impact of Driver Impatience on Stress Feeling Level 236
Kevin Bylykbashi, Ermioni Qafzezi, Phudit Ampririt, Makoto Ikeda,
Keita Matsuo, and Leonard Barolli

Mobility-Aware Narrow Routing Protocol for Underwater Wireless
Sensor Networks 245
Elis Kulla, Kuya Shintani, and Keita Matsuo

Design and Implementation of a Testbed for Delay Tolerant
Networks: Work in Progress 254
Kuya Shintani, Elis Kulla, Makoto Ikeda, Leonard Barolli,
and Evjola Spaho

Evaluation of Focused Beam Routing Protocol on Delay Tolerant
Network for Underwater Optical Wireless Communication 263
Keita Matsuo, Elis Kulla, and Leonard Barolli

A Fuzzy-Based System for Slice Service Level Agreement in 5G
Wireless Networks: Effect of Traffic Load Parameter 272
Phudit Ampririt, Ermioni Qafzezi, Kevin Bylykbashi, Makoto Ikeda,
Keita Matsuo, and Leonard Barolli

A River Monitoring and Predicting System Considering a Wireless Sensor Fusion Network and LSTM 283
Yuki Nagai, Tetsuya Oda, Tomoya Yasunaga, Chihiro Yukawa, Aoto Hirata, Nobuki Saito, and Leonard Barolli

Social Experiment of Realtime Road State Sensing and Analysis for Autonomous EV Driving in Snow Country 291
Yositaka Shibata, Akira Sakuraba, Yoshikazu Arai, Yoshiya Saito, and Noriki Uchida

A Soldering Motion Analysis System for Danger Detection Considering Object Detection and Attitude Estimation 301
Tomoya Yasunaga, Tetsuya Oda, Nobuki Saito, Aoto Hirata, Chihiro Yukawa, Yuki Nagai, and Masaharu Hirota

Performance Evaluation of a Soldering Training System Based on Haptics 308
Kyohei Toyoshima, Tetsuya Oda, Chihiro Yukawa, Tomoya Yasunaga, Aoto Hirata, Nobuki Saito, and Leonard Barolli

Performance Evaluation of WMNs by WMN-PSOHC Hybrid Simulation System Considering Two Instances and Normal Distribution of Mesh Clients 316
Shinji Sakamoto and Leonard Barolli

The Principal Dimensions Optimization of Large Ships Based on Improved Firefly Algorithm 324
Jianghao Yin and Na Deng

Improved Butterfly Optimization Algorithm Fused with Beetle Antennae Search 335
Jianghao Yin and Na Deng

A Delaunay Edge and CCM-Based SA Approach for Mesh Router Placement Optimization in WMN: A Case Study for Evacuation Area in Okayama City 346
Aoto Hirata, Tetsuya Oda, Nobuki Saito, Tomoya Yasunaga, Kengo Katayama, and Leonard Barolli

FPGA Implementation of a Interval Type-2 Fuzzy Inference for Quadrotor Attitude Control 357
Tomoaki Matusi, Tetsuya Oda, Chihiro Yukawa, Tomoya Yasunaga, Nobuki Saito, Aoto Hirata, and Leonard Barolli

Design of a Robot Vision System for Microconvex Recognition 366
Chihiro Yukawa, Tetsuya Oda, Nobuki Saito, Aoto Hirata, Tomoya Yasunaga, Kyohei Toyoshima, and Kengo Katayama

Path Control Algorithm for Weeding AI Robot   375
Misato Shiba and Hiroyoshi Miwa

**Performance Analysis of RIWM and RDVM Router Replacement
Methods for WMNs by WMN-PSOSA-DGA Hybrid Simulation
System Considering Stadium Distribution of Mesh Clients**   386
Admir Barolli, Shinji Sakamoto, and Leonard Barolli

**An Energy-Efficient Process Replication to Reduce the Execution
of Meaningless Replicas**   395
Tomoya Enokido, Dilawaer Duolikun, and Makoto Takizawa

A Byzantine Fault Tolerant Protocol for Realizing the Blockchain   406
Akihito Asakura, Shigenari Nakamura, Dilawaer Duolikun,
Tomoya Enokido, Kuninao Nashimoto, and Makoto Takizawa

**Performance Evaluation of a DQN-Based Autonomous Aerial Vehicle
Mobility Control Method in an Indoor Single-Path Environment with
a Staircase**   417
Nobuki Saito, Tetsuya Oda, Aoto Hirata, Chihiro Yukawa,
Masaharu Hirota, and Leonard Barolli

**Practical Survey on MapReduce Subgraph
Enumeration Algorithms**   430
Xiaozhou Liu, Yudi Santoso, Venkatesh Srinivasan, and Alex Thomo

**Identifying Vehicle Exterior Color by Image Processing
and Deep Learning**   445
Somayeh Abniki, Kin Fun Li, and Tom Avant

Author Index   459

Data Service Platform for Social and Community to Drive the Royal Project Foundation

Suphatchaya Autarrom[1], Kittayaporn Chantaranimi[2],
Anchan Chompupoung[1], Pichan Jinapook[1], Waranya Mahanan[3(✉)],
Pathathai Na Lumpoon[3], Juggapong Natwichai[4], Prompong Sugunsil[3],
Sumalee Sangamuang[3], Titipat Sukhvibul[5], and Pree Thiengburanathum[3]

[1] Royal Project Foundation, Chiang Mai, Thailand
[2] Data Science Consortium, Faculty of Engineering, Chiang Mai University,
Chiang Mai, Thailand
kittayaporn_c@cmu.ac.th
[3] College of Arts, Media and Technology, Chiang Mai University,
Chiang Mai, Thailand
{waranya.m,pathathai.n,prompong.sugunnasil,sumalee.sa,
pree.t}@cmu.ac.th
[4] Department of Computer Engineering, Faculty of Engineering,
Chiang Mai University, Chiang Mai, Thailand
juggapong.n@cmu.ac.th
[5] Information Technology Section, Faculty of Engineering, Chiang Mai University,
Chiang Mai, Thailand
titipat_sukhvibul@cmu.ac.th

Abstract. The Royal Project Foundation is an organization in Thailand which works on various non-profit projects since 1969. Social development, such as population structure, drug problem, educational development and community organization, is one of the foundation's focus areas. Thus, the social data has been collected with various data sources and structures. In this paper, the data service platform for the social and community is proposed to maintain the collection of various data sources and structures. The proposed architecture consists of 3 different data stores 1) data lake, 2) data staging data and data warehouse and 3) CKAN. With the proposed hybrid architecture, the various data types are allowed to be stored in the data lake. Then, the data staging area and data warehouse are used for well-defined schema data. The CKAN store the metadata and data dictionaries of the Royal Project Foundation's data. Furthermore, the prototypes of interactive business intelligence dashboards are developed in order to visualize a summary of the population, education, and public health of Royal Project Foundation's data.

© The Author(s), under exclusive license to Springer Nature Switzerland AG 2022
L. Barolli et al. (Eds.): EIDWT 2022, LNDECT 118, pp. 1–10, 2022.
https://doi.org/10.1007/978-3-030-95903-6_1

1   Introduction

1.1   The Royal Project Foundation

The Royal Project Foundation [9] is the Thai organization founded by King Bhumibol Adulyadej in 1969. This foundation was started during his visit to Doi Pui, the mountain located in northern Thailand, hill tribe's village. He discovered that the hill tribes were growing opium poppy for a living. Because of the disadvantages of opium poppy, he suggested the hill tribes to replace opium-growing with the local peach which provides a higher income. Thus, the Royal Project Development Center was started with the following objectives, help the hill tribes, reduce the destruction of natural resources, eliminate the opium poppy cultivation and preserve the soil and use the area properly. In 1972, the UN/Thai Program for Drug Abuse Control project was started. The project was supported by the UN due to the importance of alternative agriculture to replace opium poppy cultivation. After that, the foundation was supported by international organizations. The USDA/ARS funded the project for research on highland agriculture in 1973. The research studied on the suitable type of plant and the method of planting in the highland. In 1992, the Royal Project Development Center be registered as the Royal Project Foundation to provide public benefit, support the flexible work system and the better efficiency.

Currently, the Royal Project Foundation has 39 development centers in six provinces in northern Thailand. The foundation is still focusing on research and developing various works to maintain a balance in terms of economy, society and environment. The objectives of the Royal Project Foundation are

1. To research and develop knowledge that is suitable for the highlanders
2. To develop communities in the highlands for better living and self-reliance
3. To promote, restore and conserve the environment
4. To promote and develop post-harvest management and marketing of Royal Project products
5. To develop the Royal Project's learning center for highlanders.

The Royal Project Foundation works on various projects to accomplish the foundation's objectives, supporting the development of society is one of them. The foundation helps underprivileged children to have a better life and also improve the life quality of children and women by providing education and career. The foundation also educates and collects knowledge on handicrafts of various tribes. Then, restore and transfer the knowledge from seniors to younger generations. Moreover, promote the handicraft products and sell them as souvenirs to tourists for households additional income. Strengthening of self-reliance groups in various forms by bringing together groups of people to engage in various activities together such as farmers' groups, housewives' groups, savings groups and natural resource conservation groups, etc.

From the various works, the Royal Project Foundation has collected the data in terms of social development such as population structure, drug problem, educational development, community organization, information on the development of holistic health and information on the restoration and conservation of traditions, culture, local wisdom. Therefore, such data are collected in various data sources and structures.

Thus, in this paper, we study and design information systems to support social and community operations for Royal Projects Foundation by proposing a hybrid data store structure (data lake, data staging area, data warehouse and CKAN). The proposed architecture is designed to collect the various data sources and structures. We also develop a model of data linkage system and summarize the overview of social and community information in many dimensions.

The rest of this paper is organized as follows. Section 2 provides the previous works of big data, data acquisition, data lake and data warehouse. The proposed data store architecture is proposed in Sect. 3. Section 4 presents the exploratory data analysis of the data from the Royal Project Foundation. Finally, the conclusion of our work is given in Sect. 5.

2   Related Works

Big data refers to the amount and various data that grow at ever-increasing rates. According to the report by McKinsey Global Institute reports [15], Companies apply big data in their services systems, i.e., improve customer service and create personalized marketing campaigns to increase income and profits. Moreover, big data will become an essential basis of competition, underpinning new waves of productivity growth, innovation, and consumer surplus. Fundamental roles of big data in Fig. 1 are considered based on the research by [6] that consist of data acquisition, data storage such as data lake, or data warehouse, and data usage.

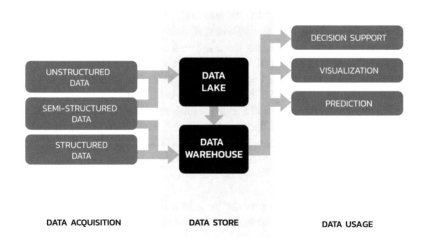

Fig. 1. Fundamental roles of big data

Data acquisition is the process of gathering, filtering, and cleaning data before putting the data into data storage. It is a necessary process and allows the data analysis tools to do their work correctly. According to Fig. 1, the type of big data can consider being structured data, semi-structured data, unstructured data. Most data analysis tools can immediately process the structured data, e.g., Spreadsheet, CSV, and data from relational databases because the structure of this data type is most fully defined. However, the data such as JSON files and XML files are structured to some extent but not complete. This structure can be defined to be semi-structured data. The semi-structured data can use for immediate analysis. Still, some parts are needed a process to extract the data according to the desired structure before it can load into the storage. Otherwise, documents containing numbers or letters which are freeform writing cannot use directly because freeform writing does not require any structure to identify. Therefore, the structure mentioned above defines unstructured data-for example, report files, video files, audio files, image files, and streaming files. The streaming [14] files are event processing which creates by social network services, e.g., Linkedin [1].

The aforementioned unstructured data can be put into the storage by the Extraction, Transformation, and Loading process or ETL [2,12,17]. Likewise, text recognition and natural language processing [13] can extract information from the unstructured data, then transforms and load it into the storage. Based on [2], the ETL's features are as follows: ease of use, the ability to apply multi-role team collaboration, significant volume performance, clustering, and job distribution, division data partitioning, functionality, automatic recovery of workflow, usability, re-usability, connections, i.e., connecting to SAP, Cloud, Hadoop, and performance.

Data lakes and data warehouses are both widely used for storing big data. A data warehouse is an information storage system for historical data that can analyze in numerous ways. Data warehouse platforms were proposed in [3,5, 10] to solve the problems of the relational database's scalability and its cost. According to [10], MapReduce has been applied in the data warehouse platform to improve efficiency in terms of computation time. Otherwise, a data lake is a repository for unstructured and semi-structured data that does still not already process for a specific purpose. In [8,11,20,21], a data lake uses as a storage source before loading data entry into the data warehouse. In addition, a distributed file system such as Hadoop Distributed File System can use to improve the complexity when the data is stored. Otherwise, CKAN is a data catalog platform that hosts two types of reference data: data dictionaries and metadata. In [4,16], the CKAN platform was used to extract and load the available data sets to the central storage.

Business intelligence is the process by which enterprises analyze current and historical data. In [18], this article aims to evaluate and publicize various Business Intelligence initiatives in the Portuguese Public Administration. Meanwhile, the incorporation of business intelligence with enterprise resource planning in SMEs was proposed in [19].

3 Architecture

As mentioned in the Sect. 1, the Royal Project Foundation has to work with various types of data. To handle the problem while maintaining a rich industrial-grade warehouse, a hybrid data store architecture has been designed as shown in Fig. 2. From the figure, the architecture consists of 3 data stores and 4 data processing pipelines, in-which will be described in Sect. 3.1 and 3.2, respectively.

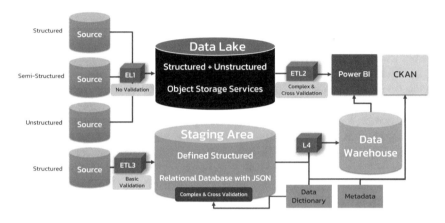

Fig. 2. Proposed data store architecture.

Note that the proposed architecture is still in the developing phase. When the implementation is finished, the efficiency of such architecture needs to be evaluated. A few example metrics to be proposed is turnaround time, duration since the queries are input to the system and results are returned. According to our requirement gathering, collecting unstructured data inside the Royal Foundation usually take a week, and it even takes more time for the paper-based documents. We also found that some data processing, which can off-load to automated data processing pipelines, are still performing manually. In conclusion, eliminating manual data collecting and data processing workloads is also a candidate to reduce the overall turnaround time of the system. In addition, the satisfaction of the data entry personals is to be measured.

3.1 Data Store

The first data store is a **data lake**. This data store allows the Royal Project Foundation to store various data types as object storage in the same place, i.e., Microsoft SharePoint. Then the Royal Project Foundation can use them as the data source in Microsoft Power BI for analytic purposes later. Since the data are in Microsoft SharePoint, employees can produce and consume the data with regular Microsoft Office solutions. Even putting every in the bucket and defining the structure later, i.e. a late-binding approach, allows starting collecting data instantaneously. However, it can have several issues in relational data validation and schema matching in the later analysis processes. Thus, we have designed a second data store for well-defined schema data.

The second data store is designed for well-defined schema data. The data store is consists of a **data staging area** and a **data warehouse**. Ingress data will be passing only basic validations then utilizing the relational database constraints before being temporarily stored in the staging area. Then, the data are interval load to the data warehouse for future analytic works. In addition, this data store also supports JSON-compatible or semi-structured data such as documented data.

The third data store is **CKAN**, a data catalog platform that hosts 2 types of reference data: **data dictionaries** and **metadata**. Therefore, 3rd party users can use those two reference data as data integration guidelines. The data dictionaries will describe the structure of published data. And on the other hand, metadata will describe miscellaneous information of each dataset such as tags, category, owner, the latest update, update interval, and so on. In addition, even we can use the CKAN as an ad-hoc data source, which 3rd party users can use CKAN as an external source directly. However, the collected social and community data contain much sensitive information. Therefore, the public ad-hoc feature is not enabled in the proposed architecture. Thus, only data dictionaries and metadata are publicly published.

3.2   Data Processing

Extract, transform, load (ETL) objectives cover extract variety data from multiple sources, transform the data into usable formats and load the data into the target database system that end-users can access. Thus, we will choose ETL tool with these key featuress: usability and performance, supported connectors and data validation.

4   Exploratory Data Analysis

4.1   User Requirements and Availability of Data

User requirements gathering is the critical process in which we understand and identify the data usage needs of organization so that we could design, develop and built the data service platform as an effective decision support tool for the implementation of the Royal Project Foundation's strategics. The techniques we use in this phase are interviewing, brainstorming and prototyping. Finally, their data usage requirements are categorized into 12 dimensions which are population, career and income, labour, health and public health, gender equality, behavior, environment, education, infrastructures, community empowerment, property and community learning center. The Royal Project Foundation's social and community data is available in both structured data such as spreadsheet and unstructured data such as hard copy documents, Portable Document Format (PDF).

4.2   Data Relationship and Relational Model

As mentioned that there are various types of data. Thus, the hybrid data store architecture with a data staging area and a data warehouse is proposed. Developing a model of a data linkage system with the goal of improving the efficiency of data management and data analysis is one of the aims of this study. For that reason, we defined data fields and their relationships from the data in each category according to the mentioned requirements. Our concepts rely on:

1. As Is state (current state) as the basis for a future: We have studied existing data to understand its structure, formats, and characteristics of data storage. We have compared what data the organization has to their requirement so that we could know what are missing pieces.
2. To Be state (future state) as the determination for a present: The aim of this process is to find out what data need to be collected, which format and characteristic are suitable for specific data. These data will be used for further analysis and dashboard development.
3. Improving state (transitional state) as the development process for data service platform: to achieve the objective in accordance with To Be state, we, therefore, need to 1.) design a relational data model and its schema with cardinality of relationship that represents the database as a collection of relations [7]. and 2) design and develop an appropriate data architecture system that covered data management process such as ETL, quality control, data governance and data sharing.

4.3   Business Intelligence Dashboard

Our business intelligence (BI) dashboard will be accessed by up to three levels of users: officer, primary executive and board of director. BI dashboards provide actionable insights in various dimensions based on the needs. These data reports will be used as a supported input to develop policies for operations that align with the Royal Project foundation strategy. BI Dashboards are designed to be able to drill down and drill up such that the users can explore data in more details. We have designed it available to drill down up to a particular sub-village level. Figure 3 shows a population dashboard prototype.

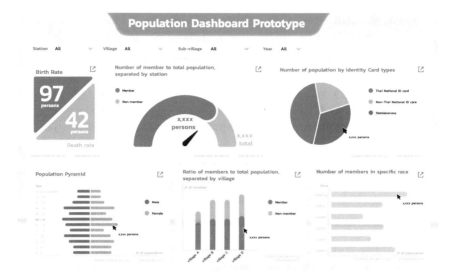

Fig. 3. Population dashboard prototype.

4.4 Data Service

Our proposed architecture, as shown in Fig. 2, also allows authorized 3rd party users to use it as a data service point in three ways. The first way is to download the processed data from each widget in the dashboards. The second way is to integrate their system into the data warehouse. The third way is to navigate the CKAN data catalog and use our curated data dictionaries and metadata as collaborative guidance with the data's owner. Besides, these three ways allow users to process data with their preferred methodology.

5 Conclusion

In summary, the paper proposed the data store architecture to maintain various data sources and data types of Royal Project Foundation data. The architecture combines the data lake, data staging area, data warehouse and CKAN together to handle such problems. Moreover, the exploratory data analysis (EDA) is presented. The EDA is implemented from the user requirement and availability of data. Also, the business intelligence dashboard prototype is designed to display the summary of Royal Project Foundation's data. For future work, the staging area, data warehouse data store and the automatically publish CKAN data dictionary and metadata will be implemented. Also, the complete business intelligence dashboard will be proposed.

Acknowledgement. We want to thank the faculty of Engineering and the College of Arts, Media, and Technology, Chiang Mai University, for supporting us in this research. Also, we are most thankful for the Royal Project Foundation that has provided financial support for the research project.

References

1. Auradkar, A., et al.: Data infrastructure at LinkedIn. In: 2012 IEEE 28th International Conference on Data Engineering, pp. 1370–1381 (2012)
2. Badiuzzaman Biplob, M., Sheraji, G.A., Khan, S.I.: Comparison of different extraction transformation and loading tools for data warehousing. In: 2018 International Conference on Innovations in Science, Engineering and Technology (ICISET), pp. 262–267 (2018)
3. Consoli, S., et al.: A smart city data model based on semantics best practice and principles. In: Proceedings of the 24th International Conference on World Wide Web, pp. 1395–1400. Association for Computing Machinery, New York (2015)
4. Correa, A.S., Correa, P.L., Silva, D.L., da Silva, F.S.C.: Really opened government data: a collaborative transparency at sight. In: 2014 IEEE International Congress on Big Data, pp. 806–807 (2014). https://doi.org/10.1109/BigData.Congress.2014. 131
5. Costa, C., Santos, M.Y.: The suscity big data warehousing approach for smart cities. In: Proceedings of the 21st International Database Engineering and Applications Symposium, pp. 264–273 (2017)
6. Desouza, K.C., Jacob, B.: Big data in the public sector: lessons for practitioners and scholars. Adm. Soc. **49**(7), 1043–1064 (2017)
7. Elmasri, R., Navathe, S.: Fundamentals of Database Systems, 6th edn. Addison-Wesley Publishing Company, USA (2010)
8. Fang, H.: Managing data lakes in big data era: what's a data lake and why has it became popular in data management ecosystem. In: 2015 IEEE International Conference on Cyber Technology in Automation, Control and Intelligent Systems, pp. 820–824 (2015)
9. Foundation, R.P.: Royal project foundation (2012). https://www. royalprojectthailand.com/
10. He, Y., Lee R., Huai, Y., Shao, Z., Jain, N., Zhang, X., Xu, Z.: RCFile: a fast and space efficient data placement structure in MapReduce-based warehouse systems. In: Proceedings-International Conference on Data Engineering, pp. 1199–1208 (2011)
11. Hedgebeth, D.: Data-driven decision making for the enterprise: an overview of business intelligence applications. **37**(4), 414–420 (2007)
12. Kimball, R., Ross, M.: The Data Warehouse Toolkit: The Definitive Guide to Dimensional Modeling. 3rd edn. (2013)
13. Kundeti, S.R., Vijayananda, J., Mujjiga, S., Kalyan, M.: Clinical named entity recognition: challenges and opportunities, pp. 1937–1945 (2016)
14. Lyko, K., Nitzschke, M., Ngonga Ngomo, A.C.: Big data acquisition. In: Cavanillas, J., Curry, E., Wahlster, W. (eds.) New Horizons for a Data-Driven Economy, pp. 39–61. Springer, Heidelberg (2016). https://doi.org/10.1007/978-3-319-21569-3_4
15. Manyika, J., et al.: Big data: the next frontier for innovation, competition and productivity. Technical report, McKinsey Global Institute (2011). https://bigdatawg. nist.gov/pdf/MGI_big_data_full_report.pdf

16. Oświecińska, K., Legierski, J.: Open data collection using mobile phones based on ckan platform. In: 2015 Federated Conference on Computer Science and Information Systems (FedCSIS), pp. 1191–1196 (2015). https://doi.org/10.15439/2015F128
17. Ponniah, P.: Data warehousing fundamentals for IT professionals (2016)
18. Ribeiro, R., Oliveira, A., Pedrosa, I.: Analysis of the impact of business intelligence in public administration. In: 2021 16th Iberian Conference on Information Systems and Technologies (CISTI), pp. 1–5 (2021). https://doi.org/10.23919/CISTI52073.2021.9476489
19. Santos, M., João, E., Canelas, J., Bernardino, J., Pedrosa, I.: The incorporation of business intelligence with enterprise resource planning in SMEs. In: 2021 16th Iberian Conference on Information Systems and Technologies (CISTI), pp. 1–6 (2021). https://doi.org/10.23919/CISTI52073.2021.9476341
20. Shvachko, K., Kuang, H., Radia, S., Chansler, R.: The hadoop distributed file system. In: IEEE 26th Symposium on Mass Storage Systems and Technologies, pp. 1–10 (2010)
21. Tudorica, B.G., Bucur, C.: A comparison between several NoSQL databases with comments and notes. In: Proceedings - RoEduNet IEEE International Conference (2011)

Implementation of a Local-Community Issues Visualization System Using Open Data and Future Population Projection

Tomoyuki Ishida[✉] and Mutsuki Kojima

Fukuoka Institute of Technology, Fukuoka 811-0295, Fukuoka, Japan
t-ishida@fit.ac.jp, s16b1019@bene.fit.ac.jp

Abstract. Currently, while the number of local governments that publish open data is steadily increasing, many services and applications that utilize open data are not zing. This study implemented a local-community issues visualization system that derives local-community issues by combining open data released by local governments with future population projection data. This system encourages users to discover local-community issues by visualizing population information, school information, and childcare facility information on WebGIS.

1 Introduction

Japan's Ministry of Economy, Trade and Industry and the Office for Promotion of Regional Revitalization, Cabinet Office, have instituted the Regional Economy Society Analyzing System (RESAS) [1] since April 2015 to support various efforts for regional revitalization. RESAS is a visualization system that aggregates big data - such as industrial structure, vital statistics, and flow of people - and effectively plans, executes, and verifies required measures. RESAS is an application programming interface that can acquire programs of published data. Additionally, application as well as policy idea contests using RESAS are conducted every year [2]. Presently, open data is operational in various fields, such as policymaking, transportation, tourism, and infrastructure management; however, it also has certain problems, which are as follows:

- Lack of platform for the easy visualization of open data
- Publishing various data in the form of open data requires considerable financial resources
- Obsolescence of data contents in open data

To address these issues, we implement a visualization system using open data from local governments and future population projection data from the National Institute of Population and Social Security Research to identify future local-community issues, such as those related to school and childcare facilities [3]. This visualization system visualizes future population projection, school, and childcare facility information for each ward using open data of Fukuoka City, Fukuoka Prefecture [4, 5]. Furthermore, it clarifies

L. Barolli et al. (Eds.): EIDWT 2022, LNDECT 118, pp. 11–20, 2022.
https://doi.org/10.1007/978-3-030-95903-6_2

future local-community issues and aids in formulating local government policies to address those issues.

The rest of the paper is organized in the followings way. The research objective is descried in Sect. 2. The system configuration is descried in Sect. 3. The local-community issues visualization system is described in Sect. 4. Section 5 evaluates the local-community issues visualization system and, finally we conclude our findings in Sect. 6.

2 Research Objective

This study uses a visualization system to identify future local-community issues, such as those related to school facilities and childcare facilities, by combining open data from Fukuoka City, Fukuoka Prefecture, with future population projection data. Currently, open data systems and applications are limited to simple visualizations. Furthermore, there have been few studies that have combined open data and future population projections to predict local-community issues. Therefore, in this study, we visualize current and predicted data on a WebGIS to clarify future local-community issues using open data published by local governments and future population projection data from the National Institute of Population and Social Security Research.

This system uses open and future population projection data to visualize past and future school and childcare facility information on WebGIS. Consequently, this system can predict the future in the following areas and aid policymaking for local governments. This study uses the OpenStreetMap API [6].

- Examining childcare projects, such as the establishment of childcare facilities, through a detailed visualization of vital data on childcare facilities for each ward.
- Examining school education projects, such as consolidation of elementary and junior high schools, through a detailed visualization of vital data on elementary and junior high schools for each ward.

3 System Configuration

The local-community issues visualization system comprises an issues solving agent, information management agent, information management server, and information database server of the local community. The configuration of this system is presented in Fig. 1.

3.1 Local-Community Information Management Agent

The local-community information management system manages open data on the local-community, such as population, school, and childcare facility information, which is stored in the local-community information database server via the local-community information management server.

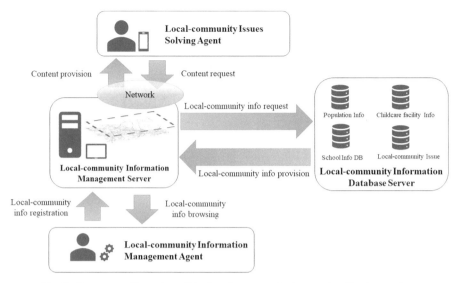

Fig. 1. System configuration of the local-community issues visualization system.

3.2 Local-Community Issues Solving Agent

The local-community issues solving agent seeks to solve local-community issues by browsing the local-community issues prediction that is visualized on WebGIS.

3.3 Local-Community Information Management Server

According to the request from the local-community issues solving agent, the local-community information management server provides future population projection data, school information, childcare facility information, and future forecasts of local-community issues that are visualized on WebGIS. Furthermore, in response to requests from the local-community information management agent, this server registers and edits various open and future population projection data stored in the local-community information database server.

3.4 Local-Community Information Database Server

The local-community information database server stores open data such as future population projection data and school and childcare facility information.

4 Local-Community Issues Visualization System

The local-community issues visualization system comprises a top, population information, school information, childcare facility information, and local-community issues screens.

A) **Top screen of the local-community issues visualization system**

Figure 2 shows the top screen of the local-community issues visualization system. The user selects visualization of information option from the menu bar of this screen, which comprises information concerning population, schools, childcare facilities, and local-community issues.

Fig. 2. Top screen of the local-community issues visualization system.

B) **Population information screen of the local-community issues visualization system**

Population information includes the population data of 2015 and 2020, projected population during 2025–2045, and estimated population in 2045, categorized as 0–14 years, 15–64 years, 65–74 years, and over 75 years. Figure 3 depicts the population visualization screen for 2045 (for each ward) based on future population projections.

C) **School information screen of the local-community issues visualization system**

School information includes data regarding the number of elementary and junior high schools, their location, and the number of students in these schools in 2020. Figure 4 shows the location information visualization screen of elementary and junior high schools.

D) **Childcare facility information screen of the local-community issues visualization system**

As shown in Fig. 5, the childcare facility information visualizes the location information of the childcare facility. When users select a pointer on WebGIS, the information (capacity) of the selected childcare facility is visualized.

E) **Local-community issue screen of the local-community issues visualization system**

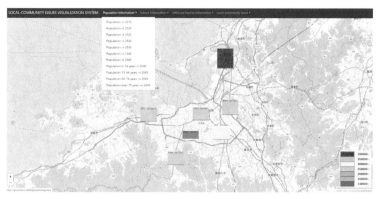

Fig. 3. Population visualization screen for 2045 (for each ward) based on future population projection.

The local-community issue information visualizes the examination of the establishment or consolidation of childcare facilities as well as that of the opening or consolidation of elementary and junior high schools, as shown in Fig. 6. On the screen, users can observe the demographic consolidation of childcare facilities and elementary schools from 2020 to 2045.

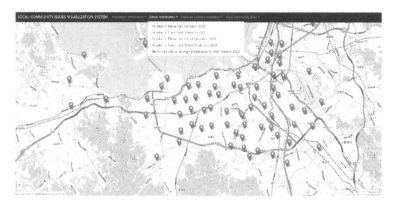

Fig. 4. Location information visualization screen of elementary and junior high schools.

5   System Evaluation

We asked five participants to use the local-community issues visualization system and conducted a questionnaire survey to assess its operability, functionality, readability, necessity, and effectiveness.

A) **Operability of the local-community issues visualization system**

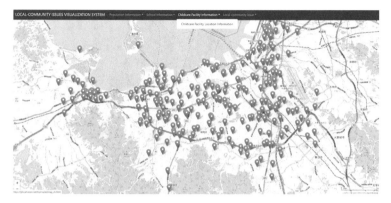

Fig. 5. Visualization screen of childcare facility information.

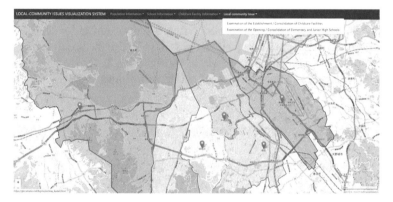

Fig. 6. Visualization screen of local-community issue information.

Figure 7 shows the operability evaluation result. Of the participants, 100% responded with "easy," and high operability could not be confirmed. Participants commented on operability, saying, "I think the operation is not too detailed and easy to use" and "I think the operation is simple and easy to understand."

B) **Functionality of the local-community issues visualization system**

Figure 8 depicts the functionality evaluation results. Of the participants, 100% responded with "satisfied" or "somewhat satisfied," and high functionality could not be confirmed. Participants commented on functionality saying, "I think it would be convenient to have a link to each childcare facility website on the blue pin of the childcare facility information."

C) **Readability of the local-community issues visualization system**

Figure 9 presents the evaluation result for readability. More than 90% of the participants responded with "easy to understand" or "somewhat easy to understand," and high readability could not be confirmed.

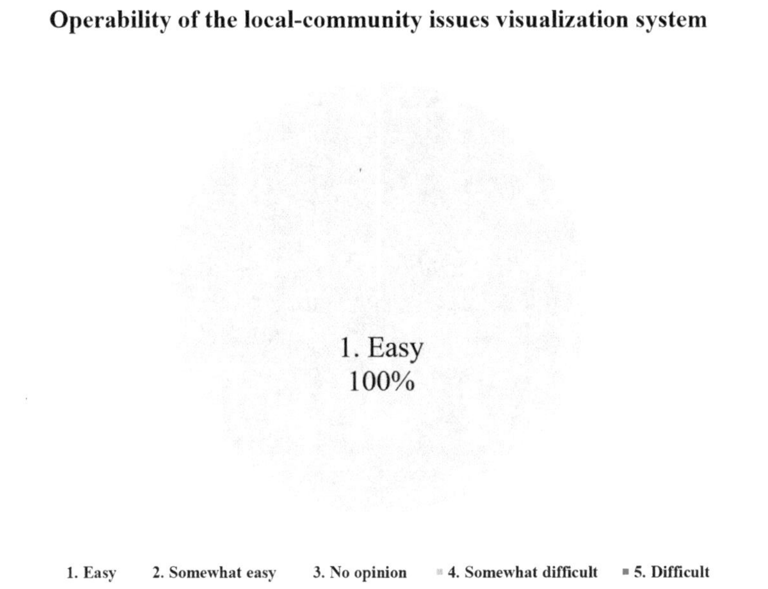
Operability of the local-community issues visualization system
1. Easy
100%
1. Easy
2. Somewhat easy
3. No opinion
4. Somewhat difficult
5. Difficult

Fig. 7. Operability of the local-community issues visualization system ($n = 10$).

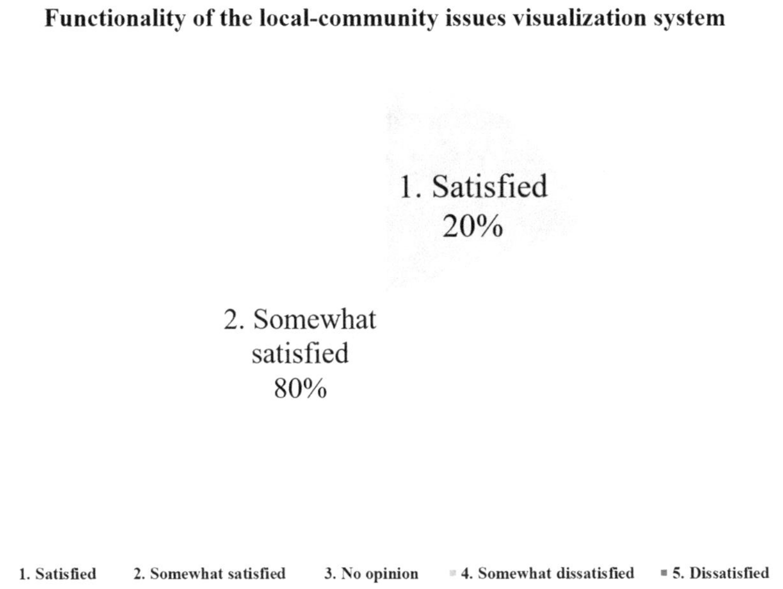
Functionality of the local-community issues visualization system
1. Satisfied
20%
2. Somewhat
satisfied
80%
1. Satisfied
2. Somewhat satisfied
3. No opinion
4. Somewhat dissatisfied
5. Dissatisfied

Fig. 8. Functionality of the local-community issues visualization system ($n = 10$).

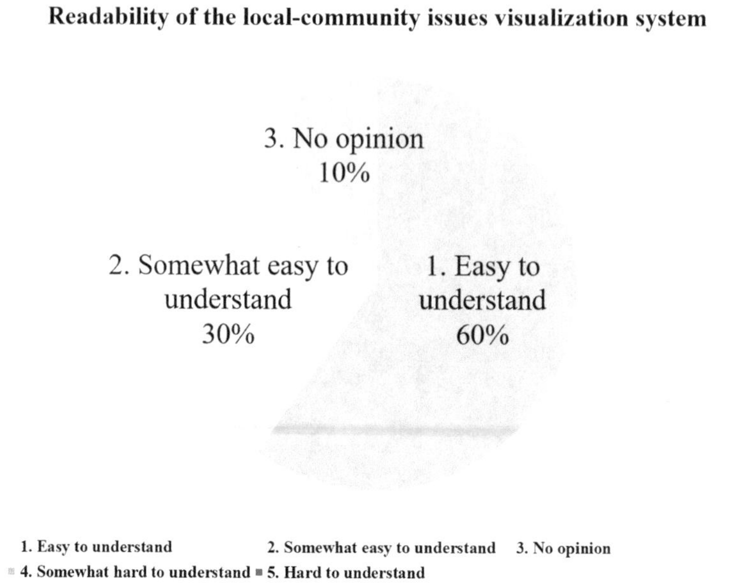

Readability of the local-community issues visualization system

3. No opinion
10%

2. Somewhat easy to
understand
30%

1. Easy to
understand
60%

1. Easy to understand   2. Somewhat easy to understand   3. No opinion
⬚ 4. Somewhat hard to understand ■ 5. Hard to understand

Fig. 9. Readability of the local-community issues visualization system ($n = 10$).

D) **Necessity of the local-community issues visualization system**

Figure 10 shows the evaluation result for necessity. More than 90% of the participants responded with "necessary" or "somewhat necessary," and high necessity was confirmed.

E) **Effectiveness of the local-community issues visualization system**

Figure 11 depicts the evaluation result for effectiveness. More than 90% of the participants responded with "effective" or "somewhat effective," and high effectiveness was confirmed.

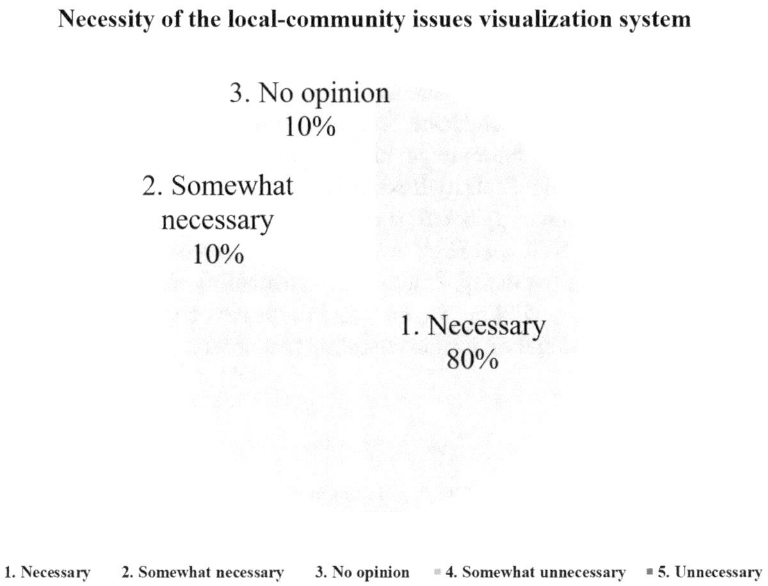

Necessity of the local-community issues visualization system

3. No opinion
10%

2. Somewhat
necessary
10%

1. Necessary
80%

1. Necessary   2. Somewhat necessary   3. No opinion   ▪ 4. Somewhat unnecessary   ▪ 5. Unnecessary

Fig. 10. Necessity of the local-community issues visualization system ($n = 10$).

Effectiveness of the local-community issues visualization system

3. No opinion
10%

2. Somewhat
effective
40%

1. Effective
50%

1. Effective   2. Somewhat effective   3. No opinion   ▪ 4. Somewhat ineffective   ▪ 5. Ineffective

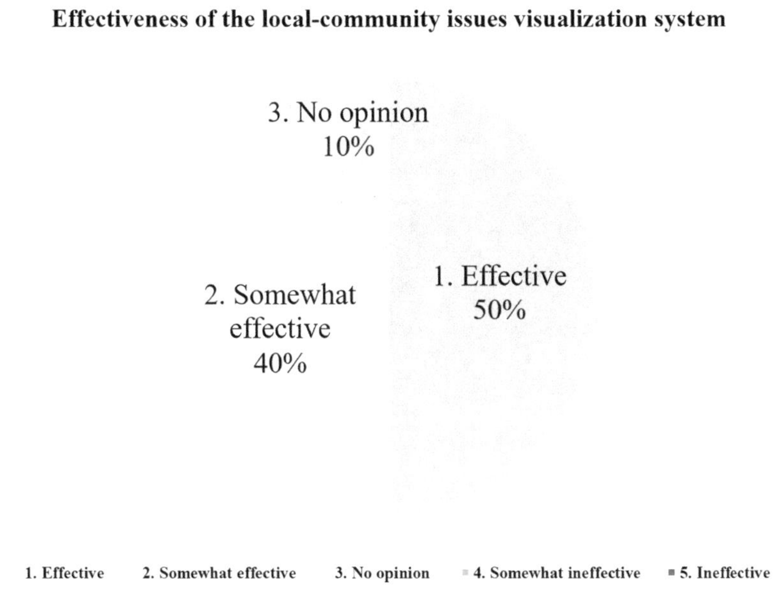

Fig. 11. Effectiveness of the local-community issues visualization system ($n = 10$).

6   Conclusion and Future Work

We used a local-community issues visualization system in this study. This system visualizes population, school, and childcare facility information by combining open data from local governments and future population projection data from the National Institute of Population and Social Security Research. Consequently, this system encourages users to identify local-community issues, such as elementary and junior high school consolidation and childcare facilities. High evaluation results were obtained in the system evaluation regarding its operability, functionality, readability, necessity, and effectiveness. As a future work, we will improve the user interface by changing the font size and font type based on the evaluation results regarding readability.

References

1. Cabinet Office and Ministry of Economy, Trade and Industry, RESAS Regional Economy Society Analyzing System, https://resas.go.jp/, last viewed November 2021
2. Cabinet Office and Ministry of Economy, Trade and Industry, RESAS API, https://opendata.resas-portal.go.jp/, last viewed November 2021
3. National Institute of Population and Social Security Research, Population & Household Projection, http://www.ipss.go.jp/site-ad/index_english/population-e.html, last viewed November 2021
4. Fukuoka City, Fukuoka City School List (Kindergarten / Elementary School / Junior High School / High School / Special Support School), https://ckan.open-governmentdata.org/dataset/school, last viewed November 2021
5. Fukuoka City, Number of Children at Fukuoka City School, https://ckan.open-governmentdata.org/dataset/401307_h26-fukuokacity-jidouseitosu, last viewed November 2021
6. OpenStreetMap, OpenStreetMap API, https://wiki.openstreetmap.org/wiki/API, last viewed November 2021

SAE+Bi-GRU Based Security Situation Prediction for Smart Grid

Lei Chen, Mengyao Zheng[✉], Zhaohua Liu, Fadong Chen, Kui Zhou, and Bin Liu

School of Information and Electrical Engineering, Hunan University of Science and Technology,
Xiangtan, China
chenlei@hnust.edu.cn

Abstract. With the degree of intelligence increases, smart grid suffers from a large number of attacks from the Internet, and urgently needs a security situation prediction method for active defense. However, traditional methods are usually unable to accurately extract deep features of smart grid, and do not jointly consider spatio-temporal features, resulting in poor situation prediction accuracy. To solve these problems, a SAE+Bi-GRU based security situation prediction algorithm is developed in this paper for smart grid, simply called SBG. Firstly, the stacked auto-encoder (SAE) is used to extract deep spatial features of smart grid at a moment. Secondly, GRU is used to predict the future situation from multiple spatial features extracted by SAE at m consecutive moments. Finally, Bi-GRU is further used to enhance the accuracy of future security situation from two directions. Extensive experiments on the public ORNL dataset prove our algorithm has a better accuracy of situation prediction.

1 Introduction

The smart grid is built on the basis of integrated, high-speed and bidirectional communication networks [1, 2]. Because the deepening connection with Internet, the smart grid suffers some cyber-attacks which can affect the stable operation of the power system [3–6]. In the *Outsmart Ing Grid Security Threats* (2018) report, more than half of power industry executives believe there will be attacked by cyber-attacks in the next five years. In 2019, the Nuclear Power Corporation of India Ltd's Intranet infected with malware. And Venezuela's power system has been attacked for three consecutive days and there have been large-scale power outages. In 2020, Brazilian Electricity Company Attacked by Sodinokibi ransomware, etc. In order to avoid the occurrence of power outages in operation of the smart grid as much as possible, security situation prediction is needed.

At present, the security situation prediction technology of the smart grid with neural network and deep learning as the technology has become a research hotspot [7–10]. For example, Li [11] based on multi-time scale state estimation, proposed a historical estimation state prediction model for multivariate time series analysis, to enhance the ability of future situation prediction. Yang J [12] based on RBF neural network find out the nonlinear mapping relationship between the front N data and the subsequent M data, and then adjust the value of N to explore the prediction results. However, the existing models have the following problems: (1) Most of them only consider the time period features of the data, without considering the changes in spatial features; (2) Extracting time period features from only one direction leads to inaccurate and incomplete feature extraction.

To overcome the above problems, we fuse a variety of deep neural networks to extract temporal and spatial features of smart grid, so as to achieve more accurate situation prediction. This paper proposed a security situation prediction algorithm based on SAE and Bi-GRU for the smart grid, namely SBG. The main contributions of this paper are as follows:

- The deep spatial features of smart grid are extracted by the SAE to realize the situation awareness;
- Bi-GRU is used to predict the future situation from two directions to enhance the accuracy and comprehensiveness of situation prediction;

The rest of this article is depicted as follows. The proposed SBG model is described in Sect. 2. The experiments are given in Sect. 3. Finally, a brief summary is presented in Sect. 4.

2   SAE+Bi-GRU Based Security Situation Prediction Algorithm for Smart Grid

2.1   SAE+Bi-GRU Based Situation Prediction Model

To realize the situation prediction more accurately, a SAE+Bi-GRU based situation prediction model is proposed for smart grid in this paper, as shown in Fig. 1. Through the combination of SAE and Bi-GRU, the model can effectively extract the spatio-temporal features of the smart grid and accurately predict its future state. This model includes two stages: situation awareness and situation prediction, and the specific idea is as follows. *The first stage is situation awareness*, where SAE is used to extract the non-linear spatial features from smart grid data at one moment. *The second stage is situation prediction*, the local spatial features extracted by SAE of m consecutive moments are taken as the input, the Bi-GRU is used to capture the periodic features which hidden in m spatial features from two different directions.

2.2   Extract Deep Features by SAE to Realize Situation Awareness

Auto encoder (AE) [13] is a neural network which often used to extract spatial features of complex data. The single-layer AE can only learn the shallow features of smart grid data, and the learning ability is limited. To solve this problem, multiple single-layer AE are stacked to form the multiple-layer SAE model [14], for getting the hidden deep spatial features of smart grid data, as shown in Fig. 2.

From the figure, it is easy to find that the SAE model contains two opposing processes. *The first process is encoding*, it is responsible for extracting deep spatial features from input data through multi-layer AE. In this process, each AE layer is used to capture the linear features from the output of the previous layer. By stacking multiple AE layers,

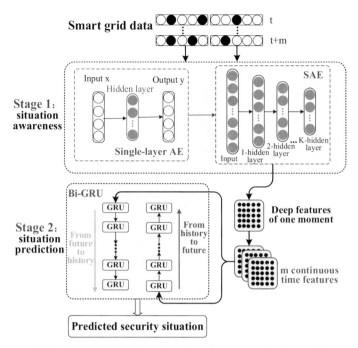

Fig.1. SAE+Bi-GRU based security situation prediction model

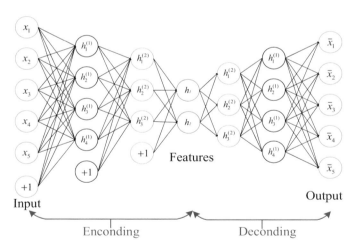

Fig. 2. The structure of SAE

multiple linear features are converted into deep nonlinear spatial features. ***The second process is decoding***, it is responsible for reconstructing the original input using the extracted deep features through multiple-layer AE. If the reconstructed input is same as the real input, it means that the SAE model is stable and can be used to accurately extract the deep spatial features. On the contrary, it means that the SAE model is not stable yet and needs to be further optimized. In this paper, the mean square loss is used as the optimization function to minimize the loss between the reconstructed input and the real input.

2.3   Use Bi-GRU to Realize Situation Prediction

Any command or cyber-attack in the smart grid both are executed for multiple consecutive moments from beginning to end. Each moment presents a certain spatial feature in the smart grid, which has been extracted by SAE. Moreover, there usually have some time periodic features hidden in the operation state of the smart grid at multiple consecutive moments. That is to say, the periodic features of smart grid can be extracted from multiple local spatial features at m consecutive moments. Based on these features, it is possible to predict the future situation of smart grid from the operating states at the previous m times. However, if you want to accurately obtain the time period features of smart grid, you must ensure that multiple local spatial features meet a fixed sequence order. To solve this problem, Bi-GRU is used in this paper to accurately extract the time periodic features from two directions.

(1) *Using GRU to predict situation from the single direction.* Firstly, m GRU units are connected to build a GRU prediction model. Secondly, m local spatial features extracted by SAE from m consecutive moments are used as input. Thirdly, SoftMax is chosen as the classifier to obtain the predicted label. Finally, the cross-entropy as the optimization function is used to minimize the difference between the real label and the predicted label at m+1 time.

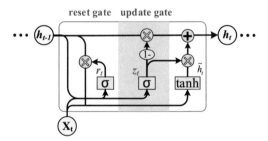

Fig. 3. The standard GRU unit

In GRU model, each GRU unit is shown in Fig. 3 [15]. From the figure, it is not difficult to find that, GRU unit uses two gate units to achieve long-term memory. *The first gate is reset gate* r_t, it determines how to combine new input information X_t with previous memories h_{t-1} for get the candidate output $\tilde{h}_t$ at the current moment, as shown in Eq. 1 and 3. The smaller the r_t, the less information from h_{t-1} is retained in $\tilde{h}_t$. *The second gate is update gate* z_t, it determines how much the memory of the previous moment h_{t-1} is saved in the memory of the current moment h_t, as shown in Eq. 2 and 4.

$$r_t = \sigma(x_t U^r + h_{t-1} W^r) \tag{1}$$

$$z_t = \sigma(x_t U^z + h_{t-1} W^z) \tag{2}$$

$$\tilde{h}_t = \tanh(x_t U^h + (h_{t-1} \circ r_t) W^h \tag{3}$$

$$h_t = (1 - z_t) \circ \tilde{h}_t + z_t \circ h_{t-1} \tag{4}$$

(2) Using Bi-GRU to enhance the accuracy of situation prediction from two directions.
Due to m local spatial features of m consecutive moments are randomly selected, this random order does not satisfy the fixed sequence order of hidden periodic features of smart grid. That is to say, based on the spatial features of m consecutive moments chosen randomly, the GRU model cannot accurately extract the periodic features of the smart grid data. To solve this problem, a Bi-GRU model is further constructed to enhance the accuracy of situation prediction, as shown in Fig. 4. The process of the Bi-GRU model is as follows: Firstly, the labeled spatial features of m consecutive moments extracted by the SAE are used as input. Secondly, the time periodic features are extracted from two directions, and are further combined to get the predicted features at time m +1. Then, the Softmax classifier is used to classify the predicted features at time m+1, and obtain its predicted label (normal or abnormal). Finally, according to the real label at time m + 1, the optimization function is used to minimize the difference between the prediction label and real label, and the stochastic gradient descent (SGD) is used to optimize the parameters of the Bi-GRU model.

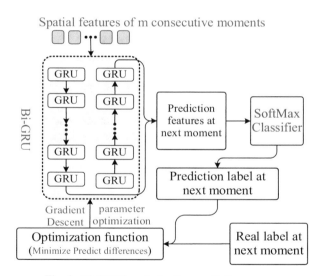

Fig. 4. Bi-GRU based situation prediction process

2.4   SAE+Bi-GRU (SBG) Algorithm

In summary, the SBG algorithm includes two aspects: model construction and situation prediction. The model construction is responsible for optimizing model parameters using

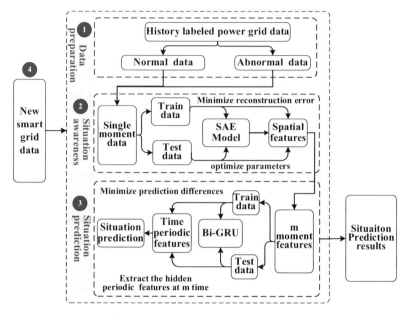

Fig. 5. SAE+Bi-GRU (SBG) algorithm

labeled test datasets, and the situation prediction is responsible for using the built stability model to predict the security situation of the smart grid.

Specifically, first, a large amount of test data are divided into normal and abnormal data. Secondly, SAE is used to extract its spatial features from the test data at a moment. Next, Bi-GRU is used to capture periodic features from the spatial features of m consecutive moments to predict the future situation. Then, based on different test data, the SAE+Bi-GRU model is repeatedly optimized until the model is stable. Finally, the stable model is used to process the new data to predict the security situation of the smart grid.

3   Experiments

3.1   Experiment Preparation

Dataset and Evaluation Metrics. The Power System Attack Dataset was used as experiments data which collected by Mississippi State University and Oak Ridge National Laboratory (referred to as ORNL dataset). It consists of 15 initial data sets, each of which contains 37 power system event scenarios. The 37 scenarios can be divided into three categories: Natural Events, No Events and Attack Events that can be regarded as label. And the ORNL dataset also contains the features of smart grid frequency, voltage magnitude and voltage phase angle, etc. Moreover, we chosen the ACC, FNR, FPR as evaluation metrics to test the algorithm performance in this paper.

3.2   Result Analysis

This section aims to verify the performance of SAE feature extractor, Bi-GRU situation predictor and the whole SBG situation prediction algorithm.

3.2.1   Performance Analysis of SAE Feature Extractor

The first experiment is to validate the performance of feature extraction of SAE on ORNL dataset from the perspectives of ACC, FPR and FNR. And GRU, Bi-GRU, Bi-LSTM, RNN and LSTM are chosen as the situation predictor, SAE, PCA and NONE are selected as feature extractor. The performance of 3 feature extractors are shown in Fig. 6.

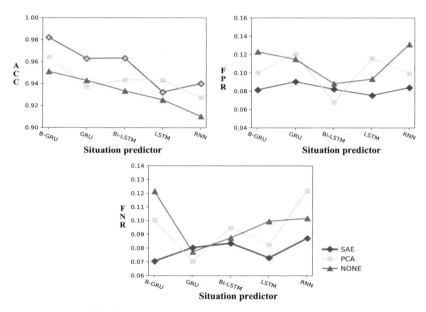

Fig. 6. Performance analysis of SAE feature extractor

In the figure, we can get the following observations. (1) The 3 feature extractors can achieve good performance in ACC, FNR and FPR. The average value of the ACC is greater than 0.93, and the average value of the FPR and FNR are less than 0.09. (2) The three feature extractors have poor stability under 5 different situation predictors, and the performance of the same feature extractor combined with different situation predictors fluctuates greatly. (3) Comparing the 3 feature extractors, SAE can extract the deep features of smart grid data better.

3.2.2   Performance Analysis of Bi-GRU Situation Predictor

The second experiment is to validate the performance of Bi-GRU as situation predictor on ORNL dataset from the perspectives of ACC, FPR and FNR. And GRU, Bi-GRU,

LSTM and RNN are chosen as the situation predictor, SAE, PCA and NONE are selected as feature extractor. The performance of 4 situation predictor is shown in Fig. 7.

From the figure, it is easy to see the following observations. (1) The 4 situation predictors show excellent performance in ACC, FNR and FPR. The average value of the ACC is greater than 0.93, and the average value of the FPR and FNR are less than 0.08. (2) The performance of 4 situation predictors is significant different under 3 feature extractors. For example, under SAE as feature extractor, Bi-GRU is the best in ACC and FNR. Under PCA as feature extractor, GRU has the best FNR. Under NONE as feature extractor, the ACC of GRU is the largest, and the FNR and FPR values of Bi-GRU are the smallest. By comprehensive comparison, Bi-GRU has the most stable performance. (3) Overall, the Bi-GRU situation predictor has the best and most stable ACC, FPR and FNR.

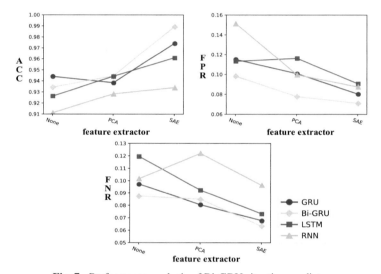

Fig. 7. Performance analysis of Bi-GRU situation predictor

3.2.3   Performance Analysis of SBG Situation Awareness Algorithm

To test the accuracy of SBG algorithm, LSTM-SAE, GRU-SAE, RNN-PCA and GUR-NONE are chosen as the comparison algorithms, and 5 groups test data of ORNL dataset are selected as experiment datasets.

Figure 8 plots the performance of 5 situation prediction algorithms. In the figure, we can get the following observations. (1) On the ACC, 5 algorithms can achieve good performance, the average value of the ACC is greater than 0.91. However, the performance of the 5 algorithms on different group of test data fluctuates sharply, with better performance on some groups, and poor performance on other groups. (2) On FPR and FNR, the SBG algorithm has better performance than other algorithms. (3) Combined with ACC, FNR and FPR, the SBG has better performance and stronger stability compared with the other 4 algorithms.

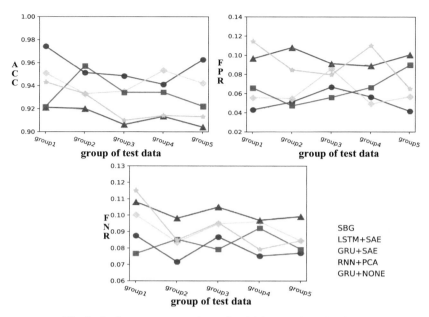

Fig. 8. Performance comparison of multiple detection algorithms

4   Conclusion

This paper proposed a SAE+Bi-GRU based security situation prediction algorithm for smart grid. The main contributions are as follows. (1) The SAE is used to convert multiple shallow linear features into deep spatial features. (2) The Bi-GRU is designed to capture the time periodic features from multiple local spatial features extracted by SAE, and then predict the future situation of smart grid. Experiments on the ORNL dataset proved that our algorithm has better ACC, lower FNR and FPR.

Acknowledgments. This work is supported by the National Natural Science Foundation of China (No.62103143); the Hunan Provincial Natural Science Foundation of China (No.2020JJ5199); the National Defense Basic Research Program of China (JCKY2019403D006); and the National Key Research and Development Program (Nos.2019YFE0105300/2019YFE0118700).

References

1. Wu, J., Kaoru, O., Dong, M.X., et al.: Big Data Analysis-based security situational awareness for smart grid. IEEE Trans. Big Data **4**, 408–417 (2016)
2. Liu, S.Y., Lin, Z.Z., Li, J.C., et al.: Review and prospect of situation awareness technologies of power system. Autom. Electr. Power Syst. **44**, 229–239 (2020)
3. Mei, S.W., Wang, Y.Y., Chen, L.J.: Overviews and prospects of the cyber security of smart grid from the view of complex network theory. High Voltage Eng. **37**, 672–679 (2011)
4. Wang, X.P., Tian, M., Dong, Z.C., et al.: Survey of false data injection attacks in power transmission systems. Power Syst. Technol. **40**, 3406–3414 (2016)

5. Kosut, O., Jia, J., Thomas, R.J., et al.: Malicious data attacks on the smart grid. IEEE Trans. Smart Grid **2**, 645–658 (2011)

6. Sakhnini, J., Karimipour, H., Dehghantanha, A.: Smart grid cyber attacks detection using supervised learning and heuristic feature selection. In: 2019 IEEE 7th International Conference on SEGE (2019)

7. Liu, Q.Y., Li, J.E., Ni, M., et al.: Situation awareness of grid cyber-physical system: current status and research ideas. Autom. Electric Power Syst. **43**, 9–21 (2019)

8. Zhang, L., Sun, W.C., Liu, X.J., et al.: The prediction algorithm of network security situation based on grey correlation entropy kalman filtering. In: Information Technology and Artificial Intelligence Conference, pp. 321–324 (2014)

9. Shahsavari, A., Farajollahi, M., Stewart, E.M., et al.: Situational awareness in distribution grid using micro-PMU data: a machine learning approach. IEEE Trans. Smart Grid **10**, 6167–6177 (2019)

10. Wang, P.Y., Govindarasu, M., et al.: Multi-agent based attack-resilient system integrity protection for smart grid. IEEE Trans. Smart Grid **11**, 3447–3456 (2020)

11. Li, Y.Z., Guo, Y.L., Peng, B., et al.: Real-time situation prediction of distribution network based on multi- time scale state estimation. Electric Power Eng. Technol. **39**, 127–134 (2020)

12. Yang, J., Li, C.H., Yu, L.S., et al.: On network security situation prediction based on RBF neural network. In: Chinese Control Conference, pp. 4060–4063 (2017)

13. Yann, L.C.: Modèles Connexionnistes de l'apprentissage. These de Doctorat. Universite P. et M. Curie (Paris 6) (1987)

14. Bengio, Y.S., Lamblin, P., Popovici, D., Larochelle, H.: Greedy layer-wise training of deep networks. In: Advances in Neural Information Processing Systems 19 (NIPS 2006)

15. Cho, K., et al.: Learning phrase representations using RNN encoder-decoder for statistical machine translation. In: Conference on Empirical Methods in Natural Language Processing, pp. 1724–1734 (2014)

Design of Identity Authentication Scheme for Dynamic Service Command System Based on SM2 Algorithm and Blockchain Technology

Jie Deng[1($\boxtimes$)], Lili Jiao[1], Lili Zhang[1], Yongjin Ren[2], and Wengang Yin[1]

[1] Police Officer College of the Chinese People's Armed Police Force, Chengdu 610213, China
[2] Armed Police Meishan Detachment, Sichuan 620010, China

Abstract. To ensure the information security of the dynamic service command system and prevent all kinds of illegal access requests, solve the privacy leakage problem of the command and decision makers in the identification process, it is necessary to conduct strong and secure identity authentication for the personnel who access the dynamic service command system to ensure the system For the security. An identity authentication scheme based on national secret algorithm and blockchain technology is proposed. The scheme creates a user's authentication block through a consensus mechanism, and uses a dynamic password to combine face recognition with a QR code for security authentication of identity. At the same time, the effectiveness and security of the scheme are studied and analyzed, which shows that the scheme can ensure the privacy of users while providing efficient identity authentication.

1 Introduction

In recent years, the armed police forces are responsible for various large-scale national security activities and handling various emergencies. Human resources alone cannot obtain timely intelligence information from various duty stations. It is necessary to establish a modern visualized dynamic service information command system. Due to the high sensitivity and confidentiality requirements of the system itself, there is an urgent need for strong security identity authentication for access to the system to ensure the security of the system, prevent all kinds of illegal access requests, and provide a strong guarantee for high-level national command and decision-making.

With the rapid development of information technology, especially the rapid development of Internet information technology in recent years, especially the rise of blockchain technology, due to its four major characteristics of distributed data storage, point-to-point transmission, consensus mechanism, and encryption algorithm, the problem of identity authentication in a non-secure environment provides a new solution. Among them, Zhang Shengnan [1] proposed the application model of blockchain in the authentication of market entities, certificate issuance, and transactions on the certificate chain. Zhang Yabing [2] proposed a cross-domain authentication scheme based on multi-layer blockchain. Zhang Jinhua [3] proposed a blockchain authentication based cross-domain and key agreement protocol in an edge computing environment. This solution makes full use of

© The Author(s), under exclusive license to Springer Nature Switzerland AG 2022
L. Barolli et al. (Eds.): EIDWT 2022, LNDECT 118, pp. 31–37, 2022.
https://doi.org/10.1007/978-3-030-95903-6_4

the blockchain technology and the provable security of the SM2 algorithm to realize a reliable and efficient armed police dynamic service command system identity solution.

2   Black-Chain Technologies

T Since its birth as a brand-new concept, blockchain has been widely studied and applied in the field of science and technology because it provides efficient and safe guarantees for the exchange and storage of data.

Blockchain technology has the characteristics of decentralization, data tamper-proof, traceability [4], making it very suitable for scenarios where high data credibility needs to be guaranteed. The Hash function, asymmetric encryption, digital signature technology, consensus algorithm and other cryptographic algorithms and encryption technologies used in blockchain technology are the main basis for ensuring the security and reliability of its data. It relies on the P2P network, a distributed database system involving multiple independent nodes, which can be understood as a distributed account book [5].

The block header contains the version number, the hash value of the previous block, and the hash value of the Merkle root node, the timestamp, the difficulty value, the random number, and the transaction record. Any transaction record is tampered with. According to the characteristics of the hash encryption algorithm, the root node of the Merkle tree of the binary tree structure will change greatly. Therefore, other nodes can clearly find the problem when verifying the hash value of the blockchain. Block chains can be divided into public chains, private chains and consortium chains. Blockchain can choose different encryption methods, such as ECC encryption algorithm, national secret algorithm, quantum encryption and other signature algorithms [6]. When designing the identity authentication protocol, this scheme writes the authentication data into the identity authentication blockchain, and realizes the identity verification of system visitors with the help of private chain [10].

3   Identity Authentication Scheme of Dynamic Service Command System

This identity authentication scheme needs to go through five processes: User's identity registration, generating identity authentication block, user login, generating electronic ID and QR code, identity authentication. The detailed design description is as follows.

3.1   User's Identity Registration

User A performs electronic identity registration on the mobile terminal through the mobile device, and the process is as follows:

① Verify the mobile phone number. Use dynamic passwords to verify the authenticity of mobile phone numbers. If the verification fails, the registration fails.
② Submit personal information. The user fills in the personal basic identity information ID_t,which is included name, ID number, gender, identity certification materials, and performs dynamic facial recognition through the camera to obtain the facial feature value ID_f, and set the user name ID_A and login password PSW_A.

③ Registration. The Electronic Identity Certification system, which is referred to as EICS, performs SM3 hash operation on the text information and picture information, and obtains the identity summary information Hash ($ID_t||ID_f$), then transmits it to the authentication service center, which is referred to as CA, and checks the identity information in Population Bank System, which is referred to as PBS. After the verification is passed, the user completes the system registration. The registration process is shown in Fig. 1.

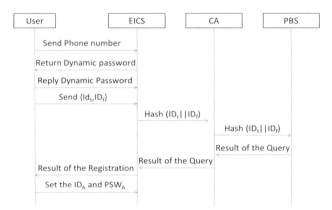

Fig. 1. Registration

3.2 Generating the Authentication Blockchain

After user A has successfully registered, a blockchain for identity authentication needs to be generated. The process is as follows:

① The service center of CA issues the user's real-name authentication information eID.
② The mobile client generates private key d_A and public key $P_A = d_A \bullet G = (x_A, y_A)$ through the SM2 algorithm, and store d_A encrypted.
③ The mobile client performs SM3 hash operation on the user's real-name information eID, identity information ID_t, facial feature value ID_f to obtain the identity hash value $e = $ Hash ($eID||ID_t||ID_f$), and generates a random number $k \in [1, n-1]$, and sets $M_0 = \{e, T_r\}$, where Tr is the registration timestamp. Then calculates $k \bullet G = (x_1, y_1)$ $r = (e +x_1)$ mod n, $r = (e +x_1)$ mod n, and set SIGr $= (r, s)$. At last, sends the message $M = \{P\_A, M0, SIGr\}$ $M = \{P_A, M_0, SIG_r\}$ to the service center of CA.
④ The service center of CA uses the SM3 Hash algorithm to calculate user A's $e' = $ Hash($eID||ID_t||ID_f$), $t = (r + s)$ mod n, $t \bullet G + P_A = (x'_1, y'_1)$, and then judges whether the formula $e' + x'_1 = r$ is established. If it is established, the verification is passed.
⑤ The service center of CA Uses SM3 algorithm to calculate $Add_A = $ Hash (P_A) as the identity authentication block identifier of user A, which is the block address of user A. In the block data structure, it is recorded the user A's Hash ($eID||ID_t||ID_f$), T_r AND SIG_r.

So as to complete the block generation through the consensus mechanism. The process is shown in Fig. 2.

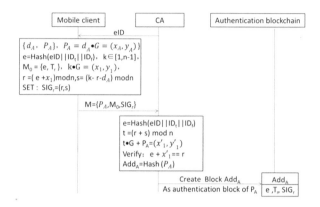

Fig. 2. Generating the Authentication block-chain.

3.3 User Login

When user A logs in to the system, he needs to input the user name ID_A, login-password PSW_A and the facial feature value ID_f by webcam to confirm that user A is logged in to the system.

3.4 Generating Electronic ID and QR Code

After user A logs into the system, the mobile client completes the following operations:

① Generate random number $k_n \in [1, n-1]$, calculate $H_{ID} =$ Hash $(eID\|ID_t\|ID_f)$.
② Set the message $M_n = \{H_{ID}, T_C\}$,
③ Using SM3 to calculate $e_n =$ Hash (M_n), $k_n \bullet G = (x_n, y_n)$, $r_n = (e_n + x_n)$ mod n, $s_n = (k_n - r_n - d_A)$ mod n. Generate user A's dynamic identity signature $SIG_{dA} = \{r_n, s_n\}$.
④ The system generates a dynamic identity QR code QR_n of user A according to the information PA, H_{ID}, Tc and $SIG_{dA} = \{r_n, s_n\}$, and the QR code is refreshed The time is set within the range of 1min. The process is shown in Fig. 3.

3.5 Identity Authentication

CA reads the user's dynamic identity QR code QRn, and obtains the message $M_{rn} = \{P_A,$ Hash $(eID\|ID_t\|ID_f)$, T_c, $SIG_{dA}\}$. Perform the following steps to complete the certification:

① Judge the timeliness of QRn through the timestamp Tc. If it expires, the authentication fails.

② Set the message $M'_n = \{\text{Hash (eID} \| \text{ID}_t \| \text{ID}_f), T_C\}$, calculate $e_n = \text{Hash }(M'_n)$ by SM3, $t_n = (r_n + s_n) \bmod n$, then calculate $t_n \bullet G + P_A = (x'_n, y'_n)$, $R_n = (e_n + x'_n) \bmod n$, and judge whether $R_n = r_n$ is established. If it is established, the verification is passed.

③ Use SM3 algorithm to calculate $\text{Add}_A = \text{Hash }(P_A)$, the block address Add_A of user A is obtained, and the identity authentication block is accessed, then compare the QR code and $e = \text{Hash (eID} \| \text{ID}_t \| \text{ID}_f)$ is or not consistent. If they are consistent, verify SIGr again.

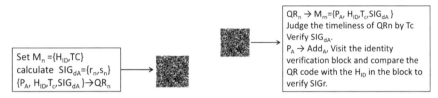

Fig. 3. Generating electronic identity stage   Fig. 4. Identity authentication stage

④ SIGr needs to be verified. Only need to calculate $t' = (r + s) \bmod n$, $t' \bullet G + P_A = (x'_1, y'_1)$, then verify: $e + x'_1 == r$ to determine whether the identity of user A is legal. The process is shown in Fig. 4.

4   System Performance Analysis

4.1   Efficiency Analysis

(1) SM2/SM3

Based on the ECC algorithm, the State Commercial Cryptography Bureau independently designed the SM2 elliptic curve public key cryptographic algorithm. The security of SM2 algorithm stems from the difficulty of solving the hash logarithm problem. Compared with RSA algorithm, its key length is shorter [7], and it has the advantages of high security, small storage space and fast signature speed. At the same time, compared to the international standard ECC algorithm, the SM2 algorithm is better in determining the correctness of decryption, the problem of plaintext encoding, the limitation of the length of the encrypted data, and the efficiency of encryption calculation [8]. Therefore, in mobile applications, the SM2 algorithm can effectively reduce the loss of operation, storage and communication, and has more advantages than RSA and the international standard ECC algorithm.

The SM3 algorithm is a cryptographic hash algorithm. The overall structure of its compression function is similar to the SHA-256 algorithm, but it can effectively avoid high-probability local collisions, resist strong collisional differential analysis, collisional linear analysis, and bit tracking methods.

In this paper, SM2 is used as a public key cryptographic algorithm to generate public and private keys to encrypt and protect user identity information, and SM3 is used as a hash algorithm for generating block addresses and hash calculations.

(2) High Facial Recognition Rate

At present, the face recognition algorithm based on convolutional neural network can achieve a recognition rate of more than 99%, and the time to recognize a single face is less than 0.05 s, and it is robust to interference such as light differences, facial expression changes, and whether there are obstructions. Improve the efficiency of registration and login.

(3) Create an Authentication Block

This article uses a private chain to construct the user's identity authentication blockchain, which is maintained and managed by a trusted authentication agency, which not only ensures the authority and security of the data, but also improves the efficiency of building blocks.

4.2 Safety Analysis

This section analyzes the security of this program from the following six aspects.

(1) Decentralized Encrypted Storage

The traditional authentication system adopts centralized management, while the blockchain adopts distributed storage. All nodes participate in the operation and maintenance of the system, and each node contains the same information record. It effectively solves the problems of disaster recovery and backup of information, anti-tampering, anti-leakage and traceability, and avoids the single point of failure and the difficulty of multi-CA trust in traditional authentication methods [1].

(2) Multi-Factor Authentication

This scheme adopts the multi-factor authentication method of dynamic password, facial recognition and identity QR code, and introduces the biometric authentication factor and dynamic password mechanism, which is more secure than the traditional two-factor authentication method of smart card and password. Even if an attacker obtains user-related information through monitoring, he cannot pass real-time facial recognition verification, so he cannot impersonate a user for authentication [9].

(3) Anti-Counterfeiting

Suppose that the attacker has the user's identity hash value and sets the message Suppose that the attacker has the user's identity hash value and sets the message M_n = {Hash (eID‖IDt‖IDf),TCn}, but it does not have the user's private key and cannot sign the message, so it cannot forge the user's dynamic The identity signature value $SIG_{dA} = \{r_n, s_n\}$, so that the user's identity QR code cannot be forged., but it does not have the user's private key and cannot sign the message, so it cannot forge the user's dynamic The identity signature value $SIGdA = \{rn, sn\}$, so that the user's identity QR code cannot be forged.

(4) Anti-Replay Attack

The timestamp and random number contained in the block header of each block can effectively prevent replay attacks in the process of creating the block. During the authentication process, a time stamp is added to each message and QR code transmitted, and the QR code is refreshed every minute. If the attacker replays the message, the receiver can judge that the message is a replay message through the timestamp and reject its authentication request.

(5) Prevent the Disclosure of Identity Information

The user's private key is encrypted and stored on the local client. Even if the attacker obtains the user's public key and calculates the identity authentication block address, the identity information obtained by the query is still in ciphertext state, and the user's identity information cannot be decrypted.

(6) Effectively Protect User Private Key

Using dynamic two-dimensional code and blockchain authentication methods, the user's identity information will not be directly transmitted online. The encrypted identity QR code will not reveal personal privacy, and has a stronger privacy protection function than a physical ID card. According to the principle of minimum information exposure, the authentication service application system will return the verification result instead of the user's identity information after successful blockchain verification, which more effectively protects user privacy.

5 Conclusion

The dynamic service system identity authentication protocol designed in this paper combines the advantages of dynamic passwords, facial recognition, two-dimensional codes and other technologies. It uses SM2/SM3 algorithm and blockchain technology to design a mobile device-based two Dimension code electronic ID card program. After analysis, the electronic identity authentication protocol proposed in this paper adopts a decentralized secret storage method and a multi-factor authentication method, which can effectively resist replay attacks, prevent attackers from cracking user identity information offline, and is effective for user privacy information.

References

1. Zhang, S., Zhang, X.: Performance optimization of the certificate trading system for renewable energy accommodation based on blockchain. Power Demand Meas. Manage. **23**(02), 10–15 (2021)
2. Yabing, Z., Bin, X.: Cross domain authentication scheme based on multi layer blockchain. Appl. Res. Comput. **38**(6), 125–131 (2021)
3. Yan, J.Z., Peng, J., Zuo, M., et al.: PKI digital certificate system based on blockchain. Telecommun. Eng. Technol. Stand. **11**, 16–20 (2017)
4. Jinshan, S., Li, R.: Survey of blockchain access control in Internet of Things. J. Softw. **30**(6), 1632–1648 (2019)
5. Biswas, S., Sharif, K., Li, F., et al.: A scalable blockchain framework for secure transactions in IoT. IEEE Internet Things J. **6**(3), 4650–4659 (2019)
6. Huang, J., Kong, L., Chen, G., et al.: Towards secure industrial IoT: blockchain system with credit-based consensus mechanism. IEEE Trans. Industr. Inf. **15**(6), 3680–3689 (2019)
7. Pengzhi, T., Mengli, Z.: A microgrid security login system based on smart card SM2. J. East China Jiaotong Univ. **37**(01), 106–112 (2020)
8. Gu, G.J.: Research and Implementation of Authentication Authorization System Based on SM2 Algorithm. Shandong University, Jinan (2013)
9. Min, X.R., Du, K.: Design of electronic license sharing platform based on block chain technology. Command Inf. Syst. Technol. **8**(2), 47–51 (2017)
10. Wang, J., Peng, W.U., Wang, X., et al.: The outlook of blockchain technology for construction engineering management. Front. Eng. Manage. **4**(1), 67–75 (2017)

Visual Authentication Codes Generated Using Predictive Intelligence

Urszula Ogiela[1]([✉]), Makoto Takizawa[2], and Marek R. Ogiela[1]

[1] AGH University of Science and Technology, 30 Mickiewicza Ave, 30-059 Kraków, Poland
uogiela@gmail.com, mogiela@agh.edu.pl
[2] Research Center for Computing and Multimedia Studies, Hosei University, 3-7-2, Kajino-cho, Koganei-shi, Tokyo 184-8584, Japan
makoto.takizawa@computer.org

Abstract. In this paper we'll describe the possibilities of using Predictive Intelligence approaches in creation of thematic-based visual security protocols, oriented for user authentication. Such protocols will use human perception abilities and cognition thresholds features evaluated for particular authorized persons.

Keywords: Cryptographic protocols · User-oriented security systems · Perception-based authentication

1 Introduction

Predictive Intelligence is one of the most important computational paradigms used in forecasting economy trends, and processes simulations. It may be also applicable for security purposes, especially for abnormal behavior detection and prediction users' web activities. It can also be useful in creation of modern security protocols especially oriented for user authentication, as well as creation of visual security procedures.

In this paper we'll describe how predictive approaches can be applied for generation of visual sequences, which can be further used in visual authentication procedures. Such visual authentication codes are extension of traditional CAPTCHA protocols, as well as cognitive CAPTCHA solutions proposed in [1, 2]. The idea of cognitive codes is based on using some personal characteristics and users' preferences in creation of authentication protocols. One of the possible extensions of such procedures is application of predictive analysis to evaluate a user-depended perception threshold, which also can be used during verification stages.

Traditional CAPTCHA can present visual patterns for users, asking questions to quickly find appropriate solution, necessary for authentication. Considering cognitive abilities and predictive analysis it is also possible to present during such verification, thematic patterns, which will be semantically connected with previous ones. So, having several already presented visual patterns, and based on personal experiences and knowledge, we can ask users to predict correctly next visual elements, which semantically will be joined with previous patterns. Such approach will be presented in following section.

2 Predictive Intelligence in Security Systems

Predictive analysis is mainly applicable for predicting future trends or users' behavior. Such analysis can also be applied for analytic task, especially in situation, in which we can obtain a great amount of data from different sources or nodes. In distributed networks is possible to register information from multiply sources as well as register personal signals from wearable sensors. Acquisition of signals from several independent sources, and analysis of such data with AI procedures is the domain of transformative computing systems. Security procedures, which are based on transformative computing technologies were also described in [3, 4]. Such procedure allows to consider different external conditions and features like position, motion, temperature etc. Of course, having security procedures, which depend of various personal or external factors, we can also try to implement in them a predictive stage in which users should guess next steps or answers, based on previous authentication elements. So finally, we can create a very universal security protocols which can work in particular situation, and be strongly depend on personal cognitive abilities or expectations. Predictive intelligence systems allow to implement several different parameters while creating security protocols. The first possible solution may operate on personal or behavioral parameters used in security procedures [5–7]. The next example may be connected with application of personal expectation or user's knowledge for security purposes. Predictive analysis allows to generate thematic patterns according user profiles, and oriented for his/her authentication, based on personal preferences. This allows to define a new visual authentication codes dedicated for personal authorization or identification. In previous works we proposed a thematic CAPTCHA codes which operate on specific knowledge and personal experiences [1]. Such solutions can be extended towards generation of visual pattern with application of predictive approaches. Having user's preferences and expertise areas it is possible to randomly generate a sequence of visual patterns, which will be semantically connected with user's knowledge or professional activities. From generated sequences of visual patterns for authentication can be selected a smaller subset with the same semantic configurations. Such selected subsets can be than use for personal authentication.

3 Security of Protocols Based on Predictive Intelligence

Traditional visual authentication codes are secure because modification introduced into the visual elements make recognition process more complicated and difficult, especially for computer systems, and pattern classification techniques. Such visual authentication procedures, which operate on images selected with predictive protocols still remain secure, because besides introduced images modifications, it requires from users to properly understand the semantic content of presented for evaluation patterns. Such patterns will be user oriented and generated according thematic or professional preferences selected for authenticated persons.

Generation of thematic visual sequences should be done especially with application of cognitive systems or transformative approaches, which additionally allow to consider some external factors, and involve them into the verification protocols.

The second important feature of predictive visual authentication approaches is possibility to establish in such procedure individual levels at which users can properly identify visual pattern. As was mentioned in paper [3] it is possible to evaluate an individual perception threshold for each user, at which he/se will be able to recognize the content of observed pattern but other users not. Such individual perception threshold may be evaluated using cognitive systems and image threshold procedures, and should be dependent also on some personal expectations and recognition experiences. After generation of such personal and thematic visual patterns it can be presented to the user with some questions connected with its content or specific features. During verification users should provide the proper answer for formulated questions and find proper semantic order of several visual parts, which fulfill requested requirements or constrains.

4   Conclusions

The concept of application predictive intelligence in creation of visual authentication codes is very promising. Visual codes like CAPTCHA usually are related to standard object recognition available for most human users, but difficult for computer systems. Replacing standard visual pattern by new sequences, which reflect personal preferences and professional expertise allow to focus verification procedure on particular person or group of users. In such protocols the main problem lays in creation of such user-oriented visual sequences, which can be generated with application of cognitive systems. Cognitive systems were described widely with relation to cognitive cryptography procedures [3]. Such systems allow to analyze semantically visual patterns, and classify them according content. Such systems imitate the natural way of human thinking, what is possible thanks the comparison of image feature with knowledge store in the database [8].

Predictive intelligence allows additionally to create a larger set of visual patterns containing additional elements, which are semantically connected with user profile or his/her expertise area. Predictive intelligence additionally allows to authorize users using blurred patterns, which can be recognized with application of personal perception thresholds. Such features allow to create a new class of security protocols which is strongly dependent and oriented for users [9, 10].

Acknowledgments. This work has been supported by the AGH University of Science and Technology research Grant No 16.16.120.773.

References

1. Ogiela, L., Ogiela, M.R.: Cognitive security paradigm for cloud computing applications. Concurr. Comput. Pract. Exp. **32**(8), e5316 (2020). https://doi.org/10.1002/cpe.5316
2. Ancheta, R.A., Reyes, F.C., Jr., Caliwag, J.A., Castillo, R.E.: FEDSecurity: implementation of computer vision thru face and eye detection. Int. J. Mach. Learn. Comput. **8**, 619–624 (2018)
3. Ogiela, L.: Transformative computing in advanced data analysis processes in the cloud. Inf. Process. Manage. **57**(5), 102260 (2020)
4. Ogiela, M.R., Ogiela, L., Ogiela, U.: Biometric methods for advanced strategic data sharing protocols. In: 2015 9th International Conference on Innovative Mobile and Internet Services in Ubiquitous Computing IMIS 2015, pp. 179–183 (2015). https://doi.org/10.1109/IMIS.2015.29

5. Ogiela, U., Ogiela, L.: Linguistic techniques for cryptographic data sharing algorithms. Concurr. Comput. Pract. Exp. **30**(3), e4275 (2018). https://doi.org/10.1002/cpe.4275
6. Ogiela, M.R., Ogiela, U.: Secure information splitting using grammar schemes. In: Nguyen, N.T., Katarzyniak, R.P., Janiak, A. (eds.) New Challenges in Computational Collective Intelligence. Studies in Computational Intelligence, vol. 244, pp. 327–336. Springer, Heidelberg (2009). https://doi.org/10.1007/978-3-642-03958-4_28
7. Ogiela, L., Ogiela, M.R., Ogiela, U.: Efficiency of strategic data sharing and management protocols. In: The 10th International Conference on Innovative Mobile and Internet Services in Ubiquitous Computing (IMIS-2016), 6–8 July 2016, Fukuoka, Japan, pp. 198–201 (2016). https://doi.org/10.1109/IMIS.2016.119
8. Guan, C., Mou, J., Jiang, Z.: Artificial intelligence innovation in education: a twenty-year data-driven historical analysis. Int. J. Innov. Stud. **4**(4), 134–147 (2020)
9. Menezes, A., van Oorschot, P., Vanstone, S.: Handbook of Applied Cryptography. CRC Press, Waterloo (2001)
10. Yang, S.J.H., Ogata, H., Matsui, T., Chen, N.-S.: Human-centered artificial intelligence in education: seeing the invisible through the visible. Comput. Educ. Artif. Intell. **2**, 100008 (2021)

Reliable Network Design Problem by Improving Node Reliability

Hiroki Yano[1], Sumihiro Yoneyama[2], and Hiroyoshi Miwa[1(✉)]

[1] Graduate School of Science and Technology, Kwansei Gakuin University, 2-1 Gakuen, Sanda-shi, Hyogo, Japan
{yano,miwa}@kwansei.ac.jp
[2] School of Science and Technology, Kwansei Gakuin University, 2-1 Gakuen, Sanda-shi, Hyogo, Japan
sumihiro@kwansei.ac.jp

Abstract. The reliability of an information network is important, since an information network such as the Internet is an important social infrastructure. Especially, when the failure of a node occurs and the information network is fragmented, services cannot be appropriately offered. In order to avoid such a situation, it is necessary to reduce the probability of network fragmentation by using the backup facility. It costs much to protect a node by backup facility; therefore, it is not practical to protect all nodes, and only critical nodes whose failures significantly degrade the performance of an information network must be protected. Since the number of these nodes must be small as much as possible, it is important to find the smallest number of nodes to be protected so that a network resulting from the failures of any non-protected nodes is not fragmented to many small connected components. In this paper, we address the problem of determining a set of the protected nodes that achieves the maximum probability that the number of nodes in the connected components of the network remaining after the failures of any non-protected nodes under cost constraint exceeds a given threshold. We formulate this problem mathematically, show its NP-hardness, and design a polynomial-time heuristic algorithm. Furthermore, we evaluate the performance of the algorithm by using the topology of actual information networks.

1 Introduction

In an information network, the high reliability which can continue the communication even in a failure is required. Since the Internet is an important social infrastructure, disruption of communication between nodes by a failure or an attack has a vast influence on service on the Internet. However, in many actual information networks, there exists the risk of communication blackout by node failures, attacks, and crash by large-scale disasters such as earthquake. Especially in a large information network, it is difficult to avoid failures and attacks. Moreover, the reference [1] pointed out the following important property about the influence of node failures on network connectivity: In a network with a scale-free property that the degree distribution satisfies the power law, when nodes are sequentially removed in descending order of degree, the average size of the connected components of the remaining network rapidly decreases. In other words, a scale-free network is fragmented to many small connected components by selective attack that nodes with large degree are selectively destroyed. Therefore, it is necessary to design a robust information network so as to keep communication among nodes against not only random failures but also selective attack. To provide a reliable network which is resistant to network failures, the design and operation of an information network is an important issue for all network service providers.

Especially, when the failure of a node occurs and the network is fragmented, services cannot be appropriately offered. In order to avoid such a situation, it is necessary to reduce the probability of network fragmentation by using the backup facility such as the duplication of communication equipment, emergency power, and strengthening of earthquake resistance and the function that, even if the node fails, it is sufficiently rapidly changed to the backup facility. If a node has the backup facility and the function, we can consider that the node does not fail, because the failure of the node is rapidly recovered and the failure cannot be detected. We call such a node a protected node. It costs much to protect a node, because it needs much network resource. Therefore, it is not practical to protect all nodes, and only critical nodes whose failures significantly degrade the performance of an information network must be protected. The number of these nodes must be small as much as possible. It is important to find the smallest number of nodes to be protected so that a network resulting from the failures of any non-protected nodes is not fragmented to many small connected components.

In this paper, we address the problem of determining a set of the protected nodes that achieves the maximum probability that the number of nodes in the connected components of the network remaining after the failures of any non-protected nodes under cost constraint exceeds a given threshold. We formulate this problem mathematically, show its NP-hardness, and design a polynomial-time heuristic algorithm. Furthermore, we evaluate the performance of the algorithm by using the topology of actual information networks.

2   Related Works

It is well known that many information network is scale-free. A scale-free network is weak against a selective node attack [1]. When nodes are sequentially removed in descending order of degree, the average size of the connected components of the remaining network rapidly decreases. In other words, a scale-free network is fragmented to many small connected components by selective attack that nodes with large degree are selectively destroyed. Furthermore, the average distance of the network increases. On the other hand, a scale-free network is robust against random failure that nodes are randomly removed; the average size of the connected components of the remaining network does not decrease very much; the average distance does not change very much.

From the viewpoint of node protection, there is the problem which determines a minimum failed node set maximizing the number of the connected components in the network resulting from a failure [2]. This problem is NP-hard and some heuristic algorithms are proposed. However, these algorithms do not always give a set of protected nodes appropriately, because this problem determines only a node set whose failure maximizes the number of the connected components.

As for link or node protection, the group of the authors has investigated some problems so far. The network design problem in [3] asks the smallest number of protected links so that the diameter of a network resulting from failures of non-protected links is less than or equal to a given threshold. This problem is NP-hard, and an approximation polynomial-time algorithm to solve the problem is proposed. The problem in [4] determines a set of protected links to keep the connectivity to a server. The reference [5] addresses the problem of determining protected links so that the master-server and all edge-servers are connected and that the capacity restrictions are satisfied. The reference [6] deals with the problem of determining protected links so that the reachability is kept within a fixed distance to multiple mirror servers even at the event of a failure. The reference [7] addresses the problem of determining protected links so that all server nodes, regardless of whether they are master or edge-servers, are connected and so that the capacity restrictions are satisfied. The reference [8] addresses the problem of determining protected links so that the master-server and all edge-servers are connected and that the capacity restrictions are satisfied and that the increase ratio of the distance in a network to the distance in the failed network does not exceed a given threshold. The reference [9] addresses the problem of determining server placement and link protection simultaneously. The problem on node protection is first formulated in [10]. This problem asks a set of protected nodes such that, even if non-protected nodes fail, the minimum size of the connected components of a remaining network is not less than a given threshold. An approximation algorithm with the approximation ratio of two for the problem restricted to the case that at most two nodes simultaneously fail is proposed.

When each link has the probability that the link is removed, the network reliability is defined so that the entire network is connected. The network design problems that maximizes the reliability or minimizes cost under reliability constraint are studied (ex. [11, 12]). Since calculating the network reliability of a given network is generally #P-complete, many previous studies designs fast approximation algorithms (ex. [13]).

There are a combination of protection and failure probability. The problem of determining the protected links to maximize the network reliability under the constraint that the number of protected links is restricted is defined in [14]. The study in [15] assumes that the failure probability of a link decreases according to the allocated cost and addresses the problem of determining cost allocation so that the failure probability of an entire network is below a threshold.

3   Vertex Protection Problem Considering Failure Probability

In this section, first, we define the problem to determine the vertices to be protected so that, even if non-protected vertices are removed, the probability that the minimum size of the connected components of the remaining network is not less than a given threshold is maximized.

Let $G = (V, E)$, where V and E are the vertex set and the edge set of G, respectively, be a connected undirected graph representing the structure of an information network. Each vertex corresponds to a node such as a router, and each edge corresponds to a link between nodes. The failure of nodes corresponds to the change of the graph from G to $G' = (V \setminus V_K, E)$ which is the induced subgraph from G by $V \setminus V_K$, where vertex set V_K corresponds to the failed nodes. When $V_P(\subseteq V)$ satisfies that, for any $V_K(\subseteq V)$ such that $|V_K| = i$ for integer $i(\geq 0)$ and such that $V_P \cap V_K = \emptyset$, the minimum size of the connected components of $G' = (V \backslash V_K, E)$ is more than or equal to positive integer L, V_P is called a (i,L)-protected vertex set. For a (i,L)-protected vertex set, V_P, let the set of $V_K(\subseteq V)$ such that $|V_K| = i$ for integer $i(\geq 0)$ and such that $V_P \cap V_K = \emptyset$ and such that the minimum size of the connected components of $G' = (V \backslash V_K, E)$ is more than or equal to positive integer L, be $R(i,L)$. From this definition, note that, even if any i vertices except the vertices in a (i,L)-protected vertex set are removed, the minimum size of the connected components of the resulting graph is L or more.

We associate the failure probability with each vertex by the failure probability function $h : V \rightarrow \mathbb{R}^+$. The failure probability of a vertex is the probability that the vertex is removed and it is independently given to the vertices.

When V_P is a vertex subset, the probability that V_P is a (i,L)-protected vertex set for all integers $i(\geq 0)$, $C(V_P)$, is defined as follows. We call $C(V_P)$ the survival probability.

$$\sum_{i=0}^{|V|} \sum_{V_K \in R(i,L)} \prod_{\forall v \in V_K} h(v) \prod_{\forall v' \notin V_K} (1 - h(v'))$$

This implies the probability that the number of nodes in the remaining connected components in an information network is larger than or equal to L for any connected component, regardless of any node failure except protected nodes, when the nodes corresponding to the protected vertex set V_P in the information network do not fail.

Problem PNPP
Input: *Network* $N = (G = (V,E), h)$, *positive integers* p, L.
Constraint: $|V_P| \leq p$
Objective: $C(V_P)$ *(maximization)*
Output: *Set of protected edges,* $V_P(\subseteq V)$.

We show an example of this problem in Fig. 1, Fig. 2, Fig. 3, and Fig. 4. We assume that $L = 4$ in this example. The value of a vertex in Fig. 1 is the failure probability of the vertex. In Fig. 2, when no vertices are protected, the all cases that the minimum size of the remaining connected components is more than or equal to positive integer L and their probability are enumerated. In Fig. 3, when vertex t_1 is protected, the all cases that the minimum size of the remaining connected components is more than or equal to positive integer L and their probability are enumerated. Similarly, in Fig. 4, when vertex t_5 is protected, the all cases that the minimum size of the remaining connected components is more than or equal to positive integer L and their probability are enumerated. In Fig. 2, $V_P = \{\emptyset\}$; in Fig. 3, $V_P = \{t_1\}$; in Fig. 4, $V_P = \{t_5\}$. In Fig. 2, $C(V_P)$ is 0.52192; in Fig. 3, $C(V_P)$ is 0.5936; in Fig. 4, $C(V_P)$ is 0.6608. These example shows that the survival probability increases by node protection. However, in both cases of Fig. 3 and Fig. 4, although the number of protected vertices is one, the survival probabilities are different. The objective of PNPP is to determine a set of protected vertices so as to maximize the survival probability under the constraint of the number of the protected vertices.

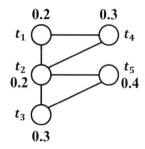

Fig. 1. Example of network and failure probability

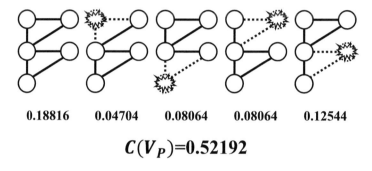

0.18816   0.04704   0.08064   0.08064   0.12544

$$C(V_P)=0.52192$$

Fig. 2. Survival probability when no vertices are protected

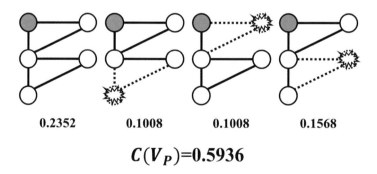

0.2352   0.1008   0.1008   0.1568

$$C(V_P)=0.5936$$

Fig. 3. Survival probability when vertex t_1 is protected

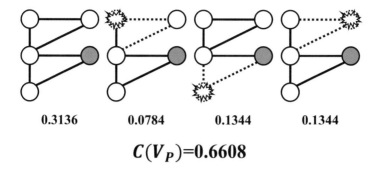

Fig. 4. Survival probability when vertex t_5 is protected

PNPP corresponds to the problem defined in [10] that is NP-hard, when we restrict PNPP so that the failure probability is one for all vertices and $p = 2$. Therefore, PNPP is NP-hard.

Theorem 1. *PNPP is NP-hard.*   □

We design a polynomial-time heuristic algorithm. We describe the basic idea as follows. First, we define the survival probability of vertex v as $c(v) = 1 - h(v)$. We solve the minimum vertex cover problem that the weight of a vertex is the survival probability of the vertex, and choose p vertices with small survival probability of vertex and large degree. In other words, the algorithm protects the vertices with large failure probability and large degree.

We used the approximation algorithm in [16] to solve the minimum vertex cover problem. The function $min\_weighted\_VC(G,c)$ in the following algorithm outputs the approximation solution of the minimum vertex cover problem for the network (G,c) whose weight function is c.

Algorithm 1: AlgorithmPNPP

Input: Network $N = (G = (V,E), h)$, a positive integer p.
Output: Set of protected vertices, V_P.

1 $c \leftarrow \{c(v) = 1 - h(v) | \forall v \in V\}$
2 $V_{MWVC} \leftarrow min\_weighted\_VC(G,c)$
3 **if** $|V_{MWVC}| \geq p$ **then**
4 $\quad$ Let V_P be a set of p vertices from V_{MWVC} in descending lexicographic order of pairs of failure probability and degree.
5 **else**
6 $\quad$ Let V_t be a set of $|V_{MWVC}| - p$ vertices from $V \backslash V_{MWVC}$ in descending lexicographic order of pairs of failure probability and degree.
7 $\quad$ $V_P \leftarrow V_{MWVC} + V_t$
8 **return** V_P

---

4   Performance Evaluation

In this section, we evaluate the proposed algorithms by using the graph structures of 18 actual ISP backbone networks provided by CAIDA (Center for Applied Internet Data Analysis) [17].

We show the results of the survival probability. In Table 1 (resp. Table 2), we assume that L is the value of 30% (resp. 50%) of the number of vertices and that the failure probability of a vertex is randomly given between 0.1and 0.3. In both tables, we show the relationship between the survival probability and the number of the protected vertices. We assume that the number of the protected vertices is 5%, 10%, and 25% of the number of vertices.

Table 1. Result of survival probability (L: 30%)

No.	n	m	L	noprotect $C(V_p)$	pnpp(5%) $C(V_p)$	V_p	pnpp(10%) $C(V_p)$	V_p	pnpp(25%) $C(V_p)$	V_p
1	5	4	2	0.879	-	-	-	-	0.973	1
2	6	5	2	0.702	-	-	1.000	0	1.000	0, 3
3	8	7	2	0.490	-	-	0.700	0	1.000	0, 4
4	8	7	2	0.980	-	-	1.000	4	1.000	4, 5
5	9	8	3	0.996	-	-	0.999	0	1.000	0, 1
6	11	10	3	0.396	0.660	2	0.730	2, 0	1.000	2, 0, 3
7	13	12	4	0.496	0.709	1	0.970	1, 8	0.970	1, 8, 10
8	13	12	4	0.406	0.580	1	0.828	1, 2	0.910	1, 2, 6
9	15	14	5	0.315	0.449	10	0.616	10, 8	0.804	10, 8, 1, 2
10	16	15	5	0.370	0.509	8	0.698	8, 6	0.899	8, 6, 2, 5
11	17	16	5	0.174	0.189	4	0.259	4, 3	0.509	4, 3, 0, 9
12	19	18	6	0.471	0.557	7	0.674	7, 10	0.797	7, 10, 1, 4, 5
13	21	20	6	0.446	0.637	1	0.910	1, 5	1.000	1, 5, 20, 3, 4
14	22	39	7	0.277	0.309	8	0.311	8, 12	0.484	8, 12, 5, 3, 4
15	23	22	7	0.252	0.359	9	0.408	9, 5	0.700	9, 5, 21, 3, 7, 14
16	27	26	8	0.258	0.368	1	0.511	1, 10, 11	1.000	1, 10, 11, 14, 15, 13, 16
17	28	56	8	0.118	0.234	6, 1	0.274	6, 1, 3	0.711	6, 1, 3, 4, 7, 5, 0
18	29	28	9	0.098	0.200	2, 15	0.285	2, 15, 6	0.828	2, 15, 6, 8, 22, 4, 7

We can observe that, according to the number of the protected vertices, the survival probability increases in all networks. When L is smaller, the connected components are more fragmented; therefore, it is desirable that L is large. However, there are few difference of the number of protected nodes in these information networks.

We show some examples of the information networks and the protected vertices in Fig. 5 and Fig. 6.

Table 2. Result of survival probability (L: 50%)

No.	$\|n\|$	m	L	noprotect $C(V_p)$	pnpp(5%) $C(V_p)$	pnpp(5%) V_p	pnpp(10%) $C(V_p)$	pnpp(10%) V_p	pnpp(25%) $C(V_p)$	pnpp(25%) V_p
1	5	4	3	0.864	-	-	-	-	0.967	1
2	6	5	3	0.701	-	-	0.999	0	1.000	0, 3
3	8	7	4	0.490	-	-	0.700	0	1.000	0, 4
4	8	7	4	0.930	-	-	0.971	4	0.989	4, 5
5	9	8	5	0.901	-	-	0.942	0	0.971	0, 1
6	11	10	6	0.395	0.659	2	0.729	2, 0	1.000	2, 0, 3
7	13	12	7	0.496	0.709	1	0.970	1, 8	0.970	1, 8, 10
8	13	12	7	0.406	0.580	1	0.828	1, 2	0.910	1, 2, 6
9	15	14	8	0.310	0.443	10	0.607	10, 8	0.786	10, 8, 1, 2
10	16	15	8	0.370	0.509	8	0.698	8, 6	0.899	8, 6, 2, 5
11	17	16	9	0.174	0.189	4	0.259	4, 3	0.509	4, 3, 0, 9
12	19	18	10	0.459	0.546	7	0.669	7, 10	0.797	7, 10, 1, 4, 5
13	21	20	11	0.446	0.637	1	0.910	1, 5	1.000	1, 5, 20, 3, 4
14	22	39	11	0.277	0.309	8	0.311	8, 12	0.484	8, 12, 5, 3, 4
15	23	22	12	0.248	0.353	9	0.408	9, 5	0.700	9, 5, 21, 3, 7, 14
16	27	26	14	0.258	0.368	1	0.511	1, 10, 11	1.000	1, 10, 11, 14, 15, 13, 16
17	28	56	14	0.118	0.234	6, 1	0.274	6, 1, 3	0.711	6, 1, 3, 4, 7, 5, 0
18	29	28	15	0.098	0.200	2, 15	0.285	2, 15, 6	0.828	2, 15, 6, 8, 22, 4, 7

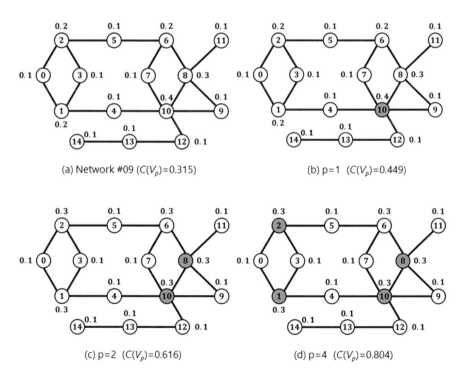

(a) Network #09 ($C(V_p)$=0.315)

(b) p=1 ($C(V_p)$=0.449)

(c) p=2 ($C(V_p)$=0.616)

(d) p=4 ($C(V_p)$=0.804)

Fig. 5. Network #09 and protected vertices

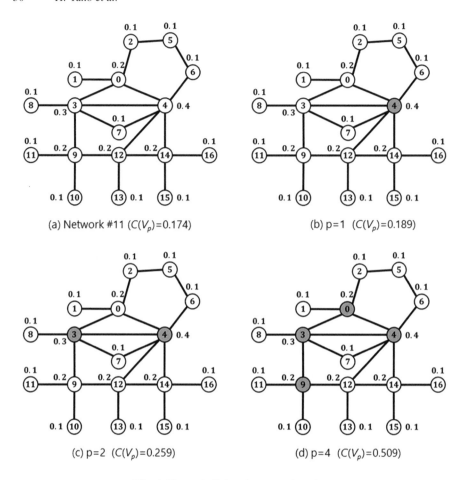

Fig. 6. Network #11 and protected vertices

5 Conclusions

In this paper, we addressed the problem of determining a set of the protected nodes that achieves the maximum probability that the number of nodes in the connected components of the network remaining after the failures of any non-protected nodes under cost constraint exceeds a given threshold.

We formulated this problem mathematically, showed its NP-hardness, and designed a polynomial-time heuristic algorithm. Furthermore, we evaluated the performance of the algorithm by using the topology of the actual information networks. In the numerical experiments, we used much larger failure probability than the actual probability; however, the survival probability of about a half of the information networks is almost one when a quarter of the number of nodes are protected.

For the future study, it remains to design more efficient approximation algorithm.

Acknowledgements. This work was partially supported by the Japan Society for the Promotion of Science through Grants-in-Aid for Scientific Research (B) (17H01742).

References

1. Albert, R., et al.: Error and attack tolerance of complex networks. Nature **406**, 378–382 (2000)
2. Arulselvan, A., Commander, C.W., Elefteriadou, L., Pardalos, P.M.: Detecting critical nodes in sparse graph. Proc. Comput. Oper. Res. **36**(7), 2193–2200 (2009)
3. Imagawa, K., Miwa, H.: Approximation algorithm for finding protected links to keep small diameter against link failures. In: Proceedings of INCoS 2011, Fukuoka, Japan, pp. 575-580, 30 November–2 December 2011 (2011)
4. Imagawa, K., Miwa, H.: Detecting protected links to keep reachability to server against failures. In: Proceedings of ICOIN 2013, Bangkok, 28–30 January 2013 (2013)
5. Irie, D., Kurimoto, S., Miwa, H.: Detecting critical protected links to keep connectivity to servers against link failures. In: Proceedings of NTMS 2015, Paris, 27–29 July 2015, pp. 1–5 (2015)
6. Maeda, N., Miwa, H.: Detecting critical links for keeping shortest distance from clients to servers during failures. In: Proceedings of HEUNET 2012/SAINT 2012, Turkey, 16–20 July 2012 (2012)
7. Fujimura, T., Miwa, H.: Critical links detection to maintain small diameter against link failures. In: Proceedings of INCoS 2010, Thessaloniki, 24–26 November 2011, pp. 339–343 (2011)
8. Yamasaki, T., Anan, M., Miwa, H.: Network design method based on link protection taking account of the connectivity and distance between sites. In: Proceedings of INCoS 2016, Ostrava, 7–9 September 2016, pp. 339–343 (2016)
9. Irie, D., Anan, M., Miwa, H.: Network design method by finding server placement and protected links to keep connectivity to servers against link failures. In: Proceedings of INCoS 2016, Ostrava, 7–9 September 2016, pp. 439–344 (2016)
10. Matsui, T., Miwa, H.: Method for finding protected nodes for robust network against node failures. In: Proceedings of INCoS 2014, Salerno, Italy, 10–12 September 2014, pp. 378–383 (2014)
11. Aggarwal, K.K., Chopra, Y.C., Bajwa, J.S.: Topological layout of links for optimising the overall reliability in a computer communication system. Microelectron. Reliab. **22**, 347–351 (1982)
12. Jan, R.-H., Hwang, F.-J., Chen, S.-T.: Topological optimization of a communication network subject to a reliability constraint. IEEE Trans. Reliab. **42**, 63–70 (1993)
13. Koide, T., Shinmori, S., Ishii, H.: The evaluations on lower bounds of all-terminal reliability by arc-packings for general networks. IEICE Trans. Fundam. Electron. Commun. Comput. Sci. **E82A**(5), 784–791 (1990)
14. Uji, K., Miwa, H.: Method for finding protected links to keep robust network against link failure considering failure probability. In: Barolli, L., Woungang, I., Hussain, O.K. (eds.) INCoS 2017. LNDECT, vol. 8, pp. 413–422. Springer, Cham (2018). https://doi.org/10.1007/978-3-319-65636-6_37
15. Morino, Y., Miwa, H.: Network design method considering cost to decrease failure probability. In: Advances in Intelligent Networking and Collaborative Systems, pp. 493–502 (2020)
16. Yehuda, R.B., Even, S.: Local-ratio theorem for approximating the weighted vertex cover problem. Ann. Discrete Math. **25**, 27–46 (1985)
17. CAIDA. http://www.caida.org/

Toward Secure K-means Clustering Based on Homomorphic Encryption in Cloud

Zheng Tu$^{(\boxtimes)}$, Xu An Wang, Yunxuan Su, Ying Li, and Jiasen Liu

Engineering University of PAP, Xi'an, China

Abstract. In recent years, the cloud computation has provided great convenience for users to outsource data for storage and computation. But when data stored in cloud, it's out of control from user, and there is a risk of private data leakage. In this paper, we propose a framework of secure K-means (CKKSKM) based on homomorphic encryption. The proposed scheme encrypts outsourcing data which can avoid to reveal the private information. We preprocess the data first and uses the Euclidean distance to calculate similarity, which can ensure the data accuracy and reduce the computing overhead. Based on CKKS homomorphic encryption, this scheme can solve the privacy security and reduce overhead of the user's outsourcing data in cloud for storage and calculation.

1 Introduction

In recent years, we have entered the era of big data. The cloud server platform helps user to manage data with sufficient storage space and powerful computing capabilities. User can upload data to a cloud server for storage, and use the computing power provided by the cloud server to process the data, which helps to reduce user overhead and improve data computation efficiency. However, users cannot control the data when the data stored in cloud. Once the data is outsourced, the sensitive information such as personal information, medical records, emails, etc., is at the risk of leakage. How to protect sensitive data stored in the cloud has been a hot research topic for a long time.

The traditional encryption algorithms cannot calculate encrypted data, but the HE (homomorphic encryption) can do this. Its decrypted result of the operation on ciphertext is equivalent to the result of the operation on plaintext. According to the types and times of operations it supports, the Homomorphic encryption is divided into the following kinds: 1.FHE (fully homomorphic encryption) which supports multiple operations and the number of operations is infinite. The representative schemes are CKKS and BGV; 2.PHE (partial homomorphic encryption) which supports some types of operations, generally addition or multiplication. And the number of operations is finite, the representative scheme is Paillier. 3.SHW (somewhat homomorphic encryption) which supports some operations and the number of operations is limited. In 2009, a fully homomorphic encryption scheme based on the ideal lattice was first proposed by Gentry[1]. Through fully homomorphic encryption, we can perform unlimited operations on the ciphertext without decrypting the data, which are equivalent to those on the plaintext. While realizing data calculation, it will not affect the confidentiality of data and avoid the

L. Barolli et al. (Eds.): EIDWT 2022, LNDECT 118, pp. 52–62, 2022.
https://doi.org/10.1007/978-3-030-95903-6_7

leakage of sensitive data, so that it can play an important role in the field of privacy protection. The interaction model is shown as Fig. 1. In 2012, Brakerski et al. proposed the BGV homomorphic encryption scheme [2]. In 2017, the CKKS scheme was proposed by Cheon et al. based on the BGV, which can support approximate operations on real numbers (complex numbers) and can realize homomorphic encryption on floating-point numbers [3]. Because CKKS solution has better computing efficiency while ensuring security, it is widely used in the industry.

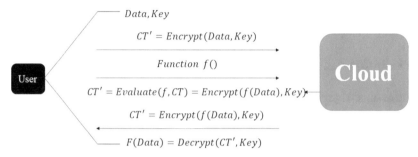

Fig. 1. Interaction model

Relying on the cloud computing platform, machine learning is developing rapidly. Common machine learning algorithms are mainly divided into two types depending on whether there is prior knowledge in the data to be processed: supervised learning and unsupervised learning. The clustering algorithm is an important part of unsupervised learning. It can cluster and compare entities or abstract things based on the similarity or dissimilarity between various classes to mine the implicit associations of big data. The K-means clustering algorithm is one of them. By calculating the similarity of sample data, it clusters samples with high similarity into one cluster. In the traditional K-means clustering scheme, the user directly transmits the unencrypted sample to the cloud server. And the cloud severs store data or return the result to the user after calculating. It helps a lot, but the security of data during interaction cannot be guaranteed. Thus, how to prevent the privacy and security of outsourcing data from being broken is an interesting problem. This paper proposes a secure K-means clustering scheme (CKKSKM) based on CKKS, which can both make use of the computing power of cloud and achieve safe and efficient clustering while protecting the security of sensitive data.

2   Related Work

Since we are in the era of big data, study of clustering algorithms has made great progress. there are many methods which can obtain the similarity between samples. For example, Euclidean distance, cosine distance, Manhattan distance, or Pearson similarity, etc. There are many classic clustering algorithms, such as K-means, KNN, DBSCAN, etc. K-means is a classic one in clustering algorithms and widely used in data mining. While enjoying the convenience brought by it, people also want to get the protection of data privacy.

Some researchers have done relate work and proposed some secure scheme of clustering algorithms [4–14]. In 2015, a K-means clustering algorithm was proposed by Rao et al., which can protect the sensitive data through Paillier [4]. In 2018, Kim et al. proposed an effective privacy protection k-Means clustering algorithm, which can quickly compare encrypted data. This algorithm gets improved performance based on Rao's scheme [5]. Wu et al. proposed an improved K-means algorithm, which acquire trustworthy sequences through global motion estimation [6]. In 2017, Nawal Almutairi et al. proposed a secure k-means through HE and updatable distance matrix [7]. In 2020, Qin et al. proposed a constant round secure multi-user k-means cluster protocol which can be used in multi-user setting [8].

3 Preliminaries

3.1 K-means Algorithm

In big data processing algorithms, clustering algorithms are generally used as the basis of other algorithms, which can cluster data, find some laws and the internal structure of the data. There are many types of clustering algorithms. K-means is one kind of partition-based clustering algorithm. Because of its simple structure, convenient and practical characteristics, it is widely used and was one of the most classic data mining algorithms. The idea is to first determine the initial cluster center in advance, and then classify the samples through the similarity between the sample and the cluster center. The higher the similarity between the samples, the more likely they are to be in the same cluster. K-means is an iterative loop algorithm, and its workflow is as follow:

a. Define the number of the clusters as K. Specify in advance that how many clusters the samples need to be clustered into. Give the cluster centers according to the number of clusters.
b. Calculate and compare the similarity between the sample wait for being clustered and each cluster center. According to the principle of maximizing similarity with the cluster center, the samples are clustered into clusters to which each cluster center belongs.
c. There are several samples in each category, and the cluster centers of various clusters in the second iteration are calculated through corresponding strategies.
d. Repeat b, c and iterate until the calculation result converges (the cluster centers are fixed or reaches the limited number of iterations), and return the result finally.

Define the samples as $[X_1, X_2, ..., X_n]$. The X is the characteristic indicators. For a given sample X, by calculating the similarity between X and the cluster centers, the samples are clustered according to the principle of the highest similarity with the cluster centers. Then, we re-determine the cluster centers, calculate the similarity again, and continue to iterate until the calculation results converge. Examples are as follows:

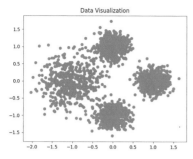

Fig. 2. Raw data distribution

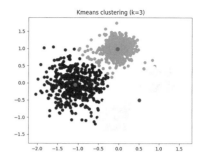

Fig. 3. Clustering result when K = 3

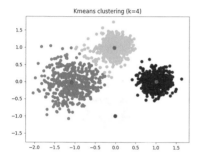

Fig. 4. Clustering result when K = 4

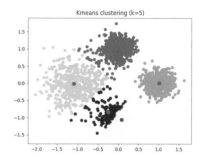

Fig. 5. Clustering result when K = 5

As shown in the Fig. 2, it is an initial sample point set generated with 4 center points. As shown in the Fig. 3, when K = 3, there are three cluster centers, and the sample point set is clustered into three clusters. As shown in the Fig. 4, when K = 4, there are four cluster centers, and the sample point set is clustered into 4 clusters. As shown in Fig. 5, when k = 5, there are five cluster centers, and the sample point set is clustered into 5 clusters.

3.2 CKKS Homomorphic Encryption Scheme

Cheon et al. proposed CKKS which can perform approximate operations on real/complex numbers [3]. The CKKSKM based on CKKS scheme and the algorithm is as follows.
Initialization:

a. Choose the secure parameter λ, the power of integers N.
b. b. Respectively choose χ_{key}, χ_{err}, χ_{enc} on $R = \mathbb{Z}[X]/(X^N + 1)$ as key distribution, error learning distribution, and encryption distribution respectively.
c. c. Define p as the basic integer, L as the level. Choose modulus on $q_l = p^l (1 \leq l \leq L)$ for the ciphertext and then randomly generate an integer P, output $pp = (N, \chi_{key}, \chi_{err}, \chi_{enc}, L, q_l)$.

1) **KeyGen** $(params) \rightarrow (pk, sk, ks, rk_r, ck)$.

    a. $s \leftarrow \chi_{key}, sk \leftarrow (1, s)$
    b. $a \leftarrow U(R_{q_l}), e \leftarrow \chi_{err}, pk \leftarrow (-as + e, a) \in R_{q_L}^2$
    c. $a' \leftarrow U(R_{q^2 l}), e' \leftarrow \chi_{err}, evk \leftarrow (-a's + e' + q_L S^2, a') \in R_{q_L^2}^2$

2) ***Encrypt***$(m, pk) \rightarrow ct$.
    $r \leftarrow \chi_{enc}, e\_0, e\_1 \leftarrow \chi\_err$.
    $ct = r \cdot pk + (m + e_0, e_1)(mod q_L)$.

3) ***Decrypt***$(ct, sk) \rightarrow m$.
    $m = < ct, sk > (mod q_l)$.

4) ***Add***$(ct, ct') \rightarrow ct_{add}$.
    $ct_{add} = ct + ct' (mod q_l)$.

5) ***Mult***$_{ks}(ct, ct') \rightarrow ct_{mult}$.
    $ct = (c_0, c_1), ct' = \left(c_0', c_1'\right) \in R_{q_l}^2$.
    $(d_0, d_1, d_2) = (c_0 c_0', c_0 c_1' + c_0' c_1, c_1 c_1')(mod q_l)$.
    $ct_{mult} = (d_0, d_1) + \left\lfloor P^{-1} \cdot d_2 \cdot sk \right\rfloor (mod \; q_1)$.

The CKKS scheme is homomorphic, so decrypted result from the operations on the encrypted data is equivalent to the result from some operations on the plaintext. While protecting the confidentiality of outsourced data, it also improves the efficiency of computation on data.

3.3 Symbol Description

In order to show the meaning of the formulas in the paper, Table 1 introduces the commonly used symbols in this paper. Lowercase bold letters are used to represent vectors and uppercase bold letters are used to represent matrices. The *enc_* before the symbol means the data is encrypted.

Table 1. Notations

Symbols	Description
X	Sample
$\tilde{X}$	Standardized sample
K	Cluster center
pk	Public key
sk	Private key
d	Euclidean distance

4 Model Analysis

4.1 CKKSKM Model

There are two main roles in the CKKSKM scheme: the USER and the cloud server provider (CSP). The CSP in the model is semi-trusted and can provide severs to the

USER, such as remote storage and outsourcing computing. The USER has a large amount of data resources locally. The data after being encrypted can safely store in cloud and calculated by CSP. The model is shown in Fig. 6:

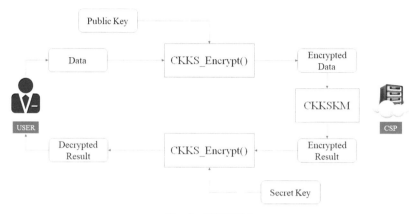

Fig. 6. CKKSKM

1) **USER**: Generate pairs of public and private key locally; encrypt data through public key and upload the encrypted data to the CSP for storage or calculation; it can receive the encrypted result returned by the CSP, and decrypt it to get the result with private key.
2) **CSP**: With large storage space and strong computing power, it provides users outsourced storage and computation. it can store the encrypted data uploaded by USER; it can perform operations on the stored encrypted data to complete the K-means clustering algorithm, and return the encrypted result to USER.

a. USER generates a pair of keys including public key and private key locally. And then sends the ciphertext to the CSP after encryption;
b. CSP receives the ciphertext sent by USER and stores it in cloud.
c. When USER needs to cluster the ciphertext of the samples stored in the CSP, USER selects the initial clustering center locally, encrypts it after standardization, and sends it to CSP.
d. CSP receives the ciphertext, calculates the similarity between samples and the clustering centers on ciphertext, and clustering the samples. When the clustering result converges, the encrypted result will be sent to USER.
e. USER receives the encrypted result returned by CSP, and obtains the result after decryption.

4.2 Security Model

In the actual interaction, CSP is untrusted or semi-trusted. And CSP will analyze the data to get useful information. Data may be intercepted by attacker during the transmission of USER and CSP. Based on the above setting, we have listed the security issues that may occur during USER's K-means clustering process:

1) CSP uses and analyzes the encrypted data stored by USER in the server, and attempts to get the key of USER. It tries to crack the encrypted data of USER to obtain the private information about plaintext.
2) During the data interaction between USER and CSP, the data may be intercepted by an attacker. The attacker attempts to crack the private keys of USER by analyzing the intercepted data and tries to obtain valuable information by cracking the encrypted data.
3) The content of the homomorphic K-means algorithm is obtained by the CSP or an attacker, and the CKKSKM algorithm is used to get the user's sensitive information.

5 System Algorithm

5.1 CKKSKM Framework

Choose the samples as $[X_1, X_2, ..., X_n]$ where $X_j = (x_1, x_2, ..., x_m)^T$. As shown in Fig. 7, the protocol consists of two parts: Initialization and Clustering. The entire process of CKKSKM is as follows:

1) **Initialization:**
   a. The feature indicators of the samples is standardized by USER locally

   $$\tilde{x}_{ij} = \frac{x_{ij} - \overline{x_j}}{\sqrt{var(x_j)}}, i \in [1, n], j \in [1, m] \tag{1}$$

   Calculate the average value of the j-th feature.

   $$\overline{x_j} = \frac{1}{n} \sum_{i=1}^{n} x_{ij}, \tag{2}$$

   Calculate standard deviation of the j-th feature

   $$var(x_j) = \frac{1}{n-1} \sum_{i=1}^{n} (x_{ij} - \overline{x_j})^2, \tag{3}$$

   Finally, we get the standardized data $[\tilde{X}_1, \tilde{X}_2, ..., \tilde{X}_n]$, where the average value is 0, the variance is 1, and it is dimensionless.
   b. USER generates (pk, sk)(public key and private key) locally; encrypts the original data of samples to obtain enc_X; encrypts the standardized data to obtain $enc\_\tilde{X}$; Both enc_X and $enc\_\tilde{X}$ are uploaded to CSP.
2) **Clustering:**
   a. USER chooses the initial clustering centers, standardizes these centers to obtain $\tilde{K}$, then encrypts $\tilde{K}$ to obtain $enc\_\tilde{K}$, and $enc\_\tilde{K}$ is uploaded to CSP.

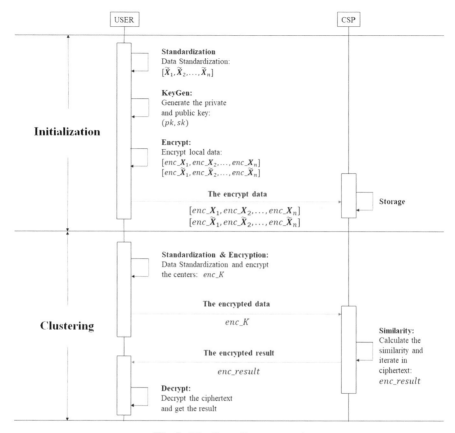

Fig. 7. The flow of our protocol

b. After receiving $\widetilde{K}$, CSP calculates the ciphertext similarity between the sample and the cluster center:

$$enc_{d\,ij} = \left(enc_{\underset{X}{\sim}} - enc_{\underset{K}{\sim}}\right)^2 = \sum_{i=1}^{n}\sum_{j=1}^{K}\left(enc_{\tilde{X}_i} - enc_{K_j}\right)^2 \tag{4}$$

Through enc_d to cluster the samples to each cluster. After clustering, we need to calculate the cluster center in the next round of clustering. Suppose the number of samples in the center of a certain type of cluster is N:

$$enc\_\tilde{K}_1' = \frac{1}{N}\sum_{x\in C_1} enc\_x \tag{5}$$

After that, we iterate continuously until the calculation result converges to get enc_result and send it back to USER.

c. USER receives $enc\_\widetilde{K}$ and obtain the clustering result through decrypting it.

5.2 Similarity Calculation

In K-means algorithm, there are many methods to obtain the similarity between the sample and each cluster center, such as Euclidean distance, cosine distance, Manhattan distance, or Pearson similarity, etc. In this paper, we use Euclidean distance to calculate the similarity of two samples $x = [x_1, x_2, ..., x_n], y = [y_1, y_2, ..., y_n]$

$$dist_{ed}(X, Y) = \|X - Y\|^2 = \sqrt{(x_1 - y_1)^2 + ... + (x_n - y_n)^2} \tag{6}$$

In the CKKSKM protocol, we use the Euclidean distance between the sample and the cluster center as the similarity. The CKKS scheme does not support direct square root operations. So there is a new method which can avoid square root. Suppose that the sample $A(x_1, y_1)$, the cluster center $K(c, d)$

$$enc_d = (enc_A - enc_K)^2 = (enc\_x_1 - enc\_c)^2 + (enc\_y_1 - enc\_d)^2 \tag{7}$$

5.3 Iterative Cluster Center Calculation

a. Calculate the Euclidean distance of each sample to each cluster centers.
b. According to the calculation results, the samples are clustered to the clusters with the highest similarity.
c. After all the sample are clustered, the cluster centers are recalculated:

The difference in the clusters can be judged by using the sum of squared errors as the objective function:

$$SSE = \sum_{i=1}^{K} \sum_{x \in C_i} (C_i - x)^2 \tag{8}$$

$$\frac{\partial SSE}{\partial C_k} = \frac{\partial}{\partial C_K} \sum_{i=k}^{K} \sum_{x \in C_i} (C_i - x)^2$$

$$= \sum_{i=k}^{K} \sum_{x \in C_i} \frac{\partial}{\partial C_K} (C_i - x)^2$$

$$= \sum_{i=k}^{K} \sum_{x \in C_i} \frac{\partial}{\partial C_K} (C_i - x)^2$$

$$= \sum_{x \in C_i} 2(C_i - x) = 0$$

$$\sum_{x \in C_i} 2(C_i - x) = 0 \rightarrow m_i C_i = \sum_{x \in C_i} x \rightarrow C_i = \frac{1}{m} \sum_{x \in C_i} x$$

When the SSE is minimum, the value of the cluster center is the mean value of the cluster. Therefore, we calculate the average value of the samples in clusters as the cluster center of the next round of clustering.

d. Repeat steps a-c when calculating each round of cluster centers until the calculation results converge.

6   Security Analysis

In CKKSKM model, the CSP is semi-trusted and will try its best to get the private information of data. The private key which can decrypt the ciphertext is only stored by USER. CSP only provides outsourced storage and calculation services, and stores and calculates the encrypted data uploaded by USER. CSP cannot obtain the private key of USER.

During the data transmission process of CKKSKM, the data of samples and cluster centers are encrypted by CKKS. The data that the CSP and the attacker can obtain are all encrypted. The security of the encrypted data is based on CKKS scheme. The probability that the CSP and the attacker get the plaintext without USER's private key is equivalent to breaking CKKS scheme. The security of the CKKS scheme based on the hard problem of LWE (learning with errors), so the probability of breaking the CKKS scheme is equivalent to solving LWE difficult problem. LWE proved to be difficult, so the probability of CSP and adversary breaking the CKKS scheme is negligible, the probability of CSP and adversary obtaining plaintext without the private key is also negligible. In summary, USER's private data is secure.

7   Conclusion

In this paper we proposed a secure K-means clustering scheme in cloud setting based on CKKS. When users outsourcing data to cloud for storage and computing, CKKSKM scheme can not only keep the security of the privacy issues but also reduce user's overhead. At the same time, we preprocess the data and use the Euclidean distance to calculate similarity, which can guarantee the data accuracy and reduce the computing overhead. In the future, we'll study the iterative calculation of the selection of K and clustering centers, try to reduce the overhead of USER and CSP and implement more secure and efficient scheme in the encrypted domain.

Acknowledgments. This work is supported by Natural Science Basic Research Plan in Shaanxi Province of China [No. 2018JM6028], the Foundation of Guizhou Provincial Key Laboratory of Public Big Data [No. 2019BDKFJJ008], Engineering University of PAP's Funding for Scientific Research Innovation Team [No. KYTD201805] and Engineering University of PAP's Funding for Key Researcher [No. KYGG202011].

References

1. Gentry, C.: Fully homomorphic encryption using ideal lattices. In: ACM STOC, pp. 169–178 (2009)
2. Brakerski, Z., Vaikuntanathan, V.: Efficient fully homomorphic encryption from (standard) LWE. SIAM J. Comput. 831–871 (2014)
3. Cheon, J.H., Kim, A., Kim, M., Song, Y.: Homomorphic encryption for arithmetic of approximate numbers. In: Takagi, T., Peyrin, T. (eds.) Advances in Cryptology. Lecture Notes in Computer Science, vol. 10624, pp. 409–437. Springer, Heidelberg (2017). https://doi.org/10.1007/978-3-319-70694-8_15

4. Rao, F., Samanthula, B.K., Bertino, E., Yi, X., Liu, D.: Privacy-preserving and outsourced multi-user K-Means clustering. In: 2015 IEEE Conference on Collaboration and Internet Computing (CIC), pp. 80–89 (2015)
5. Kim, H., Chang, J.W.: Privacy-preserving k-means clustering algorithm using secure comparison protocol and density-based center point selection. In: 2018 IEEE 11th International Conference on Cloud Computing (CLOUD), pp. 928–931 (2018)
6. Wu, M., et al.: Robust global motion estimation for video security based on improved k-means clustering. J. Ambient. Intell. Humaniz. Comput. **10**(2), 439–448 (2018). https://doi.org/10. 1007/s12652-017-0660-8
7. Almutairi, N., Coenen, F., Dures, K.: K-Means clustering using homomorphic encryption and an updatable distance matrix: secure third party data clustering with limited data owner interaction. In: Bellatreche, L., Chakravarthy, S. (eds.) DaWaK 2017. LNCS, vol. 10440, pp. 274–285. Springer, Cham (2017). https://doi.org/10.1007/978-3-319-64283-3_20
8. Hong, Q., Hao, W., Xiaochao, W., Zhihua, Z.: Secure constant-round multi-user k-means clustering protocol. J. Comput. Res. Dev. **57**, 2188–2200 (2020)
9. Vaidya, J., Clifton, C.: Privacy-preserving k-means clustering over vertically partitioned data. In: SIGKDD, pp. 206–215 (2003)
10. Kim, H.-J., Kim, H.-I., Chang, J.-W.: A privacy-preserving kNN classification algorithm using Yao's garbled circuit on cloud computing. In: 2017 IEEE 10th International Conference on Cloud Computing (CLOUD), pp. 766–769 (2017)
11. Zhang, J., Wu, G., Hu, X., Li, S., Hao, S.: A parallel K-means clustering algorithm with MPI. In: 2011 Fourth International Symposium on Parallel Architectures, Algorithms and Programming, pp. 60–64 (2011)
12. Chitrakar, A.S., Petrovic, S.: Analyzing digital evidence using parallel k-means with triangle inequality on spark. In: Proceedings of the IEEE Big Data 2nd International Workshop on Big Data Analytic for Cyber Crime Investigation and Prevention (2018)
13. Huang, Z.: Extensions to the k-means algorithm for clustering large data sets with categorical values. Data Mining Knowl. Discov. **2**, 283 (1998)
14. Liu, J., Wang, C., Tu, Z., Wang, X.A., Lin, C., Li, Z.: Secure KNN classification scheme based on homomorphic encryption for cyberspace. Secur. Commun. Netw. **2021** Article ID 8759922 (2021)

On the Insecurity of a Certificateless Public Verification Protocol for the Outsourced Data Integrity in Cloud Storage

Xu An Wang$^{(\boxtimes)}$, Xiaozhong Pan, Lixian Wei, and Yize Zhao

Engineering University of People's Armed Police, Xi'an, People's Republic of China

Abstract. Cloud auditing is an important method to check the integrity of the outsourced cloud storage. Until now, there are many cryptographic ways to implement the cloud audit protocol, such as PKI based cloud audit protocol, identity based cloud audit protocol and certificateless based cloud audit protocol. In this short paper, we focus on a certificateless based cloud audit protocol proposed by Huang et al. in 2017. Although this is a valuable protocol, we show this protocol has some flaw. Concretely the certificateless signature can be forged, and thus the audit protocol is not secure.

1 Introduction

Cloud storage is now a very popular method for outsourced data management. But due to the accidents of cloud storage, such as the data losing, the leaking of data, data deleting etc., how to ensure the integrity of cloud storage is very important. Until now, there are many proposals aiming at solving this problem, such as provable data possession, proof of retrievability, cloud auditing protocols [1–3].

Generally speaking there are three ways for implementing cloud auditing protocol: PKI based cloud audit protocol, identity based cloud audit protocol and certificateless based cloud audit protocol.

1. The first one is the PKI based cloud audit protocol. In this kind of cloud audit protocol, the data owner has its own public key and private key, and the public key is certified by the certificate authority (CA) as a certificate. The data owner uses its private key to compute the signatures of the data blocks and then outsources the data blocks and the corresponding signature to the cloud server, later the data owner can challenge the cloud server on random data blocks for the signatures. The cloud server uses the stored data blocks and signatures to give the correct proof for the data possession. Finally the data owner uses the verification equation and public keys to check the correctness of the proof. If the equation is satisfied, the data owner can ensure the cloud server storing its outsourced data well, otherwise the outsourced data has been modified.

2. The second one is the identity based cloud audit protocol. In this kind of cloud audit protocol, the data owner uses its identity as the public key, and the private key is generated by the private key generator (PKG) which is an authority trusted by all the parties, based on the identity. The concrete steps of the protocol are the same as the PKI based one, except in this setting the public key is the identity.

© The Author(s), under exclusive license to Springer Nature Switzerland AG 2022
L. Barolli et al. (Eds.): EIDWT 2022, LNDECT 118, pp. 63–67, 2022.
https://doi.org/10.1007/978-3-030-95903-6_8

3. The third one is the certificateless based cloud audit protocol. In this kind of cloud audit protocol, the data owner generates its public key and private key by following the paradigm of certificateless cryptography. In certificateless cryptography, the data owner's secret key consists of two parts, one part is generated by the private key generator and the other part is generated by itself. In this way, the private key generator can not know the whole secret key for it does not know the part generated by the data owner itself. Furthermore, the adversary who replaces the data owner's public key can not decrypt the data owner's ciphertext for it only knows the private key corresponding to the replaced public key, but do not know the part generated by the data owner itself. Thus the public key infrastructure (PKI) is no longer needed to manage the certificates. The concrete steps of the certificateless based cloud audit protocol are the same as the PKI based one, except in this setting the cryptographic mechanism is the certificateless one.

In this short paper, we focus on the certificateless based cloud audit protocol. In 2017, Huang et al. proposed a certificateless based cloud audit protocol [4]. Although this is a valuable protocol, but there are some flaws in the protocol. Concretely the certificateless signature can be forged, and thus the audit protocol is not secure.

2 Review of Huang et al.'s Signature Scheme and Cloud Auditing Protocol

In this section we review Huang et al.'s signature scheme and cloud auditing protocol. Below are the five parts of their signature scheme. Concretely they are the following:

1. Setup. System parameters and a master key are generated by this algorithm with a security parameter k. KGC runs as following.
   a. First, a multiplicative G_2 and an additive cyclic group G_1 with the prime order $q \geq 2k$ are selected.
   b. Second, $e : G_1 \times G_1 \rightarrow G_2$ is chosen. $H_1 : \{0,1\}^* \rightarrow Z_q, H_2 : G_1 \times \{0,1\}^* \rightarrow G_1$ are two hash functions.
   c. Then, $x \in Z_q$ is randomly chosen and the public key is computed as $P_{pub} = x \cdot P$.
   d. Finally, master private x is kept secretly and $Param = (G_1, G_2, q, e, P, T, H_1, H_2, P_{pub})$ is published as system parameters.
2. Partial-private-extract. x_{ID} is randomly picked by the user i with identity ID as part of his secret key, and $P_{ID} = x_{ID} \cdot P$ is computed.
3. KeyGen. The user's identity-based private key is generated by this algorithm after giving the master key x, a user's public key P_{ID}, user's identifier ID and system parameters. A random r_{ID} is chosen by KGC, and it computes $h_{ID} = H_1(ID, R_{ID}, P_{ID})$ where $R_{ID} = r_{ID} \cdot P$. Then $s_{ID} = r_{ID} + h_{ID} \cdot x$ is computed by KGC and (s_{ID}, R_{ID}) is sent to the user securely.

By verifying the equality $s_{ID}P = R_{ID} + h_{ID}P_{pub}$, the validity of s_{ID} is checked by the user with (s_{ID}, R_{ID}). Because $s_{ID} = r_{ID} + h_{ID} \cdot x$, $R_{ID} = r_{ID} \cdot P$ and $x_{ID} \cdot P = P_{ID}$. Finally, $pk_{ID} = (P_{ID}, R_{ID})$ is the public key and the pair $sk_{ID} = (s_{ID}, x_{ID})$ is the private key.

4. Sign. With inputs a message m, an identity ID, abstract index,

$$Param = (G_1, G_2, q, e, P, T, H_1, H_2, P_{pub})$$

private key $sk_{ID} = (s_{ID}, x_{ID})$, this algorithm outputs a signature. First it computes $h = H_2(m||index||ID)$ and then computes the signature as $s = x_{ID} \cdot m \cdot T + h \cdot s_{ID}$. Finally the signature and the message are uploaded by the users to the cloud server.

5. Verify. The cloud server verifies $e(s, P) = e(T, m \cdot P_{ID})e(h, R_{ID})e(h \cdot h_{ID}, P_{pub})$ holds or not where $h_{ID} = H_1(ID, R_{ID}, P_{ID})$.

Signature correctness is verified by the below equation:

$$
\begin{aligned}
e(s, P) &= e(x_{ID} \cdot m \cdot T + hs_{ID}, P) \\
&= e(x_{ID} \cdot m \cdot T, P)e(h \cdot (r_{ID} + h_{ID} \cdot x), P) \\
&= e(x_{ID} \cdot m \cdot T, P)e(h \cdot r_{ID}, P)e(h \cdot h_{ID} \cdot x, P) \\
&= e(m \cdot T, x_{ID} \cdot P)e(h, r_{ID} \cdot P)e(h \cdot h_{ID}, x \cdot P) \\
&= e(T, m \cdot P_{ID})e(h, R_{ID})e(h \cdot h_{ID}, P_{pub})
\end{aligned}
$$

Huang et al.'s certificateless public auditing scheme is constructed on the above signature scheme with the following algorithms:

1. Sign. A data file $M = \{m_1, m_2, \cdots, m_n\}$ is first divided into n blocks with abstract index. Taken the system parameters

$$Param = (G_1, G_2, q, e, P, T, H_1, H_2, p_{pub})$$

his own private key $sk_{ID} = (s_{ID}, x_{ID})$ and a message m_i as inputs, the data user computes the signature $s_i = x_{ID} \cdot m_i \cdot T + h_i s_{ID}$ where $h_i = H_2(m_i||index||ID)$ $(1 \leq i \leq n)$. The user uploads $M = (m_1, \cdots, m_n)$ and $S = (s_1, \cdots, s_n)$ to the cloud server.

2. Challenge. A c-element subset I of set $[1, n]$ and a number l is randomly chosen by the verifier to produce the challenge $chall = \{l, I\}$ and is sent to the cloud server.

3. ProofGen. The cloud server computes a c-element set $C = \{i, v_i\}$ where $i \in I$, $v_i = l^i$ with the challenge $chall = \{l, I\}$. The values $S = \sum_{i \in I} v_i \cdot s_i$ $H = \sum_{i \in I} v_i \cdot h_i$, $\mu = \sum_{i \in I} v_i \cdot m_i$ are computed by the cloud server, and the proof message $Pro = \{S, H, \mu\}$ is sent to the verifier.

4. ProofVerify. With $Pro = \{S, H, \mu\}$ from cloud server, $h_{ID} = H_1(ID, R_{ID}, P_{ID})$ is first computed by the verifier and $e(S, P) = e(T, \mu \cdot P_{ID})$ is checked. If the equality holds, the verifier will accept the proof, otherwise the proof is rejected. The correctness of the proof can be verified by the following equation:

$$e(S,P) = e(\sum_{i\in I} v_i s_i, P)$$

$$= e(\sum_{i\in I} v_i(x_{ID} \cdot m_i \cdot T + h_i \cdot s_{ID}), P)$$

$$= e(\sum_{i\in I} v_i(x_{ID} \cdot m_i \cdot T + h_i \cdot (r_{ID} + h_{ID} \cdot x)), P)$$

$$= e(\sum_{i\in I} v_i x_{ID} \cdot m_i \cdot T, P) e(\sum_{i\in I} v_i h_i \cdot r_{ID}, P) e(\sum_{i\in I} v_i h_i \cdot h_{ID} \cdot x, P)$$

$$= e(\sum_{i\in I} v_i \cdot m_i \cdot T, x_{ID} \cdot P) e(\sum_{i\in I} v_i h_i, r_{ID} \cdot P) e(\sum_{i\in I} v_i h_i \cdot h_{ID}, x \cdot P)$$

$$= e(\mu T, P_{ID}) e(\sum_{i\in I} v_i h_i, r_{ID} \cdot P) e(\sum_{i\in I} v_i h_i \cdot h_{ID}, x \cdot P)$$

$$= e(T, \mu \cdot P_{ID}) e(H, R_{ID}) e(H \cdot h_{ID}, P_{pub})$$

Note in the above equation, $S = \sum_{i\in I} v_i \cdot s_i$, $H = \sum_{i\in I} v_i \cdot h_i$, $\mu = \sum_{i\in I} v_i \cdot m_i$ and $h_{ID} = H_1(ID, R_{ID}, P_{ID})$.

3   Our Attack

Our attack is based on the following observation: after the Type I adversary who can know the partial private key generated by the KGC (Type I adversary can be the malicious KGC). Only by obtaining a signature it can forge signature, concretely the attack runs as following:

1. The adversary obtains one signature

$$s_1 = x_{ID} \cdot m_1 \cdot T + h_1 s_{ID}$$

Here $h_1 = H_2(m_1 || index || ID)$, and the adversary knows s_{ID}
2. It computes the following

$$A = s_1 - h_1 s_{ID}$$
$$= x_{ID} \cdot m_1 \cdot T$$

and

$$m_1^{-1} \cdot A = x_{ID} \cdot T$$

3. With $s_{ID}, x_{ID}T$, the adversary can forge any signature for any message m' with index $index'$ as the following:

$$s' = m' \cdot x_{ID} \cdot T + h' s_{ID}$$

where $h' = H_2(m' || index' || ID)$

The adversary can collude with the cloud server, and in this way the cloud server can modify any outsourced data block and forge the corresponding signature. And finally the cloud server can easily forge the proof to pass the verifying equation. Concretely it runs as the following:

1. The cloud server runs the above attack to forge $M = (m_1, \cdots, m_n)$ and $S = (s_1, \cdots, s_n)$ to be $M' = (m'_1, \cdots, m'_n)$ and $S' = (s'_1, \cdots, s'_n)$.
2. Challenge. A c-element subset I of set $[1,n]$ and a number l is randomly chosen by the verifier to produce the challenge $chall = \{l, I\}$ and is sent to the cloud server.
3. ProofGen. The cloud server computes a c-element set $C = \{i, v_i\}$ where $i \in I$, $v_i = l^i$ with the challenge $chall = \{l, I\}$. The values $S' = \sum_{i \in I} v_i \cdot s'_i$ $H' = \sum_{i \in I} v_i \cdot h_i$, $\mu' = \sum_{i \in I} v_i \cdot m'_i$ are computed by the cloud server, and the proof message $Pro = \{S', H', \mu'\}$ is sent to the verifier.
4. ProofVerify. With $Pro = \{S', H', \mu'\}$ from cloud server, $h_{ID} = H_1(ID, R_{ID}, P_{ID})$ is first computed by the verifier and $e(S', P) = e(T, \mu' \cdot P_{ID})$ is checked.
   It is easy to check the correctness of the proof.

4 Conclusion

In this short paper, we show that one certificateless public verification protocol for the outsourced data integrity, although being valuable, but is not secure due to the underlying signature scheme is not secure.

Acknowledgements. This work is supported by Natural Science Basic Research Plan in Shaanxi Province of China [No. 2018JM6028], the Foundation of Guizhou Provincial Key Laboratory of Public Big Data [No. 2019BDKFJJ008], Engineering University of PAPs Funding for Scientific Research Innovation Team [No. KYTD201805] and Engineering University of PAPs Funding for Key Researcher [No. KYGG202011].

References

1. Ateniese, G., et al.: Provable data possession at untrusted stores. In: CCS 2007, pp. 598–609 (2007)
2. Shacham, H., Waters, B.: Compact proofs of retrievability. In: Pieprzyk, J. (ed.) ASIACRYPT 2008. LNCS, vol. 5350, pp. 90–107. Springer, Heidelberg (2008). https://doi.org/10.1007/978-3-540-89255-7_7
3. Cash, D., Küpçü, A., Wichs, D.: Dynamic proofs of retrievability via oblivious RAM. In: Johansson, T., Nguyen, P.Q. (eds.) EUROCRYPT 2013. LNCS, vol. 7881, pp. 279–295. Springer, Heidelberg (2013)
4. Huang, L., Zhou, J., Zhang, G., Sun, J., Wang, T., Vajdi, A.: Certificateless public verification for the outsourced data integrity in cloud storage. J. Circ. Syst. Comput. 27(11), 1850181 (2018)

An Improved Density Peaks-Based Graph Clustering Algorithm

Lei Chen[(✉)], Heding Zheng, Zhaohua Liu, Qing Li, Lian Guo, and Guangsheng Liang

School of Information and Electrical Engineering, Hunan University of Science and Technology, Xiangtan, China
chenlei@hnust.edu.cn

Abstract. The density peaks algorithm is a widely accepted density-based clustering algorithm, which shows excellent performance for many discrete data with any shape, and any distribution. However, because the traditional node density and density following distance does not match the graph data, the traditional density peaks model cannot be directly applied to graph data. To solve this problem, an improved density peaks graph clustering algorithm is proposed, simply called DPGC. Firstly, a novel node density is defined for the graph data based on the aggregation of the relative neighbors with a fixed number. Secondly, a density following distance search method is designed for graph data to calculate the density following distance of each node, so as to enhance the accuracy of selecting cluster centers. Finally, an improved density peaks model is constructed to quickly and accurately cluster the complex network. Experiments on multiple synthetic networks and real networks show that our algorithm offers better graph clustering results.

1 Introduction

Graph clustering (Community detection [1]) is an important task of network analysis, its purpose is to uncover the hidden structures, the internal association of nodes, or dynamic evolution from a complex network through an unsupervised mode, to get the valuable data in the complex network. In the real world, any complex network is a graphical abstraction of a complex system, which includes a large amount of information in the development process of a complex system from a few initial nodes to thousands of nodes today. In a complex network, there are usually multiple hidden structures or associated node groups with different shapes, different sizes, and different distributions, each of which represents a core component of the complex network having a great value. However, these hidden structures are formed after a long period of gradual accumulation and evolution, and they present the characteristics of unevenness, irregularity, and dense aggregation [2]. Therefore, it is a very challenging and meaningful to accurately realize graph clustering accurately from a complex network without any supervision information.

The density peaks algorithm is an excellent clustering algorithm that is widely accepted and used in the scope of discrete data [5]. This algorithm can get excellent performance in many discrete datasets with any shape, density, and distribution. However, the density peaks model cannot be directly applied to graph data (complex network). To

L. Barolli et al. (Eds.): EIDWT 2022, LNDECT 118, pp. 68–80, 2022.
https://doi.org/10.1007/978-3-030-95903-6_9

solve this problem, the graph data is usually projected from the high-dimensional space to the low-dimensional discrete vector space [13], and then the density peaks model is used to cluster the low-dimensional vectors, to achieve graph clustering. However, the data projection operation in this method will inevitably lose the important attributes of the graph data in the high-dimensional space, and bring additional time overhead, resulting in reduced clustering accuracy and increased time complexity.

Therefore, the motivation of this paper is to analyze the success of the density peaks model in discrete data, redefine and design the new density and density following distance that is inherently related to each other, and finally form a new customized density peaks model for more accurately and quickly uncover the hidden community structures within complex networks. Base on this, an improved density peaks-based graph clustering algorithm is proposed in this paper, simply called DPGC. Firstly, a relative neighbor is defined to ensure a common number of neighbors for each node. And a new density suitable for complex networks is defined by considering the node cohesion within relative neighbor. Secondly, a density following distance search method is developed to accurately find the cluster centers. Finally, taking the new node density and density following distance as attributes, a density peaks graph clustering model is constructed to accurately and quickly discover potential unbalanced and irregular community structures in complex networks. The main contributions of this paper are as follows:

- A new local density is defined for graph data by considering node cohesion within relative neighbors.
- A distance search method is developed to obtain the density following distance of each node, and further improve the accuracy of selecting the cluster centers.
- A graph clustering algorithm, DPGC, is proposed based on the new customized density peaks model to accurately and quickly uncover the community structure of a network.

The remainder of this paper is organized as follows. The traditional density peaks model is presented in Sect. 2. Section 3 shows our improved density peaks-based graph clustering algorithm, DPGC. Extensive experiments are described in Sect. 4. Finally, Sect. 5 concludes this paper.

2 Traditional Density Peaks Model

The density peaks clustering algorithm is an excellent clustering algorithm for discrete data and was proposed in 2014 [7]. The density peaks model believes that there are multiple high-density points and low-density points in each dataset. Each high-density point has a strong leadership ability to attract multiple low-density points to move closer. After a long period of dynamic evolution, multiple uneven and irregular dense areas are formed in the dataset. And each dense area is a cluster consisting of one high-density point and multiple low-density points. Among them, the high-density point is the leader, the other low-density points surround and follow the high-density point, and the distance between the leader and any following point is small. At the same time, there is a clearer boundary among different dense areas, and the core high-density points of different dense areas often have a longer distance. Based on the above idea, the density peaks model defines two attributes for each data point: local density and density following distance. Firstly, the local density is used to determine whether the current data is a high-density point. Secondly, the density following distance is used to measure the relative distance between the current point and its following leader node. The detailed formulas for the two definitions are as follows:

Definition 1 (local density ρ). The local density ρ represents the number of other data points surrounding the current point, that is, the number of neighbors, and is defined as follows:

$$\rho_i = \sum_{j \neq i} \chi\left(d_{ij} - d_c\right). \tag{1}$$

$$\chi(x) = \begin{cases} 1 \text{ if } x < 0 \\ 0 \text{ otherwise} \end{cases}. \tag{2}$$

Where d_{ij} is the Euclidean distance between data point d_i and data point d_j, d_c is the cut-off threshold used to determine the local neighbor range of the current point.

The local density defined in Eq. (1) is suitable for the case of large-scale datasets. To adapt to the small-scale dataset, a Gaussian Kernels-based local density is defined as follows:

$$\rho_i = \sum_{i \neq j} \exp\left(-\frac{d_{ij}^2}{d_c^2}\right). \tag{3}$$

Definition 2 (density following distance δ). The density following distance δ represents the distance between the current point and the closest high-density point with greater density (Leader), and is defined as follows:

$$\delta_i = \begin{cases} \min\left(d_{ij}\right) \rho_j > \rho_i \\ \max\left(d_{ij}\right) \rho_i = \max\left(\rho_j\right) \end{cases}. \tag{4}$$

When the density of the current point is the maximum, the density following distance of the current point is the maximum distance between any two nodes in the dataset.

In the density peaks model, the local density and density following distance $< \rho, \delta >$ are set as the attributes of each data point, and the rule of "*data points with high density and large density following distance is a cluster center*" is formulated. To execute the above rule more conveniently and quickly, a cluster center evaluation index is defined as follows:

$$\gamma_i = \rho_i \cdot \delta_i \tag{5}$$

Generally, the greater the evaluation index γ, the greater the probability that the current data point is a cluster center. Through the evaluation index, the density peaks model can quickly and accurately find all cluster centers in a dataset.

3   Improved Density Peaks-Based Graph Clustering

The traditional density peaks model cannot be directly applied to graph data of complex networks. To solve this problem, an improved density peaks-based graph clustering algorithm is developed in this section, namely DPGC.

3.1   New Local Density Based on Relative Neighbors

The first reason why the traditional density peaks model cannot be directly applied to graph data is that the definition of traditional density does not match the graph data. In the traditional model, the node density is usually the number of neighbors in a fixed area centered on this node. The greater the number of neighbors, the stronger the aggregation of these nodes, and the greater the probability that they belong to the same cluster. However, in the graph data, the number of neighbors directly connected to one node is not suitable as the local density. Because the more the number of neighbors of a node does not mean the stronger the aggregation between these nodes, as shown in Fig. 1. In the figure, both the green node 1 and the blue node 2 have many neighbors, and the number of neighbors of node 2 is 10 greater than that of node 1. In the traditional density peaks model, node 2 will be selected as the high-density point and the cluster center. However, it is not difficult to find that node 2 is not suitable as a cluster center, the aggregation between node 2 and its neighbors are poor, and many neighbors of node 2 are leaf nodes. On the contrary, the number of neighbors of node 1 is less than that of node 2, but the aggregation between node 1 and its neighbors is very strong. Therefore, node 1 is more suitable as a cluster center.

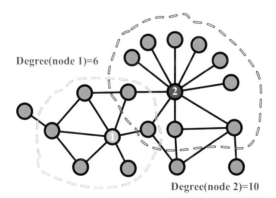

Fig. 1. Nodal degree is not suitable for measuring local density.

To solve the above problem, a new local density for graph data needs to be defined to enhance the aggregation of nodes and their neighbors. The aggregation refers to the degree to which all nodes are neighbors to each other in the set formed by this node and its neighbors. However, since each node in a complex network has a different number of neighbors, the aggregation of each node and its neighbors cannot be accurately measured. For this reason, this paper uses relative neighbors instead of direct neighbors to ensure that each node in a complex network has the same number of neighbors. Some definitions are given below.

Definition 1 (undirected graph).   Let $G = (V, E)$ be an undirected graph where V is the nodes set, E is the edges set. Each edge $e(u,v) \in E$ represents a strong connection between nodes u and v.

Definition 2 (direct neighbors of node u).   Given an undirected graph $G = (V,E)$, the neighbors of node u $N(u)$ is a node set that consists of its connected nodes, and is defined as follows:

$$N(u) = \{v \in V \,|\, \{u, v\} \in E \text{ and } v \neq u\} \tag{6}$$

Definition 3 (relative neighbors of node u).   The relative neighbors are used to set a fixed number of neighbors for each node, and is defined as follows:

$$RN(u) = top\ S\ node\ in \sum_{x \in V} minpath(x, u) \,|\, S = maxdegree \tag{7}$$

Where S is a constant, representing the fixed number of neighbors of each node. To remove the difference in node density of different networks, S is usually set to the maximum degree in a network.

The calculation process of the relative neighbors of one node is as follows. Firstly, the directly connected neighbors of the current node are selected as relative neighbors in descending order of degree. If the number of directly connected neighbors is less than S, the neighbors of all directly connected neighbors are merged to form a set. From this set, multiple nodes are chosen as relative neighbors of the current node in descending order of degree, until the number of relative neighbors is equal to S. When the relative neighbors of the nodes are selected, the local density of each node is easy to calculate.

Definition 4 (new local density).   The new local density suitable for graph data is the degree to which all relative neighbors are neighbors to each other, which is the ratio of the number of relative neighbors to the number of unions of all neighbors of the relative neighbors. The new local density of one node is defined as follows:

$$\rho_i^{new} = \frac{\|RN(v_i)\|}{\left\| \bigcup_{x \in RN(v_i)} RN(x) \right\|} = \frac{max\ degree}{\left\| \bigcup_{x \in FR(v_i)} RN(x) \right\|} \tag{8}$$

3.2   New Density Following Distance for Graph Data

The second reason why the traditional density peaks model cannot be directly applied to graph data is that the density following distance method in the traditional model is not suitable for graph data. In the traditional density peaks model, the density following distance is used to select the real cluster centers from multiple high-density points. In the traditional model, the shortest distance between the current point and the higher density points is usually used as the density following distance. However, in the graph data, the shortest path is not suitable for the density following distance. The reason is that a complex network usually has the "small world" rule, and any two nodes are interconnected by a few hops. That is to say, in a complex network, the shortest path between any two nodes is very short, which cannot be used as an index to distinguish high-density points, so that the cluster centers cannot be found accurately.

To solve the above problem, a new distance measurement method needs to be designed for the graph data to calculate the density following distance of each node, and ensure that the density following distance between high-density nodes has a greater difference. The reason why the distance between any two nodes in a complex network is usually short is that a low-density node may be connected to multiple high-density nodes at the same time. Therefore, the measurement of the new density following distance needs to avoid low-density nodes, thereby increasing the difference of the density following distance of each node. By combing the depth and breadth search, a density following distance search method is proposed in this paper, as shown in the Fig. 2.

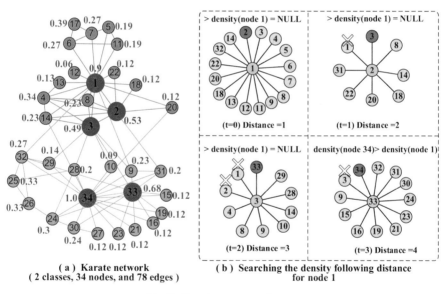

(a) Karate network
(2 classes, 34 nodes, and 78 edges)

(b) Searching the density following distance
for node 1

Fig. 2. Density-connected tree.

In the figure, a real-world network with 2 clusters (identified by different colors), 34 nodes, and 78 edges is taken as an example. In Fig. 2(a), the local density of each node is calculated according to Eq. 8, as shown by the blue numbers surrounding the nodes. It is not difficult to find that node 34, node 33, node 1, node 2, and node 3 in the Karate network are high-density nodes. To find the center of 2 clusters from 5 high-density nodes, it is necessary to set a density following distance for each node. Figure 2(b) shows the searching process of the density following distance of node 1. Firstly, the maximum density neighbor (node 2) is found from the directly connected neighbors of node 1. Since the density of node 2 is greater than that of node 1, set the density following distance of node 1 to 1, and continues searching. Secondly, the maximum density neighbor (node 3) is found from the directly connected neighbors of node 2. Since the density of node3 is greater than that of node 1, the density of node 1 increases by 1, and continue searching. Then, find the maximum density node 33 from the neighbors of node 3. Since the density of node 33 is less than that of node 1, the density of node 1 is increased by 1 again, and the search is continued. Finally, find neighbor 34 from the neighbors of node 33. Because the density of node 34 is higher than that of node 1, the search process ends, and the density following distance of node 1 is 4.

In summary, the detailed process of the density following distance search method has 3 steps. *Step 1, node ordering.* According to the local density of nodes, each node in the complex network is arranged in descending order to form a searching node set. *Step 2, start node and candidate neighbors set construction.* Choose a node from the searching node set as the start node, and construct a candidate neighbors set. Take the start node as the current node, the neighbors of the start node as the candidate neighbors set. *Step 3, repeat search.* Find the highest density node that has not been queried from the candidate neighbors set. If the density of the selected highest density node is less than that of the start node, set the selected highest density node as new current node, set the neighbors of this selected node as new candidate neighbors set, and continue searching. If the density of the selected highest density node is more than or equal than that of the start node, the process of searching ends. *Step 4, the density following distance calculation*. After the search is over, the density following distance of the starting node is the path of this search.

Based on the above method, the density following distance of each node in a complex network is very easy to calculate.

Definition 6 (new density following distance). The new density following the distance of a node is set as the searching path between this node and greater density nodes, and defined as follows:

$$\delta_i^{new} = searchpath(v_i, x) \Big|_{x \in V} \rho_i^{new} < \rho_x^{new} \tag{9}$$

3.3 Improved Density Peaks Model

According to the rule of "***data points with high density and large density following distance is a cluster center***" in the density peaks model, by combing the new local density and the density following distance, a new cluster center evaluation index is proposed for the graph data, and defined as follows:

$$\gamma_i^{new} = \rho_i^{new} \cdot \delta_i^{new} \tag{10}$$

Generally, the greater the value of the evaluation index, the greater the probability that the current node is the cluster center. Therefore, the cluster centers of a complex network can be quickly selected from all high-density points. Through the above improvements, the traditional density peaks model is naturally optimized to fit the graph data of a complex network.

3.4 Density Peaks-Based Graph Clustering (DPGC)

In summary, based on the improved density peaks model, a graph clustering algorithm is proposed in this paper, namely DPGC. The detailed process of the DPGC algorithm consists of the following 4 steps:

Step 1, Calculating the local density of each node. Taking a complex network $G = \{V, E\}$ as input, calculate firstly the relative neighbors of each node according to Eq. 7, and finally form the local density of each node ρ^{new}.

Step 2, Computing the density following distance of each node. Based on the searching method, compute the density following distance δ^{new} of each node according to Eq. 9.

Step 3, Choosing cluster centers. By merging the node density and density following distance, the cluster center evaluation index γ^{new} of each node is calculated according to Eq. 10. Based on the γ^{new}, the cluster centers of G are chosen from all high-density nodes.

Step 4, Assigning class labels for other low-density nodes. In descending order of local density, the class label of each low-density node is set to the class label with the most neighbors. If the neighbor of a low-density node does not have a class label, the node is skipped. The DPGC algorithm ends when all low-density nodes have class labels.

4 Experiments

4.1 Experimental Setup

Table 1. Comparison algorithms.

Algorithm	Full name
InfoMap [9]	Maps of random walks on complex networks reveal community structure
FastGreedy [8]	Finding community structure in very large networks
LPA [3]	Near linear time algorithm to detect community structures in large-scale networks
Louvain [4]	Fast unfolding of communities in large networks
DPC [6]	Original DPC algorithm
DPGC	An improved density peaks-based graph clustering algorithm

Comparison Algorithms. To verify the performance of the DPGC algorithm, five representative algorithms are selected to compare with it, as listed in Table 1. The InfoMap, FastGreedy, and Louvain algorithms are considered to be the best for disjoint graph clustering. The LPA algorithm offers high-speed graph clustering. The DPC algorithm is a native algorithm based on, which mapped the original graph data from a high-dimensional space to a low-dimensional vector space, and then obtained the cluster by using the traditional density peaks model. To compare the effectiveness of different graph clustering algorithms, NMI and F1-Score are chosen to evaluate their clustering results.

4.2 Synthetic Datasets

(1) Network generation

To compare the performance of various graph clustering algorithms, the LFR benchmark is used to generate 6 synthetic complex networks with different characteristics, as listed in Table 2. The generation model of the LFR benchmark is defined as LFR (Nodes, Edges, C#). In which, Nodes is the number of nodes, Edges is the number of Edges, C# is the number of communities.

Table 2. Synthetic networks.

Networks	Nodes	Edges	C#	Average degree
LFR1	64	130	2	4.062
LFR2	159	323	5	4.065
LFR3	202	397	6	3.931
LFR4	539	1105	12v	4.100
LFR5	1294	2659	20	4.110
LFR6	6109	12434	66	4.071

(2) Graph clustering performance

Figure 3 presents the clustering performance of 6 algorithms on 6 LFR networks from NIM and F1-Score metrics. In the figure, 6 algorithms have achieved good results on different networks, and the average value of the NMI and F1-Score is greater than 0.75. However, the performance of 6 algorithms changes greatly on 6 LFR networks, with better performance on some networks, and poor performance on other networks. More specifically, we get the following observations. (1) For the NMI, the performance of the 6 algorithms on the 6 LFR networks has changed drastically, and the stability is very poor. For example, the FastGreedy algorithm performs poorly on LFR4 and LFR6, but performs better on LFR3 and LFR5. Comparing the 6 algorithms, it is not difficult to find that the DPGC and Louvain algorithms have the best performance and stability. The performance of InfoMap and DPC algorithms is the worst, and the stability is not good. (2) For the F1-Score, the 6 algorithms have better stability than the NMI metric in the 6 LFR networks, and the curves of 6 algorithms are relatively stable. Comparing the 6 algorithms, DPGC and Louvain algorithms are best, InfoMap and LPA algorithms are next, FastGreedy and DPC are worst.

Table 3 further shows the computation time of 6 algorithms on the 6 LFR networks. In the table, the computation time of the LPA algorithm is the lowest, the DPGC algorithm is next, the Louvain algorithm is third, InfoMap and DPC algorithms are last. Contrasting the LPA and InfoMap algorithms, it is not difficult to find that the computation time of

InfoMap is close to 25 times that of the LPA algorithm. Moreover, focused on the DPC and DPGC algorithms, the computation time of DPC is close to 10 times that of DPGC.

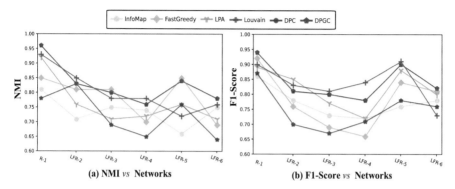

(a) NMI *vs* **Networks**   **(b) F1-Score** *vs* **Networks**

Fig. 3. Clustering performance of 6 algorithms on the LFR networks.

Table 3. Computation time of 6 algorithms on the LFR networks (ms).

Networks	LFR1	LFR2	LFR3	LFR4	LFR5	LFR6
IofoMap	8976	25931	53855	195477	201137	283959
FastGreedy	950	986	996	1995	3989	22910
Louvain	967	970	1099	3992	8975	81810
LPA	901	924	961	1062	1983	13856
DPC	9546	32885	58897	161539	366251	421895
DPGC	926	962	982	1582	2853	21963

4.3 Real-Work Datasets

(1) Network selects

5 typical real-world networks with ground truth are chosen for our experiments to further test the graph clustering performance of 6 algorithms, as listed in Table 4. Five real-worlds networks are publicly available and are easily obtained from the UCI network data repository (https://networkdata.ics.uci.edu/index.php) and Stanford large network dataset collection (http://snap.stanford.edu/data/).

(2) Graph clustering performance.

Figure 4 further presents the graph clustering accuracy of 6 algorithms on 5 real-world networks, where (a) plots the NMI results and (b) plots the F1-Score results. In the

Table 4. Real-world networks.

Networks	Nodes	Edges	Average degree	Community number
Karate	34	78	4.59	2
Dolphin	62	159	5.13	2
Polbooks	105	441	8.4	3
Adjnoun	112	425	7.59	2
Polblogs	1490	19090	22.440	2

Figure, some observations were obtained. (1) For three metrics (NMI and F1-Score), the 6 algorithms have different performances, and the performance curves of all algorithms have changed drastically. Comparing the 6 algorithms, DPGC and Louvain have the best clustering results, LPA is next, InfoMap, FastGreedy, and DPC are the worst. In additional, the stability of DPGC and Louvain algorithms is better than other algorithms. (2) Only focusing on the DPGC and DPC algorithms, the performance of the DPGC algorithm is significantly better than that of the native DPC algorithm. (3) When comparing the synthetic datasets and real-world datasets together (Fig. 3 and Figure 4), the clustering accuracy of 6 algorithms on LFR networks is better than that on real-world networks. The average of NMI and F1-Score of 6 algorithms on real networks is only 0.7, which is significantly lower than that on synthetic networks (0.8).

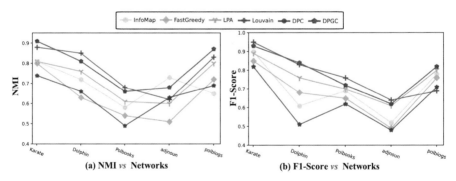

(a) NMI *vs* Networks   **(b) F1-Score *vs* Networks**

Fig. 4. Clustering performance of 6 algorithms on the real-world networks.

Table 5 further lists the time cost of 6 algorithms on the 5 real-world networks. In the table, some observations can be easily obtained. (1) For the computation time, the graph clustering time of the LPA algorithm is minimum, DPGC and Louvain algorithms are next, then is FastGreedy, DPC and InfoMap algorithms are maximum. (2) When we only compare the DPGC and DPC algorithms, with the increase of the network scale, the advantage between the time overhead of the DPGC algorithm and the DPC is getting bigger and bigger.

Table 5. Computation time of 6 algorithms on the real-world networks (ms).

Networks	Karate	Dolphin	Polbooks	Adjnoun	Polblogs
InfoMap	2959	4987	11936	12966	390935
FastGreedy	320	832	923	1023	48870
Louvain	221	816	915	992	3029
LPA	106	622	871	987	2105
DPC	3976	6917	21807	28545	413409
DPGC	180	752	923	990	3603

5   Conclusions

In this paper, an improved density peaks-based graph clustering algorithm, DPGC, is proposed for complex networks. The main contributions of this paper are as follows. (1) A new local density suitable for graph data is defined for each node by using a new relative neighbor with the fixed number and considering the agglomeration between neighbors. (2) A density following distance search method is developed to quickly obtain the density following distance of each node, and to improve the selection of cluster centers. (3) Based on the improved density peaks model, the DPGC algorithm is proposed to accurately and quickly cluster graph data. Extensive experiment results on multiple synthetic networks and real-world networks show the advantages of our algorithm.

Acknowledgments. This work is supported by the National Natural Science Foundation of China (No. 62103143); Hunan Provincial Natural Science Foundation of China (No. 2020JJ5199); the National Defense Basic Research Program of China (JCKY2019403D006); the National Key Research and Development Program (Nos. 2019YFE0105300/2019YFE0118700); and the Open Project of Key Laboratory of Intelligent Computing and Information Processing of Ministry of Education, Xiangtan University (2020ICIP06).

References

1. Mohamed, E., Agouti, T., Tikniouine, A., et al.: A comprehensive literature review on community detection: approaches and applications. Procedia Comput. Sci. **151**, 295–302 (2019)
2. Su, X., Xue, S., Liu, F., et al.: A Comprehensive Survey on Community Detection with Deep Learning. arXiv preprint arXiv, **2105**, 12584 (2021)
3. Raghavan, U.N., Albert, R., Kumara, S.: Near linear time algorithm to detect community structures in large-scale networks. Phys. Rev. E **76**, 036106 (2007)
4. Rosvall, M., Bergstrom, C.T.: Maps of random walks on complex networks reveal community structure. Proc. Natl. Acad. Sci. **105**, 1118–1123 (2008)
5. Li, Z., Tang, Y.: Comparative density peaks clustering. Expert Syst. Appl. **95**, 236–247 (2018)
6. Bhat, S.Y., Abulais, M.: OCMiner: a density-based overlapping community detection method for social networks. Intell. Data Anal. **19**, 917–947 (2015)

7. Rodriguez, A., Laio, A.: Clustering by fast search and find of density peaks. Science **344**, 1492–1496 (2014)
8. Clauset, A., Newman, M.E., Moore, C.: Finding community structure in very large networks. Phys. Rev. E **70**, 264–277 (2004)
9. Blondel, V.D., Guillaume, J.L., Lambiotte, R., et al.: Fast unfolding of communities in large networks. J. Stat. Mech. Theory Exp. **10**, 155–168 (2008)

Community Division Algorithm Based on Node Similarity and Multi-attribute Fusion

Du Tiansi[✉], Deng Na[✉], and Chen Weijie

School of Computer Science, Hubei University of Technology, Wuhan 430068, China

Abstract. In order to improve the accuracy of community division results, a community division algorithm (NSMF) based on node similarity and multi-attribute fusion is proposed. NSMF algorithm iteratively selects the node with the highest PageRank value as the initial clustering center of the community by improving the PageRank algorithm. The initial clustering node is selected through the network global information, which effectively avoids the low stability of the random selection of the initial node. Then, the node with the highest similarity with the community node is added to the community. At this time, only the local information of the network is calculated, the computational strength of the network is reduced. This paper compares with GN algorithm and FG algorithm on three real network data sets with different community structures. The results show that NSMF algorithm has the best overall performance in correct partition rate, global modularity and standardized mutual information.

1 Introduction

With the development of modern science and technology, human society is increasingly networked. As a new interdisciplinary discipline, complex network has been widely used in life science, physics, sociology, computer science and other fields, such as protein action network, communication network, interpersonal network, aviation network and so on. Through people's research on complex networks, it is found that the characteristics of complex networks are "small world", "scale-free" and "high aggregation". The nodes in complex networks are similar. Similar nodes are clustered into a community. The nodes in the same community are relatively closely connected, and the nodes among multiple communities are relatively loose. The community structure discovery of complex networks is of great significance to the study of the organizational structure and topological characteristics of complex networks.

The current community partition algorithms for complex networks are mainly divided into two categories. The first category is the Graph Segmentation Algorithm in computer science, such as the heuristic optimization K-L algorithm based on the principle of greedy algorithm proposed by Kernighan and Lin [1], and the spectral bisection method of eigenvalues and eigenvectors of Laplace matrix based on graph proposed by Pothen [2]. The second category is the hierarchical clustering algorithm in sociology. The hierarchical clustering algorithm is mainly divided into splitting algorithm and aggregation algorithm. The splitting algorithm, such as the GN algorithm proposed by

L. Barolli et al. (Eds.): EIDWT 2022, LNDECT 118, pp. 81–90, 2022.
https://doi.org/10.1007/978-3-030-95903-6_10

Girvan and Newman [3], gradually removes the edge with the largest edge intermediate number in the network to complete the community division. Because the GN algorithm needs to recalculate the edge intermediate number of each connected edge every time after removing the edge, the algorithm complexity is high, and it is not suitable for large-scale complex networks. Based on GN algorithm and greedy algorithm, Newman [4] proposed a condensation algorithm, Newman fast algorithm.

In addition, due to the similarity of nodes in complex networks, similar nodes are easy to cluster into a community. The community structure can be found by measuring the similarity of nodes. Yuan [5] and others introduced node similarity into local community mining algorithm, adopted the idea of pinch in community division, and proposed a community division algorithm based on local similarity. Ding [6] and others proposed an improved community partition algorithm by introducing mixed similarity index and community similarity index and combining the idea of hierarchical clustering. Gu [7] and others define a new similarity index according to the tightness of the common neighbors of the nodes, and cluster the nodes in combination with the improved k-means algorithm, so as to realize the community discovery of complex networks. The current measurement standards of node similarity are diverse and have room for improvement. How to improve the measurement standards of node similarity and combine with traditional clustering algorithms have a great impact on community division.

Based on the node similarity and multi-attribute fusion strategy of complex networks, taking into account the global and local information of the network, this paper combines the hierarchical clustering algorithm to divide the community of complex networks, and proposes a community division algorithm based on node similarity and multi-attribute fusion (NSMF). The algorithm iteratively selects the node with the highest PageRank value as the initial clustering center of the community by improving the PageRank algorithm. The initial clustering node is selected through the network global information, which effectively avoids the low stability of the random selection of the initial node. Then, the node with the highest similarity with the community node is added to the community. At this time, only the local information of the network is calculated, the computational strength of the network is reduced.

The content of this article is organized as follows. Some basic concepts of complex networks are described in Sect. 2, the algorithm is described in Sect. 3, the experimental results using the algorithm are described in Sect. 4, and the paper is summarized in Sect. 5.

2 Basic Concepts

2.1 PageRank Algorithm

PageRank algorithm [8] is a mathematical ranking algorithm, which determines the quality and relevance of pages through link-based methods. It is generally used in web page ranking of search engines. PageRank algorithm iteratively calculates the node weight in the complex network according to the importance of adjacent nodes. Its iterative equation in the directed graph is shown in formula (1):

$$PR(p_i) = \frac{1-\alpha}{N} + \alpha \sum_{p_j \in M(p_i)} \frac{PR(p_j)}{L(p_j)} \tag{1}$$

where: $p_1, p_2, \cdots p_n$ are nodes in complex network; N is the total number of nodes in the complex network; $L(p_j)$ is the outgoing degree of node p_j; $M(p_i)$ is the outgoing degree of node p_i; α is the damping factor, which indicates the probability that one node continues to jump to another node, which is generally set to 0.85.

In this algorithm, when PageRank algorithm is applied to undirected graph, its iterative equation is shown in formula (2):

$$PR(p_i) = \frac{1-\alpha}{N} + \alpha \sum_{p_j \in M(p_i)} \frac{w(p_{ij}) \times PR(p_j)}{d(p_j)} \tag{2}$$

where: $p_1, p_2, \cdots p_n$ are nodes in complex network; N is the total number of nodes in the complex network; $d(p_j)$ is the degree of node p_j; $M(p_i)$ is the adjacent edge set of p_j; $w(p_{ij})$ is the weight of the edge (p_i, p_j). If there is no connecting edge between the node p_i and p_j, the value of $w(p_{ij})$ is 0; α is the damping factor, which is generally set to 0.85.

The calculation process of the initial cluster center of the community is as follows: first initialize the PR values of all nodes, and then iteratively calculate and update the PR values of all nodes according to formula (2). After the kth update, when the difference between the PR value of each node and the PR value of the previous iteration is less than a certain threshold condition, the iteration is stopped.

2.2 Node Similarity

Nodes in complex networks are similar. Putting similar nodes into the same community is community division. Its core step is to constantly find the node with the highest similarity with nodes in the community and join the community. The similarity between two nodes is not only related to whether they are directly connected, but also related to their common neighbors. Therefore, the following three attributes are comprehensively considered.

Node Direct Impact. When calculating node similarity, whether two nodes are directly connected and the degree of nodes can have a direct impact on node similarity. The definition of node direct impact is shown in formula (3):

$$DI(v_i, v_j) = \frac{a(v_i, v_j)}{D(v_i)} + \frac{a(v_i, v_j)}{D(v_j)} \tag{3}$$

where: $D(v_i)$ is the degree of node v_i; $a(v_i, v_j)$ is to judge whether there is a connecting edge between nodes v_i and v_j. If there is a connecting edge, $a(v_i, v_j) = 1$, otherwise $a(v_i, v_j) = 0$.

Influence of Number of Common Neighbors. The node similarity between two nodes is also related to their common neighbors. The more common neighbors of two nodes, the higher the similarity between nodes. The definition of node common neighbor influence is shown in formula (4):

$$CN(v_i, v_j) = \frac{C(v_i, v_j)}{N(v_i) + N(v_j)} \tag{4}$$

where: $N(v_i)$ is the number of neighbor nodes of node v_i; $C(v_i, v_j)$ is the number of common neighbors of nodes v_i and v_j.

Impact of Shared Neighbor Similarity. The close relationship between the common neighbors of two nodes is also one of the criteria to measure the similarity of nodes. The closer the relationship between the common neighbors of two nodes, the higher the similarity between nodes. The definition of node shared neighbor similarity is shown in formula (5):

$$NS(v_i, v_j) = \frac{E(v_i, v_j)}{M(v_i, v_j) \times (M(v_i, v_j) - 1)/2} \tag{5}$$

where: $E(v_i, v_j)$ is the actual number of edges between nodes v_i and v_j and their common neighbor nodes; $M(v_i, v_j)$ is the number of nodes v_i and v_j and their common neighbor nodes; $M(v_i, v_j) \times (M(v_i, v_j) - 1)/2$ is the maximum number of edges that can be formed between nodes v_i and v_j and their common neighbor nodes.

To sum up, the three attributes all affect the node similarity, but their proportions are different, so different weights are assigned to the three attributes. The node similarity algorithm is shown in formula (6):

$$S(v_i, v_j) = aDI(v_i, v_j) + bCN(v_i, v_j) + cNS(v_i, v_j) \tag{6}$$

where: a, b and c are the weights of $DI(v_i, v_j)$, $CN(v_i, v_j)$ and $NS(v_i, v_j)$ respectively, and $a + b + c = 1$.

2.3   Evaluating Indicator

For the network with real known community structure, the accuracy of the algorithm can be obtained by comparing the community division results with the real community structure. Some community division evaluation indexes currently used are as follows.

Global Modularity. Global modularity [9] is an important measure of community division proposed by Newman and Girvan. The function Q is usually used to describe the modularity level of community division. Global modularity refers to the difference between the actual proportion of edge connections of internal nodes in complex networks and the proportion of edge connections of internal nodes in random networks. The definition of global modularity is shown in formula (7):

$$Q = \frac{1}{2M} \sum_{i,j} \left[\left(a_{ij} - \frac{k_i k_j}{2M} \right) \delta(\sigma_i, \sigma_j) \right] \tag{7}$$

where: a_{ij} is to judge whether there is a connecting edge between nodes v_i and v_j. If there is a connecting edge, $a_{ij} = 1$, otherwise $a_{ij} = 0$; $\delta(\sigma_i, \sigma_j)$ judge whether nodes v_i and v_j belong to the same community. When v_i and v_j belong to the same community $\sigma_i = \sigma_j$, $\delta(\sigma_i, \sigma_j) = 1$, otherwise $\delta(\sigma_i, \sigma_j) = 0$; M is the total number of edges in the complex network; k_i and k_j are the degrees of nodes v_i and v_j, respectively.

The value range of global modularity is [–1,1]. There are no more edges randomly connected between nodes in the community. The value of global modularity is negative. The better the division result of the community, the higher the degree of community, and the closer the value of global modularity is to 1. In practical complex networks, the value of global modularity is generally between [0.3, 0.7].

Local Modularity. The calculation of global modularity is based on the known global network structure, and the time complexity is high. Clauset [10] proposed local modularity to solve the problem of high time complexity in the calculation of global modularity. Generally, function R is used to refer to local modularity, and its definition is shown in formula (8):

$$R = \frac{B_{in}}{B_{in} + B_{out}} \tag{8}$$

where: B_{in} is the number of edges of internal nodes of the community; B_{out} is the number of sides between this community and external nodes of the community.

Local modularity is the ratio of the number of edges of internal nodes of the community to the number of external edges of the community. The more internal edges of a community and the fewer external edges of the community, the greater the local modularity and the better the result of community division. In this algorithm, the local modularity is used as the index of node clustering. Assuming that a node is divided into a community, if the R value increases, the node will join the community, and if the R value decreases, the node will be abandoned for the judgment of the next node.

Correct Division Rate. The correct partition rate refers to the ratio of the number of nodes correctly divided into communities to the total number of all nodes in a complex network. After any community in the community division results matches the real community, taking the real community as the evaluation standard, the nodes in any community but not in the real community and the nodes not in any community but in the real community are considered to be wrong division. The definition of correct division rate is shown in formula (9):

$$S = \frac{N_R}{N} \tag{9}$$

where: N_R is the number of nodes correctly divided into communities in the complex network; N is the total number of nodes in the network.

Normalized Mutual Information. Normalized mutual information (NMI) introduces standard common information into the comparison of community division results of complex networks to describe the correlation between community division results and real community structure. The definition of normalized mutual information is shown in formula (10):

$$I(A, B) = \frac{-2 \sum_{i=1}^{C_A} \sum_{j=1}^{C_B} m_{ij} \log\left(\frac{m_{ij}N}{m_{i.}m_{.j}}\right)}{\sum_{i=1}^{C_A} m_{i.} \log\frac{m_{i.}}{N} + \sum_{j=1}^{C_B} m_{.j} \log\frac{m_{.j}}{N}} \tag{10}$$

where: N is the total number of nodes in the complex network; C_A is the number of real societies; C_B is the number of communities obtained by the community partition algorithm; m is chaos matrix; $m_{i.}$ is the sum of row i of the chaos matrix; $m_{.j}$ is the sum of column j of the chaos matrix.

The value range of standardized mutual information is [0,1]. The higher the value of standardized mutual information, the better the result of community division.

3 NSMF Algorithm

Based on the idea of hierarchical clustering, the NSMF algorithm proposed in this paper improves the PageRank algorithm to iteratively select the node with the highest PageRank value as the initial clustering center of the community to build the initial community, calculate the node similarity in the neighbor nodes of the initial community, and find the node with the highest similarity with the community node, Calculate the local modularity of the node before and after joining the community. When Δ When the value of R is greater than 0, the node will join the community. Otherwise, the community will stop the iteration and start recalculating the PageRank value in the remaining nodes, and select the one with the highest PageRank value as the initial clustering center of the community. Repeat the above steps until all nodes are divided into the community, set a threshold (generally not greater than 3), judge whether the number of nodes in the community is less than this threshold, and if so, merge the nodes in the community into the community where the node with the largest degree is located among the neighbor nodes of the node.

The pseudo code of the algorithm is as follows:

```
Input: graph file
Output: community division result
Produce NSMF()
    while(GNumber != NULL) {
                D ← select(GNumber)
                node_p ← Find_Pro(D)
                R'p = Add_node(node_p)
                ΔR'p = R'p - Rp
                    if(ΔR'p > 0){
                            D = D + node_p
                            NSMF()}
                    else{
                            break}}
```

4 Experimental Results and Analysis

4.1 Comparison Algorithm

In order to objectively measure the rationality and effectiveness of NSMF algorithm, this experiment uses the method of comparative verification to compare the algorithm with GN algorithm and Newman fast algorithm (FG).

GN Algorithm. GN algorithm is a community partition algorithm based on edge intermediate number and splitting method. Its partition steps are as follows: calculate the edge intermediate number of all edges in the complex network, find the edge with the highest edge intermediate number and remove it, and continue to repeat the above steps until all edges in the network are removed.

Newman Fast Algorithm. Newman algorithm is a community partition algorithm based on greedy idea and aggregation method. Its partition steps are as follows: each node is a community, and constantly merge communities to maximize the modularity, so as to obtain the optimal community partition result.

4.2   Data Set

In order to ensure the authenticity of the experiment, this experiment is carried out on two real network data sets with different community structures commonly used in the field of community division.

Karate Club Network. Karate club network [11] (Zachary network) is one of the classic test networks for community division in complex networks. The network is a social network obtained by Zachary from observing the relationship between members of a karate club in American universities. The nodes are composed of club members and the edges are composed of the relationship between members. The network contains 34 nodes and 78 edges, which is divided into two communities.

American College Football Network. American college football network [3] (college football network) is one of the classic test networks for community division in complex networks. The network is a social network obtained by Girvan's observation of the game relationship between American college football leagues. The nodes are composed of teams participating in the game, while the sides are composed of games between teams. There are many teams in the same league, there are few competitions among different leagues. The network contains 115 nodes and 613 sides, which is divided into 12 communities.

4.3   Experimental Result

Division Results of Karate Club Network. The NSMF algorithm proposed in this paper divides the karate club network into four communities. The node colors of different communities are different. The division results are shown in Fig. 1.

The comparison of community division results of NSMF algorithm, GN algorithm and Newman fast algorithm (FG algorithm) on karate club network is shown in Table 1, in which a = 0.4, b = 0.4 and c = 0.2.

It can be seen from Table 1 that NSMF algorithm is higher than GN algorithm and FG algorithm in terms of global modularity and normalized mutual information. Only the division accuracy is slightly lower than FG algorithm. It can be seen that NSMF algorithm has the best overall division effect on karate club network. Although the NSMF algorithm divides the karate club network into four communities, the algorithm still accurately divides the two communities centered on node 34 and node 1. The internal nodes of the two communities centered on node 32 and node 7 are closely related and may form an independent community.

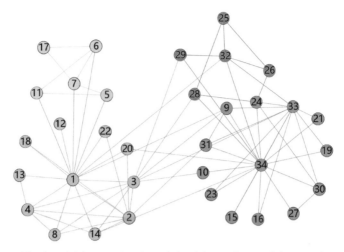

Fig. 1. Division results of nsmf algorithm on karate club network

Table 1. Community division results of different algorithms on karate club network

Algorithm	Q	NMI	P
GN	0.4013	0.5798	0.6471
FG	0.3807	0.6925	0.7353
NSMF	0.4174	0.6956	0.7059

Division Results of American College Football Network. The NSMF algorithm proposed in this paper divides the American college football network into 13 communities. The node colors of different communities are different. The division results are shown in Fig. 2.

The comparison of community division results of NSMF algorithm, GN algorithm and Newman fast algorithm (FG algorithm) on American university football network is shown in Table 2, in which a = 0.4, b = 0.4 and c = 0.2.

As can be seen from Table 2, in terms of normalized mutual information and division accuracy, NSMF algorithm is higher than GN algorithm and FG algorithm, and only the global modularity is slightly lower than GN algorithm. It can be seen that NSMF algorithm has the best overall division effect on American university football network. NSMF algorithm divides the American college football network into 13 communities, one of which is divided into two leagues. Although it is the same league, the schedule between teams is fine-tuned, resulting in not enough games between two teams in the same league.

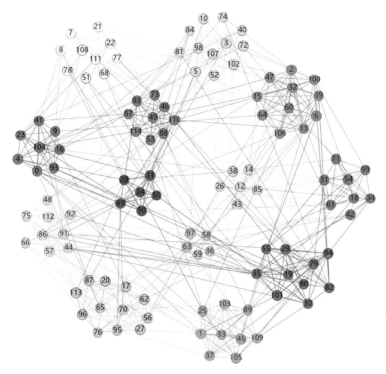

Fig. 2. Division results of nsmf algorithm on American college football network

Table 2. Community division results of different algorithms on American college football network

Algorithm	Q	NMI	P
GN	0.5996	0.9479	0.8696
FG	0.5773	0.8149	0.6261
NSMF	0.5810	0.9725	0.9217

5   Conclusion

In order to improve the accuracy of community partition results in complex networks, this paper proposes a community partition algorithm NSMF based on node similarity and multi-attribute fusion. Aiming at the low stability of community division caused by random selection of cluster centers in some clustering algorithms, NSMF algorithm iteratively selects the node with the highest PageRank value as the initial cluster center of the community by improving the PageRank algorithm, and selects the initial cluster node through the network global information, so as to effectively avoid the low stability of random selection of initial nodes. For some clustering algorithms, the number of communities needs to be set in advance. NSMF algorithm adds the node with the highest similarity with the community node to the community, and takes the reduction of local modularity as the termination condition of community clustering. Moreover, the algorithm only calculates the local information of the network, which reduces the calculation intensity of the network. In the follow-up research work, we will continue to improve the clustering method of the algorithm and reduce the time complexity of the algorithm, so that the algorithm can partition complex network communities more quickly and accurately.

Acknowledgements. The research was supported by the National Natural Science Foundation of China under grant number 61902116.

References

1. Kernighan, B.W., Lin, S.: An efficient heuristic procedure for partitioning graphs. Bell Syst. Tech. J. **49**(2), 292–307 (1970)
2. Pothen, A., Simon, H., Liou, K.P.: Partitioning sparse matrices with eigenvectors of graphs. SLAM J. Matrix Anal. Appl. **11**(3), 430–452 (1990)
3. Girvan, M., Newman, M.E.J.: Community structure in social and biological networks. Proc. Nat. Acad. Sci. **99**(12), 7821–7826 (2002)
4. Newman, M.E.J.: Fast algorithm for detecting community structure in networks. Phys. Rev. E. **69**(6), 066133 (2004)
5. Yuan, C., Chai, Y.: Local community structure mining algorithm for complex networks. Acta Autom. Sin. **40**(5), 921–934 (2014)
6. Ding, M.Z., Ma, Y.H., Li, Y.: Community division algorithm based on similarity index. Comput. Eng. **45**(2), 195–201 (2019)
7. Gu, Y.R., Chen, Y.Q.: A new community discovery algorithm based on local similarity. J. Nanjing Univ. Posts Telecommun. (Natl. Sci. Ed.) **37**(5), 48–55 (2017)
8. Page, L.: The PageRank Citation Ranking: Bringing Order to the Web, Online manuscript. Stanford Digital Libraries Working Paper **9**(1), 1 (1999)
9. Newman, M.E.J., Girvan, M.: Finding and evaluating community structure in networks. Phys. Rev. E **69**(2), 26–113 (2004)
10. Clauset, A.: Finding local community structure in networks. Phys. Rev. E Stat. Nonlinear & Soft Matter Phys. **72**(2), 026132 (2005)
11. Zachary, W.W.: An information flow model for conflict and fission in small groups. J. Anthropol. Res. **33**(4), 452–473 (1977)

Research on TCM Patent Annotation to Support Medicine R&D and Patent Acquisition Decision-Making

Du Tiansi, Deng Na[✉], and Chen Weijie

School of Computer Science, Hubei University of Technology, Wuhan 430068, China

Abstract. Traditional Chinese Medicine (TCM) patents contain abundant medical, economic and legal information. The effective analysis and mining of TCM patents is of great significance to support medicine R&D and patent acquisition decision-making. Named entity recognition is a key step in the research of TCM patents annotation. In order to solve the problem that the entity recognition in TCM patents relies on manual work in a large extent and the degree of automation is not high, in this paper, under the framework of ERNIE (Enhanced Language Representation with Informative Entities) pre-training language model, combined with the Bi-directional Gating Recurrent Unit (BiGRU) and Conditional Random Field (CRF), two important entities, herbal names and medicine effects, are recognized in TCM patent texts. Experimental results show that the performance of the method based on ERNIE-BiGRU-CRF is significantly improved compared with other models.

1 Introduction

As an important carrier of scientific and technological achievements, TCM patent can help medicine R&D personnel find blank or incomplete blind spots in medicine R&D technology to promote medicine R&D by making good use of its medical, economic and legal information. The analysis of hot medicines and hot technology information in TCM patents can provide decision support for pharmaceutical enterprises to purchase suitable and potential patents. Sun [1] and others promoted technological innovation in important fields through statistical analysis of patent information and patent analysis in the field of TCM in typical universities. Named entity recognition is a key step in the labeling of TCM patents. The herbal names and medicine effects are two important semantic entities in TCM patent. They are not only semantic extension objects of patent retrieval, but also components of patent technology efficacy matrix, and can help professionals to effectively master the relevant information of Chinese herbal medicine for medicine research and development, patent acquisition decisions, etc.

L. Barolli et al. (Eds.): EIDWT 2022, LNDECT 118, pp. 91–101, 2022.
https://doi.org/10.1007/978-3-030-95903-6_11

At present, the difficulties and challenges in the identification of herbal names and medicine effects are as follows: There are many kinds of Chinese herbal medicine and alias phenomenon is common. "ZhangDan" stone class, for example, there are "Dan", "ZhenDan", "QianHua", "QianHuang", "GuoDan", "ZhuFen", "SongDan", "ZhuDan", by more than 20 of the alias. "YuBaiFu" and "GuanBaiFu" are called "BaiFuZi", "Fen-FangJi" and "GuangFangJi" are called "FangJi", wild anise of the star Illicium family, wolfsbane of the Daphne family, aconitum of the buttercup family and nearly 40 kinds of herbs are all called gelsemium elegans. These namesake medicines sometimes have similar effects, but come from different sources, contain different chemicals, and sometimes are just two unrelated medicines. Because there is not a complete Chinese herbal medicine name dictionary, the alias phenomenon and unusual characteristics of Chinese herbal medicine name will lead to low recognition accuracy. The names of Chinese herbal medicines are generally two to five words, and the distribution is concentrated and regular. There is little difference between medicine names, while the number of words of efficacy description is usually divided into efficacy phrases of two to four words and other forms of efficacy sentences. There are great differences in language narration between some efficacy descriptions. For example, there are two to four word efficacy phrases such as "XiaoYan (diminish inflammation)", "KangGanRan (fight infection)" and "QingReJieDu (clear heat and detoxication)" in Chinese medicine patents, as well as efficacy sentences such as "JiaKuaiXinChenDaiXie (accelerate metabolism)", "ZhiLiaoXunMaZhen (treat measles)" and "YiZhiDouDouEHuaManYan (inhibit the deterioration and spread of acne)". Because the word length of efficacy description is different, it has become a difficulty in entity recognition of efficacy description.

In entity recognition, there are two traditional methods based on linguistics and statistics. However, the entity recognition methods based on linguistics and statistics have their own shortcomings. Later scholars combine the two methods, and the early representative methods are C-value method and NC-value method. Yeom [2] and others proposed a scoring method for candidate key phrases by using statistical model and graph model combined with C-value method.

The traditional entity recognition method is relatively dependent on manual, but with the rise of machine learning, more machine learning methods are gradually applied to entity recognition. There are three kinds of entity recognition methods based on machine learning: supervised, semi-supervised and unsupervised learning methods. Supervised and semi-supervised learning methods realize entity recognition through model training on the basis of annotated data. The difference between them mainly lies in the scale of annotated data. The unsupervised learning method does not annotate the data and trains the model directly. Sung [3] and others extracted phenotypic information of ischemic stroke through supervised learning and text mining techniques for structured and unstructured data from electronic medical records. Zhang [4] and others based on semi-supervised learning method, combined with non-negative matrix decomposition and harmonic function, effectively improved classification accuracy. Yenkar [5] and others used the unsupervised learning method based on gazetteers to extract locations from complaint tweets, which is of great significance for local governments in urban management.

In recent years, as a branch of machine learning, deep learning has also been widely used in natural language processing. In 2018, Devlin [6] and others proposed BERT model based on shadowing language model and deep bidirectional pre-training. In 2019, Sun [7] and others improved the BERT model and designed a new pre-training task in order to better integrate semantic and knowledge information to enhance the semantic representation ability of the model. Thus proposed ERNIE1.0 (Enhanced Language Representation with Informative Entities) model. On the basis of ERNIE1.0 model, Sun [8] and others proposed ERNIE2.0 model by constructing pre-training tasks with multi-task continuous incremental learning based on the semantic understanding pre-training framework of continuous learning. ERNIE model can be used in various application scenarios such as text classification, named entity recognition and emotion analysis. Wang [9] and others proposed a Joint training model ERNIE-Joint based on ERNIE model to achieve Chinese text classification and named entity recognition tasks. Liu [10] and others proposed a problem classification model based on ERNIE model and feature fusion to extract semantic information from a deeper level.

Support medicine R&D and patent acquisition decision-making, based on the TCM patent text for the data source, the introduction of the ERNIE training language model framework, combined with Bi-directional Gated Recurrent Unit [11] (BiGRU) to extract the context of each word semantic information, Conditional Random Field [12] (CRF) is used for sequence constraint to extract and identify two important entities of Chinese herbal names and medicine effects.

The content of this article is organized as follows. The overall construction of the model is in Sect. 2, the process of entity recognition is in Sect. 3, the experimental results using the model are in Sect. 4, and the summary of the paper is in Sect. 5.

2   Model Building

The overall structure of the ERNIE- BiGRU-CRF entity recognition model is shown in Fig. 1. The model is mainly divided into three layers: the first layer from bottom to top is ERNIE layer. The pre-training model ERNIE builds the lexical, grammatical and semantic pre-training tasks respectively to obtain lexical, grammatical and semantic information in the training data, and finally obtains the sentence-level word vector representation. The second layer is the BiGRU layer, which takes the output of the ERNIE layer as input to automatically extract contextual semantic information for each word. The third layer is CRF layer, which annotates the output of BiGRU layer. CRF is used to optimize the decoding sequence and solve the dependency between tags, so as to obtain the global optimal solution.

3   Entity Recognition Process

Through the overall analysis of TCM patents, it was found that the herbal names and medicine effects appeared in the patent would appear in the abstract in most cases, so the abstract of TCM patent was used as the data source of entity extraction in this paper. The abstract texts of 1500 TCM patents were extracted by web crawler and imported into Baidu's EasyData annotation tool. Data cleaning of the abstract was carried out through

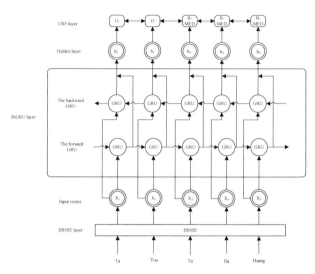

Fig. 1. The overall structure of ERNIE- BiGRU-CRF entity recognition model

the annotation tool, including the removal of repeated and empty data. The data after cleaning were manually marked with the name of Chinese herbal names and medicine effects, in which, the name of Chinese herbal names was represented by MED and medicine effects was represented by EFF. Export all annotation results from annotation tools and store them in JSON files. Figure 2 shows some of the JSON file content.

{"content": "本发明公开了一种治疗皮肤瘙痒的药物及其制备方法，它主要是由土大黄、小飞蓬、牛含水、水薏青、白首乌、白僵蚕、血满草、全蝎、土香薷、大蓟、山胡椒、小巢菜、木贼、马梢蛇、古羊藤、四方麻、地笋、麦斛、苏头、芭蕉根、赤胭、扶芳藤、羌活、沙枣叶、鸡冠苗、松木皮、制蜜、拔毒散、拦路虎、虎刺、法罗海按一定重量配比制备而成，本发明具有清热解毒、祛风、利湿、京血、理气的功能，用于治疗皮肤瘙痒见效快，疗效好。", "records": [{"span": "治疗皮肤瘙痒", "offset": [8, 13], "tag": "EFF"}, {"span": "土大黄", "offset": [29, 31], "tag": "MED"}, {"span": "小飞蓬", "offset": [33, 35], "tag": "MED"}, {"span": "牛含水", "offset": [37, 39], "tag": "MED"}, {"span": "水薏青", "offset": [41, 43], "tag": "MED"}, {"span": "白首乌", "offset": [45, 47], "tag": "MED"}, {"span": "白僵蚕", "offset": [49, 51], "tag": "MED"}, {"span": "血满草", "offset": [53, 55], "tag": "MED"}, {"span": "全蝎", "offset": [57, 58], "tag": "MED"}, {"span": "土香薷", "offset": [60, 62], "tag": "MED"}, {"span": "大蓟", "offset": [64, 65], "tag": "MED"}, {"span": "山胡椒", "offset": [67, 69], "tag": "MED"}, {"span": "小巢菜", "offset": [71, 73], "tag": "MED"}, {"span": "木贼", "offset": [75, 76], "tag": "MED"}, {"span": "马梢蛇", "offset": [78, 80], "tag": "MED"}, {"span": "古羊藤", "offset": [82, 84], "tag": "MED"}, {"span": "四方麻", "offset": [86, 88], "tag": "MED"}, {"span": "地笋", "offset": [90, 91], "tag": "MED"}, {"span": "麦斛", "offset": [93, 94], "tag": "MED"}, {"span": "苏头", "offset": [96, 97], "tag": "MED"}, {"span": "芭蕉根", "offset": [99, 101], "tag": "MED"}, {"span": "赤胭", "offset": [103, 104], "tag": "MED"}, {"span": "扶芳藤", "offset": [106, 108], "tag": "MED"}, {"span": "羌活", "offset": [110, 111], "tag": "MED"}, {"span": "沙枣叶", "offset": [113, 115], "tag": "MED"}, {"span": "鸡冠苗", "offset": [117, 119], "tag": "MED"}, {"span": "松木皮", "offset": [121, 123], "tag": "MED"}, {"span": "制蜜", "offset": [125, 126], "tag": "MED"}, {"span": "拔毒散", "offset": [128, 130], "tag": "MED"}, {"span": "拦路虎", "offset": [132, 134], "tag": "MED"}, {"span": "虎刺", "offset": [136, 137], "tag": "MED"}, {"span": "法罗海", "offset": [139, 141], "tag": "MED"}, {"span": "清热解毒", "offset": [159, 162], "tag": "EFF"}, {"span": "祛风", "offset": [164, 165], "tag": "EFF"}, {"span": "利湿", "offset": [167, 168], "tag": "EFF"}, {"span": "京血", "offset": [170, 171], "tag": "EFF"}, {"span": "理气", "offset": [173, 174], "tag": "EFF"}, {"span": "治疗皮肤瘙痒", "offset": [181, 186], "tag": "EFF"}]}

Fig. 2. Annotate the resulting JSON file contents

On the basis of the obtained JSON file, it is converted into BIO annotation text, which contains two parts: the first part is the text content, and the second part is the annotation character of BIO annotation. When annotated by BIO, "B" represents the beginning of the entity type, "I" represents the middle and end of the entity, and "O" represents the parts that are not part of the entity. The entity label types in the dataset are shown in Table 1.

Figure 3 below shows some TCM patent texts marked by BIO.

Table 1. Entity label type

Entity class	First character	Non first character
Herbal names	B-MED	I-MED
Medicine effects	B-EFF	I-EFF

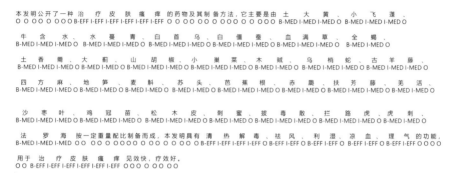

Fig. 3. Schematic diagram of some TCM patents marked by BIO

The process of converting JSON file into BIO annotation text is shown in Fig. 4. Cycle through all JSON files to determine whether the record in the JSON file being read is an entity. If it is an entity, it is marked as "Tag", otherwise it is marked as "O". Judge the keyword type of "Tag" in the record. If the keyword type of "Tag" is "MED", investigate the offset of all characters in the same array of "Tag", and mark the character with the smallest offset as "B-MED" and other characters as "I-MED". If the keyword type of "Tag" is "EFF", investigate the offset of all characters in the same array as this "Tag", and mark the character with the smallest offset as "B-EFF" and other characters as "I-EFF". Finally, store all the converted information in the text and output the text file marked by BIO.

Transfer the text marked with BIO into ERNIE-BiGRU-CRF entity recognition model for entity recognition. Take the sentence "TaYouTuDaHuang (It is made of rumex madaio)" as an example. In ERNIE-BiGRU-CRF entity recognition model, firstly, the sentence "TaYouTuDaHuang" is introduced into ERNIE layer. ERNIE layer segments the input sentence according to different granularity of words, words and entities, that is, tokenization, and then takes the token sequence and segmentation boundary of plaintext as the input of ERNIE model. In ERNIE model, the four embedding results of "TaYouTuDaHuang" such as task embedding, segment embedding, position embedding and token embedding are added as the input of transformer encoder. After being processed by transformer encoder, it is represented by the embedding vector corresponding to each token of "TaYouTuDaHuang". Then the embedding vector corresponding to each token is introduced into the BiGRU layer to extract the deep-seated features as the output. Although "TaYouTuDaHuang" can obtain the score value of each label of each word through ERNIE layer and BiGRU layer, and the label with the maximum score value is the label category of each word, there are dependencies between labels. CRF can obtain

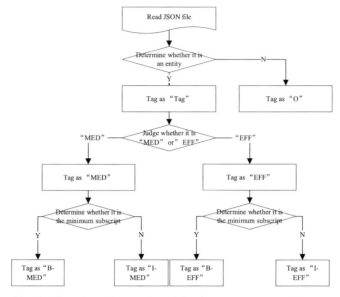

Fig. 4. Flow chart of converting JSON file into BIO annotation text

the global optimal solution by optimizing the decoding sequence and taking the dependency between tags as the constraint condition of sequence annotation. For example, the label of the first word marked on each entity must be "B-MED" or "B-EFF", and the labels of other words must be "I-MED" and "I-EFF". Moreover, different types of labels cannot appear in the same entity, such as "B-MED, I-EFF", "B-EFF, I-MED", etc. In Fig. 5, "TaYouTuDaHuang" through CRF layer to obtain three tag sequences: "O, I-MED, O, O, O", "O, O, B-MED, I-MED, I-MED", "O, O, B-MED, I-EFF, B-MED", their scores are "0.1", "0.9", "0.3" respectively. At this time, the label sequence with a score of "0.9" is the global optimal solution, that is, "TuDaHuang" It is recognized as the herbal names.

4   Experimental Results and Analysis

4.1   Experimental Data

The patent text data of this experiment is crawled from the patent database of China HowNet. Because the abstract data contains a large number of detailed Chinese herbal names and medicine effects, the abstract data is used as the experimental data source in this experiment. 1500 abstracts of TCM patents were manually marked, of which the training set accounted for 70% and the test set accounted for 30%.

The objects identified in this experiment are the following two entity categories:

One is herbal names. That is, the names of Chinese herbal medicines appearing in the abstract, such as angelica sinensis, cassia seed, ginseng, etc.

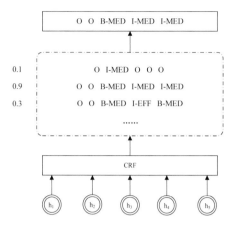

Fig. 5. CRF sequence annotation diagram

The second is medicine effects. That is, the phrases and words that describe the functions of medicines and Chinese herbal medicine appear in the abstract, such as dredging meridians, brightening eyes, relaxing tendons and relieving pain, etc.

4.2 Experimental Results and Analysis

In order to verify the effectiveness and generalization of ERNIE-BiGRU-CRF model, the model was compared with six mainstream models: BiLSTM-CRF, BiGRU-CRF, BERT-BiLSTM, BERT-BiGRU, BERT-BiLSTM-CRF and BERT-BiGRU-CRF. Among them, the evaluation indicators are precision, recall and F1-score.

In order to intuitively reflect the influence of epoch parameters on ERNIE-BiGRU-CRF model, Fig. 6 shows the influence of epoch parameters on ERNIE-BiGRU-CRF model. When epoch is 9, the maximum F1 value of herbal names entity recognition is 95.0%. Figure 7 shows the influence of epoch parameters on ERNIE-BiGRU-CRF model in medicine effects entity recognition. When epoch is 8, the maximum F1 value of medicine effects entity recognition is 87.6%. Figure 8 shows the influence of epoch parameters on ERNIE-BiGRU-CRF model in herbal names and medicine effects entity recognition. When epoch is 11, the maximum F1 value of herbal names and medicine effects entity recognition is 91.4%.

Figure 9 shows the accuracy, recall rate and F1 value of herbal names and medicine effects of these seven different entity recognition models. The epoch value of the first six models is 5. Where, "M" represents the herbal names, "E" represents the medicine effects, and "ME" represents the overall description of the herbal names and medicine effects.

The F1 value of herbal names entity recognition obtained after the training of ERNIE-BiGRU-CRF model proposed in this paper reaches 95.0%, the F1 value of medicine effects entity recognition reaches 87.6%, and the overall F1 value reaches 91.4%. Compared with the BiLSTM-CRF model, the accuracy rate, recall rate and F1 value are significantly improved. Compared with the BiGRU-CRF model, the recall rate and F1 value are significantly improved, indicating that Ernie pre training model can better train

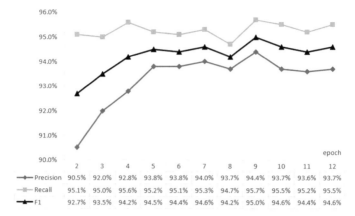

Fig. 6. Influence of epoch parameters on ERNIE-BiGRU-CRF model in herbal names entity recognition

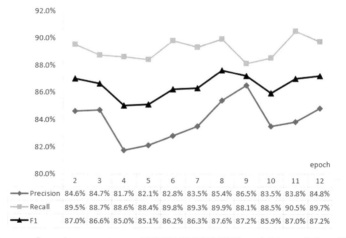

Fig. 7. Influence of epoch parameters on ERNIE-BiGRU-CRF model in medicine effects entity recognition

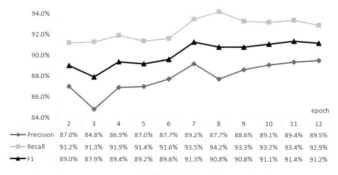

Fig. 8. Influence of epoch parameters on ERNIE-BiGRU-CRF model in overall entity recognition

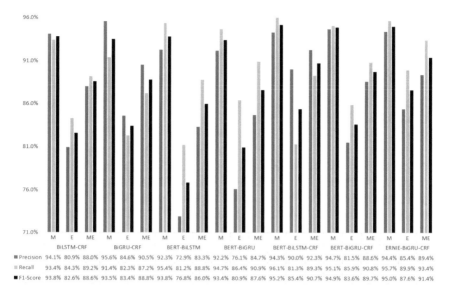

	M	E	ME	M	E	ME	M	E	ME	M	E	ME	M	E	ME	M	E	ME	M	E	ME
	BiLSTM-CRF			BiGRU-CRF			BERT-BiLSTM			BERT-BiGRU			BERT-BiLSTM-CRF			BERT-BiGRU-CRF			ERNIE-BiGRU-CRF		
Precision	94.1%	80.9%	88.0%	95.6%	84.6%	90.5%	92.3%	72.9%	83.3%	92.2%	76.1%	84.7%	94.3%	90.0%	92.3%	94.7%	81.5%	88.6%	94.4%	85.4%	89.4%
Recall	93.4%	84.3%	89.2%	91.4%	82.3%	87.2%	95.4%	81.2%	88.8%	94.7%	86.4%	90.9%	96.1%	81.3%	89.3%	95.1%	85.9%	90.8%	95.7%	89.9%	93.4%
F1-Score	93.8%	82.6%	88.6%	93.5%	83.4%	88.8%	93.8%	76.8%	86.0%	93.4%	80.9%	87.6%	95.2%	85.4%	90.7%	94.9%	83.6%	89.7%	95.0%	87.6%	91.4%

Fig. 9. Experimental results of entity recognition with different models

through the lexical structure, grammatical structure and semantic information in the text data, so as to find valuable information and better semantic information in the training data. Compared with BERT-BiLSTM model and BERT-BiGRU model, the accuracy, recall and F1 value are significantly improved, indicating that CRF layer can carry out sequence labeling to obtain the global optimal solution, which significantly improves the experimental effect. Compared with the BERT-BiLSTM-CRF model, the recall rate and F1 value are significantly improved. Compared with the BERT-BiGRU-CRF model, the accuracy rate, recall rate and F1 value are significantly improved, mainly because ERNIE adopts a better knowledge masking strategy, which can mask all words of the entity, so that the model can better understand the semantic information between entities and better identify entities.

Experiments show that the recall rate of this model is as high as 93.4% while taking into account the F1 value, indicating that this model can accurately identify the herbal names and medicine effects entity, and can better support for medicine R&D and patent acquisition decision-making. The trained model can better extract the Chinese herbal names and medicine effects entity in the new text data, as shown in Fig. 10.

EFF:治肺脾亏损型鼻粘膜变异;MED:桑叶;MED:柴胡;MED:地龙;MED:板蓝板;MED:薏苡仁;MED:桑叶;MED:柴胡;MED:地龙;MED:板蓝板;MED:薏苡仁;EFF:治肺脾亏损型鼻粘膜变异;EFF:清热解毒;EFF:健脾化湿;
EFF:治疗风湿骨病;MED:松香;MED:虫白蜡;MED:雪猪油;MED:松红梅;MED:托玛琳粉;MED:砭石粉;MED:冰片;MED:樟脑;MED:寻骨风;MED:瑞香植;MED:细辛;MED:赤芍;MED:竹叶椒;MED:防己;MED:海桐皮;MED:川乌;MED:芦子藤;MED:马钱子;MED:梧桐根;MED:麝香;MED:没药;MED:干姜;MED:人参;MED:三七;MED:辣椒;EFF:祛风除湿;EFF:活血祛瘀;EFF:消肿止痛;EFF:温经散寒;EFF:舒筋活络;EFF:清热解毒;
EFF:治风热型急性咽炎;MED:桔梗;MED:射干;MED:蓝根;MED:马勃;MED:桔梗;MED:射干;MED:蓝根;MED:马勃;EFF:治风热型;EFF:疏风清热;EFF:解毒利咽;
EFF:治疗慢性阻塞性肺病;MED:生地黄;MED:麦冬;MED:梅花;MED:蒲公英;MED:知母;MED:桃仁;MED:白芷;MED:荷叶;MED:生地黄;EFF:养阴生津;MED:麦冬;EFF:润肺清心;MED:梅花;EFF:疏肝解郁;MED:生津化痰;MED:蒲公英;MED:知母;EFF:清热解毒泻火;EFF:清肺火;MED:桃仁;EFF:活血化瘀;MED:白芷;EFF:散风消肖;MED:荷叶;EFF:升发清阳;EFF:养阴润肺;EFF:清热解毒;EFF:慢性阻塞性肺病;
EFF:治热毒炽盛证急性扁桃体炎;MED:马勃;MED:金莲花;MED:升麻;MED:桔梗;MED:桔梗;MED:马勃;MED:金莲花;MED:升麻;MED:桔梗;EFF:治热毒炽盛证;EFF:急性扁桃体炎;EFF:清热解毒;

Fig. 10. The model extracts entities from new text

5 Conclusion

In this paper, in order to support medicine R&D and patent acquisition decision-making, this paper proposes ERNIE-BiGRU-CRF model, which solves the problems that traditional methods largely rely on manual annotation and low degree of automation in entity recognition of Chinese herbal names and medicine effects, and the disadvantage that BERT model is difficult to model high-level semantic knowledge. The model obtains the sentence level vector representation through ERNIE pre-training model, obtains the prior semantic knowledge of the entity by using the knowledge masking strategy, extracts the feature through BiGRU, obtains the context information of each word, and transmits the obtained feature matrix into the CRF layer for sequence annotation to obtain the global optimal solution. Experiments show that this method is effective for entity recognition of Chinese herbal names and medicine effects in TCM patent text.

However, due to the diversity of entity text types, the extraction effect needs to be further improved. The future research direction will focus on the extraction of entity relationships between Chinese herbal names and medicine effects entities.

Acknowledgments. The research was supported by the National Natural Science Foundation of China under grant number 61902116.

References

1. Sun, C.L., Sun, L.B.: Analysis of technological innovation in the field of traditional Chinese medicine in colleges and universities from the perspective of patent. Chin. J. New Drugs **30**(10), 915–920 (2021)
2. Yeom, H., Ko, Y., Seo, J.: Unsupervised-learning-based key phrase extraction from a single document by the effective combination of the graph-based model and the modified C-value method. Comput. Speech Lang. **58**, 304–318 (2019)
3. Sung, S.F., Lin, C.Y., Hu, Y.H.: EMR-based phenotyping of ischemic stroke using supervised machine learning and text mining techniques. IEEE J. Biomed. Health Inform. **24**(10), 2922–2931 (2020)
4. Zhang, F., Li, L.: Semi-supervised learning method based on nonnegative matrix decomposition and harmonic function. Stat. Decis. Making **35**(22), 71–73 (2019)
5. Yuan, N., Lu, K.Z., Yuan, Y.H.: Extraction of phenotypic named entities of symptoms in TCM medical records based on depth representation. World Sci. Technol. Modernization Tradit. Chin. Med. **20**(03), 355–362 (2018)
6. Devlin, J., Chang, M.W., Lee, K.: Bert: pre-training of deep bidirectional transformers for language understanding. In: Proceedings of 2019 Conference of the North American Chapter of the Association for Computational Linguistics: Human Language Technologies, Minneapolis, USA, 2–7 June 2019. Stroudsburg, PA: ACL, pp. 4171–4186 (2019)
7. Sun, Y., Wang, S., Li, Y.: ERNIE: enhanced representation through knowledge integration. arXiv:1904.09223 (2019)
8. Sun, Y., Wang, S., Li, Y.: ERNIE 2.0: a continual pre-training framework for language understanding. In: Proceedings of 34th AAAI Conference on Artificial Intelligence, NY, USA, 7–12 Feb 2020. Palo Alto, CA: AAAI press, pp. 8968–8975 (2020)
9. Wang, Y., Sun, Y., Ma, Z.: An ERNIE-Based joint model for Chinese named entity recognition. Appl. Sci. **10**(16), 5711 (2020)

10. Liu, G., Yuan, Q., Duan, J., Kou, J., Wang, H.: Chinese question classification based on ERNIE and feature fusion. In: Zhu, X., Zhang, M., Hong, Yu., He, R. (eds.) NLPCC 2020. LNCS (LNAI), vol. 12431, pp. 343–354. Springer, Cham (2020). https://doi.org/10.1007/978-3-030-60457-8_28

11. Cho, K., Merrienboer, B.V., Gulcehre, C.: Learning phrase representations using RNN encoder-decoder for statistical machine translation. In: Proceedings of 2014 Conference on Empirical Methods in Natural Language Processing, Doha, Qatar, 25–29 October 2014. Stroudsburg, PA: ACL, pp. 1724–1734 (2014)

12. Lafferty, J., Mccallum, A., Pereira, F.: Conditional random fields: probabilistic models for segmenting and labeling sequence data. In: Proceedings of 18th International Conference on Machine Learning, Williamstown, USA, 28 June–1 July 2001. San Francisco: Morgan Kaufmann, pp. 282–289 (2001)

An Algorithm for GPS Trajectory Compression Preserving Stay Points

Shota Iiyama[1(✉)], Tetsuya Oda[2], and Masaharu Hirota[3]

[1] Graduate School of Informatics, Okayama University of Science (OUS),
1-1 Ridaicho, Kita-ku, Okayama 700-0005, Japan
`i21im01cs@ous.jp`
[2] Department of Information and Computer Engineering,
Okayama University of Science (OUS), 1-1 Ridaicho, Kita-ku,
Okayama 700-0005, Japan
`oda@ice.ous.ac.jp`
[3] Department of Information Science, Okayama University of Science (OUS),
1-1 Ridaicho, Kita-ku, Okayama 700-0005, Japan
`hirota@mis.ous.ac.jp`

Abstract. There has been a significant increase in the amount of available trajectory data due to the widespread use of mobile devices equipped with the global positioning system. The increase in trajectory data, however, comes with problems, such as increased storage costs and difficulty in analyzing the data. These issues can be addressed by using compression methods. In this study, we propose an offline compression method that preserves the features of the trajectory data, such as stay points and trajectory shapes. When users stay at a particular location for a certain period, such as waiting at traffic lights or bus stops, the recorded positioning points are redundant. The proposed method compresses the trajectory data using the stay point information. We evaluated the performance of the proposed method using experimental results and a trajectory dataset.

1 Introduction

The widespread use of mobile devices equipped with the global positioning system (GPS) has given rise to a large volume of recorded trajectory data. Trajectory data are an important source of information for analyzing human behavior patterns [11,13], recommendation of travel routes for tourists [5,6], and next location prediction [2,12].

Consequently, this has led to several problems. First, the amount of traffic from a mobile device to the server increases. Second, the storage cost of the trajectory data increases. Finally, the processing speed for analysis using the trajectory data decreases. A solution to these problems is trajectory compression. Trajectory compression is a method that reduces data size by removing redundant positioning points from the original trajectory data.

L. Barolli et al. (Eds.): EIDWT 2022, LNDECT 118, pp. 102–113, 2022.
https://doi.org/10.1007/978-3-030-95903-6_12

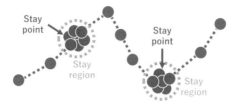

Fig. 1. Stay points and stay regions in trajectory data.

Many methods have been proposed to compress trajectory data, such as the Douglas-Peucker (DP) [3] and Spatial QUalIty Simplification Heuristic (SQUISH) [8] methods. Such trajectory compression methods preserve the shape of the trajectory and/or movement speed but do not preserve the stay point information. Stay point information is used to analyze the point where a user stays after moving through the tourist area and the stay of the associated traffic flow. Therefore, it is important to preserve the stay points of the original trajectory when applying a trajectory compression method.

Next, we provide the definition of stay point information. Figure 1 shows an example of trajectory data containing stay points. The red points surrounded by the blue circles represent the stay points. Moreover, the blue circles represent the stay region. A stay region is where a user stays at a particular location for a certain period, such as a bus stop or traffic lights. In addition, stay points are the positioning points included in a stay region.

Although the compression results using the existing trajectory compression methods could preserve the stay points, they do not necessarily preserve both the start and end points of the stay points; thus, stay point information cannot be used in the analysis of the compression results. Furthermore, such methods can adjust the degree of compression of the trajectory using parameters. When the degree of compression is reduced, the existing compression methods may preserve both the start and end of stay points. However, the application of compression results in achieving many redundant points other than the start and end points that should be compressed, and thus the compression rate gets reduced. By contrast, when the degree of compression increases, the compression rate also increases, but the stay points are removed. Therefore, it is difficult to apply one of the existing methods and obtain a high compression rate while preserving the stay points.

In this study, positioning points, where the trajectory shape and movement speed have significantly changed, and stay points are collectively referred to as feature points. The performance of the analysis may be degraded compared with the original trajectory when a trajectory, whose feature points are lost owing to the trajectory compression, is used in the analysis. Therefore, we propose a trajectory compression method that preserves the feature points of the trajectory at a high compression rate. The proposed method uses SQUISH - Extended

(SQUISH-E (μ)) [9], which preserves the trajectory shape, and the method proposed in [14] (we refer to as SP), which finds stay points. Finally, we combine the compression results of the SQUISH-E(μ) and SP methods to remove redundant compression points. Consequently, the compression results produced by the proposed method permanently preserve the start and end points of the stay points.

2   Related Work

Both offline and online compression methods are used for trajectory compression. Offline compression methods compress the trajectory data using the entire positioning points. Online compression methods compress the trajectory data using only the acquired positioning and a few other points. Therefore, offline compression methods can be applied only after the trajectory data are complete. However, an online compression method can be applied during the positioning process. The advantage of offline compression methods is their high performance because they use all the positioning point information for compression. Moreover, the advantage of the online compression methods is that they compress data sequentially; thus, the amount of data preserved is less.

The DP method [3] preserves the shape of trajectories. DP recursively computes the compression points and splits a trajectory based on the perpendicular Euclidean distance (PED). Top-down time ratio (TD-TR) [7] is an extension of the DP method that uses the synchronized Euclidean distance (SED), which considers the timestamp of positioning points. We elaborate on SED in Sect. 3.1. The advantage of this method is that it preserves the trajectory shape and speed of movement. However, the compression time of TD-TR is slower than that of DP because the computation time required to calculate SED is more than that of PED. The TD-TR Reduce method [4] is an extension of TD-TR to reduce the compression time. This method extracts feature points from trajectory data, whose movement direction and speed have changed, and applies TD-TR to the extracted points.

Dead reckoning [10], a fast trajectory compression method, uses the speed of an object to predict the next moving point and its distance from the actual positioning point to determine whether it is a compression point. The critical point (CP) [1] is a trajectory compression method that focuses on the direction of movement. In this method, the compression point is a location where the direction of movement of the user has changed. The SQUISH method [8] is a trajectory compression method that uses a priority queue. This method adds positioning points until the buffer is complete and removes the points with a low priority. The rest points in the buffer are regarded as compression points.

3   Proposed Method

In our proposed method, we use both the SQUISH-E(μ) and SP methods. Figure 2 shows an overview of the proposed method. First, we apply SQUISH-E(μ) and SP to a trajectory $T = \{p_0, p_1, \cdots, p_l\}$. Here, the positioning point

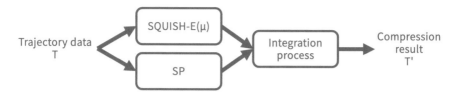

Fig. 2. Overview of proposed method.

$p_i = (x_i, y_i, t_i)$ represents the longitude, latitude, and timestamp. Subsequently, we integrate the compression points extracted by each method, thereby yielding the compression result T'.

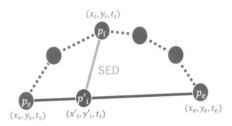

Fig. 3. Synchronized euclidean distance.

3.1 Synchronized Euclidean Distance

First, we describe the SED that was used in SQUISH-E(μ). Figure 3 illustrates an example of calculating the SED of these points. SED is the Euclidean distance between the positioning point $p_i(x_i, y_i, t_i)$ and pseudo point $p'_i(x'_i, y'_i, t_i)$. The pseudo point p'_i is the point approximated on the line between the two positioning points $\{p_s(x_s, y_s, t_s)|s < i\}$ and $\{p_e(x_e, y_e, t_e)|i < e\}$. In this study, when we calculate the SED between p_i and p'_i based on p_s and p_e, we denote it as $SED(p_i, p_s, p_e)$. The $SED(p_i, p_s, p_e)$ is calculated as follows:

$$x'_i = x_s + \frac{t_i - t_s}{t_e - t_s}(x_e - x_s)$$

$$y'_i = y_s + \frac{t_i - t_s}{t_e - t_s}(y_e - y_s) \tag{1}$$

$$SED(p_i, p_s, p_e) = \sqrt{(x_i - x'_i)^2 + (y_i - y'_i)^2}$$

3.2 SQUISH-E(μ)

SQUISH-E(μ) is a trajectory compression method that uses a priority queue. Priority is the degree of change in the shape of the trajectory data when a positioning point p_i is dequeued from the priority queue. In SQUISH-E(μ), the shape differences in the trajectory before and after the dequeue does not exceed threshold μ. Therefore, the advantage of this method is that it preserves the shape of trajectory data.

SQUISH-E(μ) consists of the following two steps:

step 1: Enqueue all positioning points of the trajectory data T into the priority queue Q.
step 2: Dequeue the low-priority points from Q.

Step 1 enqueues the positioning point p_i of the trajectory data T to the priority queue Q. Here, we denote the priority of p_i as $priority(p_i)$. When enqueuing p_i to Q, the method initializes $priority(p_i) \leftarrow \infty$ and $\pi(p_i) \leftarrow 0$. Here, the variable $\pi(p_i)$ is used to maintain the neighborhood's maximum priority of the removed points in the next step. When enqueuing a positioning point other than the first point ($i \geq 2$), we change the priority $priority(p_i)$ as follows:

$$priority(p_{i-1}) = SED(p_{i-1}, pred(p_{i-1}), succ(p_{i-1})), \tag{2}$$

where $pred(p_i)$ and $succ(p_i)$ are p_i's closest predecessor and successor among the points in Q. Hence, the priorities of the start and end points of Q are $priority(p_s) \leftarrow \infty, priority(p_e) \leftarrow \infty$, which prevent them from being removed from Q. After we finish inquiring all the positioning points to Q, the method moves to step 2.

In step 2, the method dequeues the positioning points with low-priority values in Q and compresses the trajectory using the following procedure:

1. We find the lowest priority positioning point p_j in Q.
2. We update $\pi(pred(p_j))$ and $\pi(succ(p_j))$ based on $priority(p_j)$. If $\pi(pred(p_j)) < priority(p_j)$, we set $\pi(pred(p_j)) \leftarrow priority(p_j)$. Furthermore, if $\pi(succ(p_j)) < priority(p_j)$, then $\pi(succ(p_j)) \leftarrow priority(p_j)$.
3. We dequeue p_j from Q and update the priorities of $pred(p_j)$ and $succ(p_j)$. Here, we let p_k be either $pred(p_j)$ or $succ(p_j)$. If p_k is the first or last point in Q, then the priority is not updated. Otherwise, $priority(p_k) = \pi(p_k) + SED(p_k, pred(p_k), succ(p_k))$.
4. We repeat the process until the minimum priority value in Q is larger than μ.

After the completion of step 2, the points in Q are the result of the SQUISH-E(μ) compression R_{sq}.

3.3   Extraction of Stay Points

Here, we explain the process of extracting stay points from trajectory data using the SP method.

In our proposed method, we used SP [14] to extract stay points from the trajectory data. This method defines the points in the trajectory data that satisfy all the following conditions as stay points:

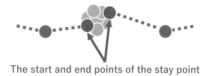

The start and end points of the stay point

Fig. 4. Example of the stay points extracted using our method.

$$m < i \leq n$$
$$Dist(p_m, p_i) \leq T_{distance}$$
$$Dist(p_m, p_{n+1}) > T_{distance} \tag{3}$$
$$Int(p_m, p_n) \geq T_{time}$$

$Dist(p_m, p_i)$ denotes the geospatial distance between the positioning points p_m and p_i. $T_{distance}$ is the threshold of the distance between the two positioning points. $Int(p_m, p_n)$ is the difference in the timestamps between the positioning points p_m and p_n. T_{time} is the difference threshold. After the stay points are extracted, we regard each consecutive set of stay points in the original trajectory data as a stay region.

The method presented in [14] defines the average latitude and longitude, and the time of the first and last points included in an extracted stay region as the information of a stay point. By contrast, our proposed method defines the first and last points included in an extracted stay region as stay points because it uses stay point information for trajectory compression. The stay points extracted using our proposed method are demonstrated in Fig. 4.

We define the start and end points of the stay points as R_{sp}. In addition, the stay region is indicated by R_{sr}. The information of the stay points and stay region are used together with the results of applying the SQUISH-E(μ) method.

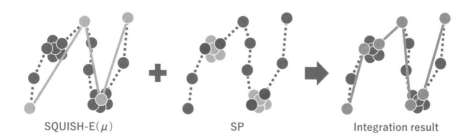

SQUISH-E(μ)   SP   Integration result

Fig. 5. Integration process.

3.4 Integration

The last process in the proposed method integrates the compression results of SQUISH-E(μ) and SP. Figure 5 presents an example of the integration process. First, we remove all the compression points from the compression result R_{sq} of SQUISH-E(μ) that overlap with the stay region R_{sr} extracted using the SP method. Subsequently, we integrate R_{sq} and the compression result of stay points R_{sp} into the compression result of the proposed method. As a result, we can compress the trajectory and maintain its features while permanently preserving the start and end points of the stay regions.

4 Experiments

Here, we evaluate whether the compression results using the proposed method can preserve the feature points of a trajectory.

4.1 Experimental Conditions

We used the Microsoft GeoLife dataset [15–17] as the experimental data. We used 171 trajectories traveled by foot or bicycle in Beijing, China. The mean and median numbers of the positioning points for each trajectory were approximately 1,016 and 540, respectively. The mean and median sampling rates of each trajectory were approximately 24 and 5 s, respectively.

We used four evaluation criteria: SED error, speed error, compression rate, and compression time. The SED error evaluates the shape differences of the trajectory before and after the data are compressed. The smaller the value of the SED error, the better preserved is the trajectory's original shape. The speed error evaluates the different in movement speed before and after the data are compressed. The smaller the value of the speed error, the smaller is the compression result owing to the loss of speed information. The compression rate evaluates the proportion of positioning points removed by the applied method. The higher the compression rate, the smaller is the compression result, and lesser is the trajectory data size. Compression time is the time required to compress the trajectory data. The smaller the compression time, the faster the method compresses the trajectory data.

We used the SQUISH-E(μ) method to compare the SED error, speed error, and compression rate. In addition, we used the SQUISH-E(μ) and SP methods to compare the compression time. The parameters of the SQUISH-E(μ) method varied from 10 to 200 m in steps of 10 m. We fixed the distance threshold at 30 m, while the stay time threshold for the SP parameters was 30 s.

4.2 Experiment Results

Figure 6 shows a comparison between the performance of the proposed and comparison methods. The red, blue, and green lines represent the performance of the proposed, SQUISH-E(μ), and SP methods. The vertical and horizontal axes represent each evaluation criterion and threshold value, respectively.

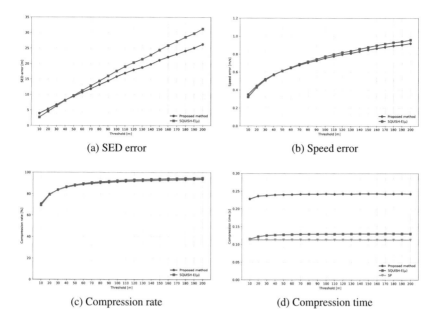

(a) SED error

(b) Speed error

(c) Compression rate

(d) Compression time

Fig. 6. Experiment results.

4.2.1   SED Error

Figure 6(a) illustrates the SED error yielded by applying the proposed and SQUISH-E(μ) methods. Our method yielded a larger SED error compared with the SQUISH-E(μ) method when the threshold is 40 m or less because the compression result obtained by our method includes the start and end points of the stay points. Moreover, when the threshold is 50 m or higher, our method yielded a smaller SED error than that of the SQUISH-E(μ) method, and the difference increased as the threshold increased because our method preserves the stay points that were not preserved by the SQUISH-E(μ) method. When the threshold value was small, the compression result yielded by applying the SQUISH-E(μ) method had many compression points; thus, the result is similar to the original trajectory shape because this method tends to overpreserve the stay points. In the proposed method, the number of compression points is less than that of the SQUISH-E(μ) method because the compression result obtained by applying this method had only two stay points. Therefore, the SED error obtained by applying our method was slightly larger. However, the SQUISH-E(μ) method does not necessarily preserve the start and end points of the stay points, whereas our method has the advantage of preserving the start and end points of the stay points.

The SQUISH-E(μ) method with a large threshold tends to decrease the number of compression points, and thus the shape is hardly preserved. Moreover, the compression result obtained by applying our method preserves the stay points extracted by the SP method, which is not affected by the threshold. Therefore, the compression result obtained by our method can preserve the stay points removed by the SQUISH-E(μ) method. Consequently, our method can preserve the trajectory shape better than the SQUISH-E(μ) method at large thresholds.

4.2.2 Speed Error

Figure 6(b) shows a comparison between the speed error obtained by applying the proposed and SQUISH-E(μ) methods. When the threshold was 40 m or less, the proposed method yielded a larger speed error compared with the SQUISH-E(μ) method. Moreover, when the threshold was 50 m or higher, the proposed method yielded a smaller speed error than that of the SQUISH-E(μ) method. In addition, the difference between the speed errors obtained by applying the proposed and SQUISH-E(μ) methods was small throughout the entire process because our method employs the SQUISH-E(μ) method, and majority of the compression points are similar. However, SQUISH-E(μ) tends to overcompress the trajectory data when the threshold becomes large, whereas our method preserves the compression points using the SP method, which can reduce the adverse effect of overcompression by the SQUISH-E(μ) method. Therefore, as the threshold increases, our method performed slightly better than SQUISH-E(μ) in preserving the information of the trajectory movement speed.

4.2.3 Compression Rate

Figure 6(c) illustrates a comparison between the compression rates obtained by applying the proposed and SQUISH-E(μ) methods. When the threshold was 20 m or less, our method yielded a higher compression rate compared with the SQUISH-E(μ) method. Moreover, when the threshold was 30 m or higher, our method yielded a lower compression rate than that of the SQUISH-E(μ) method. In addition, the difference between the compression rate obtained by applying the proposed and SQUISH-E(μ) methods is small. When the threshold was small, the SQUISH-E(μ) method compressed the stay points in the trajectory data; however, our method replaced the compression points with stay points and thus reduced the number of compression points and increased the compression rate. When the threshold value was large, the compression result obtained by applying our method preserved not only the compression points extracted by the SQUISH-E(μ) method but also the stay points extracted by the SP method. Therefore, applying our method increases the number of compression points and decreases the compression rate. Consequently, our method can improve the compression rate using the stay point information when the threshold is small.

4.2.4 Compression Time

Figure 6(d) demonstrates a comparison between the compression times obtained by applying the proposed, SQUISH-E(μ), and SP methods. Here, the compression time of SP is constant because the SP method is not affected by the threshold value. The compression time of the proposed and SQUISH-E(μ) methods did not change significantly when the threshold was 60 m. In addition, the compression time of the proposed method was approximately equal to the sum of the execution times of the SQUISH-E(μ) and SP methods because our method employs both the SQUISH-E(μ) and SP methods. In future studies, we will attempt to reduce the compression time by determining whether the point is a stay point during the execution of the SQUISH-E(μ) method or not.

4.2.5   Discussion

Based on the SED error, speed error, and compression rate, the performance of our proposed method was slightly lower than that of SQUISH-E(μ) when the threshold of the methods was large. Moreover, when the threshold was small, the proposed method showed better results. This is because the proposed method modifies the compression results obtained by applying the SQUISH-E(μ) method using the stay point information. Therefore, a trade-off exists between the proposed and SQUISH-E(μ) methods in terms of the evaluation criteria, depending on the threshold value. In addition, as our method preserves the stay points extracted by the SP method, the compression result obtained by applying our method included stay point information.

5   Conclusion

We proposed a trajectory compression method that preserves the shape of the trajectory data and information of the stay points. The proposed method uses the stay points extracted from the trajectory data and modifies the compression result using the SQUISH-E(μ) method. Furthermore, based on several evaluation criteria, we compared the performance of the proposed method with that of the SQUISH-E(μ) and SP methods. The experimental results proved that the use of stay points for trajectory compression was helpful in trajectory compression.

In the future, we intend to extend this study in the following directions: In this study, we propose an offline compression method that employs two online compression methods. Therefore, we will first improve the proposed method so that it can be applied online to preserve the features of trajectory data. Second, we will improve the execution time of the proposed method. The compression time of the proposed method was approximately equal to the sum of that of the SQUISH-E(μ) and SP methods because we employed both the SQUISH-E(μ) and SP methods to the trajectory data and combined the results. Therefore, it is necessary to develop a method that can simultaneously extract the stay points during the execution of SQUISH-E(μ). Third, we intend to validate the performance of the proposed method using other datasets.

Acknowledgment. This work was supported by JSPS KAKENHI Grant Number JP19K20418 and Grant for Promotion of OUS Research Project (OUS-RP-20-3).

References

1. Barbeau, S., et al.: Dynamic management of real-time location data on GPS-enabled mobile phones. In: 2008 The Second International Conference on Mobile Ubiquitous Computing, Systems, Services and Technologies, pp. 343–348 (2008)
2. Chen, M., Zuo, Y., Jia, X., Liu, Y., Yu, X., Zheng, K.: CEM: a convolutional embedding model for predicting next locations. IEEE Trans. Intell. Transp. Syst. **22**(6), 3349–3358 (2021)
3. Douglas, D.H., Peucker, T.K.: Algorithms for the reduction of the number of points required to represent a digitized line or its caricature. Cartogr.: Int. J. Geogr. Inf. Geovis. **10**(2), 112–122 (1973)
4. Hansuddhisuntorn, K., Horanont, T.: Improvement of TD-TR algorithm for simplifying GPS trajectory data. In: 2019 First International Conference on Smart Technology Urban Development (STUD), pp. 1–6 (2019)
5. Kurashima, T., Iwata, T., Irie, G., Fujimura, K.: Travel route recommendation using geotags in photo sharing sites. In: Proceedings of the 19th ACM International Conference on Information and Knowledge Management, CIKM 2010, pp. 579–588 (2010)
6. Memon, I., Chen, L., Majid, A., Lv, M., Hussain, I., Chen, G.: Travel recommendation using geo-tagged photos in social media for tourist. Wirel. Pers. Commun. **80**(4), 1347–1362 (2015)
7. Meratnia, N., de By, R.A.: Spatiotemporal compression techniques for moving point objects. In: Bertino, E., et al. (eds.) EDBT 2004. LNCS, vol. 2992, pp. 765–782. Springer, Heidelberg (2004). https://doi.org/10.1007/978-3-540-24741-8_44
8. Muckell, J., Hwang, J.H., Patil, V., Lawson, C.T., Ping, F., Ravi, S.S.: SQUISH: an online approach for GPS trajectory compression. In: Proceedings of the 2nd International Conference on Computing for Geospatial Research & Applications, COM.Geo 2011 (2011)
9. Muckell, J., Olsen, P.W., Hwang, J.H., Lawson, C.T., Ravi, S.S.: Compression of trajectory data: a comprehensive evaluation and new approach. GeoInformatica **18**(3), 435–460 (2014)
10. Trajcevski, G., Cao, H., Scheuermanny, P., Wolfsonz, O., Vaccaro, D.: On-line data reduction and the quality of history in moving objects databases. In: Proceedings of the 5th ACM International Workshop on Data Engineering for Wireless and Mobile Access, MobiDE 2006, pp. 19–26. Association for Computing Machinery (2006)
11. Yang, L., Wu, L., Liu, Y., Kang, C.: Quantifying tourist behavior patterns by travel motifs and geo-tagged photos from flickr. ISPRS Int. J. Geo-Inf. **6**(11), 345 (2017)
12. Yao, D., Zhang, C., Huang, J., Bi, J.: SERM: a recurrent model for next location prediction in semantic trajectories. In: Proceedings of the 2017 ACM on Conference on Information and Knowledge Management, CIKM 2017, pp. 2411–2414 (2017)
13. Yuan, Y., Medel, M.: Characterizing international travel behavior from geotagged photos: a case study of flickr. PLoS One **11**(5), 1–18 (2016)
14. Zheng, V.W., Zheng, Y., Xie, X., Yang, Q.: Collaborative location and activity recommendations with GPS history data. In: Proceedings of the 19th International Conference on World Wide Web, WWW 2010, pp. 1029–1038. Association for Computing Machinery (2010)

15. Zheng, Y., Li, Q., Chen, Y., Xie, X., Ma, W.Y.: Understanding mobility based on GPS data. In: Proceedings of the 10th International Conference on Ubiquitous Computing, UbiComp 2008, pp. 312–321. Association for Computing Machinery (2008)
16. Zheng, Y., Xie, X., Ma, W.: GeoLife: a collaborative social networking service among user, location and trajectory. IEEE Data Eng. Bull. **33**(2), 32–39 (2010)
17. Zheng, Y., Zhang, L., Xie, X., Ma, W.Y.: Mining interesting locations and travel sequences from GPS trajectories. In: Proceedings of the 18th International Conference on World Wide Web, WWW 2009, pp. 791–800. Association for Computing Machinery (2009)

Blockchain for Islamic HRM: Potentials and Challenges on Psychological Work Contract

Olivia Fachrunnisa[1]([⊠]) and Fannisa Assyilah[2]

[1] Department of Management, Faculty of Economics, UNISSULA, Semarang, Indonesia
`olivia.fachrunnisa@unissula.ac.id`
[2] School of Design, University of Western Australia, Perth, Australia
`22368055@student.uwa.edu.au`

Abstract. Blockchain technology has evolved and has now entered the non-finance function since some organizations have begun to use blockchain to facilitate human resources management practices. One of them is to make psychological work contract formation process more efficient and effective. This paper analyzes the potential and challenges of using Blockchain-based application for Islamic human resources management practices. Drawing from the current literature review existence, the paper discussed various interesting mechanism to support the formation and maintenance of psychological work contract that can bring some potentials and benefits. The paper also underlined the challenges facing in applying blockchains in Islamic HRM and some strategies to face the challenges. We argue that blockchain has the ability to harmonize differing sharia-compliance in work contract or work agreement between employee and employers.

1 Introduction

A work contract is a legally binding agreement between an employee and their employer outlining the employee's and employer's obligations and responsibilities. The purpose of a labor contract is to create a relationship between the two parties regarding their respective rights and duties. Work contracts offer workers the certainty that they are working for professional organizations which defined duties and commitments in all employment circumstances. In addition, the employment contract serves as a safeguard for employers and their key stakeholders' intellectual property protection. Some fundamental aspect of the necessity of work contract for both employees and employers are as follows: employee job safeguards and workforce stability for the employer, employee-employer mutual awareness of each other's responsibilities; the necessity of protecting workers' rights; and a work contract is the most efficient way for an employer to safeguard a trade mark or maintain confidentiality, and employees end up leaving [1, 2]. Work contracts, in core allow ongoing growth for the employment relationship, propelling it forward. Several studies have also investigated the significance of work contracts, for example [3], offering employment relationships and small business HRM practices in the form of AAC (Awareness, Action, Comprehensiveness, and Excellence) model framework to increase small business commitment sustainability. The investing in the work relationship then contributes to the accomplishment of economic, social, and environmental sustainability [4].

L. Barolli et al. (Eds.): EIDWT 2022, LNDECT 118, pp. 114–122, 2022.
https://doi.org/10.1007/978-3-030-95903-6_13

Meanwhile, as a result of the technical advancements brought about IR 4.0, the nature of the business model underwent a transition. The deployment of automated robots, artificial intelligence (AI), and the Internet of Things (IoT) in business organizations will result in increased efficiency, blurring the lines between the real world and virtual environments, which will be linked by the Cyber-Physical Production System (CPPS) [5].

In this era, digital work platforms such as Rappi, Gojek, Uber, Upwork, etc. increase workers' agency to contract with choses clients/employers, rates, tasks, and methods and times of working [6]. They are free to choose several jobs with different income levels and within certain time frame (temporary time), and even tend to quickly move from one digital work platform to another. Thus, short-term contract mediated by digital platforms are becoming a trend for employee-employer relationships. These short-term work contracts enable for flexible work schedules, freedom to choose job, and the ability to work whenever and wherever you want. Employees have the option of joining other platforms to generate additional revenue due to a lack of other career possibilities. Unfortunately, working circumstances on this platform are hard.

These short-term work contracts enable for flexible work schedules, the independence to choose tasks, and the ability to work whenever and wherever you want. Employees have the option of joining other platforms to generate additional revenue due to a lack of other career possibilities. Unfortunately, working circumstances on this platform are hard. Insufficient salaries and erratic revenue, arbitrary dismissal of employment and social security, and hurdles to multi-homing (joining to some more than one platform), are examples of these. Indeed, algorithmic control impacts employment procedures and performances in manners that reduce the autonomy of platform workers. They wind up working long hours to make a reasonable livelihood, which has an influence on their work-life and therefore can cause anxiety.

Nevertheless, economists, psychology, and management theory have all argued that platforms are more than just employment traders who link people with jobs in exchange for the payment. Platforms, in particular, strategically manage needs, coordination issues, information, and contractual budget [7]. Employers, on the other hand, can lose because employment conditions brimming with gig workers become less advantageous when single specialists fail to develop skills through valuable social associations. Employers, like employees, suffer when the physical distribution of formation comes at a cost of the accessible good of mutual understanding and learning [8, 9].

These problems require a technology or HRM practice approach that emphasizes human value as the core system, to balance the welfare of employees and employers. So that a mutually beneficial, sustainable employee-employer relationship can actually create decent work. While an employment contract is a written legal agreement, a psychological contract is formed by the easily repeatable, declarations, and promises of one side of the transaction and how they are accepted by other. Several research, for example [10, 11]; and [12], have demonstrated the relevance of psychological work contracts to the job relationship. In answer to alterations in the digital-work platform, it is necessary to adhere to the two aspects of the psychological work contract, both transactional and relational.

Transactional psychological work contract is based on precisely defined, monetisable transactions that take place between the employee and the employer over a predetermined and often brief period of time. Transactional distinguished by the employee's restricted personal participation in the workplace and diminished organizational civic activity. Employees are driven by extrinsic motivation under this contract, such as career fat monitoring and extra benefits. Meanwhile, the relational psychological work contract is based on social exchange theory [13], which states that an exchange of effort and loyalty on the part of the employee exchange for job stability and career advancement on the part of the employer. Social exchange theory stresses reciprocal advantages between employee-employer, with the hope that the other party would reciprocate. Employees who receive support from their employers, such as career progression, will feel obligated to reciprocate, maybe by giving exceptional performance or commitment. Employers with relational work contract tend to adopt a paternalistic role, demonstrating concern for employees' well-being and taking on most of the responsibility for managing their careers. As a result, employee-employer interactions are often borderless, long-tem, and open-ended. Employee motivation in this contract focuses on intrinsic motivation, which includes job security, a comfortable/friendly work environment, personal support, flexible work environment, personal support, flexible working circumstances, possibilities for personal and career growth, and so on. The primary goal of the relational psychological work contract is to meet the requirement of both employee-employer [11, 14].

Meanwhile, related to the use of digital technology advances, efforts to implement Blockchain into work contracts for have been carried out. However, it turns out that the role of this technology is still limited to building smart contracts for psychological work contract fulfilment in a transactional context, for example [15–20]. Dolzhenko 2021 [21], introduced the role of Blockchain in smart contracts to maintain the stability of the employment relationship, but it is still a discourse. Furthermore, [22] states that the application of digital labor contracts still requires at least determining the conditions for the existence of such a system and ways to protects its elements (employer, digital platform developer, and employee) from changes and shocks. Several studies have demonstrated Blockchain's role in achieving a balance between transactional and non-transactional data. However, it is not been able to seek the achievement of a relational psychological work contract.

The relational psychological work contract is related to social exchange theory and emphasizes intrinsic motivation. Therefore, to achieve this relational employment contract, it is necessary to incorporate religious values into HRM practices, one of which is Islamic values.

2   Islamic HRM

The incorporation of Islamic values into HRM practices consist of recruitment, selection, training and development, remuneration, performance appraisal and employee engagement. In Islamic HRM, recruitment is premises on the basis of fairness for all job seekers (Holy Qur'an. 4: 58–59). Because a Muslims's main obligation at work is not only to handle out the mission of the organization/employer, but also to their God. Selection decisions in Islamic HRM are based on fairness, good deeds, and wisdom (knowledge and understanding). In terms of selection, Islamic ideals include justice, honesty, trust, and accountability (Holy Qur'an. 28: 26). Islamic HRM training and development focuses not only on expanding intellectual competence, but also on personal virtues rooted in spirituality or religion (*Etqan, Ihsan,* and *al-Falah*). In Islamic value, compensation is founded on the notion of fairness (Holy Qur'an. 83: 1–3). The Islamic value of compensation is due to two factors: material advantages, such as salaries and earnings, are only a means of subsistence, whilst the actual return of work is with Allah since it is regarded an act of worship [23]. Furthermore, performance evaluation may be described as the process of determining an employee's performance, interaction with them, and advising and preparing for improvement. In an Islamic context, performance evaluation is based on ethical guidance and the practice of the Prophet Muhammad (SAW) and his immediate four Caliphs. The ethical domain is disclosed in Qur'anic commands, and it consists of contractual agreements, self-responsibility and control, and obligation to God, who holds the top position for performance evaluation [24].

In terms of work contract, Islam regards a person's employment as the constant reminder of an obligatory relationship between the business and the individual. Both the employer and the employee have expectation that must be satisfied in full. According to the Qur'an, each commitment or engagement is subject to a contract that must be fulfilled by the partners (Holy Qur'an. 17: 34) *"And fulfill engagement (promise), for the engagement will be enquired into."* More precisely, the Islamic approach teaches duties in mutual connections between them, including of Social Exchange theory, LMX theory, and reciprocity norms and golden rule, Islamic Work Ethic (Maslaha as mutual welfare), taqwa as framework, and looking for halal as deeds [25].

3 Psychological Work Contract

The initial concept of the application of the psychological work contract in the workplace was developed on 1960 by Argyris. *"Since the foreman realize the employees in this system will tend to produce optimally under passive leadership, and since the employees in this system will tend to produce optimally under passive leadership, and since the employees agree, a relationship may be hypothesized to evolve between the employees and the foreman which might be called the 'psychological work contract'. The employee will maintain high production, low grievances, etc., if the foreman guarantee and respect the norms of the employee informal culture (i.e., let the employees alone, make certain they make adequate wages, and have secure jobs). This is precisely what the employees' need* [26]. Furthermore, Rousseau 1995 [27] developed, *"The term psychological contract refers to an individual's belief regarding the terms and conditions of a reciprocal exchange agreement between the focal person and another party. Key issues here include the belief that a promise has been made and a consideration offered in exchange for it, binding the parties to some set of reciprocal obligations.'* Meanwhile Morrison & Robinson 1997 [28], explain that *"An employee's beliefs about the reciprocal obligations between that employee and his or her organization, where these obligations are based perceived promises and are not necessarily recognized by agents of the organization."* In other terms, if employees believe that employer respects their right to grow and improve and utilize their own ingenuity, they will respect the organization's right to change. Employees are likely to perform considerably better if employers do not intervene too much with them, respect their culture or the conventions of the employee group, and let them to get on with the profession.

Psychological work contract characteristics, consisting of the beliefs containing the implied nature, the personal nature, perceived arrangement, continuous exchange among employee and employer, the party members to the psychological contracts, and formed by the organizations. As previously stated, the psychological work contract is made up of two components: transactional and relational. Whereas the motive for completing the transactional psychological work contract is extrinsic, the motivation for completing the relational psychological work contract is intrinsic motivation. Transactional contracts related to material purposes are rooted in economic exchange (monetizable exchange, career tracking, salary, incentives, etc.). Meanwhile, the relational contract focuses non-material successes that result in commitment and loyalty for mutual benefit or that suit the demands of the employee-employer relationship.

The link between psychological contract and HRM practices begins with the process of employer promises, progresses through employer conduct, generates employee promises, shapes employee behavior, and finally returns to the beginning. Psychological work contract fulfilment happens when the employee has faith in the employer to keep commitments and meet expectations, and vice versa. Employees that fulfill the psychological contract demonstrate to their coworkers that they are devoted to them, appreciate their contributions to the business, and aim to keep the connection going. The organization's availability of open and honest communication, being allocated interesting and challenging work, and having decent management [29] are all elements that impact psychological work contracts.

4   Blockchain Technology and Work Contract

Blockchain is related with connectivity, decentralization, ledger dispersion, and good thermal stability, and it may aid in the documentation of transactions and asset monitoring on a trade or business [30]. The word "decentralization" refers to the fact that many more than one entity or agent administers the complete database. Data and information kept on numerous host systems, over (one silo database), on the other hand, might be extremely vulnerable to hackers. On the most basic level, this technology enables the user community to have safe access to shared information. Like the world wide web, blockchain as a whole is neither owned or managed by a single person [31]. Blockchain is essentially a "journey" of "blocks," or a string of structured "blocks" in a network of individuals. This route is important for document checking, entry security, and system maintenance. A block comprises data, hash, and hash from the preceding block, making a chain-like pattern [32]. A block is strong enough to hold a variety of items, including currency, votes in a election, and data. While "hash" a value generated by a string of ext. Hashes can be understood as fingerprints, with each providing a specific, as a form of identification to the block. A content in the block evolves, so does the hash.

There are three variations of privacy in Blockchain Technology: totally public, closed with small set of users, and hybrid public private. A totally public blockchain is accessible for all users to observe, access, and contribute to. According to [33] the next form of blockchain is limited to a certain set of users, for illustration (companies, banking communities, or agents). With contrast, in a hybrid public-private blockchain, those with privileged access can see certain data in full, while the public can only see a portion of it. Constructing chains in this form of blockchain is not possible for everyone. When a blockchain is added, it becomes immutable, which means that information cannot be erased or modified. Blockchain's underlying premise, based on the original concept of financial ledgers. If a user wants to change something on the blockchain, they won't be able to rebuild it. The requested modifications, on the other hand, will be recorded in a new block noting the data change and the period of the adjustment. If it is inevitable the content that has already been entered to be modified, an information path that cannot be omitted will be generated. This blockchain technology could allow users to store, track, and transmit data. Additionally, blockchain improves operational activities, notably the effectiveness and efficiency of human resource management activities.

Initially, blockchain technology was best recognized for its use in crypto currencies and Bitcoin. Nevertheless, it is predicted that this technology will benefit all aspect of the business organization. Many recent studies have investigated the importance of blockchain technological advancements in business management, notably in management of human resources.

In the context of work contracts from HRM practice, several studies have pointed to the role of blockchain in building a contract system to Gig-Workers in Gig-Economy. Wang et al. 2017 [15] introduced blockchain technology as a smart-contract model built and maintained by an engineering team. This contract is founded on the assumption that all acts made by individuals are motivated by their self-interest, and that important information is asymmetric between firms and employees. In blockchain, the trust connection across the stem is automatically formed and does not require a special organization to grasp. Michaelides [16] discusses smart-contracts as a framework for verifying educational and job record (e.g., APPI and TechnoJobs). A blockchain can collect and store data. As a result, it can monitor a person from one job to the next, giving recruiters and HR officials with information on the candidate's experience and talents. Other important documents include: length of employment in previous jobs, pay, bonuses, and objectives stated for former companies. Specific computer protocols can be included in blockchain to govern the delivery of digital contracts. All of these factors can have an impact on benefits, salary, retirement packages, and bonuses. At the same time, some precautions must be in place about how the data is gathered in order to avoid the undesirable effects of trespassing on one's privacy.

Moreover, Ni et al. 2019 [17] suggested a smart-contract as a re-emerging technology with the growth of the blockchain. A smart-contract is a collection of scenario-response programing rules and logic that is put on the blockchain as decentralized, trustworthy shared program code. Smart-contract share many of the same features as blockchain data, such as distributed recording, storage, verification, and non-tampering. The suggested smart-contract blockchain is linked to organizational staff behavior, which guides planning and analysis. The technique's elements include: assessment strategy (meet employee and employer demands, design a realistic individual and community index system to qualitatively analyze individual and community behavioral characteristics); reward and punishment strategy (for determining what kind of feedback is given to individuals or groups of organizations). Employee incentives for transparent, fair, tamper-resistant, and human intervention are achieved as a result.

The framework and mechanism of smart-contract-driven blockchain provide a wide range of business integration possibilities. The biggest potential benefits stem from blockchain platforms include immutable entries, great transparency, digitally automated operations, and decentralized control mechanisms [18]. More specifically for temporary workers, [19] designed a smart-contract focused to remuneration fulfilment and based on hours worked exclusively for temporary employees. Meanwhile, [20] explored smart-contracts, which contain agreements including blockcahain technology. The conditions of the deal and the penalties that would apply if the agreements were broken. There are a few examples of smart-contracts in action, disrupting and transforming traditional business models. First, when an employee completes the contractual hours, which are recorded in blockchain, the employee is immediately compensated. A new recruit might be part of a second implementation. Where a new recruit successfully completes the required documentation, such as background checks or terms and conditions agreements, they can be granted access to the company's system. It should be emphasized, however, that the diverse natures of occupations my necessitate different processing durations and result in varying levels of success. As a result, knowing job type categorization is necessary in order to identify the completion deadline, material needs, and conditions agreed. Although the concept of smart-contracts appears to be a simple and effective solution to improve corporate activities, it is not without pitfalls. In principle, they are accident-proof, which may be both an advantage and a risk. Because it cannot be readily modified, the programming underlying the contract must be flawless. Smart-contract payments are immutable, so if a mistake is made, there is no way to reverse it. According to recent study, the state has already recognized the prospect of implementing digital work contracts and electronic work books. The next phase is to create digital passports based on blockchain technology, to which any information and transactions relevant to market themes may be linked. However, in order to achieve this goal, it is important to first define the prerequisites for the creation of such a system as well as methods to safeguard its constituents from shifts and disruptions [22].

5   Conclusion

Based on the extensive the literature review, it can be concluded that employee-employer relationships may be maintained by incorporating Blockchain-based Islamic HRM into Psychological Work Contract. Blockchain technology serves as a developer of smart-contract systems, which are more inclined to the implementation of psychological work contracts needing an HRM strategy by incorporating Islamic principles into HRM practices in the context of work contracts. In psychological work contracts, Islamic HRM consists of a promise to fulfill commitments between employee and employer (QS. 17: 34), and also for God, who has the highest grade in performance appraisal [24]. The Islamic approach teaches duties in reciprocal relationships between employer and employee, which include Social Exchange theory, LMX theory, the reciprocity norm and golden rule, Islamic Work Ethic (Maslaha as mutual welfare), taqwa as framework, and pursuing halal as an action [25]. Employers are responsible for three things:

ensuring timely payroll to employees, ensuring employees have enough time to perform they obligatory religious prayers; treating employees with respect and burdening them according to their abilities; and ensuring the safety of their employees at work. Then, workers' obligations to their employers include: executing their task honestly, rigorously sticking to the job timings agreed upon in the work contract; and completing their duties as promises in the employment contract [34].

References

1. Elton, W.: The Importnce of an Employment Contract, ZEGAL Articles HR, [Online] (2021). https://zegal.com/blog/post/importance-of-employment-contract/
2. Schwarzkopf Professional, The Importance of an Employment Contract, Business (2021). https://www.schwarzkopf-professional.se/en/home/education/ask/business/0614/the-import ance-of-an-employment-contract.html. Accessed 02 Dec 2021
3. Maheshwari, M., Samal, A., Bhamoriya, V.: Role of employee relations and HRM in driving commitment to sustainability in MSME firms. Int. J. Product. Perform. Manag. **69**(8), 1743–1764 (2020)
4. Rincon-Roldan, F., Lopez-Cabrales, A.: The impact of employement relationships on firm sustainability. Empl. Relations, vol. ahead-of-p, no. ahead-of-print (2021)
5. Szabó-Szentgróti, G., Végvári, B., Varga, J.: Impact of industry 4.0 and digitization on labor market for 2030-verification of Keynes' prediction. Sustain **13**(14), 1–19 (2021)
6. Charlton, E.: What is the gig economy and what's the deal for gig workers?," World Economic Forum (2021). https://www.weforum.org/agenda/2021/05/what-gig-economy-wor kers/. Accessed 05 Dec 2021
7. Wood, A.J., Lehdonvirta, V.: Antagonism beyond employment: how the 'subordinated agency' of labour platforms generates conflict in the remote gig economy. Socio-Econ. Rev. **19**(4), 1369–1396 (2021)
8. Connelly, C.E., Fieseler, C., Matej, Č., Giessner, R.: Working in the digitized economy : HRM theory & practice. Hum. Resour. Manag. Rev. **31**(1) pp. 1–7 (2020)
9. Goswami, M.: Revolutionizing employee employer relationship via gig economy. Mater. Today Proc. (2020)
10. Mohamed Jaafar, S., Nik Mat, N.H.: A conceptual analysis of effective gig works system for sustainable employment in challenging times. J. Contemp. Issues Bus. Gov. **27**(1), 1427–1440 (2021)
11. Chinyamurindi, W.T.: The Dynamism of psychological contract and workforce diversity: implications and challenges for industry 4.0 HRM. In: Coetzee, M., Deas, A., (eds.) Redefining the Psychological Contract in the Digital Era: Issues for Research and Practice, no. April, Springer (2021)
12. Braganza, A., Chen, W., Canhoto, A., Sap, S.: Productive employment and decent work: the impact of AI adoption on psychological contracts, job engagement and employee trust. J. Bus. Res. **131**, 485–494 (2021)
13. Blau, P.M.: Exchange and Power in Social Life. Wiley, New York, NY (1964)
14. Donohue, R., Tham, T.L.: Chapter 3 : psychological contracts essential reading. In: Contemporary HRM Issues in the 21st Century, vol. 44, no. 2, pp. 299–306 (2006)
15. Wang, X., Feng, L., Zhang, H., Lyu, C., Wang, L., You, Y.: Human resource information management model based on blockchain technology. In: Proceedings - 11th IEEE International Symposium on Service-Oriented System Engineering, SOSE 2017, pp. 168–173 (2017)
16. Michaelides, M.: The challenges of AI and blockchain on HR recruiting practices. Cyprus Rev. **30**(1), 169–180 (2018)

17. Ni, X., Yuan, Y., Wang, F.Y.: Behavioral management for employees based on blockchain and smart contracts. In: Proceedings of - IEEE International Conference Service Operation Logistics Informatics 2019, SOLI 2019, pp. 248–252 (2019)
18. Peisl, T., Shah, B.: The impact of blockchain technologies on recruitment influencing the employee lifecycle. In: Walker, A., O'Connor, R.V., Messnarz, R. (eds.) EuroSPI 2019. CCIS, vol. 1060, pp. 695–705. Springer, Cham (2019). https://doi.org/10.1007/978-3-030-28005-5_54
19. Pinna, A., Ibba, S.: A blockchain-based decentralized system for proper handling of temporary employment contracts. In: Arai, K., Kapoor, S., Bhatia, R. (eds.) SAI 2018. AISC, vol. 857, pp. 1231–1243. Springer, Cham (2019). https://doi.org/10.1007/978-3-030-01177-2_88
20. Soules, C.: Blockcahin Technology as a Disruptive Innovator in Human Resource Management, WCBT Student Publications, [Online] (2020). https://digitalcommons.sacredheart.edu/wcob_sp/. Accessed: 23 Jun 2021
21. Lai, J.: The application prospects of blockchain technology in human resource management. Mod. Manag. Forum **4**(4), 167–171 (2020)
22. Dolzhenko, R.: Blockchain as an imperative of labor relations digitalizing. SHS Web Conf. **93**, 1–8 (2021)
23. Ahmad, K.: Leadership and work motivation from the cross cultural perspective. Int. J. Commer. Manag. **19**(1), 72–84 (2009)
24. Habib Rana, M., Shaukat Malik, M.: Human resource management from an Islamic perspective: a contemporary literature review Int. J. Islam. Middle East. Financ. Manag. **9**(1), 109–124 (2016)
25. Ahmed Haj Ali, A.R., Bin Noordin, K., Achour, M.: The Islamic approach of obligations in mutual relations between employee and employer. Int. J. Ethics Syst. **34**(3), 338–351 (2018)
26. Argyris, C.: Understanding Organizational Behavior. Dorsey Press, Homewood, IL (1960)
27. Rousseau, E.F.: Psychological Contracts in Organizations: Understanding Writen and Unwriten Agreements. SAGE Publications, Thousand Oaks, CA (1995)
28. Morrison, E.W., Robinson, S.L.: When employees feel betrayed: a model of how psychological contract violation develops. Acad. Manag. Rev. **22**, 226–256 (1997)
29. Conway, N., Briner, R.B.: Understanding Psychological Contracts at Work: A Critical Evaluation of Theory and Reseaerch. Oxford University Press (2006)
30. Casino, F., Dasaklis, T.K., Patsakis, C.: A systematic literature review of blockchain-based applications: Current status, classification and open issues. Telemat. Inform. **36**, 55–81 (2019)
31. Gamage, H.T.M., Weerasinghe, H.D., Dias, N.G.J.: A survey on blockchain technology concepts, applications, and issues. SN Comput. Sci. **1**(2), 1–15 (2020). https://doi.org/10.1007/s42979-020-00123-0
32. Yaga, D., Mell, P., Roby, N., Scarfone, K.: Blockchain technology overview (2018)
33. Akgiray, V.: Blockchain technology and corporate governance: technology, markets, regulation, and corporate governanc (2018)
34. Mirza, M.O.N.: Employer-employee relationships in Islam: a normative view from the perspective of orthodox islamic scholars. Int. J. Bus. Manag. **11**(4), 59 (2016)

Human-Value Orientation as Center for Business Transformation Model in Digital Era

Ardian Adhiatma$^{(\boxtimes)}$ and Nurhidayati

Department of Management, Faculty of Economics, UNISSULA, Semarang, Indonesia
{ardian,nurhidayati}@unissula.ac.id

Abstract. Human-Technology Interaction (HTI) is an intensive interaction between humans and technology to achieve organizational goals. In the digital era, the interaction between humans and technology is becoming more massive due to its efficient and effective ways to meet the work requirements. Therefore, redesigning the business model through the Human-Value orientation is needed. Human value orientation applies to the principle that welfare, health and humanities are priorities in running a business. The human ecosystem in the organization consisting of customers, workforce, and partners becomes the new focus of business strategy. Human comfort and welfare are a top priority, even though it must coexist with the obstacles and take advantage of advances in information technology. The ability to unite to improve the collective capabilities of human resources and the use of technology will provide significant changes for business sustainability. This paper presents several methods and techniques for implementing the transformation model with the human value orientation approach.

Keywords: Business transformation · Human-technology interaction · Religious values

1 Introduction

Currently, the global industrial world is entering a new era known as the Industrial Revolution 4.0 or also known as the digital era. In general, Indonesia has actively implemented a new era marked by the movement of various sectors of life towards fully automated digital. We can see this phenomenon with evidence that there are more and more digital-based business people around us. Call it Google, Facebook, Youtube to messaging application services or messenger. The main feature of the digital era is how data becomes important. Of all the uploaded data, there is artificial intelligence which then translates it into an algorithm. This algorithm then becomes data that can be maximized to help businesses, including reading trends. Although computer technology has been around for decades, the concept of digital transformation is relatively new. This concept was present in the 1990s with the introduction of the internet. Since then, the ability to transform traditional forms of media (such as documents and photos) has waned amid the importance that digital technology brings to society. Today, digitization touches every part of our lives, influencing the way we work, shop, travel, educate, manage and live. Transformation practices are usually used in a business context. The introduction of digital technology has sparked the creation of new business models and revenue streams. Emerging technologies such as artificial intelligence are accelerating transformation, while basic technologies such as data management and analytics are required to analyze the huge amount of data that results from digital transformation. Digital transformation is not just about technology. It takes place at the intersection of people, business and technology, guided by broader business strategy. Success exists when organizations can effectively use data created by or through technology in ways that enable business change to occur dynamically. In the digital era, the interaction between humans and technology is increasingly massive because it is an efficient and effective way to fulfill work needs. Therefore, it is necessary to redesign the business model through a Human-Value orientation.

Technological advances are one of the biggest challenges in the difficulty of building models that integrate humans with technology. To create new habits and management practices related to how humans adapt, behave and work with existing technology, in addition to meeting human needs such as the search for meaning, connection and well-being in the workplace as well as maximizing employee potential through capacity building and protecting ethical values. This is because of the focus on technology and how humans can interact with technology.

Religiosity (the importance of religion to an individual) is associated with honesty (Arrunada 2010), high ethical standards (Weaver and Agle 2002), and greater risk aversion (Diaz 2000). Personal religious beliefs influence behavioureven among individuals who have not internalized societal beliefs and values (Spicer and Bailey 2007). Consistent with this, it shows that risk-taking and behaviourbased on religious values can make a person have more superior values with work orientation not only looking at material aspects, but also aspects of social relationships, interactions, and a balanced unity between technology and humans.

Business transformation is very necessary in responding to various challenges in today's world, not only looking at the aspects of changes that occur in the digital era but also seeing how interactions between humans and technology can become a single unit that can build human values that instill a religious character.

2   Human – Technology Interaction

As defined in the 2018 resource trends report, the "new social contract" proposes an increase in relationships that are more human-centered than those between individuals and organizations and organizations and society (Dimple et al. 2018). Since 2018, we have seen the speed and scale of change continue to increase. With the advancement of technology brings bigger and bolder changes in less time. But as new technologies and digital transformation dominate the conversation in the boardroom, human concerns are seen as separate, indirectly at odds with technological advances.

Technological progress that is getting faster and faster has certainly made many changes to a country, including Indonesia. Taking advantage of the progress made during the crisis and planning future goals as changes in behaviourto keep running activities trigger companies to rely more on technology.

Currently, the use of applications for long-distance communication is still high to stay connected with colleagues and clients. Focusing solely on returning to work is not the right choice, as organizations take advantage of everything they have experienced and learned over the past few months. Companies are required to adapt to technological advances that can support work. Organizations must embrace the perspective of New York Times columnist Thomas Friedman that humans wishing to adapt in an age of acceleration must develop dynamic stability. Rather than trying to stop the storm of inevitable change, it is better to encourage leaders to build an eye that moves with the storm, drawing energy from it, but creating a dynamic platform of stability within (Zach 2017).

Thus, the concept of the relationship between humans and technology provides a conclusion that technological developments can contribute significantly to making people's daily lives healthier, safer, more independent, fun and comfortable. New technologies can also provide new means of communication and entertainment and contribute to solving challenges, such as saving energy or improving health and well-being. For technology to be successful, people need to trust, accept and use it naturally. Technology produces rapid progress and leads to major changes in society. Access to new technologies can also improve consumers' quality of life exponentially (Scott 2019).

3   Business Transformation

Today's business transformation into a digital business refers to the process and strategy of using digital technology to drastically change the way businesses operate and serve customers. This expression has become common in the era of digitalization. That's because every organization regardless of size or industry is increasingly reliant on data and technology to operate more efficiently and deliver value to customers. Companies throughout Indonesia continue to survive with the changes in the digitalization era. Of course the company is trying to survive and find the right moment to bounce back. So that companies can survive in facing the challenges that exist in this digitalization era, like it or not, business transformation must be carried out. Business transformation is a comprehensive change process that requires companies to position themselves with the aim that companies can become better at responding to new business challenges that face or face a rapidly changing business environment. Changes made by companies in undergoing the digitalization era can be carried out comprehensively and continuously, consisting of changes to the company's perspective, mindset, and pattern of action. Changes are made because human health and safety is currently a priority. Changes made by companies in undergoing the digitalization era can be carried out comprehensively and continuously, consisting of changes to the company's perspective, mindset, and pattern of action. Changes are made because human health and safety is currently a priority. Changes made by companies in undergoing the digitalization era can be carried out comprehensively and continuously, consisting of changes to the company's perspective, mindset, and pattern of action. Changes are made because human health and safety is currently a priority.

Changes can also be made through business strategy, organizational capabilities and corporate culture. Business transformation cannot be separated from the current popular trends. This transformation can be carried out in various forms of business, including business model transformation, strategy transformation, structural transformation, operational transformation and corporate culture transformation. Transformation of business models and strategies, namely choosing the main source of business income that is adapted to society in the new normal era, such as technical sales and delivery of goods, in structural transformation repositioning the organization according to the needs of the company's model and strategy. Operational transformation is carried out by preparing infrastructure which includes sales promotion facilities and health protocols in the office.

Business transformation requires changes in organizational and management aspects that are aligned with changes in all systems and management aspects such as planning management, operations management, human resources, marketing management, financial management, and others. The implementation and business transformation steps that can be carried out by the company go through several stages, including the identification stage, the implementation stage and the implementation evaluation stage. Identification in this case is to identify the needs of the community. Continued at the identification stage of the availability of raw materials, production equipment and human resource capabilities. Then the identification stage of the standard operation procedure (SOP) is also carried out. Identification of the selection of a good promotional tool is also carried out so that the promotion that is carried out is able to hit the target. And the last is to identify the pattern and distribution strategy to consumers. After the identification stage, operational activities are carried out and do not forget to evaluate the results of the implementation of activities.

Transformation can be interpreted as an effort to accelerate business by involving technological tools. The process is not changing all manual business models to digital, but trying to see opportunities that can help business processes with digital (Hidayat et al. 2020). The company's business transformation has enormous benefits for the company, among others, the company can focus more on businesses that are more promising and financially profitable, as well as being able to improve organizational capabilities so that they have strong support power. After carrying out a business transformation, it is expected that the company in acting has new rules, operations, ways of working and strategies. There is a lot to consider to make a transformation for a better business in the future. Strong leaders who place the digital and customer experience at the core of the business model drive successful transformation efforts. These leaders must ensure their digital transformation strategy addresses the cultural gap and that everyone understands where the organization is going in the future.

4   Religious Values

Religious belief is one of the most influential determinants of individual behavior. Tiliouine and Belgoumidi (2009) assert that religious practice has become a source of strength for entrepreneurs because it has a tendency to influence them. As a result, the application of religious practices in a company provides a force that can influence personality traits and positive attitudes towards the success of Muslim entrepreneurs' companies. The influence of religious factors on moral values is considered crucial on the grounds that it can increase the religious values of business visionaries. In addition, they can improve their business performance better (Kotey and Meredith 1997). Based on the positive characteristics contained in religious values, including savings and productivity, precision and time saving, have a sense of pride in the work done, commitment and loyalty to the organization, the need for high achievement, honesty, have a high internal. self-control, seeing ambition and success as signs of God's help. According to Adnan and Mohamad Dahlan (2002), doing business is considered as worship because it is an economic activity that allows people to earn a living and the most common component of any society is religion, thus stating that historically it has shown that religion has become a force in business. Ethical issues that occur in the business world can actually be overcome if universal religious values can be adopted as a source of ethics, especially in Indonesia, where people tend to be religious. In doing business without a religious foundation, it will only be an idea to do good but humans do not have a strong urge to do so. He also explained that the application of religion in a company with religious diversity to its employees and stakeholders is not something difficult. There are universal values found in all religions. For example, the values of honesty, fairness, responsibility, and so on. All religions should give birth to positive ethics that can be applied in a company. For example the values of honesty, fairness, responsibility, and so on. All religions should give birth to positive ethics that can be applied in a company. For example the values of honesty, fairness, responsibility, and so on. All religions should give birth to positive ethics that can be applied in a company.

Table 1. Visualization human-value orientation concepts

No	Concepts	Definition
1	Human – technology interaction	A new social resource trend that proposes an increase in relationships that are more human-centered than the relationships between individuals and organizations and organizations and society associated with digital interactions (Dimple et al. 2018)
2	Business transformation	Efforts to accelerate business by involving technological tools. The process is not changing all manual business models to digital, but trying to see opportunities that can help business processes with digital (Hidayat et al. 2020)
3	Religious values	Religious beliefs reflect individual behaviouras a source of power that tends to influence Tiliouine and Belgoumidi (2009)

Based on the table the concept of visualization of human value will be created In the digital era supported by the interaction between humans and technology will make a job more efficient and effective. Therefore, it is necessary to redesign the business model through a Human-Value orientation. The orientation of human values with the principle that welfare, health and humanity are priorities in running a business, which is balanced with strong religious values. So that human value will be superior and have a positive effect on a business.

5   Conclusion

The globalization trend of the digital era facilitated by worldwide communication and Internet technology has influenced the formation of identity and human values. A person's religiosity can describe his moral values to make a choice between good and bad. religious factors and business performance found that honesty and kindness, trust, truth and justice and equality showed a positive relationship with business performance and one of the factors that influenced the application of the Islamic work ethic was religious practice. Work ethics regulated by Islam can form individuals or employees with a high sense of self-discipline. They will try to comply with every rule set in an organization such as doing work according to procedures, punctuality, mutual respect and integrity. High discipline is a civilized way of life in which employees become more positive for the organization in terms of commitment, dedication, job satisfaction, cooperation, creativity and work improvement. So that doing business transformation is no longer an aspiration that focuses on the future, but a reality in the present.

The company's business transformation in the digital era has enormous benefits for the company, including the company being able to focus more on more promising businesses and increasing the organization's ability to have strong support. Rather than trying to stop the storm of inevitable change, it is better to encourage leaders to build decisions that move with the storm, draw energy from the storm and create a platform of dynamic stability within it based on religious values.

References

Arrunada, B.: Protestants and catholics: similar work ethic, different social ethic. Econ. J. **120**(547), 890–918 (2010)

Liedtka, J., Ogilvie, T.: Designing for Growth: a Design Thinking Tool Kit for Managers. New York: Columbia Business School Publishing (Book 2), Columbia University Press (2011)

Mazzoni, M.: 15 companies retooling their operations to fight COVID-19, Triple Pundit, 1 May 2020

Kolakowski, N.: COVID-19 burnout growing among remote workers, Dice Insights, 5 May 2020

Smith, K.: Pandemic fuels burnout among nearly half of US workers, Orange County Register, 16 April 2020

Ryan, T.: Trudeau announces wage top-ups for front-line workers, but details unclear, 7 May 2020

Scott, D., Saariluoma, P., Cañas, J.J., Leikas, J.: Designing for life: a human perspective on technology development [Book review]. IEEE Trans. Prof. Commun. **62**(4), 400–401 (2019). https://doi.org/10.1109/tpc.2019.2946941

Vechakul, J., Shrimali, B.P., Sandhu, J.S.: Human-centered design as an approach for place-based innovation in public health: a case study from Oakland. Matern. Child Health J. **19**(12), 2552–2559 (2015)

Weaver, G.R., Agle, B.R.: Religiosity and ethical behaviourin organizations: a symbolic interactionist perspective. Acad. Manage. Rev. **27**, 77–97 (2002)

Wyche, S., Olson, J., Karanu, M.: Redesigning agricultural hand tools in Western Kenya redesigning agricultural hand tools in Western Kenya: considering human-centered design in ICTD. Inf. Technol. Int. Dev. **15**, 97–112 (2019)

Zach, S.L.: Thomas Friedman on human interaction in the digital age, Aspen Institute, 10 January 2017

An Energy-Efficient Algorithm to Make Virtual Machines Migrate in a Server Cluster

Dilawaer Duolikun[1(✉)], Tomoya Enokido[2], Leonard Barolli[3],
and Makoto Takizawa[1]

[1] Research Center for Computing and Multimedia Studies, Hosei University,
Tokyo, Japan
makoto.takizawa@computer.org
[2] Faculty of Business Administration, Rissho University, Tokyo, Japan
eno@ris.ac.jp
[3] Department of Information and Communications Engineering,
Fukuoka Institute of Technology, Fukuoka, Japan
barolli@fit.ac.jp

Abstract. Applications can take advantage of virtual computation services independently of heterogeneity and locations of servers by using virtual machines in clusters. Here, a virtual machine on an energy-efficient host server has to be selected to perform an application process. In this paper, we propose an SMI (Simple MI) algorithm to estimate the energy consumption of a server to perform application processes. Then, we propose an SMIM (SMI Migration) algorithm to make a virtual machine migrate from a host server to a guest server to reduce the total energy consumption of the servers by using the SMI algorithm. In the evaluation, we show the energy consumption of servers in a cluster can be reduced in the SMIM algorithm compared with other algorithms.

Keywords: Server selection algorithm · Migration of virtual machines · Green computing systems · SMI algorithm · SMIM algorithm

1 Introduction

Scalable information systems like the IoT (Internet of Things) [31] are composed of millions of computers and devices and consume huge amount of energy [27]. We have to reduce the energy consumption of clusters [2–5,17] to realize green computing systems [27]. Virtual machines [1–6,27] are widely used to provide applications with virtual services on computation resources like CPUs independently of heterogeneity and locations of servers. Once a client issues a request to a cluster, a virtual machine on a host server is selected to perform an application process to handle the request. Algorithms [7–9,13–16,24,26] are proposed to select an energy-efficient host server. In addition to selecting virtual machines,

L. Barolli et al. (Eds.): EIDWT 2022, LNDECT 118, pp. 130–141, 2022.
https://doi.org/10.1007/978-3-030-95903-6_15

virtual machines migrate to other servers which consume smaller energy in the migration approach [7,9,17–21,25]. Algorithms to make a system more reliable by replicating processes on virtual machines are proposed [10,11,19].

Power consumption and computation models of servers [2–5,13,16] are proposed. The power consumption models of fog nodes of the IoT are also proposed [30,31]. By using the models, the execution time and energy consumption of a server to perform processes are obtained by simulating the computation steps of the processes in the SM (SiMulation) algorithm [13–16]. However, it is difficult to *a priori* know the computation residue of processes and it takes time to simulate the computation of processes on a server while the execution time and energy consumption can be precisely estimated. In the SP (SimPle) algorithm [17,18,28,29], only the number of processes on a server is utilized to do the estimation. The MI (Monotonically Increasing) [22] and MMI (Modified MI) [23] algorithms are proposed to more precisely estimate the execution time and energy consumption of a server. Here, the total computation residue RS_t of n_t active processes p_{t1}, ..., p_{t,n_t} on a server s_t is assumed to be known. The computation residue RP_{ti} of each process p_{ti} is assumed to be $RP_{t1} \cdot \alpha^{i-1}$ $(\alpha \geq 1)$ and $RP_{t1} = RS_t \cdot (1 - \alpha)/(1 - \alpha^{n_t})$. However, it takes time to calculate the computation residue of each process.

In this paper, we newly propose an SMI (Simple MI) algorithm where the computation residue RP_{ti} of each process p_{ti} is assumed to be $RP_{t1} \cdot i$. Then, we propose an SMIM (SMI Migration) algorithm where a virtual machine migrates to reduce the total energy consumption of the host and guest servers by using the SMI algorithm. In the evaluation, we show the energy consumption of servers can be reduced in the SMIM algorithm than other algorithms.

In Sect. 2, we present the computation and power consumption models of a server. In Sect. 3, we discuss the SMI algorithm. In Sect. 4, we propose the SMIM algorithm. In Sect. 5, we evaluate the SMIM algorithm.

2   System Model

Each server s_t is equipped with np_t homogeneous CPUs. Each CPU supports pc_t homogeneous cores and each core supports tn_t threads. The server s_t totally supports nc_t $(= np_t \cdot pc_t)$ cores and nt_t $(= nc_t \cdot tn_t)$ threads. Application processes issued by clients are performed on a virtual machine vm_k of a server s_t. A process being performed is *active*. Let $SP_t(\tau)$ and $VP_k(\tau)$ be numbers of active processes on a server s_t and a virtual machine vm_k at time τ, respectively. A server s_t and virtual machine vm_k are *active* if and only if (iff) $|SP_t(\tau)| > 0$ and $|VP_k(\tau)| > 0$, respectively. A virtual machine vm_k is *smaller* than vm_h iff $|VP_k(\tau)| < |VP_h(\tau)|$. A virtual machine vm_k on a host server s_h can migrate to a guest server s_g in a live manner [1]. Here, active processes on vm_k does not terminate but are suspended during the migration.

In this paper, a term *process* stands for an application process which uses CPU resources [3]. Time is modeled to be a sequence of time units [tu]. The *minimum execution time* $minT_{ti}$ [tu] shows the execution time of a process p_i on a server s_t where only the process p_i is performed without any other process.

Let $minT_i$ be $\min(minT_{1i}, \ldots, minT_{mi})$. The amount of computation of each process p_i is defined to be $minT_i$. The thread computation rate TCR_t of a server s_t is $minT_i/minT_{ti}$ (≤ 1) for any process p_i. On a server s_t where n_t processes are active, each process is performed at rate $NPR_t(n_t)$ in the MLC (Multi-Level Computation) model [13,15,16]:

[MLC Model]

$$NPR_t(n_t) = \begin{cases} TCR_t & \text{if } 0 < n_t \leq nt_t. \\ nt_t \cdot TCR_t/n_t & \text{if } n_t > nt_t. \end{cases} \tag{1}$$

The computation residue RP_i of each active process p_i is decremented by the computation rate $NPR_t(n_t)$ for one time unit. The total computation rate $NSR_t(n_t)$ ($\leq nt_t \cdot TCR_t$) of a server s_t is $NPR_t(n_t) \cdot n_t$ for $n_t > 0$.

The power consumption $NE_t(n_t)$ [W] of a server s_t with n_t active processes is given as follows:

[MLPCM (Multi-Level Power Consumption) Model] [13,15–17]

$$NE_t(n_t) = \begin{cases} minE_t & \text{if } n_t = 0. \\ minE_t + n_t \cdot (bE_t + cE_t + tE_t) & \text{if } 1 \leq n_t \leq np_t. \\ minE_t + np_t \cdot bE_t + n_t \cdot (cE_t + tE_t) & \text{if } np_t < n_t \leq nc_t. \\ minE_t + np_t \cdot bE_t + nc_t \cdot cE_t + n_t \cdot tE_t & \text{if } nc_t < n_t < nt_t. \\ maxE_t(= minE_t + np_t \cdot bE_t + nc_t \cdot cE_t + nt_t \cdot tE_t) & \\ & \text{if } n_t \geq nt_t. \end{cases} \tag{2}$$

An *idle* server s_t just consumes the minimum power $minE_t$ [W], where no process is performed, i.e. $n_t = 0$. Once a CPU, core, and thread are activated on a server s_t, the power consumption NE_t of a server s_t increases by bE_t, cE_t, and tE_t [W], respectively. A server s_t consumes the maximum power $maxE_t$ if every thread is active, i.e. $n_t \geq nt_t$. The electric power $E_t(\tau)$ [W] consumed by a server s_t to perform n_t ($= |SP_t(\tau)|$) processes at time τ is $NE_t(|SP_t(\tau)|)$. Energy consumed by a server s_t from time st [tu] to et is $\sum_{\tau=st}^{et} NE_t(|SP_t(\tau)|)$ [W tu].

Let C_t, T_t, and E_t be variables denoting a set of active processes, active time, and energy consumption of each server s_t, respectively. Variables RP_i and T_i show the computation residue and execution time of each process p_i in the set C_t, respectively. τ shows time. When a process p_i starts on a server s_t, the computation residue RP_i is $minT_i$. At each time τ, RP_i is decremented by the process computation rate $NPR_t(n_t)$. If $RP_i \leq 0$, p_i terminates. E_t and T_t give the total energy consumption and execution time of the server s_t, respectively.

[Computation Model of Processes on a Server s_t]

1. Initially, $E_t = 0$; $C_t = \phi$; $T_t = 0$; $\tau = 1$;
2. while ()
   (a) for each process p_i which starts on a server s_t at time τ,
       i. $C_t = C_t \cup \{p_i\}$; $RP_i = minT_i$; $T_i = 0$;
   (b) $n_t = |C_t|$;   /* number of active processes */
       $E_t = E_t + NE_t(n_t)$;   /* energy consumption */
       $T_t = T_t + 1$; /* server is active at time τ */

(c) for each process p_i in C_t,
    i. $RP_i = RP_i - NPR_t(n_t)$; /* residue is decremented */
    ii. $T_i = T_i + 1$;
    iii. if $RP_i \leq 0$, $C_t = C_t - \{p_i\}$; /* p_i terminates */
(d) $\tau = \tau + 1$;

3 An SNI (Simple MI) Algorithm

Suppose n_t processes p_{t1}, ..., p_{t,n_t} are active on a server s_t at time τ, i.e. $SP_t(\tau) = \{p_{t1}, \ldots, p_{t,n_t}\}$. RS_t denotes the total computation residue $RP_{t1} + \ldots + RP_{t,n_t}$ of active processes $p_{t1}, \ldots, p_{t,n_t}$ on a server s_t. In this paper, we assume each active process p_{ti} in the set $SP_t(\tau)$ has the computation residue $RP_{ti} \cdot i$ at current time τ ($i = 1, \ldots, n_t$). The total computation residue RS_t of the server s_t is $RP_{t1} \cdot (1 + 2 + \ldots + n_t) = RP_{t1} \cdot n_t \cdot (n_t + 1)/2$. RP_{t1} is $2 \cdot RS_t/(n_t \cdot (n_t + 1))$ given as the following function $RF_t(RS_t, n_t)$:

$$RF_t(r, n) = 2 \cdot r/(n \cdot (n + 1)). \tag{3}$$

The total computation residue RS_t of a server s_t of time τ is assumed to be proportional to the number n_t of active processes, i.e. $RS_t = c \cdot aT \cdot n_t$ where aT is the average one of minimum execution time of all the processes and c is a constant $1/2$. In addition, the total number n_t of active processes is assumed to be obtained at each time τ.

First, each process p_{ti} is performed at the computation rate $NPR_t(n_t)$. An active process whose computation residue is the smallest is referred to as *top*. In the set $SP_t(\tau)$, p_{t1} is a top process since the computation residue RP_{t1} is the smallest. The computation residue RP_{ti} of each process p_{ti} is decremented by $RP_{t1} = RF_t(RS_t, n_t)$ ($i = 1, \ldots, n_t$). The execution time is given as a function $T_{t1}(RS_t, n_t) = RF_t(RS_t, n_t)/NPR_t(n_t)$. If $n_t \geq nt_t$, $T_{t1}(RS_t, n_t) = 2 \cdot RS_t/((n_t+1) \cdot nt_t \cdot TCR_t)$. Totally the computation $RF_t(RS_t, n_t) \cdot n_t$ of the n_t processes $p_{t1}, \ldots, p_{t,n_t}$ is performed for $T_{t1}(RS_t, n_t)$ time units [tu]. Then, p_{t1} terminates since RP_{t1} gets zero. The server s_t consumes the energy $E_{t1}(RS_t, n_t) = NE_t(n_t) \cdot T_{t1}(RS_t, n_t)$ [W tu] for $T_{t1}(RS_t, n_t)$ [tu]. Then, the $(n_t - 1)$ processes $p_{t2}, \ldots, p_{t,n_t}$ are performed where p_{t2} is the top process. Here, the execution time $T_{t2}(RS_t, n_t)$ to terminate the top process p_{t2} is $RF_t(RS_t, n_t)/NPR_t(n_t - 1)$ [tu]. The computation $RF_t(RS_t, n_t) \cdot (n_t - 1)$ of the $(n_t - 1)$ processes is performed. The server s_t consumes the energy $E_{t2}(RS_t, n_t) = NE_t(n_t - 1) \cdot T_{t2}(RS_t, n_t)$ [W tu] for $T_{t2}(RS_t, n_t)$ [tu]. Here, p_{t2} terminates. Thus, the execution time $T_{ti}(RS_t, n_t)$ is $RF_t(RS_t, n_t)/NPR_t(n_t - i + 1)$ and the energy consumption $E_{ti}(RS_t, n_t)$ is $NE_t(n_t - i + 1) \cdot T_{ti}(RS_t, n_t)$ where p_{ti} is a *top* process, i.e. $(i - 1)$ processes $p_{t1}, \ldots, p_{t,i-1}$ are terminated and $(n_t - i + 1)$ processes $p_{ti}, \ldots, p_{t,n_t}$ are still active ($i = 1, \ldots, n_t$). As presented in the MLC model (1), the computation rate $NPR_t(n_t - i + 1)$ is $nt_t \cdot TCR_t/(n_t - i + 1)$ if $n_t - i + 1 \geq nt_t$, otherwise TCR_t. By using the following function D_t, the execution time $T_{ti}(RS_t, n_t)$ is $D_t(RS_t, n_t) \cdot (n_t - i + 1)$ if $n_t - i + 1 \geq nt_t$, otherwise $D_t(RS_t, n_t) \cdot nt_t$.

$$D_t(r, n) = RF_t(r, n)/(nt_t \cdot TCR_t). \tag{4}$$

Suppose $n(\leq n_t)$ processes $p_{t,n_t-n+1}, \ldots, p_{t,n_t}$ are active and the others $(n_t - n)$ $p_{t1}, \ldots, p_{t,n_t-n}$ are terminated. The execution time $T_{t,n_t-n+1}(RS_t, n_t)$ for a top process p_{t,n_t-n+1} is the following function $TN_t(RS_t, n_t, n)$:

$$TN_t(r, n_t, n) = \begin{cases} D_t(RS_t, n_t) \cdot n & \text{if } n \geq n_t - nt_t \text{ and } n_t \geq nt_t. \\ D_t(RS_t, n_t) \cdot nt_t & \text{otherwise.} \end{cases} \quad (5)$$

We consider the execution time $ET_t(RS_t, n_t, n) = \sum_{i=n}^{n_t} TN_t(RS_t, n_t, n)$ to terminate $(n_t - n)$ processes $p_{t1}, \ldots, p_{t,n_t-n}$ $(n \leq n_t)$ in the set $SP_t(\tau)$. Here, n processes $p_{t,n_t-n+1}, \ldots, p_{t,n_t}$ are still active. First, suppose $n_t \geq nt_t$. The execution time is $D_t(RS_t, n_t) \cdot (n_t + (n_t - 1) + \ldots + n)$ for $n \geq nt_t$. If $n < nt_t$ and $n_t \geq nt_t$, the execution time is $D_t(RS_t, n_t) \cdot ((n_t + (n_t - 1) + \ldots + nt_t) + nt_t \cdot (n - nt_t))$. Next, suppose $n_t \leq nt_t$. Here, since at most one process is performed on each thread, the execution time is $D_t(RS_t, n_t) \cdot nt_t \cdot n$.

$$ET_t(r, n_t, n) = \begin{cases} D_t(r, n_t) \cdot (-n^2 + n + n_t^2 + n_t)/2 \\ \quad \text{if } n \leq n_t - nt_t \text{ and } n_t > nt_t. \\ D_t(r, n_t) \cdot (nt_t \cdot n + (n_t^2 + n_t - 3 \cdot nt_t^2 + nt_t)/2) \\ \quad \text{if } n > n_t - nt_t \text{ and } n_t > nt_t. \\ D_t(r, n_t) \cdot nt_t \cdot n \text{ if } n_t \leq nt_t. \end{cases} \quad (6)$$

The total execution time $TET_t(RS_t, n_t)$ to perform all the processes in the set $SP_t(\tau)$ is $ET_t(RS_t, n_t, n_t)$.

A server s_t consumes the total energy $TES_t(RS_t, n_t)$ to perform n_t processes $p_{t1}, \ldots, p_{t,n_t}$ in the set $SP_t(\tau)$:

$$TES_t(r, n_t) = \sum_{n=1}^{n_t} (NE_t(n) \cdot TN_t(r, n_t, n)). \quad (7)$$

By using the equation $tm = ET_t(RS_t, n_t, n)$ (6), the number n of active processes to be performed at time $\tau + tm$ [tu] is obtained as $EN_t(RS_t, n_t, tm)$:

$$EN_t(r, n_t, tm) = \begin{cases} (1 + \sqrt{1 + 4 \cdot (n_t^2 + n_t) - 8 \cdot tm/D_t(r, n_t)})/2 \\ \quad \text{if } n_t > nt_t \text{ and } tm \leq (n_t^2 + n_t - nt_t^2 + nt_t). \\ (tm/D_t(r, n_t) + (n_t{}^2 + n_t - 3 \cdot nt_t^2 + nt_t)/2)/nt_t \\ \quad \text{if } n_t > nt_t \text{ and } tm > (n_t^2 + n_t - nt_t^2 + nt_t). \\ tm/(D_t(r, n_t) \cdot nt_t) \quad \text{if } n_t \leq nt_t. \end{cases} \quad (8)$$

We consider the computation residue $TRS_t(RS_t, n_t, tm)$ of the server s_t of time $\tau + tm$ [tu]. First, the number np_1 of processes active at time $\tau + tm$ is $\lfloor EN_t(RS_t, n_t, tm) \rfloor$. Let np_2 be $np_1 - 1$. Let t_1 and t_2 be the execution time $ET_t(RS_t, n_t, np_1)$ and $ET_t(RS_t, n_t, np_2)$, respectively. Here, $t_1 < t_2$. If $np_1 < nt_t$, $TRS_t(RS_t, n_t, tm)$ is $RF_t(RS_t, n_t) \cdot ((np_1 + 1) \cdot np_1/2 + np_2 \cdot (t_2 - tm)/(t_2 - t_1))$. Next, suppose $np_1 \leq n_t - nt_t$. The amount of computation of the server s_t from time τ to time t_1 is $RF_t(RS_t, n_t) \cdot (n_t + (n_t - 1) + \ldots + np_1) = RF_t(RS_t, n_t) \cdot (n_t + np_1) \cdot (n_t - np_1 + 1)/2$. Hence, $TRS_t(RS_t, n_t, tm)$ is $RS_t - RP_{t1} \cdot [(n_t + np_1) \cdot (n_t - np_2 + 1)/2 + np_2 \cdot (tm - t_1)/(t_2 - t_1)]$. The total computation residue $TRS_t(RS_t, n_t, tm)$ of time $\tau + tm$ is as follows:

$$TRS_t(r, n_t, tm) = \begin{cases} RF_t(RS_t, n_t) \cdot ((np_1 + 1) \cdot np_1/2 \\ \quad + np_2 \cdot (t_2 - tm)/(t_2 - t_1)) & \text{if } np_1 \leq nt_t. \\ RS_t - RF_t(RS_t, n_t) \cdot ((n_t + np_1) \cdot (n_t - np_1 + 1) \\ \quad + 2 \cdot np_2 \cdot (tm - t_1)/(t_2 - t_1)) & \text{if } np_1 > nt_t. \end{cases} \quad (9)$$

The energy consumption $TEE_t(RS_t, n_t, tm)$ of the server s_t for tm time units [tu] from current time τ is given as follows, where $np_1 = \lfloor EN_t(RS_t, n_t, tm) \rfloor$:

$$TEE_t(r, n_t, tm) = \begin{cases} \sum_{n=np_1}^{n_t} (NE_t(n) \cdot TN_t(r, n_t, n)) + NE_t(np_1 - 1) \cdot \\ TN_t(r, n_t, np_1 - 1) \cdot (tm - t_1)/(t_2 - t_1). \end{cases} \quad (10)$$

4 Migration of Virtual Machines

We discuss the total energy consumption of a pair of a host server s_h and a guest server s_g where a virtual machine migrates. First, suppose no virtual machine migrates from the server s_h to s_g. Here, the servers s_h and s_g consume the energy $NE_h = TES_h(RS_h, n_h)$ and $NE_g = TES_g(RS_g, n_g)$ [W tu] for $NT_h = TET_h(RS_h, n_h)$ and $NT_g = TET_g(RS_g, n_g)$ time units [tu], respectively.

Next, suppose a virtual machine vm_k with $nv_k (= |VP_k(\tau)|)$ active processes on a host server s_h migrates to a guest server s_g at current time τ and restarts at time $\tau + tm$. Let VRS_k be the total computation residue of the nv_k processes on the virtual machine vm_k. VRS_k is assumed to be $aT \cdot nv_k/2$. Suppose n_g $(= |SP_g(\tau)|)$ processes are active. The total computation residue RS_g of the server s_g is assumed to be $aT \cdot n_g/2$. The guest server s_g consumes the energy $NEE_g = TEE_g(RS_g, n_g, tm)$ from time τ to $\tau + tm$. At time $\tau + tm$, $nn_g = EN_h(RS_g, n_g, tm)$ processes are active and the total computation residue NRS_g is $TRS_g(RS_g, n_g, tm)$. Now, nv_k processes on the virtual machine vm_k restart on the guest server s_g. Totally $(nn_g + nv_k)$ processes are performed and the total computation residue of the server s_g is $NRS_g + VRS_k$. The execution time MTT_g of the active processes is $TET_g(NRS_g + VRS_k, nn_g + nv_k)$ and the energy consumption MEE_g is $TES_g(NRS_g + VRS_k, nn_g + nv_k)$. The total execution time MT_g is $tm + MTT_g$. The guest server s_g totally consumes the energy $ME_g = NEE_g + MEE_g$ from time τ to $\tau + MT_g$ Fig. 1(a).

Next, suppose no process is active at time $\tau + tm$, i.e. every process in the set $SP_g(\tau)$ terminates before $\tau + tm$, i.e. $tm > NT_g$. Here, only the nv_k processes on the virtual machine vm_k are active. Hence, the total execution time MT_g is $tm + TET_g(VRS_k, nv_k)$. The guest server s_g consumes the energy $ME_g = NE_g + (tm - NT_g) \cdot minE_g + TES_g(VRS_k, nv_k)$ from time τ to $\tau + MT_g$ as shown in Fig. 1(b).

At time τ, n_h processes are active on the host server s_h. Since nv_k processes on the virtual machine vm_k leave the server s_h, $(n_h - nv_k)$ processes are active on the server s_h at time τ. The computation residue RS_h of the server s_h decreases to $RS_h - VRS_k$. The execution time MT_h is $TET_h(RS_h - VRS_k, n_h - nv_k)$ and the energy consumption ME_h is $TES_h(RS_h - VRS_k, n_h - nv_k)$.

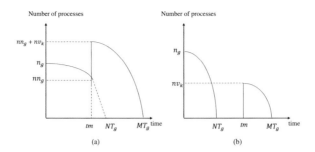

Fig. 1. Number of processes.

If $NE_h + NE_g < ME_h + ME_g$, the total energy to be consumed by the servers s_h and s_g might be reduced if a virtual machine vm_k migrates. Even an idle server s_t consumes the power $minE_t$ as presented in the MLPC model (2). This means, even if every process terminates on one server, say s_h, the server consumes the power until the other server, say s_g, gets idle. Let t_h and t_g be a pair of the execution time of the servers s_h and s_g, respectively. In order to take into consideration, the energy consumption of an idle server, we introduce the following function:

$$mE_{hg}(t_h, t_g) = \begin{cases} minE_h \cdot (t_g - t_h) \text{ if } t_h \leq t_g. \\ minE_g \cdot (t_h - t_g) \text{ if } t_h > t_g. \end{cases} \tag{11}$$

Let NEE_{hg} be the total energy consumption $NE_h + NE_g + mE_{hg}(NT_h, NT_g)$ of the servers s_h and s_g from time τ to $\tau + \max(NT_h, NT_g)$ where no virtual machine migrates. On the other hand, $MEE_{hg:k}$ shows the total energy consumption $ME_h + ME_g + mE_{hg}(MT_h, MT_g)$ of the servers s_h and s_g from time τ to $\tau + \max(MT_h, MT_g)$ where a virtual machine vm_k migrates from the host server s_h to the guest server s_g.

The migration (MG) condition for a virtual machine vm_k on a host server s_h and a guest server s_g is as follows:

[MG (Migration) Condition] $NEE_{hg} > MEE_{hg:k}$
If the MG condition is satisfied, the total energy to be consumed by the host and guest servers s_h and s_g can be reduced by making the virtual machine vm_k migrate from s_h to s_g.

[SMIM Algorithm]

1. Let S and $AS(\subseteq S)$ be sets of all the servers and active servers in a cluster;
2. For every server s_t, let n_t be the number of active processes and RS_t be $aT \cdot n_t/2$;
3. **while** $(AS \neq \phi)$ **do**
   (a) Select an active host server s_h in AS whose energy $TES_h(RS_h, n_h)$ is the largest;
   (b) Select a smallest active virtual machine vm_k on the host server s_h;

(c) Select a guest server s_g in S where the MG condition holds and $MEE_{hg:k}$ is the smallest;

(d) If the guest server s_g is found, the virtual machine vm_k migrates from the host server s_h to the guest server s_g;

(e) Otherwise, s_h is removed from the set AS;

In paper [25], if a server s_h is selected to be a host one and guest one, the server is not selected as a guest server and a host server of another server, respectively. By this algorithm, the number of migrations of virtual machines can be reduced. However, each server s_t cannot send virtual machines to other guest servers once the server s_t is taken as a host or guest one. In the SMIM algorithm, each active server s_h can make multiple virtual machines migrate as long as the MG condition is satisfied.

5   Evaluation

We evaluate the SMIM algorithm in terms of the execution time TT and energy consumption TE of servers and the average execution time AT of processes compared with the non-migration RD (random), RB (round robin), SP [18], and SMI algorithms and the migration SPM [28] algorithm. In each algorithm, once a process is issued by a client, a host server s_t and a virtual machine vm_k hosted by the server s_t are selected to perform the process. In the RD algorithm, a virtual machine vm_k is randomly selected. In the RB algorithm, a host server s_h is selected in the round-robin way. In the SP and SPM algorithms, a host server s_h where $n_t/NSR_t(n_t)$ is maximum is selected. In the SMI and SMIM algorithms, a host server s_h is selected where the total energy consumption $TES_t(RS_t, n_t)$ is the largest. In the RB, SP, SPM, SMI, and SMIM algorithms, a smallest virtual machine vm_k is selected on the selected host server s_h. In the SPM and SMIM algorithms, virtual machines migrate to guest servers. Here, the migration time tm is one [tu]. Every ten time units [tu], the MG condition is checked.

We consider four servers $s_1, \ldots, s_4$ ($m = 4$) whose parameters are shown in Table 1. Each server initially hosts twenty virtual machines. Totally n processes $p_1, \ldots, p_n$ are issued from time 1 to $xtime$, where $xtime$ is 1,000 [tu]. For each process p_i, the starting time $stime_i$ is randomly taken from 1 to $xtime$ [tu]. The minimum execution time $minT_i$ of each process p_i is randomly taken from 1 to 20 [tu] so that the average value aT of $minT_1, \ldots, minT_n$ is 10 [tu].

Figure 2 shows the total energy consumption TE [k W tu] of the servers $s_1, \ldots, s_4$ for number n of processes. As shown in Fig. 2, the total energy consumption TE of the SMIM algorithm is smaller than the other algorithms while several percent smaller than the SPM one. The TE of the SMI algorithm is about 5 to 10 [%] smaller than the SP algorithm. This means, the energy consumption of the servers can be more precisely estimated in the SMI estimation algorithm. In addition, the TE of the servers can be reduced by making virtual machines migrate. For example, the TE of the SMIM and SPM algorithms is about 20 to 30 [%] smaller than the SMI and SPO algorithms for $n = 6,000$.

Table 1. Parameters of servers

sid	TCR	nb	nc	nt	minE[W]	bE[W]	cE[W]	tE[W]	maxE[W]
1	1	1	8	16	250	25	12	4	435
2	0.9	1	8	16	200	20	10	2	332
3	0,7	1	4	8	180	15	8	1	235
4	0.5	1	2	4	100	10	6	1	126

Figure 3 shows the total execution time, i.e. active time TT [tu] of the servers for the number n of the processes. The TT of the SMIM and SPM algorithms is almost the same and is smaller than the other algorithms for $n > 3,500$.

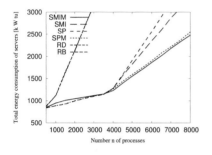

Fig. 2. Total energy consumption of servers.

Fig. 3. Total execution time of servers.

Figure 4 shows the average execution time AT of n processes. For $n < 3,500$, the TT is abut aT, i.e. 10 [tu] since about one process is performed on each thread of the servers, The AT of the SMI, SP, SMIM, and SPM algorithms linearly increases as n increases for $n > 3,500$ as shown in Fig. 4. The more number of processes, the more number of migrations of each virtual machine. Each time a virtual machine migrates, processes on the virtual machine is suspended for tm time units, i.e. one [tu].

In Fig. 5, SMIM and SPM show the average numbers of migrations of each virtual machine in the SMIM and SPIM algorithms, respectively. Virtual machines migrate in the SMIM algorithm a little bit more times than the SPM algorithm. SMIM-NM and SMP-NM indicate the average numbers of migrations of each process in the SMIM and SPIM algorithms, respectively. Processes migrate in the SPM algorithm a little bit more times than the SMIM algorithm. Each process migrates at most one time for $n < 3,500$. The number of migrations linearly increases as n increases for $n > 3,500$. For example, each process migrates about two time for $n = 5,500$ and three times for $n = 7,000$. Here, the execution time of each process increases by two and three [tu] since the processes are suspended during the migration.

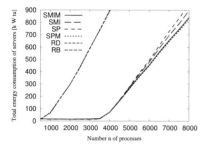

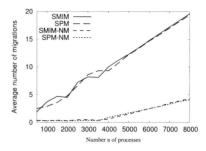

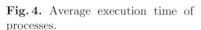

Fig. 4. Average execution time of processes.

Fig. 5. Average number of migrations of a virtual machine.

6   Concluding Remarks

It is critical to reduce the energy consumption of servers to realize green societies. In this paper, we first proposed the SMI (Simple MI) algorithm to more simply estimate the execution time and energy consumption of a server only by using the number of active processes on a server. We proposed the SMIM algorithm where virtual machines on host servers migrate to guest servers by using the SMIM algorithm so that the total energy consumption of the servers can be reduced. In the evaluation, we showed the energy consumption of servers can be reduced in the SMIM algorithm compared with the other algorithms.

References

1. KVM: Main Page - KVM (Kernel Based Virtual Machine) (2015). http://www. linux-kvm.org/page/Main_Page
2. Enokido, T., Aikebaier, A., Takizawa, M.: Process allocation algorithms for saving power consumption in peer-to-peer systems. IEEE Trans. Ind. Electron. **58**(6), 2097–2105 (2011)
3. Enokido, T., Aikebaier, A., Takizawa, M.: A model for reducing power consumption in peer-to-peer systems. IEEE Syst. J. **4**(2), 221–229 (2010)
4. Enokido, T., Aikebaier, A., Takizawa, M.: An extended simple power consumption model for selecting a server to perform computation type processes in digital ecosystems. IEEE Trans. Ind. Inform. **10**(2), 1627–1636 (2014)
5. Enokido, T., Takizawa, M.: Integrated power consumption model for distributed systems. IEEE Trans. Ind. Electron. **60**(2), 824–836 (2013)
6. Enokido, T., Takizawa, M.: An energy-efficient load balancing algorithm to perform computation type application processes for virtual machine. In: Proceedings of the 18th International Conference on Network-Based Information Systems (NBiS-2016), pp. 32–39 (2015)
7. Enokido, T., Takizawa, M.: Power consumption and computation models of virtual machines to perform computation type application processes. In: Proceedings of the 9th International Conference on Complex, Intelligent, and Software Intensive Systems (CISIS-2015), pp. 126–133 (2015)

8. Enokido, T., Duolikun, D., Takizawa, M.: An energy efficient load balancing algorithm based on the active time of cores. In: Barolli, L., Xhafa, F., Conesa, J. (eds.) BWCCA 2017. LNDECT, vol. 12, pp. 185–196. Springer, Cham (2018). https://doi.org/10.1007/978-3-319-69811-3_16

9. Enokido, T., Duolikun, D., Takizawa, M.: The energy consumption laxity-based algorithm to perform computation processes in virtual machine environments. Int. J. Grid Util. Comput. **10**(5), 545–555 (2019)

10. Enokido, T., Duolikun, D., Takizawa, M.: The improved redundant active time-based (IRATB) algorithm for process replication. In: Barolli, L., Woungang, I., Enokido, T. (eds.) AINA 2021. LNNS, vol. 225, pp. 172–180. Springer, Cham (2021). https://doi.org/10.1007/978-3-030-75100-5_16

11. Enokido, T., Duolikun, D., Takizawa, M.: The redundant active time-based algorithm with forcing meaningless replica to terminate. In: Barolli, L., Yim, K., Enokido, T. (eds.) CISIS 2021. LNNS, vol. 278, pp. 206–213. Springer, Cham (2021). https://doi.org/10.1007/978-3-030-79725-6_20

12. Enokido, T., Duolikun, D., Takizawa, M.: The improved redundant active time-based algorithm with forcing termination of meaningless replicas in virtual machine environments. In: Barolli, L., Chen, H.-C., Enokido, T. (eds.) NBiS 2021. LNNS, vol. 313, pp. 50–58. Springer, Cham (2022). https://doi.org/10.1007/978-3-030-84913-9_5

13. Kataoka, H., Duolikun, D., Enokido, T., Takizawa, M.: Energy-efficient virtualisation of threads in a server cluster. In: Proceedings of the 10th International Conference on Broadband and Wireless Computing, Communication and Applications (BWCCA-2015), pp. 288–295 (2015)

14. Kataoka, H., Duolikun, D., Sawada, A., Enokido, T., Takizawa, M.: Energy-aware server selection algorithms in a scalable cluster. In: Proceedings of IEEE the 30th International Conference on Advanced Information Networking and Applications (AINA-2016), pp. 565–572 (2016)

15. Kataoka, H., Sawada, A., Dilawaer, D., Enokido, T., Takizawa, M.: Multi-level power consumption and computation models and energy-efficient server selection algorithms in a scalable cluster. In: Proceedings of the 19th International Conference on Network-Based Information Systems (NBiS-2016), pp. 210–217 (2016)

16. Kataoka, H., Nakamura, S., Duolikun, D., Enokido, T., Takizawa, M.: Multi-level power consumption model and energy-aware server selection algorithm. Int. J. Grid Util. Comput. **8**(3), 201–210 (2017)

17. Duolikun, D., Enokido, T., Takizawa, M.: Static and dynamic group migration algorithms of virtual machines to reduce energy consumption of a server cluster. In: Nguyen, N.T., Kowalczyk, R., Xhafa, F. (eds.) Transactions on Computational Collective Intelligence XXXIII. LNCS, vol. 11610, pp. 144–166. Springer, Heidelberg (2019). https://doi.org/10.1007/978-3-662-59540-4_8

18. Duolikun, D., Enokido, T., Takizawa, M.: Simple algorithms for selecting an energy-efficient server in a cluster of servers. Int. J. Commun. Netw. Distrib. Syst. **21**(1), 1–25 (2018)

19. Duolikun, D., Enokido, T., Hsu, H.H., Takizawa, M.: Asynchronous migration of process replicas in a cluster. In: Proceedings of IEEE the 29th International Conference on Advanced Information Networking and Applications (AINA-2015), pp. 271–279 (2015)

20. Duolikun, D., Watanabe, R., Enokido, T., Takizawa, M.: An eco migration algorithm of virtual machines in a server cluster. In: Proceedings of IEEE the 32nd International Conference on Advanced Information Networking and Applications (AINA-2015), pp. 189–196 (2018)

21. Duolikun, D., Enokido, T., Takizawa, M.: Energy-efficient group migration of virtual machines in a cluster. In: Barolli, L., Takizawa, M., Xhafa, F., Enokido, T. (eds.) AINA 2019. AISC, vol. 926, pp. 144–155. Springer, Cham (2020). https://doi.org/10.1007/978-3-030-15032-7_12

22. Duolikun, D., Enokido, T., Barolli, L., Takizawa, M.: A monotonically increasing (MI) algorithm to estimate energy consumption and execution time of processes on a server. In: Barolli, L., Chen, H.-C., Enokido, T. (eds.) NBiS 2021. LNNS, vol. 313, pp. 1–12. Springer, Cham (2022). https://doi.org/10.1007/978-3-030-84913-9_1

23. Duolikun, D., Enokido, T., Barolli, L., Takizawa, M.: An energy-efficient algorithm to make virtual machines migrate in a server cluster. In: Barolli, L. (ed.) BWCCA 2021. LNNS, vol. 346, pp. 25–36. Springer, Cham (2022). https://doi.org/10.1007/978-3-030-90072-4_3

24. Inoue, T., Aikebaier, A., Enokido, T., Takizawa, M.: Algorithms for selecting energy-efficient storage servers in storage and computation oriented applications. In: Proceedings of IEEE the 26th International Conference on Advanced Information Networking and Applications (AINA-2016), pp. 920–927 (2016)

25. Noaki, N., Saitto, T., Duolikun, D., Enokido, T., Takizawa, M.: An energy-efficient algorithm for virtual machines to migrate considering migration time. In: Proceedings of the 15th International Conference on Broadband and Wireless Computing, Communication and Applications (BWCCA-2020), pp. 341–354 (2020)

26. Noguchi, K., Saito, T., Duolikun, D., Enokido, T., Takizawa, M.: An algorithm to select a server to minimize the total energy consumption of a cluster. In: Barolli, L., Takizawa, M., Yoshihisa, T., Amato, F., Ikeda, M. (eds.) 3PGCIC 2020. LNNS, vol. 158, pp. 18–28. Springer, Cham (2021). https://doi.org/10.1007/978-3-030-61105-7_3

27. Natural Resources Defense Council (NRDS): Data center efficiency assessment - scaling up energy efficiency across the data center industry: evaluating key drivers and barriers (2014). http://www.nrdc.org/energy/files/data-center-efficiency-assessment-IP.pdf

28. Watanabe, R., Duolikun, D., Enokido, T., Takizawa, M.: An eco model of process migration with virtual machines. In: Proceedings of the 19th International Conference on Network-Based Information Systems (NBiS-2016), pp. 292–297 (2016)

29. Watanabe, R., Duolikun, D., Takizawa, M.: Simple estimation and energy-aware migration models of virtual machines in a server cluster. Concurr. Comput. Pract. Exp. **30**(21), e4771 (2018)

30. Oma, R., Nakamura, S., Duolikun, D., Enokido, T., Takizawa, M.: An energy-efficient model for fog computing in the internet of things (IoT). Internet Tings **1–2**, 14–26 (2018)

31. Oma, R., Nakamura, S., Enokido, T., Takizawa, M.: A tree-based model of energy-efficient fog computing systems in IoT. In: Barolli, L., Javaid, N., Ikeda, M., Takizawa, M. (eds.) CISIS 2018. AISC, vol. 772, pp. 991–1001. Springer, Cham (2019). https://doi.org/10.1007/978-3-319-93659-8_92

Energy Consumption Model of a Device Supporting Information Flow Control in the IoT

Shigenari Nakamura[1(✉)], Tomoya Enokido[2], and Makoto Takizawa[3]

[1] Tokyo Metropolitan Industrial Technology Research Institute, Tokyo, Japan
nakamura.shigenari@iri-tokyo.jp
[2] Rissho University, Tokyo, Japan
eno@ris.ac.jp
[3] Hosei University, Tokyo, Japan
makoto.takizawa@computer.org

Abstract. In the CBAC (Capability-Based Access Control) model to make the IoT (Internet of Things) secure, subjects are issued capability tokens, i.e. a set of access rights on objects in devices, by device owners. Objects are data resource in a device which are used to store sensor data and action data. Through manipulating objects of devices, data are exchanged among subjects and objects. Here, since the illegal information flow and the late information flow occur, subjects can get data which the subjects are not allowed to get. In our previous studies, the OI (Operation Interruption) and the TBOI (Time-Based OI) protocols are implemented to interrupt operations implying both illegal and late types of information flows. In addition, two types of capability token selection algorithms are proposed to make the protocols more useful. In this paper, an electric energy consumption model of a device supporting the protocols is proposed. Based on the model, it is shown that the electric energy consumption in the OI and the TBOI protocols with capability token selection algorithms is smaller than the conventional OI and TBOI protocols.

Keywords: IoT (Internet of Things) · Device security · CBAC (Capability-Based Access Control) model · Information flow control · Capability token selection algorithm · Electric energy consumption model

1 Introduction

It is widely recognized that access control models [5] are useful to make information systems such as database systems [4] secure. For the IoT (Internet of Things) [6, 19], the CBAC (Capability-Based Access Control) model named "CapBAC model" [7] is proposed. Here, subjects are issued capability tokens which are collections of access rights. Only the authorized subjects can manipulate objects of devices only in the authorized operations. Data are exchanged among subjects and objects through manipulating objects of devices. Therefore, a subject might get data via other subjects and objects even if the subject is granted no access right to get the data, i.e. illegal information flow might occur [5, 15]. In addition, a subject might get data generated out of the validity

© The Author(s), under exclusive license to Springer Nature Switzerland AG 2022
L. Barolli et al. (Eds.): EIDWT 2022, LNDECT 118, pp. 142–152, 2022.
https://doi.org/10.1007/978-3-030-95903-6_16

period of a capability token to get the data. Even if the time τ is not within the validity period of the capability token, the subject sb_i can get the data generated at time τ. The data are older than the subject sb_i expects to get, i.e. the data come to the subject sb_i *late* [16]. Hence, it is critical to prevent the illegal information flow and the late information flow to make the IoT secure.

In the papers [10, 11, 13], types of protocols to prevent illegal information flow in database systems and P2PPSO (Peer-to-Peer Publish/Subscribe with Object concept) systems [13] are proposed and discussed. In order to solve both types of illegal and late information flow problems in the IoT, the OI (Operation Interruption) and the TBOI (Time-Based OI) protocols are implemented in our previous studies [15, 16]. In the OI protocol, operations implying illegal information flow are interrupted, i.e. not performed, at devices. On the other hand, in the TBOI protocol, both types of illegal and late information flows are prevented by interrupting operations implying them.

The OI and the TBOI protocols are implemented in a Raspberry Pi3 Model B+ [1] with Raspbian [2] which is used as an IoT device. The device and a subject are a CoAP (Constrained Application Protocol) server and a CoAP client, respectively, in CoAPthon3 [21]. The OI and the TBOI protocols are evaluated in terms of the request processing time which is the period between when a device receives a request and when the device sends the answer of the request. Here, it is shown that the request processing time gets longer as the number of capability tokens whose signatures are verified in devices increases in the OI and the TBOI protocols. Hence, the MRCTSD (Minimum Required Capability Token Selection for Devices) algorithm to reduce the number of capability tokens used is proposed [14]. In the MRCTSD algorithm, minimum required capability tokens to prevent the illegal information flow and the late information flow are selected and used in devices. In the evaluation, the OI and the TBOI protocols supporting the MRCTSD algorithm are implemented in a Raspberry Pi 3 Model B+.

In the MRCTSD algorithm, capability tokens used to make authorization decisions are selected by devices. However, the more number of capability tokens are sent from subjects to devices, the more complex the capability token selection is. Since the devices support just low processing power and smaller size of memories, it is important to avoid concentrating loads in devices. Hence, the MRCTSS (MRCTS for Subjects) algorithm to select capability tokens sent from subjects to devices is proposed [18]. Here, the minimum required capability tokens are selected by subjects and sent to devices. In the evaluation, the OI and the TBOI protocols supporting the MRCTSS algorithm are implemented in a Raspberry Pi 3 Model B+.

In this paper, an electric energy consumption model of a device realized in a Raspberry Pi 3 Model B+ [1] equipped with Raspbian [2] supporting the protocols is proposed. In the protocols, the authorization process can be modeled to be a computation process which use CPU resources like scientific computation. In the paper [9], the MLPC (Multi-Level Power Consumption) and the MLC (ML Computation) models of a device to perform computation processes are proposed. Here, devices are mainly characterized in terms of the number of cores and threads of its CPU. This means, the power consumption of a device depends on the numbers of active cores and threads. The MLPC and the MLC models are validated by measuring the electric energy consumption of a device to perform computation processes. Based on the MLPC and the

MLC models, the electric energy consumption model of a device supporting the protocols is proposed. By using the electric energy consumption model, it is shown that the total electric energy consumption in the OI and the TBOI protocols with capability token selection algorithms is smaller than the conventional OI and TBOI protocols.

In Sect. 2, the system model and types of information flow relations are discussed. In Sect. 3, the protocols to prevent both types of information flows and the capability token selection algorithms are discussed. In Sect. 4, the electric energy consumption model of a device supporting the protocols is proposed.

2   System Model

Suppose an IoT is composed of a pair of sets D and SB which consist of the number dn of devices $d_1, \ldots, d_{dn}$ $(dn \geq 1)$ and the number sbn of subjects $sb_1, \ldots, sb_{sbn}$ $(sbn \geq 1)$, respectively. In order to make the IoT secure, the CBAC (Capability-Based Access Control) model is discussed [7]. Here, each device d_k holds the number on^k of objects $o_1^k, \ldots, o_{on^k}^k$ $(on^k \geq 1)$. A term, "object o_m^k" stands for a component object in the device d_k. Subjects manipulate data of objects in devices. Here, each subject sb_i is issued a set CAP^i which consists of the number cn^i of capability tokens $cap_1^i, \ldots, cap_{cn^i}^i$ $(cn^i \geq 1)$. A capability token cap_g^i is designed as shown in the papers [15, 16]. Each capability token cap_g^i includes the public keys of an issuer and a subject. A signature generated with the private key of the issuer is also included in the capability token cap_g^i. These keys and signatures are generated in the ECDSA (Elliptic Curve Digital Signature Algorithm) [8] and then encoded into Base64 form. The validation period of the capability token cap_g^i is defined with its fields $cap_g^i.NB$ and $cap_g^i.NA$ (NB: Not Before, NA: Not After). The capability token cap_g^i is valid at time τ where $cap_g^i.NB < \tau < cap_g^i.NA$.

If a subject sb_i tries to manipulate data of an object o_m^k in a device d_k in an operation op, the subject sb_i sends an access request with a capability token cap_g^i to specify the subject sb_i is allowed to manipulate the object o_m^k in the operation op to the device d_k. The access request is accepted, i.e. the operation op is performed on the object o_m^k if the device d_k confirms that the subject sb_i is allowed to manipulate the object o_m^k in the operation op. Otherwise, the access request is rejected. Since the device d_k just checks the capability token cap_g^i to authorize the subject sb_i, it is easier to adopt the CBAC model to the IoT than the ACL (Access Control List)-based models such as RBAC (Role-Based Access Control) [20] and the ABAC (Attribute-Based Access Control) [22] models.

Let a pair $\langle o, op \rangle$ be an access right. Subjects issued a capability token including an access right $\langle o, op \rangle$ is allowed to manipulate data of an object o in an operation op. A set of objects whose data a subject sb_i is allowed to get is $IN(sb_i)$ i.e. $IN(sb_i) = \{o_m^k \mid \langle o_m^k, get \rangle \in cap_g^i \wedge cap_g^i \in CAP^i\}$.

Through manipulating data of objects in devices, the data are exchanged among subjects and objects. For example, if a subject sb_i puts data got from an object o_n^l to another object o_m^k, the data of the object o_n^l flow into the subject sb_i and the object o_m^k. Objects whose data flow into entities such as subjects and objects are referred to as *source* objects for these entities. Let $sb_i.sO$ and $o_m^k.sO$ are sets of *source* objects of a subject sb_i and an object o_m^k, respectively, which are initially ϕ. In this example, $sb_i.sO = o_m^k.sO = \{o_n^l\}$.

A capability token cap_g^i is valid from the time $cap_g^i.NB$ to the time $cap_g^i.NA$. Let a pair of times $gt^i.st(o_m^k)$ and $gt^i.et(o_m^k)$ be the start and end time when a subject sb_i is allowed to get data from the object o_m^k. The time when data of an object o_m^k are generated is referred to a generation time. Let $minOT_m^k(o_n^l)$ and $minSBT^i(o_n^l)$ be the earliest generation times of data of an object o_n^l which flow to an object o_m^k and a subject sb_i, respectively.

Based on the CBAC model, we define types of information flow relations on objects and subjects as follows [12, 17]:

Definition 1. An object o_m^k *flows* to a subject sb_i ($o_m^k \rightarrow sb_i$) iff (if and only if) $o_m^k.sO \neq \phi$ and $o_m^k \in IN(sb_i)$.

Definition 2. An object o_m^k *legally flows* to a subject sb_i ($o_m^k \Rightarrow sb_i$) iff $o_m^k \rightarrow sb_i$ and $o_m^k.sO \subseteq IN(sb_i)$.

Definition 3. An object o_m^k *illegally flows* to a subject sb_i ($o_m^k \mapsto sb_i$) iff $o_m^k \rightarrow sb_i$ and $o_m^k.sO \not\subseteq IN(sb_i)$.

Definition 4. An object o_m^k *timely flows* to a subject sb_i ($o_m^k \Rightarrow_t sb_i$) iff $o_m^k \Rightarrow sb_i$ and $\forall o_n^l \in o_m^k.sO$ ($gt^i.st(o_n^l) \leq minOT_m^k(o_n^l) \leq gt^i.et(o_n^l)$).

Definition 5. An object o_m^k *flows late* to a subject sb_i ($o_m^k \mapsto_l sb_i$) iff $o_m^k \Rightarrow sb_i$ and $\exists o_n^l \in o_m^k.sO \neg(gt^i.st(o_n^l) \leq minOT_m^k(o_n^l) \leq gt^i.et(o_n^l))$.

3   Information Flow Control

3.1   Protocols

In the CBAC model for the secure IoT, data are exchanged among subjects and objects. Here, a subject sb_i may get data of an object o_m^k flowing to another object o_n^l by accessing the object o_n^l even if the subject sb_i is not issued a capability token cap_g^i to get the data from the object o_m^k, i.e. illegal information flow from the object o_m^k to the subject sb_i occurs. In addition, although no illegal information flow occurs, a subject sb_i may get data from an object o_m^k generated out of validity period of a capability token cap_g^i to get the data. Here, the data are older than the subject sb_i expects to get, i.e. information comes to the subject sb_i *late*. In our previous studies, the OI (Operation Interruption) [15] and the TBOI (Time-Based OI) [16] protocols are implemented to prevent illegal information flow and both illegal and late types of information flows, respectively.

In order to prevent the illegal information flow, sets of *source* objects are manipulated in the OI protocol. Here, if data of an object o_m^k flow into an entity, the object o_m^k is added to a *source* object set of the entity. The sets of *source* objects are updated as follows:

1. Initially, $sb_i.sO = o_m^k.sO = \phi$ for every subject sb_i and object o_m^k;
2. If a device d_k generates data by sensing events occurring around itself and stores the data to its object o_m^k, $o_m^k.sO = o_m^k.sO \cup \{o_m^k\}$;

3. If a subject sb_i gets data from an object o_m^k, $sb_i.sO = sb_i.sO \cup o_m^k.sO$;
4. If a subject sb_i puts data to an object o_m^k, $o_m^k.sO = o_m^k.sO \cup sb_i.sO$;

On the other hand, in the TBOI protocol, the earliest generation time of data of every *source* object is also updated as follows:

1. Initially, $sb_i.sO = o_m^k.sO = \phi$ for every subject sb_i and object o_m^k;
2. If a device d_k generates data by sensing events occurring around itself and stores the data to its object o_m^k at time τ.
   a. If $minOT_m^k(o_m^k) = $ NULL, $minOT_m^k(o_m^k) = \tau$;
   b. $o_m^k.sO = o_m^k.sO \cup \{o_m^k\}$;
3. If a subject sb_i gets data from an object o_m^k.
   a. For each object o_n^l such that $o_n^l \in (sb_i.sO \cap o_m^k.sO)$, $minSBT^i(o_n^l) = min(minSBT^i(o_n^l), minOT_m^k(o_n^l))$;
   b. For each object o_n^l such that $o_n^l \notin sb_i.sO$ but $o_n^l \in o_m^k.sO$, $minSBT^i(o_n^l) = minOT_m^k(o_n^l)$;
   c. $sb_i.sO = sb_i.sO \cup o_m^k.sO$;
4. If a subject sb_i puts data to an object o_m^k.
   a. For each object o_n^l such that $o_n^l \in (sb_i.sO \cap o_m^k.sO)$, $minOT_m^k(o_n^l) = min(minOT_m^k(o_n^l), minSBT^i(o_n^l))$;
   b. For each object o_n^l such that $o_n^l \notin o_m^k.sO$ but $o_n^l \in sb_i.sO$, $minOT_m^k(o_n^l) = minSBT^i(o_n^l)$;
   c. $o_m^k.sO = o_m^k.sO \cup sb_i.sO$;

Based on the sets of *source* objects and the earliest generation time of data of every object, the illegal information flow and the late information flow are detected. The OI and the TBOI protocols perform as follows to prevent illegal information flow and both illegal and late types of information flows, respectively:

[OI protocol] A *get* operation on an object o_m^k issued by a subject sb_i is interrupted if $o_m^k \mapsto sb_i$.

[TBOI protocol] A *get* operation on an object o_m^k issued by a subject sb_i is interrupted if $o_m^k \mapsto_l sb_i$.

3.2 MRCTSD (Minimum Required Capability Token Selection for Devices) Algorithm

In our previous studies [15, 16], the OI and the TBOI protocols are implemented on a Raspberry Pi 3 Model B+ [1] whose operating system is Raspbian [2]. In the evaluation, a subject sb_i issues operations to a device d_k to manipulate data of an object o_m^k in the device d_k. If the subject sb_i issues a *get* operation on the object o_m^k to the device d_k, the device d_k has to confirm the subject sb_i is issued capability tokens to get data from not only the object o_m^k but also the other objects in $o_m^k.sO$. Here, the request processing time which is the period between when a device receives a request and when the device sends the answer of the request increases as the number of capability tokens whose signatures are verified increases. Most part of the request processing time is time to verify signatures. Hence, the number of capability tokens used should be reduced to shorten the request processing time.

In the paper [14], the MRCTSD (Minimum Required Capability Token Selection for Devices) algorithm is proposed for the OI and the TBOI protocols. Suppose a subject sb_i issues a *get* operation on an object o_m^k to the device d_k. Here, the device d_k selects minimum required capability tokens to confirm the subject sb_i is allowed to get data from not only the object o_m^k but also the *source* objects in $o_m^k.sO$. The OI-MRCTSD protocol and the TBOI-MRCTSD protocol are proposed and implemented. In the evaluation, it is shown that the request processing times of the OI-MRCTSD and the TBOI-MRCTSD protocols are smaller than the OI and the TBOI protocols.

3.3 MRCTSS (MRCTS for Subjects) Algorithm

In the MRCTSD algorithm [14], capability tokens used to make authorization decisions are selected in devices. However, the more number of capability tokens are sent from subjects to devices, the more complex the capability token selection is. Since the devices support just low processing power and smaller size of memory, it is important to avoid concentrating loads in devices.

In the paper [18], the MRCTSS (MRCTS for Subjects) algorithm to select capability tokens sent from subjects to devices is proposed. Suppose a subject sb_i tries to send a *get* operation on an object o_m^k to a device d_k. In the OI and the TBOI protocols, the subject sb_i has to indicate that the subject sb_i is allowed to get data from not only the object o_m^k but also the other *source* objects of the object o_m^k. However, when the subject sb_i starts to send a *get* operation, the subject sb_i dose not know which *source* objects are in the object o_m^k. Hence, the subject sb_i selects minimum required capability tokens to indicate the subject sb_i is allowed to get data from every object in $IN(sb_i)$ in the MRCTSS algorithm. The OI-MRCTSS protocol and the TBOI-MRCTSS protocol are proposed and implemented. In the evaluation, it is shown that the size of a UDP datagram in a *get* access request in the OI-MRCTSS and the TBOI-MRCTSS protocols are smaller than the OI, the TBOI, the OI-MRCTSD, and the TBOI-MRCTSD protocols.

4   Electric Energy Consumption Model

The IoT is more scalable than the cloud computing systems because huge number and various types of nodes are included. Hence, it is required to reduce the electric energy consumption by not only servers but also devices. In this paper, an electric energy consumption model of a device supporting the protocols to prevent the illegal information flow and the late information flow is proposed. Here, a Raspberry Pi 3 Model B+ [1] equipped with Raspbian [2] is regarded as a device d_k. The device d_k is manipulated by a subject sb_i which is implemented in Python. Both the device d_k and the subject sb_i are realized as a CoAP (Constrained Application Protocol) server and a CoAP client in CoAPthon3 [21], respectively.

In the OI and the TBOI protocols, a capability token is designed as shown in the papers [15,16]. Suppose the subject sb_i issues a *get* operation on an object o_m^k in the device d_k. In every protocol, a device d_k supports the authorization process composed of the following steps.

Step 1 The device d_k confirms which objects the subject sb_i is allowed to get data from.

Step 2 The device d_k verifies a signature with the public key of the subject sb_i.

Step 3 The device d_k verifies a signature of every capability token with every issuer's public key.

If the access request is accepted, the device d_k sends a CoAP response to the subject sb_i.

In all the steps, the CPU resources of the device d_k are used. Let p_m^k be a process composed of all the steps for the object o_m^k on the device d_k. Here, a process p_m^k is modeled to be a computation process which uses the CPU resources like scientific computation.

4.1 MLPC (Multi-Level Power Consumption) Model

In the paper [9], the MLPC (Multi-Level Power Consumption) model of a device to perform computation processes is proposed. Devices are mainly characterized in terms of the numbers of cores and threads of their CPUs. A CPU is characterized in terms of the number nc^k of cores and the number nct^k of threads on each core. Let nt^k be the total number of threads in a device d_k, i.e. $nt^k = nc^k \cdot nct^k$. A device d_k provides nt^k (≥ 1) threads $th_0^k, th_1^k, \ldots, th_{nt^k-1}^k$ on nc^k cores $c_0^k, c_1^k, \ldots, c_{nc^k-1}^k$. A thread th_i^k is *active* iff at least one process is performed on the thread th_i^k in a device d_k. A core c_j^k is *active* iff at least one thread of the core c_j^k is active. The device d_k consumes the electric power bC^k [W] if at least one core is active. The power consumption of a device d_k increases by values cPW^k [W] and tPW^k [W] as the numbers of active cores and active threads increase by one, respectively. Cores and threads which are not active are *idle*. If no process is performed on a device, every core and thread is idle. Here, the device d_k consumes the minimum electric power $minPW^k$ [W]. The power consumption $PW^k(\tau)$ [W] of a device d_k to perform processes at time τ is given as follows:

$$PW^k(\tau) = minPW^k + \gamma(\tau) \cdot [bC^k + \sum_{j=0}^{nc^k-1} \alpha_j(\tau) \cdot \{(cPW^k + \beta_j(\tau) \cdot tPW^k)\}]. \quad (1)$$

If at least one core is active at time τ, $\gamma(\tau) = 1$. Otherwise, $\gamma(\tau) = 0$, i.e. an idle device d_k consumes minimum electric power $minPW^k$. If a core c_j^k is active on the device d_k at time τ, $\alpha_j(\tau) = 1$. Otherwise, $\alpha_j(\tau) = 0$. $\beta_j(\tau)$ ($\leq nct^k$) is the number of active threads on a core c_j^k at time τ. If $\alpha_j(\tau) = 1$, $1 \leq \beta_j(\tau) \leq nct^k$. '$\gamma(\tau) = 1$' means $\alpha_j(\tau) = 1$ for some core c_j^k.

Generally, processes are allocated to threads in the RR (Round-Robin) algorithm. First, a first process is allocated to a thread th_0^k. Then, a second process is allocated to a thread th_1^k. If a greater number pn of processes than the total number nt^k of threads are concurrently performed ($pn \geq nt^k$), the device d_k consumes the maximum electric power $maxPW^k$ [W] since every thread is active.

4.2 MLC (Multi-Level Computation) Model

Let $CP^k(\tau)$ be a set of processes which are being performed on a device d_k at time τ. In the paper [9], the MLC (ML Computation) model is proposed. Here, a process

p_m^k is how much processed by a device d_k at time τ is referred to as 'computation rate $cr_m^k(\tau)$'. Let $maxC^k$ be the maximum computation rate of a core in the device d^k. The computation rate $cr_m^k(\tau)$ of a process p_m^k in a device d_k at time τ is given as follows:

$$cr_m^k(\tau) = \begin{cases} maxC^k & (\text{if } |CP^k(\tau)| \le nt^k). \\ nt^k \cdot maxC^k/|CP^k(\tau)| & (\text{if } |CP^k(\tau)| > nt^k). \end{cases} \tag{2}$$

For $|CP^k(\tau)| > nt^k$, the computation rate $cr_m^k(\tau)$ for the process p_m^k decreases as the number of processes concurrently performed on the device d_k, i.e. $|CP^k(\tau)|$, increases. Hence, for $|CP^k(\tau)| > nt^k$, the execution time of the process p_m^k increases as $|CP^k(\tau)|$ increases. The total electric energy $EE^k(st, et)$ [J] of a device d_k from time st to time et is given as follows.

$$EE^k(st, et) = \sum_{\tau=st}^{et} PW^k(\tau). \tag{3}$$

4.3   Experiments

The power consumption of a device d_k, Raspberry Pi 3 Model B+ equipped with Raspbian, is measured by using the power meter UWmeter [3]. The CPU of the device d_k is 'Broadcom BCM2837B0, Cortex-A53' where $nc^k = 4$, $nct^k = 1$, and $nt^k = 4$. In the experiment, multiple processes are concurrently performed on the device d_k. First, a process is forked to pn child processes p_1, ..., p_{pn} ($pn \ge 1$). Here, the same computation is performed on every child process. If no other process is performed on the device d_k, it takes about 10 s to perform one child process. Child processes are bound to some number of threads in a CPU by using the system call *taskset*. Each child process waits to start until specified time so that every child process can start simultaneously [9].

Figure 1 shows the power consumption of the device d_k. The label 'idle' means no computation process is performed on the device d_k. For $pn \le 4$, if the number of processes increases by one, the numbers of cores and threads also increase by one, respectively. The more numbers of cores and threads are active, the more power is consumed in the device d_k. For $pn > 4$, even if the number of processes increases, the power consumption does not change.

Figure 2 shows the average execution time of a process performed on the device d_k. For $pn \le 4$, the average execution time does not change. For $pn > 4$, the average execution time linearly increases as the number pn of processes increases.

As a result, the parameters to decide the power consumption of the device d_k is made clear. Here, $minPW^k$, $maxPW^k$, bC^k, and $cPW^k + tPW^k$ are 2.55, 4.55, 0.20, and 0.45 [W], respectively.

Example 1. Suppose a device d_k receives four access requests to get data of objects o_1^k to o_4^k from subjects. Here, a computation process p_1^k to p_4^k are performed on the device d_k from time st to time et, i.e. $CP^k(\tau) = \{p_m^k \ (m = 1, \dots 4)\}$ ($st \le \tau \le et$). In each access request, a subject sb_i attempts to send a set of five capability tokens such that $\{cap_g^i \mid \langle o_g^k, get \rangle \in cap_g^i (g = 1, ..., 4)\} \cup \{cap_5^i \ (= \{\langle o_1^k, get \rangle, \dots, \langle o_4^k, get \rangle\})\}$ to the device d_k. The set of *source* objects of the object o_m^k ($m = 1, \dots 4$) includes o_1^k to o_4^k. In our previous studies [14], the execution time of the authorization process is

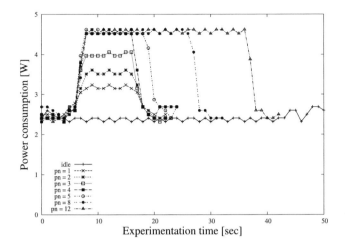

Fig. 1. Power consumption for concurrent executions.

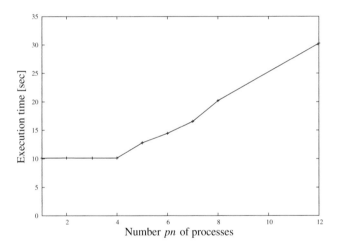

Fig. 2. Average execution time of a process.

measured. In the OI and the TBOI protocols, the capability tokens cap_1^i to cap_4^i are used for each access request and the execution time of every process is about 100[ms]. Hence, the total electric energy $EE_k(st, et) = 0.1 \cdot (2.55 + 0.20 + 0.45 \cdot 4) = 0.46$ [J]. On the other hand, in the OI-MRCTSD, the TBOI-MRCTSD, the OI-MRCTSS, and the TBOI-MRCTSS protocols, only the capability token cap_5^i is selected for each access request and the execution time of every process is about 40 [ms]. Hence, the total electric energy $EE_k(st, et) = 0.04 \cdot (2.55 + 0.20 + 0.45 \cdot 4) = 0.18$ [J].

5   Concluding Remarks

In the CBAC (Capability-Based Access Control) model proposed to make the IoT (Internet of Things) secure, capability tokens are issued to subjects as access rights. Since data are exchanged among subjects and objects through manipulating objects, two types of illegal and late information flows occur. In our previous studies, the OI (Operation Interruption) and the TBOI (Time-Based OI) protocols were proposed to prevent both types of illegal and late information flows. In addition, the MRCTSD (Minimum Required Capability Token Selection for Devices) and the MRCTSS (MRCTS for Subjects) algorithms were proposed to make the OI and the TBOI protocols useful. In this paper, an electric energy consumption model of a device realized in a Raspberry Pi 3 Model B+ equipped with Raspbian supporting the protocols is proposed. Based on the model, it is shown that the electric energy consumption in the OI and the TBOI protocols with capability token selection algorithms is smaller than the conventional OI and TBOI protocols.

Acknowledgements. This work was supported by Japan Society for the Promotion of Science (JSPS) KAKENHI Grant Number JP20K23336.

References

1. Raspberry pi 3 model b+. https://www.raspberrypi.org/products/raspberry-pi-3-model-b-plus/
2. Raspbian, version 10.3, 13 February 2020. https://www.raspbian.org/
3. Uwmeter (2011). http://www.metaprotocol.com/UWmeter/Features.html
4. Date, C.J.: An Introduction to Database Systems, 8th edn. Addison Wesley, Boston (2003)
5. Denning, D.E.R.: Cryptography and Data Security. Addison Wesley, Boston (1982)
6. Hanes, D., Salgueiro, G., Grossetete, P., Barton, R., Henry, J.: IoT Fundamentals: Networking Technologies, Protocols, and Use Cases for the Internet of Things. Cisco Press, Indianapolis (2018)
7. Hernández-Ramos, J.L., Jara, A.J., Marín, L., Skarmeta, A.F.: Distributed capability-based access control for the internet of things. J. Internet Serv. Inf. Secur. **3**(3/4), 1–16 (2013)
8. Johnson, D., Menezes, A., Vanstone, S.: The elliptic curve digital signature algorithm (ECDSA). Int. J. Inf. Secur. **1**(1), 36–63 (2001)
9. Kataoka, H., Nakamura, S., Duolikun, D., Enokido, T., Takizawa, M.: Multi-level power consumption model and energy-aware server selection algorithm. Int. J. Grid Util. Comput. **8**(3), 201–210 (2017)
10. Nakamura, S., Duolikun, D., Aikebaier, A., Enokido, T., Takizawa, M.: Read-write abortion (RWA) based synchronization protocols to prevent illegal information flow. In: Proceedings of the 17th International Conference on Network-Based Information Systems, pp. 120–127 (2014)
11. Nakamura, S., Duolikun, D., Enokido, T., Takizawa, M.: A read-write abortion protocol to prevent illegal information flow in role-based access control systems. Int. J. Space-Based Situated Comput. **6**(1), 43–53 (2016)
12. Nakamura, S., Enokido, T., Takizawa, M.: Information flow control based on the CapBAC (capability-based access control) model in the IoT. Int. J. Mob. Comput. Multimed. Commun. **10**(4), 13–25 (2019)

13. Nakamura, S., Enokido, T., Takizawa, M.: Information flow control in object-based peer-to-peer publish/subscribe systems. Concurr. Comput.: Pract. Exp. **32**(8), e5118 (2020)

14. Nakamura, S., Enokido, T., Takizawa, M.: A capability token selection algorithm for lightweight information flow control in the IoT. In: Barolli, L., Chen, H.-C., Enokido, T. (eds.) NBiS 2021. LNNS, vol. 313, pp. 23–34. Springer, Cham (2022). https://doi.org/10.1007/978-3-030-84913-9_3

15. Nakamura, S., Enokido, T., Takizawa, M.: Implementation and evaluation of the information flow control for the internet of things. Concurr. Comput.: Pract. Exp. **33**(19), e6311 (2021)

16. Nakamura, S., Enokido, T., Takizawa, M.: Information flow control based on capability token validity for secure IoT: implementation and evaluation. Internet Things **15**, 100, 423 (2021)

17. Nakamura, S., Enokido, T., Takizawa, M.: Time-based legality of information flow in the capability-based access control model for the internet of things. Concurr. Comput.: Pract. Exp. **33**(23), e5944 (2021)

18. Nakamura, S., Enokido, T., Takizawa, M.: Traffic reduction for information flow control in the IoT. In: Barolli, L. (ed.) BWCCA 2021. LNNS, vol. 346, pp. 67–77. Springer, Cham (2022). https://doi.org/10.1007/978-3-030-90072-4_7

19. Oma, R., Nakamura, S., Duolikun, D., Enokido, T., Takizawa, M.: An energy-efficient model for fog computing in the internet of things (IoT). Internet Things **1–2**, 14–26 (2018)

20. Sandhu, R.S., Coyne, E.J., Feinstein, H.L., Youman, C.E.: Role-based access control models. IEEE Comput. **29**(2), 38–47 (1996)

21. Tanganelli, G., Vallati, C., Mingozzi, E.: CoApthon: easy development of CoAP-based IoT applications with python. In: IEEE 2nd World Forum on Internet of Things (WF-IoT 2015), pp. 63–68 (2015)

22. Yuan, E., Tong, J.: Attributed based access control (ABAC) for web services. In: Proceedings of the IEEE International Conference on Web Services (ICWS 2005), p. 569 (2005)

A Fuzzy-Based System for Assessment of QoS of V2V Communication Links in SDN-VANETs

Ermioni Qafzezi[1(✉)], Kevin Bylykbashi[2], Phudit Ampririt[1], Makoto Ikeda[2], Keita Matsuo[2], and Leonard Barolli[2]

[1] Graduate School of Engineering, Fukuoka Institute of Technology (FIT), 3-30-1 Wajiro-Higashi, Higashi-Ku, Fukuoka 811–0295, Japan
{bd20101,bd21201}@bene.fit.ac.jp
[2] Department of Information and Communication Engineering, Fukuoka Institute of Technology (FIT), 3-30-1 Wajiro-Higashi, Higashi-Ku, Fukuoka 811-0295, Japan
makoto.ikd@acm.org, {kt-matsuo,barolli}@fit.ac.jp

Abstract. Vehicular Ad hoc Networks (VANETs) are characterized by high mobility of vehicles and dynamic topology changes, which lead to intermittent connectivity of V2V links. Data communication links among vehicles must satisfy certain requirements to induce a successful transmission. Therefore the evaluation of Quality of Service (QoS) of communication links set up among adjacent vehicles is an important parameter in VANETs. In this paper, we propose a fuzzy-based system to assess QoS in Software Defined VANETs (SDN-VANETs). Our proposed system, called Fuzzy System for Assessment of Quality of Service (FSAQoS), determines whether a communication link is reliable and satisfies certain needs in terms of data exchange. FSAQoS is implemented in the vehicles equipped with SDN modules and in SDN controllers. When a vehicle needs additional resources, it makes a request to use the available resources of neighboring vehicles. However, for a successful data exchange, the communication link between a pair of vehicles should be reliable and provide low latency. In addition, the proposed system takes into consideration the interference caused by transmissions from other vehicles, and the beacon messages distributed in the network that inform vehicles for their neighbor's condition and their location. The output of FSAQoS decides the QoS of a certain data link. We evaluate FSAQoS by computer simulations. QoS of the communication links is high when interference is permissible, data exchange is reliable and it is provided with low latency, and when beacon messages are generated with moderate frequency throughout the network.

1 Introduction

The long distances separating homes and workplaces/facilities/schools as well as the traffic present in these distances make people spend a significant amount of time in vehicles. Thus, it is important to offer drivers and passengers ease of driving, convenience, efficiency and safety. This has led to the emerging of Vehicular Ad hoc Networks (VANETs), where vehicles are able to communicate and share important information among them. VANETs are a relevant component of Intelligent Transportation System (ITS), which offers more safety and better transportation.

VANETs are capable to offer numerous services such as road safety, enhanced traffic management, as well as travel convenience and comfort. To achieve road safety, emergency messages must be transmitted in real-time, which stands also for the actions that should be taken accordingly in order to avoid potential accidents. Thus, it is important for the vehicles to always have available connections to infrastructure and to other vehicles on the road. On the other hand, traffic efficiency is achieved by managing traffic dynamically according to the situation and by avoiding congested roads, whereas comfort is attained by providing in-car infotainment services.

The advances in vehicle technology have made possible for the vehicles to be equipped with various forms of smart cameras and sensors, wireless communication modules, storage and computational resources. While more and more of these smart cameras and sensors are incorporated in vehicles, massive amounts of data are generated from monitoring the on-road and in-board status. This exponential growth of generated vehicular data, together with the boost of the number of vehicles and the increasing data demands from in-vehicle users, has led to a tremendous amount of data in VANETs [9]. Moreover, applications like autonomous driving require even more storage capacity and complex computational capability. As a result, traditional VANETs face huge challenges in meeting such essential demands of the ever-increasing advancement of VANETs.

The integration of Cloud-Fog-Edge Computing in VANETs is the solution to handle complex computation, provide mobility support, low latency and high bandwidth. Each of them serves different functions, but also complements eachother in order to enhance the performance of VANETs. Even though the integration of Cloud, Fog and Edge Computing in VANETs solves significant challenges, this architecture lacks mechanisms needed for resource and connectivity management because the network is controlled in a decentralized manner. The prospective solution to solve these problems is the augmentation of Software Defined Networking (SDN) in this architecture.

The SDN is a promising choice in managing complex networks with minimal cost and providing optimal resource utilization. SDN offers a global knowledge of the network with a programmable architecture which simplifies network management in such extremely complicated and dynamic environments like VANETs [8]. In addition, it will increase flexibility and programmability in the network by simplifying the development and deployment of new protocols and by bringing awareness into the system, so that it can adapt to changing conditions and requirements, i.e., emergency services [3]. This awareness allows SDN-VANET to make better decisions based on the combined information from multiple sources, not just individual perception from each node.

In a previous work [7], we have proposed an intelligent approach to manage the Cloud-Fog-Edge resources in SDN-VANETs using Fuzzy Logic (FL). We have presented a Cloud-Fog-Edge layered architecture which is coordinated by an intelligent system that decides the appropriate resources to be used by a particular vehicle (hereafter will be referred as *the vehicle*) in need of additional computing resources. The main objective is to achieve a better management of these resources. In another work [6], we focused only on the edge layer resources. The Fuzzy System for Assessment of Neighboring Vehicle Processing Capability (FS-ANVPC) assesses the processing capability for each neighboring vehicle separately, hence helpful neighboring vehicles could be discovered and a better assessment of available edge resources can be made.

In this work, we propose a Fuzzy System for Assessment of Quality of Service (FSAQoS) to determine the QoS of communication links between vehicles, in order to offer efficient and reliable communication.

The remainder of the paper is as follows. In Sect. 2, we present an overview of Cloud-Fog-Edge SDN-VANETs. In Sect. 3, we describe the proposed fuzzy-based system. In Sect. 4, we discuss the simulation results. Finally, conclusions and future work are given in Sect. 5.

2   Cloud-Fog-Edge SDN-VANETs

While cloud, fog and edge computing in VANETs offer scalable access to storage, networking and computing resources, SDN provides higher flexibility, programmability, scalability and global knowledge. In Fig. 1, we give a detailed structure of this novel VANET architecture. It includes the topology structure, its logical structure and the content distribution on the network. As it is shown, it consists of Cloud Computing data centers, fog servers with SDNCs, Road-Side Units (RSUs), RSU Controllers (RSUCs), Base Stations (BS) and vehicles. We also illustrate the Infrastructure-to-Infrastructure (I2I), Vehicle-to-Infrastructure (V2I), and Vehicle-to-Vehicle (V2V) communication links. The fog devices (such as fog servers and RSUs) are located between vehicles and the data centers of the main cloud environments.

The safety applications data generated through in-board and on-road sensors are processed first in the vehicles as they require real-time processing. If more storing and computing resources are needed, *the vehicle* can request to use those of the other adjacent vehicles, assuming a connection can be established and maintained between them for a while. With the vehicles having created multiple virtual machines on other vehicles, the virtual machine migration must be achievable in order to provide continuity as one/some vehicle may move out of the communication range. However, to set-up virtual machines on the nearby vehicles, multiple requirements must be met and when these demands are not satisfied, the fog servers are used.

Cloud servers are used as a repository for software updates, control policies and for the data that need long-term analytics and are not delay-sensitive. On the other side, SDN modules which offer flexibility and programmability, are

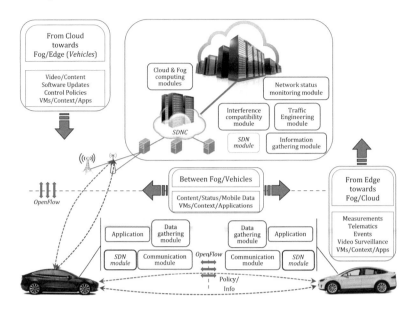

Fig. 1. Logical architecture of Cloud-Fog-Edge SDN-VANET with content distribution.

used to simplify the network management by offering mechanisms that improve the network traffic control and coordination of resources. The implementation of this architecture promises to enable and improve the VANET applications such as road and vehicle safety services, traffic optimization, video surveillance, telematics, commercial and entertainment applications.

3   Proposed Fuzzy-Based System

In this section, we present our proposed fuzzy based system. A *vehicle* that needs storage and computing resources for a particular application can use those of neighboring vehicles, fog servers or cloud data centers based on the application requirements. For instance, for a temporary application that needs real-time processing, *the vehicle* can use the resources of adjacent vehicles if the requirements to realize such operations are fulfilled. Otherwise, it will use the resources of fog servers, which offer low latency as well. Whereas real-time applications require the usage of edge and fog layer resources, for delay tolerant applications, vehicles can use the cloud resources as these applications do not require low latency.

To fully exploit the available edge layer resources, it is necessary to use reliable V2V communication links. The proposed system (FSAQoS) assesses the QoS of the link between two vehicles. FSAQoS is implemented in the SDNC and in the vehicles which are equipped with SDN modules. If a vehicle does not have an SDN module, it sends the information to SDNC which sends back its decision. The system uses the beacon messages received from the adjacent vehicles to extract information, such as their current position, velocity, direction, etc., and based on the received data, the performance of each communication link among adjacent vehicles is decided.

The structure of the proposed system is shown in Fig. 2. For the implementation of our system, we consider four input parameters: Link Latency (LL), Radio Interference (RI), Effective Reliability (ER), and Update Information for Vehicle Position (UIVP) to determine the Quality of Service (QoS) of the communication links between vehicles.

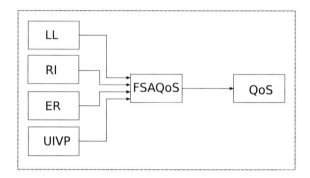

Fig. 2. Proposed system structure.

LL: Link latency is the total amount of time needed for a data packet to travel from the sender to the destination. It is important to provide a low link latency, since VANETs are very dynamic and suffer from intermittent communication.

RI: Radio Interference indicates the unwanted signals that come from transmissions of nearby vehicles and deteriorate the reception of information. The unwanted energy can cause low data speed transmission and even complete loss of information.

ER: We define effective reliability as the capacity of the network to successfully deliver messages to its destination. There are many factors that influence ER such as: the bandwidth of transmission medium, number of transmission errors, the buffer size, and so on.

UIVP: Beacon messages are transmitted periodically throughout the network to inform other vehicles for their position and other attributes. It is necessary to have the coordinates of surrounding vehicles in order to detect a dangerous situation or to monitor traffic. However, too many beacon packets will increase occupied bandwidth in the network, whereas a few beacon packets will decrease the possibility of obtaining the accurate position of vehicles.

QoS: The output parameter values consist of values between 0 and 1, with the value 0.5 working as a border to determine if a certain communication link is capable to exchange information successfully through V2V communications.

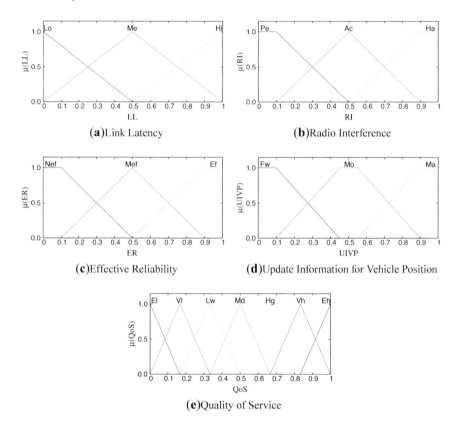

Fig. 3. Membership functions.

We consider FL to implement the proposed system because our system parameters are not correlated with each other. Having three or more parameters which are not correlated with each other results in a non-deterministic polynomial-time hard (NP-hard) problem and FL can deal with these problems. Moreover, we want our system to make decisions in real time and fuzzy systems can give very good results in decision making and control problems [1, 2, 4, 5, 10, 11].

The input parameters are fuzzified using the membership functions shown in Fig. 3(a)–3(d). In Fig. 3(e) are shown the membership functions used for the output parameter. We use triangular and trapezoidal membership functions because they are suitable for real-time operation. The term sets for each linguistic parameter are shown in Table 1. We decided the number of term sets by carrying out many simulations. In Table 2, we show the Fuzzy Rule Base (FRB) of our proposed system, which consists of 81 rules. The control rules have the form: IF "conditions" THEN "control action". For instance, for Rule 1: "IF LL is Lo, RI is Pe, ER is Nef, and UIVP is Fw, THEN QoS is Hg" or for Rule 60: "IF LL is Hi, RI is Pe, ER is Mef, and UIVP is Ma, THEN QoS is El".

Table 1. System parameters and their term sets.

Parameters	Term sets
Link Latency (LL)	Low (Lo), Medium (Me), High (Hi)
Radio Interference (RI)	Permissible (Pe), Acceptable (Ac), Harmful (Ha)
Effective Reliability (ER)	Not Effective (Nef), Medium Effective (Mef), Effective (Ef)
Update Information for Vehicle Position (UIVP)	Few (Fw), Moderate (Mo), Many (Ma)
Quality of Service (QoS)	Extremely Low (El), Very Low (Vl), Low (Lw), Moderate (Md), High (Hg), Very High (Vh), Extremely High (Eh)

Table 2. FRB of FSAQoS.

No	LL	RI	ER	UIVP	QoS	No	LL	RI	ER	UIVP	QoS	No	LL	RI	ER	UIVP	QoS
1	Lo	Pe	Nef	Fw	Hg	28	Mef	Pe	Nef	Fw	Vl	55	Hi	Pe	Nef	Fw	El
2	Lo	Pe	Nef	Mo	Eh	29	Mef	Pe	Nef	Mo	Lw	56	Hi	Pe	Nef	Mo	Vl
3	Lo	Pe	Nef	Ma	Hg	30	Mef	Pe	Nef	Ma	Vl	57	Hi	Pe	Nef	Ma	El
4	Lo	Pe	Mef	Fw	Vh	31	Mef	Pe	Mef	Fw	Lw	58	Hi	Pe	Mef	Fw	El
5	Lo	Pe	Mef	Mo	Eh	32	Mef	Pe	Mef	Mo	Md	59	Hi	Pe	Mef	Mo	Lw
6	Lo	Pe	Mef	Ma	Vh	33	Mef	Pe	Mef	Ma	Lw	60	Hi	Pe	Mef	Ma	El
7	Lo	Pe	Ef	Fw	Eh	34	Mef	Pe	Ef	Fw	Md	61	Hi	Pe	Ef	Fw	Vl
8	Lo	Pe	Ef	Mo	Eh	35	Mef	Pe	Ef	Mo	Hg	62	Hi	Pe	Ef	Mo	Md
9	Lo	Pe	Ef	Ma	Eh	36	Mef	Pe	Ef	Ma	Md	63	Hi	Pe	Ef	Ma	Vl
10	Lo	Ac	Nef	Fw	Md	37	Mef	Ac	Nef	Fw	El	64	Hi	Ac	Nef	Fw	El
11	Lo	Ac	Nef	Mo	Hg	38	Mef	Ac	Nef	Mo	Lw	65	Hi	Ac	Nef	Mo	El
12	Lo	Ac	Nef	Ma	Md	39	Mef	Ac	Nef	Ma	El	66	Hi	Ac	Nef	Ma	El
13	Lo	Ac	Mef	Fw	Hg	40	Mef	Ac	Mef	Fw	Vl	67	Hi	Ac	Mef	Fw	El
14	Lo	Ac	Mef	Mo	Eh	41	Mef	Ac	Mef	Mo	Md	68	Hi	Ac	Mef	Mo	Vl
15	Lo	Ac	Mef	Ma	Hg	42	Mef	Ac	Mef	Ma	Vl	69	Hi	Ac	Mef	Ma	El
16	Lo	Ac	Ef	Fw	Vh	43	Mef	Ac	Ef	Fw	Lw	70	Hi	Ac	Ef	Fw	El
17	Lo	Ac	Ef	Mo	Eh	44	Mef	Ac	Ef	Mo	Hg	71	Hi	Ac	Ef	Mo	Lw
18	Lo	Ac	Ef	Ma	Vh	45	Mef	Ac	Ef	Ma	Lw	72	Hi	Ac	Ef	Ma	El
19	Lo	Ha	Nef	Fw	Lw	46	Mef	Ha	Nef	Fw	El	73	Hi	Ha	Nef	Fw	El
20	Lo	Ha	Nef	Mo	Md	47	Mef	Ha	Nef	Mo	Vl	74	Hi	Ha	Nef	Mo	El
21	Lo	Ha	Nef	Ma	Lw	48	Mef	Ha	Nef	Ma	El	75	Hi	Ha	Nef	Ma	El
22	Lo	Ha	Mef	Fw	Md	49	Mef	Ha	Mef	Fw	Vl	76	Hi	Ha	Mef	Fw	El
23	Lo	Ha	Mef	Mo	Hg	50	Mef	Ha	Mef	Mo	Lw	77	Hi	Ha	Mef	Mo	El
24	Lo	Ha	Mef	Ma	Md	51	Mef	Ha	Mef	Ma	Vl	78	Hi	Ha	Mef	Ma	El
25	Lo	Ha	Ef	Fw	Hg	52	Mef	Ha	Ef	Fw	Lw	79	Hi	Ha	Ef	Fw	El
26	Lo	Ha	Ef	Mo	Vh	53	Mef	Ha	Ef	Mo	Md	80	Hi	Ha	Ef	Mo	El
27	Lo	Ha	Ef	Ma	Hg	54	Mef	Ha	Ef	Ma	Lw	81	Hi	Ha	Ef	Ma	El

4   Simulation Results

The simulations were conducted using FuzzyC and the results are shown for three scenarios. Figure 4, Fig. 5, and Fig. 6 show the results for low, medium and high link latency, respectively. We show the relation between QoS and UIVP for different ER values while keeping LL and RI values constant.

Figure 4(a), shows the scenario when a low communication latency is supported for data transmission and the interference caused by transmissions of

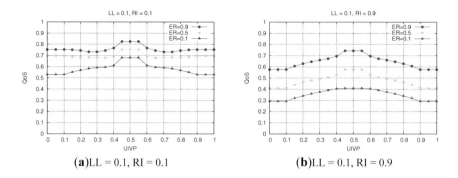

(**a**)LL = 0.1, RI = 0.1   (**b**)LL = 0.1, RI = 0.9

Fig. 4. Simulation results for LL = 0.1.

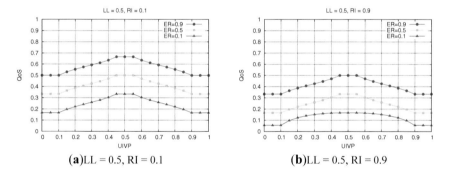

(**a**)LL = 0.5, RI = 0.1   (**b**)LL = 0.5, RI = 0.9

Fig. 5. Simulation results for LL = 0.5.

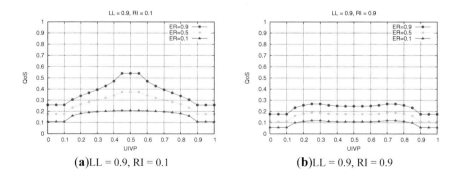

(**a**)LL = 0.9, RI = 0.1   (**b**)LL = 0.9, RI = 0.9

Fig. 6. Simulation results for LL = 0.9.

other vehicles is permissible. Due to this, we see that all V2V communication links provide very high QoS, despite the reliability values. However, harmful interference will degrade the links' performance, as seen in Fig. 4(b). In the case of high RI and not effective ER, the communication will lead to loss of information, which is indicated by the low value of QoS. With the increase of LL, the link performance will deteriorate even more, as seen in Fig. 5. The results for high LL are shown in Fig. 6. A high latency means that more time is needed for the information to be transmitted between two vehicles. In Fig. 6(a), only the links with high ER and few UIVP are considered as capable of offering high performance. Whereas, when a harmul RI is present (see Fig. 6(b)), none of links is considered successful.

Throughout the scenarios, we see that QoS reaches peak values when the UIVP values are moderate. The performance of the link communication does not increase when too many UIVP packets are transmitted because more packets occupy more bandwidth in the network.

5   Conclusions

In this paper, we proposed a fuzzy-based system to assess QoS of V2V link communication in a layered Cloud-Fog-Edge architecture for SDN-VANETs. Our proposed system decides appropriate communication links between vehicles in order to exchange data packets successfully, and help a *vehicle* that needs additional resources to accomplish different tasks. The proposed FL-based system is implemented in the SDN module and decides the QoS of link connections based on LL, RI, ER, and UIVP. After assessing the performance of each V2V link, our previous proposed FS-ANVPC system [6] evaluates the processing capability of edge layer in SDN-VANETs and decides whether the edge layer is appropriate for data processing. We evaluated our proposed system by computer simulations. From the simulations results, we conclude as follows.

- The best V2V link performance is reached when LL is low, RI is Permissible, ER of packet delivery is high, and the amount of beacon messages transmitted in the network is moderate.
- In all scenarios, UIVP gives the best performance when it has a moderate value.
- The increase of LL and RI values, deteriorates the QoS of data links in the edge layer.

In the future, we would like to make extensive simulations to evaluate the proposed system and compare the performance with other systems.

References

1. Kandel, A.: Fuzzy Expert Systems. CRC Press, Inc., Boca Raton (1992)
2. Klir, G.J., Folger, T.A.: Fuzzy Sets, Uncertainty, and Information. Prentice Hall, Upper Saddle River (1988)

3. Ku, I., Lu, Y., Gerla, M., Gomes, R.L., Ongaro, F., Cerqueira, E.: Towards software-defined VANET: architecture and services. In: 13th Annual Mediterranean Ad Hoc Networking Workshop (MED-HOC-NET), pp. 103–110 (2014)
4. McNeill, F.M., Thro, E.: Fuzzy Logic: A Practical Approach. Academic Press Professional, Inc., San Diego (1994)
5. Munakata, T., Jani, Y.: Fuzzy systems: an overview. Commun. ACM **37**(3), 69–77 (1994)
6. Qafzezi, E., Bylykbashi, K., Ampririt, P., Ikeda, M., Matsuo, K., Barolli, L.: A QoS-aware fuzzy-based system for assessment of edge computing resources in SDN-VANETs. In: Barolli, L., Woungang, I., Enokido, T. (eds.) AINA 2021. LNNS, vol. 225, pp. 63–72. Springer, Cham (2021). https://doi.org/10.1007/978-3-030-75100-5_6
7. Qafzezi, E., Bylykbashi, K., Ikeda, M., Matsuo, K., Barolli, L.: Coordination and management of cloud, fog and edge resources in SDN-VANETs using fuzzy logic: a comparison study for two fuzzy-based systems. Internet of Things **11**, 100169 (2020)
8. Truong, N.B., Lee, G.M., Ghamri-Doudane, Y.: Software defined networking-based vehicular adhoc network with fog computing. In: 2015 IFIP/IEEE International Symposium on Integrated Network Management (IM), pp. 1202–1207 (2015)
9. Xu, W., et al.: Internet of vehicles in big data era. IEEE/CAA J. Autom. Sinica **5**(1), 19–35 (2018)
10. Zadeh, L.A., Kacprzyk, J.: Fuzzy Logic for the Management of Uncertainty. Wiley, New York (1992)
11. Zimmermann, H.J.: Fuzzy control. In: Zimmermann, H.J. (ed.) Fuzzy Set Theory and Its Applications, pp. 203–240. Springer, Dordrecht (1996). https://doi.org/10.1007/978-94-015-8702-0_11

Reliable and Low-Cost Digital Transformation Technology Using Progressive Web Apps in Fog Computing Architecture for Small and Medium Industries in Indonesia

Zulkifli Tahir[✉], Amil Ahmad Ilham, Ais Prayogi Alimuddin,
Muhammad Zulfadly A. Suyuti, and Charina

Department of Informatics, Hasanuddin University, Makassar, Indonesia
{zulkifli,amil,aisprayogi}@unhas.ac.id, {suyutimza19d,
charina15d}@student.unhas.ac.id

Abstract. In the Industry 4.0 era, industries require connectivity and interaction through digital transformation. It has become a concern in Indonesia, especially in the Small and Medium Industries (SMIs) sector. However, the implementation of the digital system in Indonesian SMIs is still relatively low. Most problems are Internet access and the amount of capital owned by the SMIs. Therefore, this study applies an innovative web solution using a Progressive Web Apps (PWA) that runs on a Fog Computing architecture. The system can work quickly and reliably with the capability to manage resources offline by applying PWA integrated with Service Worker, and performs the computing process at the end of the network closest to the activities of SMIs by using the low-cost middleware minicomputers in Fog Computing architecture. The integrated system innovation can become one of the best solutions for implementing digital transformation to support the sustainability of SMIs in Indonesia.

1 Introduction

The main activity of the fourth industrial revolution is digital transformation. The industries require connectivity and interaction through integrated digital Information and Communication Technology (ICT) that can be utilized throughout the manufacturing chain to achieve efficiency and improve product quality. The revolution will reshape fundamental processes and ways of working in Industry [1]. Many modern technologies are applied to bring about a revolution by connecting all components in industries. Indonesia has made a roadmap to realize the industrial era 4.0, which will then produce Indonesia 4.0 [2]. The sector that plays the most role for economic growth in developing countries including Indonesia is the Small and Medium Industries (SMIs) [3]. Indonesia has more than 62.9 million of SMIs or equivalent to 99.99% of the total types of Industrial units. However, the contribution from the micro sector only reached 30.1%, still less than the large sector which reached 42.9% [4]. The quick implementation of digital transformation is expected to increase the effectiveness of the industry which will indirectly improve the economy.

L. Barolli et al. (Eds.): EIDWT 2022, LNDECT 118, pp. 163–174, 2022.
https://doi.org/10.1007/978-3-030-95903-6_18

The availability of fast and reliable technology through internet services and web technology is crucial in industrial activities [4]. It is estimated that around 25.2 billion types of products will be connected to the 5.8 billion human population via the Internet by 2025 [5]. Based on an announcement from The Ministry of Cooperatives and SMIs in Indonesia, around 3.79 million SMIs have used online applications to market their products [6]. However, when compared to the number of SMIs, this value is still measly. This condition shows that there is still a gap from SMIs in Indonesia to the desire to use the online system. Some of the main problems are inadequate Internet access in several regions in Indonesia [7], and the small capital of SMIs so that they rarely invest in technology implementation [8]. On the other hand, new technology is needed that can provide more affordable, faster, and better service capabilities [9].

In our previous research, web technology has been implemented for SMIs which has a standard of reliability with faster performance than conventional web technology and can also work even in conditions of slow Internet or no access at all [10]. The current study aims to develop the implementation of the web technology on a network architecture that is more suitable for SMIs, which in this case uses the Fog Computing architecture. The reason is the industries are mostly run locally process, so if we connect to server systems such as cloud systems or web services via the Internet, it will result in high latency. It can also add additional tasks such as security consideration and complex data configuration [11]. Fog Computing is expected to be suitable because it has a middleware system that provides processes close to industrial activities. The middleware can be adapted to inexpensive devices to run web server processes that act like micro datacenters. Such tools are likely to be implemented in SMIs, most of which have limited capital.

In this study, we analyzed the use of minicomputer as middleware in the Fog Computing architecture. The type of minicomputer analyzed is the Raspberry Pi. The purpose of this analysis is to determine whether this device is suitable and can be applied in SMIs compared to more expensive devices such as computers that are generally used as local web servers. The minicomputer will act as a web server that runs the same system like a computer server device as a web server. The system uses the concept of a Progressive Web Apps (PWA) that integrates Service Worker, which is the result of our previous research [10]. This web system is designed so that it has faster than conventional web systems, and is considered reliable in terms of availability because it can work both online and offline. There are several important parameters that become the benchmark for analysis, namely the average response time, processing time in the browser, and the percentage of Central Processing Unit (CPU) utility. Both types of servers will receive data from the client computers, process the data, then give the appropriate responses such as saving the data to a database or returning the responses data to the client computers.

2   Digital Transformation Technology for SMIs

2.1   Small and Medium Industries (SMIs)

The definition of industry based on the Law of the Republic of Indonesia is distinguished based on the criteria for net worth and annual sales results. Based on net worth, the criteria for the small industries are from 50 million to 500 million rupiahs and medium industries from 500 million to 10 billion rupiahs. Then based on annual sales results, small industries are from 300 million to 2.5 billion rupiahs, and medium industries are from 2.5 billion to 50 billion rupiahs [12]. Based on Indonesia's central statistical agency, Industries are grouped based on the number of workers, small industries between 5 to 19 workers, and medium industries ranging from 20 to 99 workers [13]. SMIs have several characteristics, i.e., most of them are individually owned, the capital owned is relatively small, the financial corporation comes from the owner, and transactions are usually direct to the owner [14].

The Indonesian government strongly supports SMIs as one of the most important factors in economic growth [15]. The factor is evidenced by as much as 20.51% of Gross Domestic Product (GDP) comes from industrial production [16]. Synergy is needed between government, industrial entrepreneurs and researchers in developing technological innovation to build appropriate systems in SMIs. The National Economic and Industrial Committee of Indonesia has drawn up a road map for industrial development until 2045. The road map focuses on industrial development from 2007 to 2025, focuses on expansion from 2026–2035, and accelerates industry from 2036–2045 [17]. This road map must be supported by all parties with their respective roles.

2.2 Digital Transformation

Transformation is important to get a better response and improve the performance of the business. In industry 4.0, digital technology becomes dominant where the industries must carry out the digital transformation of all activities. Digital transformation mostly uses technological innovations and new business models that are applied to all aspects [18]. It most uses both software and hardware technologies, including the use of ICT and embedded systems, which cover many fields, including data analysis, cloud services, machine learning, artificial intelligence, microdevices with sensors and actuators, and so on [19]. By using these technologies, almost all industrial activities can be carried out quickly and accurately, including transactions, business management, profit and loss reports, or even monitoring of machine work processes, and much more [20]. In a pandemic situation, digital transformation is the most effective way for SMIs to survive. Covid-19 does not have to be a barrier for SMIs to increase their activities, because they can continue to run their business through a digital system, and most likely will continue to use it in the future [21]. SMIs that implement digital transformation have the same goals, those are improving the quality of products and services [22].

Based on a Google Temasek report, Indonesia has a very fast growth in digital transformation. It was recorded that in 2020 there were 40 million new digital users. This shows a high percentage increase reaching 37% above the average in Southeast Asia. This is also influenced by the pandemic situation which applies lockdown rules so that many new people are forced to find solutions to make it easier to get the necessities of life. This rapid increase through digital transformation is seen by the increase in Gross Merchandise Value (GMV) growth in the online system sector by 54%. Overall, it is predicted that by 2025 the GMV in Indonesia will reach 23% to US$ 124 billion [23].

2.3 Progressive Web Apps (PWA)

The new web concept which is a collection of several existing technology features with capabilities such as other installed applications including being able to be added to the home screen and can be used offline, but running on browsers and being cross-platform, are some of the introductions of PWA [24]. Technically, PWA run with a manifest, an Application Shell, and a Service Worker. Manifest is JSON-based metadata that will cache in the background system [25]. The Application Shell is a Graphical User Interface (GUI) and programming logic that reads the app cache and is executed when the system is offline. This application shell is rendered and directly reads the app cache so that it can make access times faster than non-PWA websites [26]. Then the one that handles network proxies, caches the logic, and makes PWAs run in an offline state is the task of the Service Worker [27].

Our previous research has implemented an offline base web system as a machine maintenance information system for SMIs by using several features, namely React JS, CouchDB, PouchDB, RxDB, RxJS, indexedDB [10]. The system has been proven to work quickly and reliably, but it still runs on a common server computer which is quite difficult to implement in SMIs due to price issues. The current research uses PWA-based React JS and is implemented on inexpensive Raspberry Pi as a micro server. Raspberry Pi is widely used as an edge computing device in Fog Computing architecture.

2.4 Fog Computing Architecture

There are many paradigms to Fog Computing related to edge computing, including multi-access edge computing (MEC), mist computing, and cloudlets, and how they can work together with Cloud Computing and Software-Defined Networking (SDN). Fog Computing can be interpreted as the form of computing systems, storage, networking, and data management on nodes that are close to users or devices, which is not only processed on the cloud system [28]. The process of computing and storage can be carried out at the end of the network through microdata centers [29]. Because of the proximity to users or devices, the latency of edge computing can be lower than directly to the cloud system. Based on research from 1000 popular websites, edge computing can reduce response time based on locations around the world, except in North America and Europe because of adequate network technology [30].

The price of infrastructure for the implementation of Fog Computing is also relatively cheaper. Based on the researcher's evaluation, sellers of edge device infrastructure devices can earn 5–15% more revenue, and buyers can pay at least 20% less [31]. The formula for calculating price estimates for fog services based on CPU, memory, storage, and bandwidth has been described by research in [32]. Moreover, to calculate the number of utilities in the fog nodes based on the number of workloads, we can look at the analytical model by researchers in [33] and metrics by researchers in [34]. Based on the advantages described, of course, industrial systems can get many benefits when using the appropriate Fog Computing architecture, especially in SMIs.

3   PWA in Fog Computing Architecture

The system created is based on an interview with one of the SMI in Indonesia. The SMI is still carrying out the manual recording of financial income and expenditure and does not yet have an information system. This system is adapted to the needs of SMI, although it is not too complex, it is sufficient to run business processes on the SMI. This system is also tested by connecting one type of industrial machine. The machine is a type of robotic arm that is monitored by several sensors including camera, vibration, rotation, sound, motion sensors, and so on. These sensors are used to monitor work errors or damage to the machine. The goal is the machines in the industry can be handled quickly so that it does not hamper the production process and reduce SMIs losses.

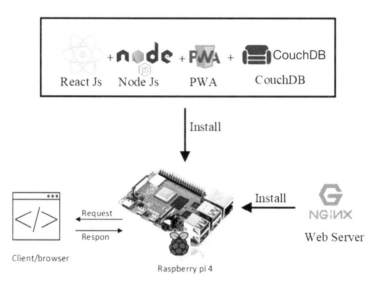

Fig. 1. Progressive Web Apps in Edge Device of Raspberry Pi.

An overview of the technology used is shown in Fig. 1. The technology is built on a Raspberry Pi 4 device, with a specification of a 1.5GHz 64-bit quad-core processor, up to 4 GB of RAM. React JS, Node JS, PWA, and CouchDB are installed to build the system. Then the webserver uses Nginx technology. This web technology uses an offline-first app model that utilizes Service Worker. The system was developed with React JS library, CouchDB, Local Storage, and IndexedDB. CouchDB serves to manage databases when online, and IndexedDB to be accessed locally when offline. Both databases will synchronize data when there is an Internet connection. CouchDB is supported by browsers to replicate data to offline devices and handle data synchronization when devices are online again. Because the information system for SMI is not too large and complex, the NoSQL database with local storage is suitable to be implemented.

The workflow of the system is shown in Fig. 2. The system will first send a request. Then the Service Worker running on the client-side will check the status of the online or offline connection. If the connection is online, the request will be forwarded to the

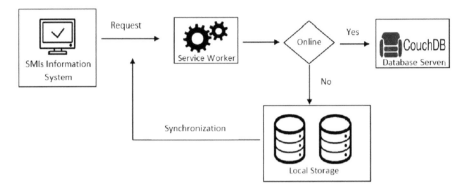

Fig. 2. The system workflow.

system and database on the server-side. After processing, the response is returned to the appropriate client browser. The database server synchronization will also be done automatically with the local system and database. If the connection is offline, the system will provide a client response via data stored on the local system and database.

The installation of the entire system on the Fog Computing architecture can be seen in Fig. 3. The architecture is divided into several layers that have different functions.

Layer 1 is the layer closest to users and machines. Users can interact directly with the devices they use via the local network or the Internet network. While the machines can interact through the machine monitoring system using sensors that can send the desired data to the web server system on the Raspberry Pi or other computing tools.

Layer 2 consists of edge devices which are usually mini computers. This is the layer where the webserver is run and the PWA system is installed. The mini-computer can also be used as a place for computing the data sent by the sensors. The mini-computer

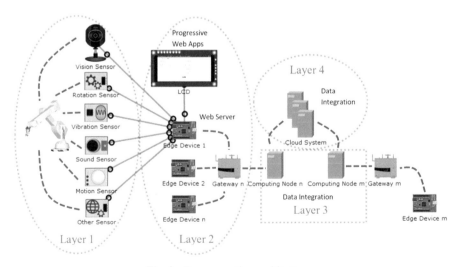

Fig. 3. Fog computing architecture.

can be considered to represent a system in SMI. Several systems can be connected at this layer, depending on the business processes of each SMI.

Layer 3 is a layer computing node that connects devices at layer 2. This third layer can be applied later to a larger level of information systems, for example for local governments that monitor the development of SMIs processes in their area.

Layer 4 is a centralized data center that receives large computing from all devices at layer 3. This central server runs mostly on Cloud Computing. Data from all SMIs in a large area can be obtained for greater use as well. The data can also be used by the government as an effort to improve services to SMIs. Data analysis technologies such as data science, machine learning, data mining, big data, and so on are usually processed at this layer.

To process the PWA-based information system that was created, SMI is sufficient to provide devices at the 1st and 2nd layers. These layers are similar to the commonly used client-server systems. Development for layers 3 and 4 can also be implemented if needed, but we still need to consider the issue of cost. Here, cooperation with the local government should be carried out to produce an integrated system for the benefit of economic development in the managed area.

4   Testing and Analysis

The design of the test scenario is carried out to find out and get performance results from the SMI Information System web server application with Progressive Web Apps (PWA) technology which is applied to the Raspberry Pi platform. The first test of the PWA concept used the Lighthouse extension on Google Chrome, to test the PWA concept with aspects of performance, accessibility, best practices, SEO, and PWA. Then the second test calculates the response time using the Apache Jmeter application, and by paying attention to the percentage of use of the Central Processing Unit (CPU) based on the addition of the number of threads representing the number of users accessing the server. The second test analysis also shows the browser processing time when the system is accessed for the first time compared to the second, third, and so on of the access times with differences in the availability of cache in the browser.

4.1   PWA Test and Analysis

Fig. 4. Lighthouse test without Service Worker.

In this test, two scenarios were carried out. The first scenario used the "Clear Cache" mode with the intention that all requests made by the application were sent to the server without the role of a Service Worker. While the second scenario will use preserve cache which utilizes application storage on Service Worker that can manage every request that must be sent to the server or cache local storage. In the first test, the performance index value obtained is 89 with a total page access time of 6 s, the accessibility index value is 100, the best practices index value is 92 and the SEO index value is 100. Figure 4 below is the results of the assessment using Google Lighthouse without Service Worker. There is one criterion that is orange, but it has reached a value of 89, which means that the total page access time performance is also almost in a green category.

Fig. 5. Lighthouse test with Service Worker.

Furthermore, in the second test, there was a significant increase in the performance index value which increased to 100 with a total value of 0.5 s. For other criteria, the index value remains the same, namely the accessibility criteria with an average value of 100, the best practices criteria with a score of 92, and the SEO criteria with a value of 100 as shown in Fig. 5. From the two test scenarios, it can be seen that using a Service Worker to manage every request that must be sent to the server or cache storage, it allows faster performance than without a Service Worker.

4.2   Response Time Test and Analysis

These tests are recorded to represent the online system situation. There are two parameters tested using Apache Jmeter. The first is the response time test, which means calculating the value of the time interval when the request is made until the response is received by the user who made the first request to the webserver. The second test also shows the percentage of CPU usage on the server-side. Testing the response time and CPU percentage is recorded by giving one to 500 requests at once from the client to the server, with each experiment ten times. After that, the average response time is calculated in milliseconds and the percentage change before and after requests CPU usage is calculated in percentage interval before and after requests. In addition to using the Raspberry Pi, it also shows the calculation results from the server computing device for comparison.

From the test results in Table 1, it can be seen that with the addition of the number of threads, all the parameters tested also experienced an increase. On the Raspberry Pi server device, the average response time of less than 1 s is up to 200 threads. This

Table 1. Test results.

No	Number of threads	Average Rasp Pi response time	Average computer response time	Average Rasp Pi CPU	Average computer CPU
1	1	289	54	1	0.7
2	50	562	64	6.5	2.8
3	100	660	139	12.5	5.5
4	150	700	261	14.5	6.5
5	200	998	282	16	7.7
6	250	1295	342	18.2	9.6
7	300	1469	408	19.6	10
8	350	1725	422	23.4	13.3
9	400	1811	452	26.8	13.8
10	450	2002	513	30.9	16.5
11	500	2448	662	31.8	20.6

shows that the performance of the device is very fast with the number of users under approximately 200. At that time, the increase in the percentage of performance on the server CPU was only about 16%. In experiments up to 500 threads, the response time is not too fast with a delay of 2,448 s. The addition of CPU performance also reached 31.8%, and from testing, it was found that several packet losses indicated that the Raspberry Pi server was working almost at its maximum. Indeed, the comparison with ordinary server computers still looks very fast even though up to 500 users access simultaneously. But it is quite rare for a system in SMI to be accessed by more than 100 concurrent users. So that the use of the Raspberry Pi to be used as a server on SMI can be considered still very fast and suitable for implementation. Plus, in this research situation, the price of a computer device that matches the specifications used can reach 3–5 times the price of a Raspberry Pi device.

The test in Table 1 is a case scenario when the user directly requests the server, where there is no cache role in the client browser. With the help of the process from the client browser to display the system, the response time to the user will be faster. The experiments for this situation are recorded on is shown in Fig. 6. The time used for the web page to successfully retrieve the existing data as a whole from the server to the browser then is also displayed in the record as the browser average response time. A rather high browser average response time is only obtained when the web page is accessed for the first time, because at that time in addition to responding to the user, the system also sends data to be cached on the browser on the client-side to speed up further response times. It is proven that when accessed the second time, the third time, and so on, it shows a fast response time that is under 1 s. Based on the test results, it can be concluded that the system is considered to be able to meet the needs of SMIs.

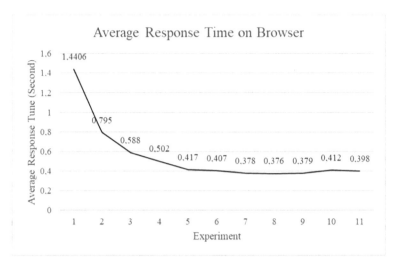

Fig. 6. Average Access Time on Browser.

5   Conclusion

This research has succeeded in developing a web system with PWA technology that is run on edge devices from the Fog Computing architecture using Raspberry Pi. This system is proven to be reliable for users in SMIs, which can process up to 200 users simultaneously in an online state and can still work reliably in an offline state. By implementing using the integration of these technologies it is much cheaper when compared to using the typical servers. Based on the results of the analysis, this technology is very suitable to be implemented for SMIs with the advantages of reliability and lower prices, not only for Indonesia, but also for all developing countries.

Acknowledgments. All authors would like to thank for Institute for Research and Community Service, Hasanuddin University for assistance with the research and funding process.

References

1. Ghobakhloo, M.: Industry 4.0, digitization, and opportunities for sustainability. J. Clean. Prod. **252**, 119869 (2020)
2. Hidayatno, A., Destyanto, A.R., Hulu, C.A.: Industry 4.0 technology implementation impact to industrial sustainable energy in Indonesia: a model conceptualization. Energy Proc. **156**, 227–233 (2019)
3. Arshad, M., Arshad, D.: Internal capabilities and SMEs performance: a case of textile industry in Pakistan. Manage. Sci. Lett. **9**(4), 621–628 (2019)
4. Budhi, M., Lestari, N., Suasih, N., Wijaya, P.: Strategies and policies for developing SMEs based on creative economy. Manage. Sci. Lett. **10**(10), 2301–2310 (2020)
5. Stryjak, J., Sivakumaran, M.: The Mobile Economy 2019. GSM Association (2019)
6. Ministry of Communication and Information of Republic of Indonesia: 3.79 Million SMEs Go Online (3,79 Juta UKM Sudah Go Online) (2017). www.kominfo.go.id

7. Batubara, B.M.: The problems of the world of education in the middle of the Covid-19 pandemic. Budapest Int. Res. Crit. Inst. (BIRCI-J.) Human. Soc. Sci. **4**(1), 450–457 (2021)
8. Tunggal, H.P., Joesron, T.S.: Technical efficiency analysis of Indonesian small and micro industries: a stochastic frontier. Approach. No. 201903. Department of Economics, Padjadjaran University (2019)
9. Putra, N.E.: The development of the digital economy in Indonesia in terms of trends, opportunities, and challenges (Perkembangan Ekonomi Digital di Indonesia dari Segi Tren, Peluang, dan Tantangan) (2018). http://www.feb.ui.ac.id/blog/2018/11/23/perkembangan-ekonomi-digital-di-indonesia-dari-segitren-peluang-dan-tantangan/
10. Tahir, Z., Dasmito, A.R., Niswar, M.: A reliable offline web system for small and medium industries. In: MATEC Web of Conferences, vol. 331. EDP Sciences (2020)
11. Aazam, M., Zeadally, S., Harras, K.A.: Deploying fog computing in industrial internet of things and industry 4.0. IEEE Trans. Industr. Inf. **14**(10), 4674–4682 (2018)
12. Suci, Y.R.: The development of MSMEs (micro, small and medium enterprises) in Indonesia. (Perkembangan UMKM (Usaha mikro kecil dan menengah) di Indonesia). J. Ilmiah Cano Ekonomos **6**(1), 51–58 (2017)
13. Adam, M., Ibrahim, M., Ikramuddin, I., Syahputra, H.: The role of digital marketing platforms on supply chain management for customer satisfaction and loyalty in small and medium enterprises (SMEs) at Indonesia. Int. J. Supply Chain Manage. **9**(3), 1210–1220 (2020)
14. Nasution, D.P., Lubis, I.: The development of demand for small and medium industries in Indonesia. Development **4**(10) (2019)
15. Febrian, A.F., Maulina, E., Purnomo, M.: The influence of social capital and financial capability on sustainable competitive advantage through entrepreneurial orientation: empirical evidence from small and medium industries in Indonesia using PLS-SEM. Adv. Soc. Sci. Res. J. **5**(12), (2018)
16. Zagloel, T.Y.M., Ardi, R., Poncotoyo, W.: Six sigma implementation model based on critical success factors (CSFs) for Indonesian small and medium industries. In: MATEC Web of Conferences, vol. 218, p. 04017. EDP Sciences (2018)
17. Tosida, E.T., Andria, F., Wahyudin, I., Jatna, T.: Development strategy for telematics small and medium industries in Indonesia. Jurnal Penelitian Pos dan Informatika **9**(1), 37–52 (2019)
18. Mihardjo, L.W.W., Rukmana, R.A.N.: Does digital leadership impact directly or indirectly on dynamic capability: Case on Indonesia telecommunication industry in digital transformation? J. Soc. Sci. Res., 832–841 (2018)
19. Ebert, C., Duarte, C.H.C.: Digital Transformation. IEEE Softw. **35**(4), 16–21 (2018)
20. Rahmatullah, R., Inanna, I., Sahade, S., Nurdiana, N., Azis, F., Bahri, B.: Utilization of digital technology for management effectiveness micro small and medium enterprises. Int. J. Sci. Technol. Res. **9**(04), 1357–1362 (2020)
21. Indriastuti, M., Fuad, K.: Impact of covid-19 on digital transformation and sustainability in small and medium enterprises (SMES): a conceptual framework. In: Conference on Complex, Intelligent, and Software Intensive Systems, pp. 471–476. Springer, Cham (2020). https://doi.org/10.1007/978-3-030-50454-0_48
22. Fachrunnisa, O., Adhiatma, A., Lukman, N., Majid, M.N.A.: Towards SMEs' digital transformation: the role of agile leadership and strategic flexibility. J. Small Bus. Strateg. **30**(3), 65–85 (2020)
23. Google, Temasek and Bain & Company: e-Conomy SEA 2020 report. Google e-Conomy (2020). https://economysea.withgoogle.com
24. Majchrzak, T.A., Biørn-Hansen, A., Grønli, T.: Progressive web apps: the definite approach to cross-platform development? In: Proceedings of the 51st Hawaii International Conference on System Sciences (2018)
25. LePage, P., Steiner, T., Beaufort, F.: Add a web app manifest (2018). https://web.dev/add-manifest/

26. Osmani, A.: The app shell model (2019). https://developers.google.com/web/fundamentals/architecture/app-shell
27. Gaunt, M.: Service workers: an introduction (2021). https://developers.google.com/web/fundamentals/primers/service-workers
28. Yousefpour, A., et al.: All one needs to know about fog computing and related edge computing paradigms: a complete survey. J. Syst. Architect. **98**, 289–330 (2019)
29. OpenEdgeConsortium: About - the who, what, and how. http://openedgecomputing.org/.Technical Report, OpenEdge Computing
30. Le Tan, C.N., Klein, C., Elmroth, E.: Location-aware load prediction in edge data centers. In: 2017 Second International Conference on Fog and Mobile Edge Computing (FMEC), pp. 25–31. IEEE (2017)
31. Prasad, A.S., Arumaithurai, M., Koll, D. and Fu, X.: Raera: a robust auctioning approach for edge resource allocation. In: Proceedings of the Workshop on Mobile Edge Communications, pp. 49–54 (2017)
32. Giang, N.K., Blackstock, M., Lea, R., Leung, V.C.: Developing IoT applications in the fog: A distributed dataflow approach. In: 2015 5th International Conference on the Internet of Things (IOT), pp. 155–162, IEEE (2015)
33. El Kafhali, S., Salah, K.: Efficient and dynamic scaling of fog nodes for IoT devices. J. Supercomput. **73**(12), 5261–5284 (2017). https://doi.org/10.1007/s11227-017-2083-x
34. Malandrino, F., Kirkpatrick, S., Chiasserini, C.F.: How close to the edge? delay/utilization trends in MEC. In: Proceedings of the 2016 ACM Workshop on Cloud-Assisted Networking, pp. 37–42 (2016)

A Low-Cost Solution for Smart-City Based on Public Bus Transportation System Using Opportunistic IoT

Evjola Spaho[1(✉)] and Andrea Koroveshi[2]

[1] Department of Electronics and Telecommunication, Faculty of Information Technology, Polytechnic University of Tirana, Mother Teresa Square, No. 4, Tirana, Albania
[2] Faculty of Information Technology, Polytechnic University of Tirana, Tirana, Albania
andrea.koroveshi@fti.edu.al

Abstract. Internet of Things (IoT) is a heterogeneous network of interconnected things where different devices and technologies collaborate to offer services. Recently, smart city projects are gaining a lot of attention from the governments. Different applications based on IoT devices are used to improve the services for citizens. Opportunistic Networks (OppNets) are a category of Disruption Tolerant Networks (DTNs) characterized by intermittent connectivity between the nodes. Opportunistic IoT (OppIoT) network is a blend of smart devices and human mobility. Smart devices equipped with interfaces and buffers (pedestrian smart phones, buses on-board units) using the human mobility will enable the communication and send sensors data to the gateways. Gateways will be connected to Internet with cloud servers. The data received from the sensors and than send to the gateways should be for non time-critical applications. In this work, we propose the usage of OppIoT where bus transportation system, their mobility, pedestrians mobility create opportunistic contacts to build a smart city with a low cost. In order to reduce the cost, devices will use opportunistic communication and not the subscription cost to connect to the Internet and only the gateways will be connected. So, the optimization of the number of gateway nodes and their placement is needed. We conder a case study of Tirana city, Albania. Based on Geo-referenced public transportation data published from the municipality of Tirana city, we evaluated the minimal number of gateways and their position using different algorithms. The results shows that the proposed algorithm for minimal number of gateways performs better and three gateways are needed to offer coverage for all bus lines in Tirana.

1 Introduction

Smart cities [1] are proposed to facilitate citizens needs and improve their quality of life. The main goal of a smart city is to manage better the city infrastructures and resources, and to offer better services to the citizens.

L. Barolli et al. (Eds.): EIDWT 2022, LNDECT 118, pp. 175–182, 2022.
https://doi.org/10.1007/978-3-030-95903-6_19

Many IoT applications generate delay tolerant content [2–6]. For example, IoT equipment deployed at a remote experimental site, collecting data for statistical purposes. In such applications, the data can be stored for a period of time before being transferred to the place where they will be analyzed and processed.

Since the number of sensors for different applications in smart cities [7] for example: environmental monitoring, network monitoring [8], traffic monitoring and control [9], parking, water quality, waste collection and management [10], street lighting, noise monitoring and air quality etc. is huge, cellular networks with high subscription fees are an expensive solution for data delivery.

A solution to reduce the cost of implementation is to use OppIoT with opportunistic communication between buses of transportation systems using mobility and usage of a small number of gateways that will be connected with Internet.

Smart cities need smart transportation systems. Intelligent Transportation System can help managing traffic and mobility in smart cities by using different communication technologies and minimizing pollution, increasing security, comfort and safety of drivers and travelers.

Public bus transportation system and its mobility using opportunistic communication can be used to deliver the IoT sensor data to the gateways connected to the Cloud servers where data will be processed. Collecting and processing data can help in creating new strategies to improve the quality of life.

Optimizing the number of gateways and their positions will help to reduce the implementation and maintenance cost.

In this paper, we propose the usage of bus transportation system, their mobility, pedestrians mobility and OppIoT with opportunistic contacts to build a smart city with a low cost and use different algorithms to choose the right number of gateways and their placement. We consider a case study of Tirana city, Albania. Real data about time-tables of the buses of public transportation system in Tirana city are used to evaluate different algorithms and their performance.

The rest of the paper is organized as follows: In Sect. 2 are presented different algorithms for gateways placement in a smart-city. In Sect. 3, we present the proposed algorithm. Section 4 presents evaluation scenarios and results. Section 5 concludes the paper and suggests some insights for the future work.

2 Gateway Placement Algorithms

The main elements of a low-cost smart city are explained in the following. Sensors nodes sense data from the environment. Buses and bus lines move in predefined routes, have stopping stations, timetables and that do the movement in a periodic way. Buses are equipped with very large buffers to store the data until the gateway is contacted. Pedestrians equipped with smart phones or smart devices will receive the data from the sensors then buffer for a while until it will encounter a bus and forward to it. Gateways are the final destination of the data received from the sensors. Since they will play a key role to send the data to the cloud, they are fixed devices placed near continuous power. These elements and their mobility using opportunistic communication will deliver IoT data to the gateways.

In this work, we address the problem of choosing a minimal number of gateways to be used for reducing the cost and find their position for covering all bus lines of Tirana city. The most significant bus stops should be found and the gateways will be positioned in these stops. The idea is to find a good position of gateways in order to have a direct path from each stop to the gateway. In this way all data received from sensors will be transmitted to the gateway using public transportation. There are different algorithms to choose the gateways placement in order to cover all the bus lines. In the following is a short explanation of two algorithms.

Degree centrality is a simple algorithm that counts the total number of immediate connections a given node has in the network. For a node, the degree centrality is the number of edges it has. The node is more central if it has the higher number of edges.

Betweenness centrality is an algorithm used to detect the amount of influence a node has in a graph. This algorithm calculates unweighted shortest path between all pairs of nodes in the graph and each node receives a score based on the number of shortest paths that pass through it. Nodes that lie more frequently on the shortest paths will have a higher betweenness centrality score.

3   Proposed Algorithm

The proposed algorithm, for minimal number of gateways, considers the set of lines R and the set of stops L. We need to find the minimal number of stops that cover all bus lines. Each line Ri is a subgraph of set R. The algorithm starts with analyzing the first and the second line by finding the common stops and creating the subset of Li stops. This process is repeated for all 14 lines. In this way, we can find the most important stop that covers most of the bus lines and the first gateway will be located there. For the lines that do not have any common stops, a new gateway is needed. When there are lines not covered from the first gateway, a similar process of finding the common stops and decide where to place the other gateway is repeated for the remaining lines. The flowchart of the proposed algorithm is shown in Fig. 1.

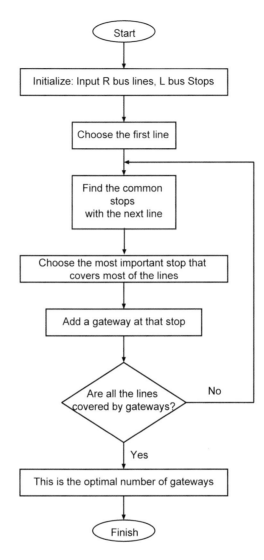

Fig. 1. Flowchart of the proposed algorithm.

4   Evaluation Scenarios and Results

Tirana city area is 41.8 square kilometers. The urban road network includes 482 roads with a total length of 1,101 km. Tirana public bus transportation system consists of 14 lines and 451 bus stops and the distance between stops is from 250 to 400 m. Real data for the bus lines of the public transportation system in Tirana city are received from Open data [11]. These data includes: bus lines itineraries, time-tables and stops for the third trimester of 2020. We used GeoJSON data (an open standard format for encoding a variety of geographic data structures) published by Tirana municipality Open data for

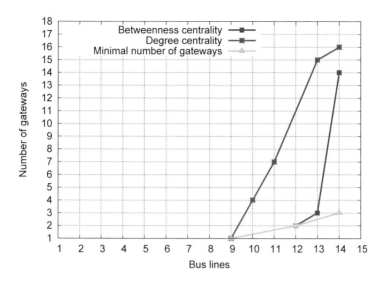

Fig. 2. Results for number of gateways vs. number of lines covered for three algorithms.

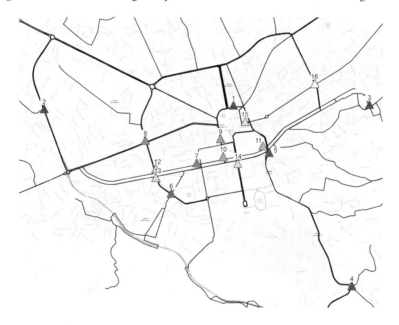

Fig. 3. Gateway placement using degree centrality algorithm.

public transportation. The analysis is done by using GIS (Geographic Information System) softwares and Geo-referenced data and also Global Mapper. We used Geographic Resources Analysis Support System (GRASS GIS), a geographic information system

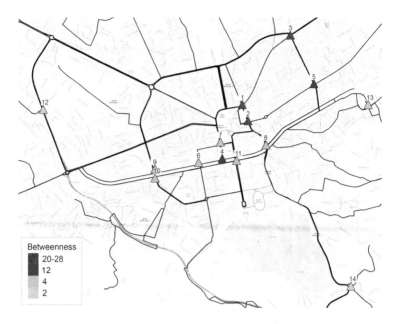

Fig. 4. Gateway placement using betweenness centrality algorithm.

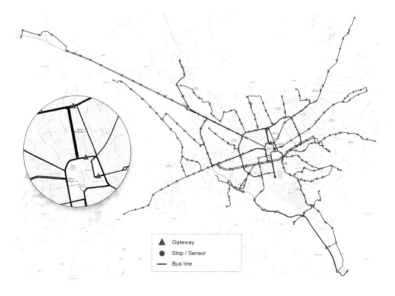

Fig. 5. Gateway placement using proposed minimal number of gateways algorithm.

software suite used for geo-spatial data management and analysis, image processing, producing graphics and maps, spatial and temporal modeling, and visualizing.

In Fig. 2 are presented the results of number of gateways vs. number of covered bus lines for three algorithms explained in the previous section. From the results, the proposed algorithm for minimal number of gateways performs better compared with Betweenness centrality and Degree centrality algorithm, because only 3 gateways can cover all bus lines. While, for Betweenness centrality algorithm and Degree centrality algorithm are needed 14 and 16 gateways, respectively to cover all the bus lines.

Figure 3 shows 16 gateways located across bus stops when Degree centrality algorithm is used. From the ranking of stops according to their centrality value for each line, it results that 14 gateways are required to cover all public transport lines in Tirana.

Figure 4 shows the location of the gateways in separate bus stops on the map of Tirana, according Betweenness centrality algorithm. 14 gateways are placed across bus stops according to their grade value, from the most important station with value 1 (with more weight) to the station with the lowest value (14).

Finally, Fig. 5 shows the gateway placement according to the proposed algorithm for minimal number of gateways. According to this algorithm, only three gateways are enough to cover all bus lines. Two gateways are located in the center and very close to each other while the third gateway is located few km away. In the city center there are two gateways because some lines such as line 4, 7 and 11 do not have a station there. The third gateway located few km away covers the line that does not pass through the city center. With some minimal itinerary changes of lines 4, 7 and 11, Tirana public transport lines can be covered with only two gateways.

5   Conclusions and Future Work

This paper discussed the possibility of the usage of OppIoT and the existing infrastructure of public bus transportation system and pedestrians mobility to deliver data from different sensors to the gateways and build a low-cost smart city. We considered as a case study Tirana city, Albania. The performance of three algorithms is compared to find the minimal number of gateways and the gateway positioning. Best results are obtained from the algorithm for minimal number of gateways where only three gateways are needed and the implementation cost is lower compared with Betweenness centrality and Degree centrality algorithms which need 14 and 16 gateways, respectively.

In the future, we would like to propose and evaluate the performance of new optimization algorithms and consider also social-based routing schemes, new strategies and points of interests to increase the pedestrian mobility to deliver data in a smart city.

References

1. Spaho, E.: Usage of DTNs for low-cost IoT application in smart cities: performance evaluation of spray and wait routing protocol and its enhanced versions. Int. J. Grid Util. Comput. **12**(2), 173–177 (2021)
2. Uchida, N., Ishida, T., Shibata, Y.: Delay tolerant networks-based vehicle-to-vehicle wireless networks for road surveillance systems in local areas. Int. J. Space-Based Situated Comput. **6**(1), 12–20
3. Spaho, E., Dhoska, K., Bylykbashi, K., Barolli, L., Kolici, V., Takizawa, M.: Performance evaluation of routing protocols in DTNs considering different mobility models. In: Barolli, L., Takizawa, M., Xhafa, F., Enokido, T. (eds.) WAINA 2019. AISC, vol. 927, pp. 205–214. Springer, Cham (2019). https://doi.org/10.1007/978-3-030-15035-8_19

4. Bylykbashi, K., Spaho, E., Barolli, L., Xhafa, F.: Impact of node density and TTL in vehicular delay tolerant networks: performance comparison of different routing protocols. Int. J. Grid Util. Comput. **7**(3), 136–144 (2017)
5. Spaho, E., Dhoska, K., Bylykbashi, K., Barolli, L., Kolici, V., Takizawa, M.: Performance evaluation of energy consumption for different DTN routing protocols. In: Barolli, L., Kryvinska, N., Enokido, T., Takizawa, M. (eds.) NBiS 2018. LNDECT, vol. 22, pp. 122–131. Springer, Cham (2019). https://doi.org/10.1007/978-3-319-98530-5_11
6. Spaho, E., Dhoska, K., Barolli, L., Kolici, V., Takizawa, M.: Enhancement of binary spray and wait routing protocol for improving delivery probability and latency in a delay tolerant network. In: Barolli, L., Hellinckx, P., Enokido, T. (eds.) BWCCA 2019. LNNS, vol. 97, pp. 105–113. Springer, Cham (2020). https://doi.org/10.1007/978-3-030-33506-9_10
7. Spaho, E., Biberaj, A., Tahiraga, A.: LoRaWAN for an IoT-based environmental monitoring application in Tirana city. Pollack Periodica **16**(2), 92–97 (2021)
8. Valeri, A., Musalimov, V.: MATLAB-based graphic user interface for monitoring and control of wireless sensor networks. Int. J. Innov. Technol. Interdiscip. Sci. **2**(2), 181–191 (2019)
9. Opoku, D., Kommey, B.: FPGA-based intelligent traffic controller with remote operation mode. Int. J. Innov. Technol. Interdiscip. Sci. **3**(3), 490–500 (2020)
10. Spaho, E., Dhoska, K.: Proposal of a LoRaWAN-based IoT system for food waste management. Int. J. Innov. Technol. Interdiscip. Sci. **3**(3), 474–479 (2020)
11. Open Data, Tirana Municipality. https://opendata.tirana.al. Accessed 28 Mar 2021

A ML-Based System for Predicting Flight Coordinates Considering ADS-B GPS Data: Problems and System Improvement

Kazuma Matsuo[1], Makoto Ikeda[2(✉)], and Leonard Barolli[2]

[1] Graduate School of Engineering, Fukuoka Institute of Technology, 3-30-1 Wajiro-higashi, Higashi-ku, Fukuoka 811-0295, Japan
mgm21107@bene.fit.ac.jp
[2] Department of Information and Communication Engineering, Fukuoka Institute of Technology, 3-30-1 Wajiro-higashi, Higashi-ku, Fukuoka 811-0295, Japan
makoto.ikd@acm.org, barolli@fit.ac.jp

Abstract. The development of a low-cost aircraft surveillance system based on Automatic Dependent Surveillance-Broadcast (ADS-B) has attracted significant attention and there are many applications. The ADS-B signals have many data about the aircraft and we are particularly interested in the idea of utilizing this data to develop flight predictions. In this paper, we present an ML-based system for predicting three-dimensional flight location coordinates by using route classification from ADS-B. The evaluation results show that our proposed system can predict three-dimensional flight coordinates, but the accuracy is not high because of the GPS fluctuations.

Keywords: ADS-B · Flight prediction · ML · GPS fluctuation

1 Introduction

To realize carbon neutrality, a variety of strategies are required, including improving aircraft operations, boosting eco-airports and developing biojet fuel technology. Automated Air Traffic Management (ATM) systems are very important for improving navigation, communications, surveillance and passenger safety [17]. They will be the core of Automatic Dependent Surveillance-Broadcast (ADS-B) technology [12,18,22,24, 25,27].

Machine Learning (ML) has been utilized to predict the arrival time of several modes of transportation, including bus, rail, and aircraft [1–3,5,11,15,16]. The data for a routed operation, such as a daily operation can be gathered and used as training data for ML and neural networks [8,9,14,20]. In this research, we focus on aircraft signals and collect them to predict the aircraft's future position coordinates.

In our previous work [10], we proposed an ML-based flight prediction system considering two-dimensional location coordinates from our gathered data of an ADS-B receiver. However, we did not consider the altitude and the data set is very small.

In this research, we present an ML-based approach for predicting three-dimensional flight location coordinates from ADS-B data. The proposed method is evaluated using

ADS-B data over the northern Kyushu area, which is classified into different route classes. We prepare the training data by filtering the received ADS-B data. Although the ADS-B receiver may receive the aircraft's identification number, altitude, speed, and location coordinates, only the time and location coordinates of the received data are used to predict the aircraft's flight location.

The following is the overall structure of the paper. Section 2 contains a description of the related works. In Sect. 3, we describe the flight prediction system design. Section 4 provides the evaluation results. Finally, conclusions and future work are given in Sect. 5.

2   Related Works

2.1   Overview

An Airport Surveillance Radar (ASR) is composed of two components: Primary Surveillance Radar (PSR) [21] and Secondary Surveillance Radar (SSR) [23]. The installed radar systems are determined by the size and function of the airport. The PSR is a fully integrated radar system consisting of a transmitter and a receiver. The PSR does not transmit altitude data and aircraft identification. The SSR transmitter sends a question signal to an aircraft's transponder, which responds with a reply signal. When a transponder is damaged, the communication is disrupted and monitoring becomes impossible. The ADS-B is more sophisticated aircraft surveillance system than the SSR. Aircraft location, altitude and other data are continually transmitted by the ADS-B. Anyone can receive this signal from aircraft using an ADS-B receiver [19].

It is possible to use PSR in the passive bistatic and multi-static primary surveillance radars [4,6,7,13,26]. They consist of separate transmitter and receiver. It is expected that by receiving signals from several locations, the monitoring coverage area will be extended and the update frequency will increase.

2.2   Problems with Three Weeks Data

In prior work [10], we proposed an ML-based approach for predicting two-dimensional flight location coordinates from a subset of ADS-B data. The training dataset was used for one week. We encountered some problems while comparing one-week and three-week data collecting periods to classify the aircraft movement prediction in two dimensions. The accuracy results in Table 1 show that the three-week data set is very low. The accuracy of latitude is 94.47% and 9.23%, respectively.

Table 1. Accuracy results of proposed two-dimension aircraft prediction system.

Term	Number of data	Accuracy (Latitude)
One Week	973	94.47
Three Weeks	1764	9.23

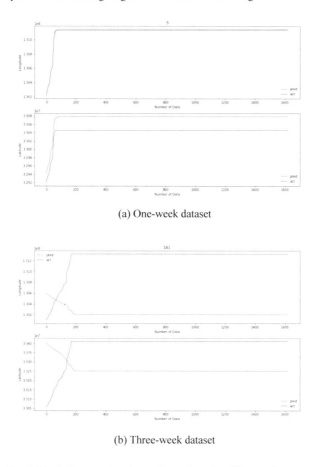

(a) One-week dataset

(b) Three-week dataset

Fig. 1. Prediction results of two dimensions for different datasets.

In Fig. 1(a), we present the results of a one-week data set with a slight variation between predicted and actual paths. On the other hand, Fig. 1(b) confirms that the difference is larger for the three-week data set. The upper figure indicates longitude and the bottom one indicates latitude. To address this problem, we classify flight paths during the learning phase.

3   ML-Based Flight Prediction System

3.1   System Overview

We have constructed a Raspberry Pi-based system to gather 1,090 MHz signal from an ADS-B receiver. The ADS-B received data consists of eight messages, the third of which contains the location coordinates. Each message contains 22 fields. We used the fields containing the longitude, latitude, altitude and time of message transmission as training data for ML.

	ICAO	Lat	Lon	Alt	TC	Date	Datetime
0	86EC40	34.240	131.119	17850	11	2020-02-07	2020-02-07 21:00:00
6	86EC40	34.240	131.119	17825	11	2020-02-07	2020-02-07 21:00:00
93	86EC40	34.235	131.109	17625	11	2020-02-07	2020-02-07 21:00:07
96	86EC40	34.235	131.109	17625	11	2020-02-07	2020-02-07 21:00:07
109	86EC40	34.234	131.107	17575	11	2020-02-07	2020-02-07 21:00:08
...	...	...	...	...	...	...	...
4771	86EC40	33.609	130.435	300	11	2020-03-27	2020-03-27 21:06:46
4797	86EC40	33.607	130.436	250	11	2020-03-27	2020-03-27 21:06:49
4805	86EC40	33.607	130.436	250	11	2020-03-27	2020-03-27 21:06:49
4818	86EC40	33.606	130.437	250	11	2020-03-27	2020-03-27 21:06:50
4905	86EC40	33.601	130.440	125	11	2020-03-27	2020-03-27 21:06:58

Fig. 2. ML's table data format.

In Fig. 2, we show an example of utilizing Jupyter Notebook to export a part of ML table data. The International Civil Aviation Organization (ICAO) specifies a mode S code of the ICAO aircraft address, which implies that the same aircraft's data (hex: 86EC40) is extracted. We prepare the retrieved data from an ADS-B message, which is sorted by timestamp. The second message denotes latitude, the third message denotes longitude, the fourth column denotes altitude, and the fifth column denotes Type Code (TC). The following messages are date and received time (t_n). The difference between the first received time t_0 and t_n for each aircraft identification code is represented by T_{diff}.

$$T_{diff} = t_0 - t_n$$

The measured data is stored in CSV format and is organized according to the aircraft identification code. These measurements were taken using an ADS-B receiver located at the Fukuoka Institute of Technology. The receiver's latitude and longitude are 33.695289 and 130.440596, respectively.

3.2 Classification and Data Processing

To handle three-dimensional data and long-term data, we attempt to improve the accuracy by categorizing each route. We are importing CSV data into an array for ML using Python. Then, the padding technique is applied to the end of the data, which is aligned with the longest data length available for training. To deal with time series data using ML, we adopt random forest regression. As the training data for this system, we measured the data for about three months. The dataset is received on a periodic basis at an interval of approximately one second. These data are from the period before the COVID-9 epidemic and aircraft flight restrictions in Japan. The longitude, latitude and altitude are predicted as performance indicators.

4   Evaluation Results

In Fig. 3, we show the prediction results of three dimensional flight coordinates. We observe that the oscillation is present in both the actual and predicted results for longitude. This problem is caused by the GPS module's accuracy. The location data are incorrectly detected with the GPS module. Our ML-based system is greatly impacted by even this seemingly insignificant oscillation.

The location data is incorrectly detected by the GPS module. We think that the preprocessing with TC is required in order to increase the learning accuracy. The process could be modified according to GPS accuracy information. We also believe that moving average and smoothing techniques are one of the keys to solve this GPS fluctuation. When computing moving averages, it would also be a good idea to include speed and altitude as weighting variables.

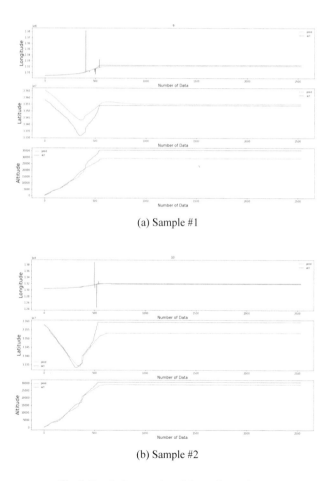

(a) Sample #1

(b) Sample #2

Fig. 3. Prediction results of three dimensions.

5   Conclusions

In this paper, we discussed the ML-based approach for predicting three-dimensional flight location coordinates from categorized route patterns from ADS-B data. We used random forest regression for training and predicted three-dimensional location. From the evaluation results, we found some problems for the proposed system to predict three-dimensional flight position coordinates, and discussed several ways to improve the system. It is important to explore additional parameters and algorithms in future work in order to adapt the proposed system to a variety of situations.

Acknowledgment. The ADS-B data are supported by the Electronic Navigation Research Institute (ENRI) with which we have research collaboration. The authors would like to thank ENRI for their assistance.

References

1. Cai, Q., Alam, S., Duong, V.N.: A spatial-temporal network perspective for the propagation dynamics of air traffic delays. Engineering **7**(4), 452–464 (2021). https://www.sciencedirect.com/science/article/pii/S2095809921000485
2. Choi, S., Kim, Y.J., Briceno, S., Mavris, D.: Prediction of weather-induced airline delays based on machine learning algorithms. In: 2016 IEEE/AIAA 35th Digital Avionics Systems Conference (DASC), pp. 1–6 (2016)
3. Duan, Y., Yisheng, L.V., Wang, F.Y.: Travel time prediction with LSTM neural network. In: 2016 IEEE 19th International Conference on Intelligent Transportation Systems (ITSC), pp. 1053–1058 (2016)
4. Edrich, M., Schroeder, A.: Design, implementation and test of a multiband multistatic passive radar system for operational use in airspace surveillance. In: 2014 IEEE Radar Conference, pp. 12–16 (2014)
5. Gui, G., Liu, F., Sun, J., Yang, J., Zhou, Z., Zhao, D.: Flight delay prediction based on aviation big data and machine learning. IEEE Trans. Veh. Technol. **69**(1), 140–150 (2020)
6. Honda, J., Otsuyama, T., Watanabe, M., Makita, Y.: Study on multistatic primary surveillance radar using DTTB signal delays. In: 2018 International Conference on Radar (RADAR), pp. 1–4 (2018)
7. Honda, J., Otsuyama, T.: Feasibility study on aircraft positioning by using ISDB-T signal delay. IEEE Antennas Wirel. Propag. Lett. **15**, 1787–1790 (2016)
8. Kim, Y.J., Choi, S., Briceno, S., Mavris, D.: A deep learning approach to flight delay prediction. In: 2016 IEEE/AIAA 35th Digital Avionics Systems Conference (DASC), pp. 1–6 (2016)
9. Martínez-Prieto, M.A., Bregon, A., García-Miranda, I., Álvarez Esteban, P.C., Díaz, F., Scarlatti, D.: Integrating flight-related information into a (big) data lake. In: 2017 IEEE/AIAA 36th Digital Avionics Systems Conference (DASC), pp. 1–10 (2017)
10. Matsuo, K., Ikeda, M., Barolli, L.: A machine learning approach for predicting 2D aircraft position coordinates. In: Barolli, L., Chen, H.-C., Enokido, T. (eds.) NBiS 2021. LNNS, vol. 313, pp. 306–311. Springer, Cham (2022). https://doi.org/10.1007/978-3-030-84913-9_30
11. Moreira, L., Dantas, C., Oliveira, L., Soares, J., Ogasawara, E.: On evaluating data preprocessing methods for machine learning models for flight delays. In: 2018 International Joint Conference on Neural Networks (IJCNN), pp. 1–8 (2018)

12. Nijsure, Y.A., Kaddoum, G., Gagnon, G., Gagnon, F., Yuen, C., Mahapatra, R.: Adaptive air-to-ground secure communication system based on ADS-B and wide-area multilateration. IEEE Trans. Veh. Technol. **65**(5), 3150–3165 (2016)
13. O'Hagan, D.W., Baker, C.J.: Passive bistatic radar (PBR) using FM radio illuminators of opportunity. In: 2008 New Trends for Environmental Monitoring Using Passive Systems, pp. 1–6 (2008)
14. Olive, X., et al.: OpenSky report 2020: analysing in-flight emergencies using big data. In: 2020 AIAA/IEEE 39th Digital Avionics Systems Conference (DASC), pp. 1–10 (2020)
15. Pamplona, D.A., Weigang, L., de Barros, A.G., Shiguemori, E.H., Alves, C.J.P.: Supervised neural network with multilevel input layers for predicting of air traffic delays. In: 2018 International Joint Conference on Neural Networks (IJCNN), pp. 1–6 (2018)
16. Peters, J., Emig, B., Jung, M., Schmidt, S.: Prediction of delays in public transportation using neural networks. In: International Conference on Computational Intelligence for Modelling, Control and Automation and International Conference on Intelligent Agents, Web Technologies and Internet Commerce (CIMCA-IAWTIC'06), vol. 2, pp. 92–97 (2005)
17. Post, J.: The next generation air transportation system of the United States: vision, accomplishments, and future directions. Engineering **7**(4), 427–430 (2021). https://www.sciencedirect.com/science/article/pii/S209580992100045X
18. Schäfer, M., Strohmeier, M., Lenders, V., Martinovic, I., Wilhelm, M.: Bringing up OpenSky: a large-scale ADS-B sensor network for research. In: IPSN-14 Proceedings of the 13th International Symposium on Information Processing in Sensor Networks, pp. 83–94 (2014)
19. Sciancalepore, S., Alhazbi, S., Di Pietro, R.: Reliability of ADS-B communications: novel insights based on an experimental assessment. In: Proceedings of the 34th ACM/SIGAPP Symposium on Applied Computing. SAC '19, pp. 2414–2421. Association for Computing Machinery, New York (2019). https://doi.org/10.1145/3297280.3297518
20. Shi, Z., Xu, M., Pan, Q., Yan, B., Zhang, H.: LSTM-based flight trajectory prediction. In: 2018 International Joint Conference on Neural Networks (IJCNN), pp. 1–8 (2018)
21. Skolnik, M.I.: Introduction to Radar System, 3rd edn. Mcgraw-Hill College, New York (1962)
22. Smith, A., Cassell, R., Breen, T., Hulstrom, R., Evers, C.: Methods to provide system-wide ADS-B back-up, validation and security. In: 2006 IEEE/AIAA 25th Digital Avionics Systems Conference, pp. 1–7 (2006)
23. Stevens, M.C.: Secondary Surveillance Radar. Artech House, Norwood (1988)
24. Strohmeier, M., Lenders, V., Martinovic, I.: On the security of the automatic dependent surveillance-broadcast protocol. IEEE Commun. Surv. Tutor. **17**(2), 1066–1087 (2015)
25. Strohmeier, M., Schäfer, M., Lenders, V., Martinovic, I.: Realities and challenges of NextGen air traffic management: the case of ADS-B. IEEE Commun. Mag. **52**(5), 111–118 (2014)
26. Willis, N.J.: Bistatic Radar, 2nd edn. Artech House, Norwood (1995)
27. Yang, A., Tan, X., Baek, J., Wong, D.S.: A new ADS-B authentication framework based on efficient hierarchical identity-based signature with batch verification. IEEE Trans. Serv. Comput. **10**(2), 165–175 (2017)

Fault Detection from Bend Test Images of Welding Using Faster R-CNN

Shigeru Kato[1]($\boxtimes$), Takanori Hino[1], Hironori Kumeno[1], Shunsaku Kume[1], Tomomichi Kagawa[1], and Hajime Nobuhara[2]

[1] Niihama College, National Institute of Technology, Niihama, Japan
{s.kato,t.hino,h.kumeno,t.kagawa}@niihama-nct.ac.jp
[2] University of Tsukuba, Tsukuba, Japan
nobuhara@iit.tsukuba.ac.jp

Abstract. The human visual inspection to find defects from welding joints is very tough. The examiners have to inspect many bend test fragments carefully. The present study aims to build an automatic detection system capable of finding cracks from bend test fragments. This paper describes the automatic detection method employing Faster R-CNN to detect crack regions. First, we introduce our achievement and explain the focused issue. Second, the structure of the proposed Faster R-CNN is explained, and then the present paper shows the experiment of automatic detection using web-camera working in real-time. Finally, conclusions and future works are discussed.

1 Introduction

In Japan, the shortage of welding technicians and the aging of skilled workers capable of providing technical guidance has become a pressing issue. To compensate for this problem, several attempts to pass on welding skills and train beginners are being made at an engineering level [1–3].

The number of candidates for licenses in the field of gas tungsten arc welding [4], which is applied to a wide range of construction, is increasing yearly not only among Japanese trainees but also among overseas technical trainees from Southeast Asia. One issue is the need for examiners to exercise caution in their judgment when visually inspecting the appearance of the plates welded by candidates. To this end, we construct FLATWE, a welding plate evaluation system that uses convolutional neural networks (CNN) [5–7]. As shown in Fig. 1, FLATWE evaluates the appearance of installed welded plates.

© The Author(s), under exclusive license to Springer Nature Switzerland AG 2022
L. Barolli et al. (Eds.): EIDWT 2022, LNDECT 118, pp. 190–200, 2022.
https://doi.org/10.1007/978-3-030-95903-6_21

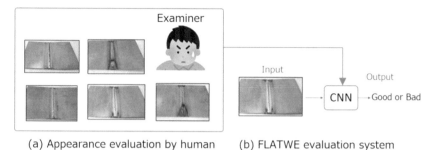

(a) Appearance evaluation by human   (b) FLATWE evaluation system

Fig. 1. Welding plate appearance evaluation.

Welded test plates that have passed the visual inspection under FLATWE are then subjected to the bend test to determine whether they pass or fail under the Japanese Industrial Standards (JIS) welding technician evaluation test [8]. As shown in Fig. 2, in addition to the rejected specimens that rupture during the bend test [9], pass/fail decisions are made by visually inspecting the weld joints with a loupe to evaluate the presence, number, and the size of microdefects, such as cracks and voids. Currently, a considerable amount of manpower is needed to deal with a large number of specimens, and generating educational benefits by providing trainees with feedback on judgment results is difficult.

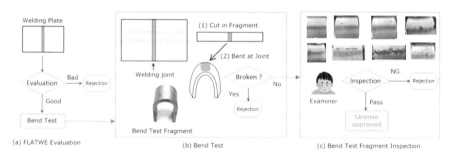

Fig. 2. Bend test for welding joint firmness.

Therefore, we construct an R-CNN-based evaluation system called BENTWE (Fig. 3) [10]. However, in R-CNN [11], detecting regions of interest (ROIs) is a very time-consuming process and may lead to defects in image processing called selective search. As very large numbers of ROIs are proposed using selective search from a single image, the method is not suitable for real-time evaluation.

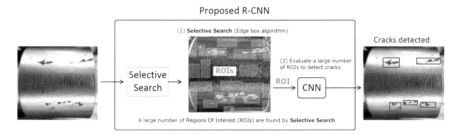

Fig. 3. Previous system (BENTWE).

Therefore, in this work, we describe the construction of a system that introduces Faster R-CNN [12] (Fig. 4). Using CNN, the system generates a feature vector of an image called the feature map from the input image. Then, ROIs can be predicted as cracks by the region proposal network (RPN) at high speeds. Crack region estimation is then performed in the later stage of the neural network (NN). In this work, we have described a prototype of a real-time automatic defect detection system for cracks and holes in bend specimens that uses a USB camera. We also discuss its future development.

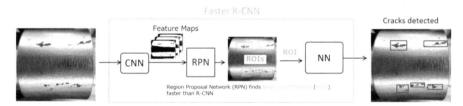

Fig. 4. Proposed faster R-CNN.

There have been various remarkable studies [13–18] related to detection of welding defect using CNN and machine learning, however our study focuses on beginner-level technician works' evaluation to judge whether the Japanese welding license is qualified or not.

The bend test pieces in this paper were a joint made by arc-welding carbon steel 9 mm thickness plates.

2 System for Crack Detection Using Faster R-CNN

Figure 5 shows the structure of the proposed system. The input layer comprises 600 × 800 RGB image data obtained from a web camera. The feature map is generated using the segments from the vgg16 [19] CNN and outputted from relu5_3 followed by conv5_3. The structure of the CNN for feature map generation is shown in Table 1. The RPN estimates the location of the object (Crack) from the feature map. Then, in the ROI pooling layer, the ROI feature vector of the object is sent to the NN in the later stage to estimate a more precise location of the crack.

Table 1. Structure of vgg16 for generating feature maps.

Layer name	Operation	Filters Num	Filter	Stride	Zero padding	Output
input	-	-	-	-	-	$600 \times 800 \times 3$
conv1_1	Conv+ReLU	64	$3 \times 3 \times 3$	1×1	$1 \times 1 \times 1 \times 1$	$600 \times 800 \times 64$
conv1_2		64	$3 \times 3 \times 64$	1×1	$1 \times 1 \times 1 \times 1$	$600 \times 800 \times 64$
pool 1	maxpooling	-	2×2	2×2	-	$300 \times 400 \times 64$
conv2_1	Conv+ReLU	128	$3 \times 3 \times 64$	1×1	$1 \times 1 \times 1 \times 1$	$300 \times 400 \times 128$
conv2_2		128	$3 \times 3 \times 128$	1×1	$1 \times 1 \times 1 \times 1$	$300 \times 400 \times 128$
pool 2	maxpooling	-	2×2	2×2	-	$150 \times 200 \times 128$
conv3_1		256	$3 \times 3 \times 128$	1×1	$1 \times 1 \times 1 \times 1$	$150 \times 200 \times 256$
conv3_2	Conv+ReLU	256	$3 \times 3 \times 256$	1×1	$1 \times 1 \times 1 \times 1$	$150 \times 200 \times 256$
conv3_3		256	$3 \times 3 \times 256$	1×1	$1 \times 1 \times 1 \times 1$	$150 \times 200 \times 256$
pool 3	maxpooling	-	2×2	2×2	-	$75 \times 100 \times 256$
conv4_1		512	$3 \times 3 \times 256$	1×1	$1 \times 1 \times 1 \times 1$	$75 \times 100 \times 512$
conv4_2	Conv+ReLU	512	$3 \times 3 \times 512$	1×1	$1 \times 1 \times 1 \times 1$	$75 \times 100 \times 512$
conv4_3		512	$3 \times 3 \times 512$	1×1	$1 \times 1 \times 1 \times 1$	$75 \times 100 \times 512$
pool 4	maxpooling	-	2×2	2×2	-	$37 \times 50 \times 512$
conv5_1		512	$3 \times 3 \times 512$	1×1	$1 \times 1 \times 1 \times 1$	$37 \times 50 \times 512$
conv5_2	Conv+ReLU	512	$3 \times 3 \times 512$	1×1	$1 \times 1 \times 1 \times 1$	$37 \times 50 \times 512$
conv5_3		512	$3 \times 3 \times 512$	1×1	$1 \times 1 \times 1 \times 1$	$37 \times 50 \times 512$

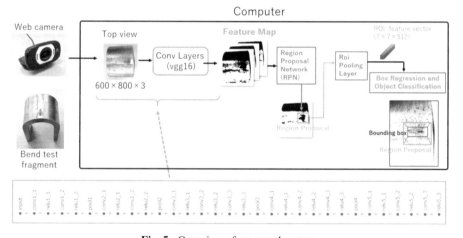

Fig. 5. Overview of proposed system.

As shown in Fig. 6, the input image is divided into a 37×50 grid, which is the same size as the feature map. Three anchor boxes are placed in each grid, with the anchor at its center. The RPN calculates the misalignment between each anchor box and the object and outputs the Region Proposal to the ROI pooling layer. The ROI pooling layer sends the ROI feature vector of the proposed region to the later stage. Finally, in the output layer of the Faster R-CNN, the bounding box is adjusted so that the region surrounds the object more precisely.

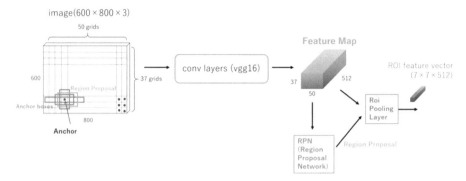

Fig. 6. Concept of anchor box and ROI feature vector generation.

Three types of anchor boxes with [height, width] of [47, 197], [18, 30], and [40, 62] are used.

3 Training Faster R-CNN Detector

Faster R-CNN training is performed. The image and gTruth in [10] are used for training. However, all the images and rectangle representing gTruth are resized to a resolution of 600 × 800 before use. An example of the resizing is shown in Fig. 7. The label is called "Crack", and 186 images have 1,149 gTruths.

Fig. 7. gTruth for training.

To avoid overfitting, this study randomly uses four versions of the original image, namely, as is, rotated left/right, rotated up/down, and rotated left/right/upside down. In the training, the entire RPN and Faster R-CNN are repeated on an alternate basis, as appropriate. Figure 8 shows the RPN and Faster R-CNN output sections. The [dx, dy, dw, dh] in the RPN is the amount of adjustment of the Anchor Box, which is used to calculate the Region Proposal area [x1, y1, x2, y2]. The output of boxDeltas is used for the box bounding of "Crack."

In the training, the amount of adjustment [dx, dy, dw, dh] for the Anchor Box from gTruth is tuned so that loss is minimized [12] in the training data. For this training

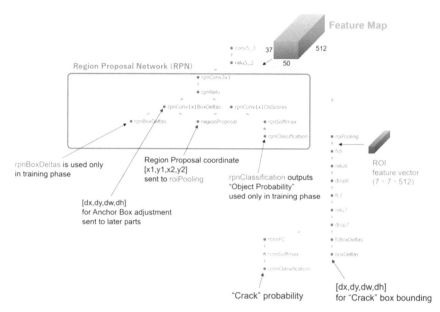

Fig. 8. RPN configuration and output part of Faster R-CNN.

rpnBoxDeltas layer in the RPN is employed. As well, boxDeltas for adjusting Region Proposal is tuned so as to minimize the box bounding error in the training phase.
Table 2 shows the training conditions.

Table 2. Condition of training Faster R-CNN.

	method/value
Solver	SGDM
Learning Rate	0.001
Max Epoch	30
Negative Overlap Range	[0, 0.3]
Positive Overlap Range	[0.6, 1.0]

Windows 10 64 bit is the operating system used, and the system was developed using MathWorks by MATLAB. The CPU is an Intel Core i9-10980XE (18 cores, 3.0 GHz), the main memory is 128 GB, and the GPU is Nvidia GeForce RTX 3090 (24 GB VRAM, 10496 cores). Table 3 summarizes the training. As shown in Figs. 9 (a) and 9 (b), as the training progresses, the accuracy for the training data improves. As depicted in Fig. 9 (c), the overall loss in the network also decreases.

Table 3. Training details.

	Value
Initial Mini-batch loss	1.6641
Initial Mini-batch accuracy	51.96 %
Initial RPN Mini-batch accuracy	67.19 %
Final Mini-batch loss	0.0854
Final Mini-batch accuracy	99.79 %
Final RPN Mini-batch accuracy	100 %
Total Iterations for training	5580
Total Time	28 min : 25 sec

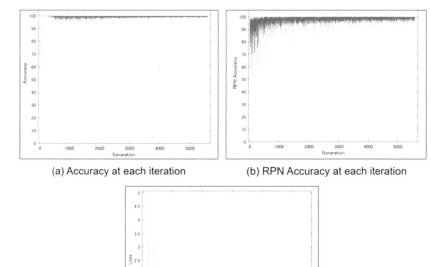

(a) Accuracy at each iteration   (b) RPN Accuracy at each iteration

(c) Loss at each iteration

Fig. 9. Training progress.

4   Validation of Faster R-CNN Detector

We evaluated the results by utilizing 77 images that are not used for the training. In some cases, as shown in Fig. 10 (a), cracks were detected accurately. However, in other cases, cracks were overlooked (Fig. 10 (b)), and letters in the background of the newspaper were misidentified as cracks (Fig. 10 (c)). Approximately 5 to 7 images contained noticeable oversights.

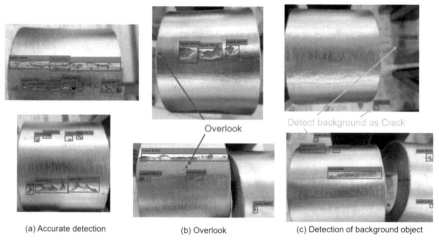

(a) Accurate detection   (b) Overlook   (c) Detection of background object

Fig. 10. Test result.

Table 4 shows the number of Gt (Ground truth, i.e., correct crack region) and Bb (Bounding box, i.e., detected regions by Faster R-CNN.) The total Gt included in all test images is 241, and the entire Bb is 372.

Table 4. Validation result of Faster R-CNN detector.

Number of Test image	77
Number of Ground Truth	241 (AOR Score : 83.37 %)
Number of Bounding Box	372 (AOR Score: 65.63 %)

The AOR (Accurate Overlap Rate) defined in Eq. (1) indicates the overlapping score of each ground truth for the bounding boxes, and the AOR is numerical value ranges [0, 1].

$$AOR(Gt_j) = \max_{i \in \{1,2\cdots,372\}} \left(\frac{area(Gt_j) \cap area(Bb_i)}{min[area(Gt_j), area(Bb_i)]} \right), where\, j \in \{1, 2 \ldots, 241\} \quad (1)$$

Equation (2) is a numerical score representing that each Bb overlaps the ground truth.

$$AOR(Bb_j) = \max_{i \in \{1,2\cdots,241\}} \left(\frac{area(Gt_i) \cap area(Bb_j)}{\min[area(Gt_i), area(Bj)]} \right), where\ j \in \{1, 2 \ldots, 372\} \quad (2)$$

Since the average of AORs in Eq. (1) was 0.8337 (83.37%), it is found that the cracks were detected with high accuracy. Contrary, the average of AORs of bounding boxes was 0.6563 (65.63%); thus, several regions except crack areas were detected by mistake (e.g., background objects such as the characters of newspapers), or ground truth areas were overlooked in some cases. Figure 11 shows the example of results in which both ground truth and detected cracks rectangles are drawn.

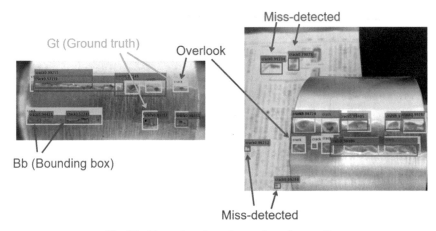

Fig. 11. Examples of test image detection result.

The results of the real-time web camera detection of the test specimens not used for training are shown in Fig. 12. Here (a1) and (b1) are the detection results, and (a2) and (b2) visualize the activity of the feature map. The visualization method was based on the MathWorks website [20] using Grad-CAM [21].

As shown in Figs. 12 (a1) and (b1), the crack parts are accurately detected. (a2) and (b2) are a colorful representation of the activity level of the relu5_3 layer, which outputs the feature map. The area surrounded by the highly active region was clearly recognized as having "Cracks." The detection process was very fast, and the time lag was not a problem in the discussions, including the group evaluations.

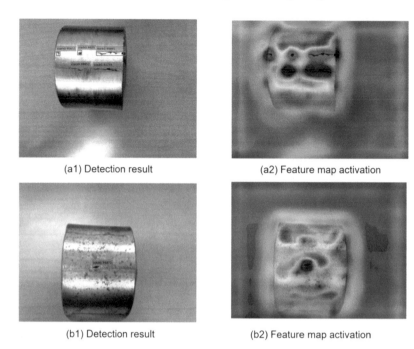

(a1) Detection result	(a2) Feature map activation
(b1) Detection result	(b2) Feature map activation

Fig. 12. Real time detection using web camera.

5 Conclusion

In this work, we have presented a prototype of a system for detecting cracks from RGB images of the surface of a welded bend specimen using Faster R-CNN and a web camera. An example of the execution of the prototype was also discussed. The results showed that the proposed Faster R-CNN could detect prominent cracks. In the future, we aim to develop a system that will allow multiple people to discuss the overall evaluation of cracks by using a projector to expand the output results of this system. Our end goal is to build a system that can automatically evaluate test specimens using artificial intelligence.

Acknowledgments. The authors would like to thank Crimson Interactive Pvt. Ltd. (Ulatus) - www.ulatus.jp for their assistance in manuscript translation and editing, and Ueno in MathWorks for technical advice. This work was supported by a Grant-in-Aid from JWES (The Japan Welding Engineering Society).

References

1. Asai, S., Ogawa, T., Takebayashi, H.: Visualization and digitation of welder skill for education and training. Weld. World **56**, 26–34 (2012)
2. Byrd, A.P., Stone, R.T., Anderson, R.G., Woltjer, K.: The use of virtual welding simulators to evaluate experimental welders. Weld. J. **94**(12), 389–395 (2015)

3. Hino, T., et al.: Visualization of gas tungsten arc welding skill using brightness map of backside weld pool. Trans. Mat. Res. Soc. Jpn. **44**(5), 181–186 (2019)
4. Niles, R.W., Jackson, C.E.: Weld thermal efficiency of the GTAW process. Weld. J. **54**, 25–32 (1975)
5. Kato, S., Hino, T., Yoshikawa, N.: Fundamental study on evaluation system of beginner's welding using CNN. In: Barolli, L., Hellinckx, P., Natwichai, J. (eds.) 3PGCIC 2019. LNNS, vol. 96, pp. 821–827. Springer, Cham (2020). https://doi.org/10.1007/978-3-030-33509-0_77
6. Kato, S., Hino, T., Kumeno, H., Kagawa, T., Nobuhara, H.: Automatic detection of beginner's welding joint. In: Proceedings of 2020 Joint 11th International Conference on Soft Computing and Intelligent Systems and 21st International Symposium on Advanced Intelligent Systems, pp. 465–467 (2020)
7. Kato, S., et al.: Evaluation for angular distortion of welding plate. In: Arai, K. (eds.) IntelliSys 2021. LNNS, vol. 294, pp. 344–354. Springer, Cham (2022). https://doi.org/10.1007/978-3-030-82193-7_23
8. The Japanese Welding Engineering Society. http://www.jwes.or.jp/en/qualification.html. Accessed 3 Sept 2021
9. Wan, Y., Jiang, W., Li, H.: Cold bending effect on residual stress, microstructure and mechanical properties of Type 316L stainless steel welded joint. Eng. Fail. Anal. **117**, 104825 (2020)
10. Kato, S., Hino, T., Kume, S., Nobuhara, H.: Crack detection from weld bend test images using R-CNN. In: Barolli, L., (eds.) 3PGCIC 2021. LNNS, vol. 343, pp. 289–298. Springer, Cham (2022). https://doi.org/10.1007/978-3-030-89899-1_31
11. Girshick, R., Donahue, J., Darrell, T., Malik, J.: Rich feature hierarchies for accurate object detection and semantic segmentation. In: CVPR 2014 Proceedings of the 2014 IEEE Conference on Computer Vision and Pattern Recognition, pp. 580–587 (2014)
12. Ren, S., He, K., Girshick, R., Sun, J.: Faster R-CNN: towards real-time object detection with region proposal networks. Adv. Neural Inf. Process. Syst. **28** (2015)
13. Park, J.-K., , An, W.-H., Kang, D.-J.: Convolutional neural network based surface inspection system for non-patterned welding defects. Int. J. Precis. Eng. Manuf. **20**(3), 363–374 (2019)
14. Dung, C.V., Sekiya, H., Hirano, S., Okatani, T., Miki, C.: A vision-based method for crack detection in gusset plate welded joints of steel bridges using deep convolutional neural networks. Autom. Constr. **102**, 217–229 (2019)
15. Zhang, Z., Wen, G., Chen, S.: Weld image deep learning-based on-line defects detection using convolutional neural networks for Al alloy in robotic arc welding. J. Manuf. Process. **45**, 208–216 (2019)
16. Dai, W., et al.: Deep learning assisted vision inspection of resistance spot welds. J. Manuf. Process. **62**, 262–274 (2021)
17. Abdelkader, R., Ramou, N., Khorchef, M., Chetih, N., Boutiche, Y.: Segmentation of x-ray image for welding defects detection using an improved Chan-Vese model. Mater. Today Proc. **42**, Part 5, 2963–2967 (2021)
18. Zhu, H., Ge, W., Liu, Z.: Deep learning-based classification of weld surface defects. Appl. Sci. **9**(16), 3312 (2019)
19. Simonyan, K., Zisserman, A.: Very deep convolutional networks for large-scale image recognition. arXiv preprint arXiv:1409.1556 (2014)
20. MathWorks: https://jp.mathworks.com/help/deeplearning/ug/investigate-network-predictions-using-class-activation-mapping.html?lang=en. Accessed 8 Sept 2021
21. Selvaraju, R.R., Cogswell, M., Das, A., Vedantam, R., Parikh, D., Batra, D.: Grad-CAM: visual explanations from deep networks via gradient-based localization. In: IEEE International Conference on Computer Vision (ICCV), pp. 618–626 (2017)

An Efficient Local Search for the Maximum Clique Problem on Massive Graphs

Kazuho Kanahara[✉], Tetsuya Oda, Elis Kulla, Akira Uejima, and Kengo Katayama

Graduate School of Engineering, Okayama University of Science, 1-1 Ridaicho, Kita-ku, Okayama-shi 700-0005, Japan
t20sd01kk@ous.jp, katayama@ice.ous.ac.jp

Abstract. The Maximum Clique Problem (MCP) is one of the most important combinatorial optimization problems that has many practical applications such as community search in social networks. Since the MCP is known to be NP-hard, much effort has been devoted to the development of metaheuristic algorithms to find a high quality clique (solution) within reasonable running times. The *Multi-start k-opt Local Search* incorporating *k*-opt local search (MKLS) is well known as a simple and effective metaheuristic for MCP. However it takes long time to search the high-quality solution for difficult massive graphs such as real world social networks, because the search space is too large. In the case of applying metaheuristic algorithms for massive sparse graphs, adequate process such as reduction process is necessary to focus on promising search space. In this paper, we present a *Multi-start k-opt Local Search with graph Reduction process* (MKLS-R), for solving the maximum clique problem on massive graphs. MKLS-R is evaluated on difficult massive graphs of Network-Repository graphs. The experimental results showed that the graph reduction process in MKLS-R contributes to the improvement of the search performance of MKLS for the difficult massive graphs.

1 Introduction

The Maximum Clique Problem (MCP) is to find a complete subgraph of maximum cardinality in a general graph. The MCP includes important applications in different domains: coding theory, geometry, fault diagnosis, bioinformatics, wireless networks and telecommunications, and more recently, social network analysis [3,20].

Since the MCP is known to be NP-hard [9], much effort has been devoted to the development of metaheuristic algorithms to find a high quality clique (solution) within reasonable running times. Among the most well known metaheuristic algorithms such as Genetic Algorithm, Simulated Annealing, Iterated Local Search, and Multi-start Local Search (MLS) [3,11,20], the simplest and powerful metaheuristic algorithm is MLS. It provides the most basic and simple framework in the metaheuristic research field that mainly consists of two processes: generating initial solutions and Local Search. MLS repeatedly applies the local search operator to an initial solution generated randomly or greedily until the stopping condition is satisfied, and finally returns the best solution. The *Multi-start k-opt Local Search* (MKLS) incorporating *k*-opt local search [12] is well known as a simple and effective metaheuristic for MCP.

L. Barolli et al. (Eds.): EIDWT 2022, LNDECT 118, pp. 201–211, 2022.
https://doi.org/10.1007/978-3-030-95903-6_22

The above metaheuristic algorithms have a common drawback that takes a long time to search the high-quality solution for difficult massive graphs such as real world social networks, because the search space is too large. In the case of applying metaheuristic algorithms for massive graphs, additional process such as reduction process is necessary to reduce the search space (graph size) and consequently the search time.

In this paper, we present a *Multi-start k-opt Local Search with graph Reduction process* (MKLS-R for short), for the maximum clique problem on massive graphs. MKLS-R provides a very simple framework based on the general MLS, mainly consisting of local search process (KLS), process of generating initial solution, and graph reduction process. KLS is an effective local search based on *variable depth search* (VDS) [13, 15]. For the graph reduction process, we reduce the search space by dropping vertices that do not consist of larger cliques than the size of the best clique that is obtained in the search. In process of generating initial solution, we randomly select a single vertex $v \in V$ in the (reduced) graph G. MKLS-R is evaluated on difficult massive graphs of Network-Repository graphs [17]. The experimental results showed that the graph reduction process contributes to the improvement of the search performance of MKLS for the difficult massive graphs.

This paper is organized as follows. In Sect. 2, we define MCP and explain the basic procedure of local search. In Sect. 3, we introduce our proposed MKLS with graph reduction. Experimental results are shown in Sect. 4. Finally, we draw our conclusions in Sect. 5.

2   Maximum Clique Problem and Basic Local Search Procedure

Let $G = (V, E)$ be an arbitrary undirected graph where V is the set of n vertices and $E \subseteq V \times V$ is the set of edges in G. For a subset $S \subseteq V$, let $G(S) = (S, E \cap S \times S)$ be *the subgraph* induced by S.

A graph $G = (V, E)$ is *complete* if all its vertices are pairwise adjacent, i.e., $\forall i, j \in V$ with $i \neq j$, $(i, j) \in E$. A *clique* C is a subset of V such that the induced subgraph $G(C)$ is complete. The objective of the maximum clique problem (MCP) is to find a clique of maximum cardinality in G.

The MCP is one of the most fundamental problems in combinatorial optimization which has various real applications, such as mobile networks, computervision, cluster analysis, coding theory, tiling, fault diagnosis, biological analysis, social network analysis, design of quantum circuits, etc.; see the recent survey [20] and lecture notes [18] on the MCP, which also contains an extensive bibliography. The MCP is known to be NP-hard [9] for arbitrary graphs, and strong negative results have been shown in [10, 14, 20]. Therefore, much effort has been devoted to the development of efficient heuristic and meta heuristic algorithms to find near-optimal solutions to denser and large graphs within reasonable times. These promising heuristic and metaheuristic algorithms can be found in [3, 11, 20].

Most of the meta heuristic approaches to the MCP are based on the principle of *local search* (or local improvement). The basic heuristical principle prefers vertices of higher degrees to ones of lower degrees in order to obtain a larger clique. The basic procedure is as follows: Given a current clique CC having a single vertex $v \in V$, one of the vertices in vertex set PA of the highest degree given by $deg_{G(PA)}$ is repeatedly added to expand CC until PA is empty, where PA denotes the vertex set of possible additions, i.e., the vertices that are connected to all vertices of CC, and $deg_{G(S)}(v)$ stands for the degree of a vertex $v \in S$ in the subgraph $G(S)$, where $S \subseteq V$. This method is called *1-opt local search* in [12] because at each iteration t, a single vertex $v \in PA^{(t)}$ is moved to $CC^{(t)}$ to obtain a larger clique $CC^{(t+1)} := CC^{(t)} \cup \{v\}$.

The 1-opt local search has been generalized in [12] by moving multiple vertices, instead of a single vertex, at each iteration, in order to obtain a better clique. The generalized local search is called *k-opt local search* (KLS for short). In KLS, variable, not fixed, k vertices are moved to or from a current clique simultaneously at each iteration by applying a sequence of add and drop move operations. The idea of the sequential moves in KLS is borrowed from *variable depth search* (VDS) [13,15]. The detailed procedure of KLS is reviewed in the Sect. 3.1

3 Multi Start *K*-opt Local Search with Graph Reduction Process

In this section we show an Multi start Local Search incorporating KLS and graph size Reduction process, called *Multi start k-opt Local Search with Reduction process* (*MKLS-R* for short), for the MCP.

Given a local optimum, each iteration of MKLS-R consists of *Local Search* process in which KLS is employed as a dedicated local search and generate initial solution processes. As the additional process is performed the case of new best clique C_{best} is found, we use *Reduction* that reduces the search graph by dropping vertices that do not consist of a clique larger then the best clique C_{best}.

The top level flow of MKLS-R is shown in Fig. 1. At first we initialize the best clique C_{best}. In the main loop (Lines 2–9), we generate a feasible solution (clique) C that contains a single vertex selected from V at random. We then compute the associated PA, and OM. KLS is applied to C, and the resulting local optimum is stored as the best clique C_{best}, and we reduce graph size (Lines 4–7). The main loop is repeated until the stopping condition (Line 9) is satisfied.

```
procedure Multi-start-k-opt-Local-Search-with-Reduction
input: graph G = (V,E);
output: best clique Cbest in G;
begin
1    Cbest := ∅;
2    repeat
3        generate C, compute PA and OM;
4        C := Local-Search(C,PA,OM);
5        if |C| > |Cbest| then
6            Cbest := C;
7            G := Reduction(G, Cbest);
8        endif
9    until terminate = true;
10   return Cbest;
end
```

Fig. 1. The pseudo code of multi start *k*-opt local search with graph reduction process for MCP

3.1 Local Search Process

The work of *Local Search* process is to find local optima in a given graph G. In the process we use KLS shown in the previous section at line 4 in Fig. 1. In this section, we simply review the algorithm KLS, based on *variable depth search* (VDS) [13, 15], for the MCP.

The basic procedure of KLS is given as follows. Given an initial feasible solution (clique) CC with PA ($\neq \emptyset$), in each iteration KLS starts from the solution $CC^{(0)}$ (equal to the initial clique CC) with $PA^{(0)}$ (we here set $t = 0$). The first move from $CC^{(0)}$ to $CC^{(1)}$ is randomly performed by the add move[1] Thus, the solution $CC^{(1)}$ can be obtained by $CC^{(0)} \cup \{v\}$, where vertex v is selected randomly from $PA^{(0)}$, and $PA^{(0)}$ is updated to be $PA^{(1)}$. Notice that the other information, OM, is also updated, where OM denotes, given a current clique CC, the vertex set of one edge missing, i.e., the vertices that are connected to $|CC| - 1$ vertices of CC, provided that $CC \subseteq OM$. Such an add phase is repeated until $PA = \emptyset$.

We assume that $PA = \emptyset$ after some add moves (e.g., of l times). It is obvious that the add move is impossible for $CC^{(l)}$. However, the k-opt neighborhood search tries to find better cliques after dropping one or several vertices from $CC^{(l)}$.

The drop phase starts from $CC^{(l)}$ with no possible additions. This phase is continued until no vertex can be dropped from the current clique, or several vertices (at least one vertex) can be added in the add phase. During the drop phase we choose a vertex $v \in CC^{(l)}$ that leads up to the largest set of possible additions (i.e., $|PA^{(l+1)}|$ is maximized) that are connected to the member of $CC^{(l)} \setminus \{v\}$. Such a vertex v can be found by checking the number of all nodes that are lacking the edge to v in $OM^{(l)}$.

Let us assume that a vertex v to be dropped from the current clique is found. The vertex v is dropped from CC. Then we update the information of PA, etc. If a vertex addition to CC is possible, then the add phase immediately starts again; otherwise the drop phase continues. The k-opt neighborhood search which consists of the two phases is repeated until the termination condition is satisfied. The information of PA and OM is updated as in [2] whenever a single vertex is moved.

At the final stage in each iteration, t solutions, i.e., $CC^{(1)}, \ldots, CC^{(t)}$, obtained by the sequence of the add-and-drop moves to the given initial solution are obtained. From the t solutions, we select the best one $CC^{(k)}$ ($1 \leq k \leq t$). The best solution becomes a new initial feasible solution setting $CC^{(0)} := CC^{(k)}$ for the next k-opt neighborhood search. KLS is repeated until no better (variable k-opt) solution is found.

Figure 2 shows the pseudo code of KLS used in this paper for MCP. We here suppose that CC, PA, OM and P are given in advance. In the figure, we use a *gain* denoted by g (see line 2), where g is defined as the difference between the clique size of a given initial solution and that of a current one during the search.

KLS has inner and outer loops. In the inner loop (lines 3–15), a sequence of the add-and-drop moves for a given initial solution is obtained, and the best (k-opt) solution found by the sequence is selected. In the outer loop (lines 1–17), the best solution

---

[1] Note that the add phase of KLS in this paper randomly selects a vertex v from PA, although the original KLS [12] selects a vertex with highest degree. Therefore, calculating the degree of vertices is not performed in the addphase of KLS.

selected in the inner loop is evaluated by checking the value of the maximum gain g_{max} (line 16) whether it is better than the previous one ($|CC_{prev}|$).

The termination condition for the inner loop (line 15) is that the set D is empty. Initially D is set to the vertex set of the initial clique CC (i.e., CC_{prev}) at line 2 before the search in the inner loop starts. A vertex v, which is dropped in the drop phase, is removed from D if v is in CC_{prev} at line 12. It is notable that *no* parameter setting is required in KLS.

3.2 Reduction Process

We here show a graph reduction process based on upper bound for MCP. The graph reduction process is often applied to exact algorithm [8, 19] and heuristics algorithm for large sized combinatorial optimization problems [4] such as MCP [6, 7, 16]. In the many graph size reduction processes, lower and upper bounds are used to reduce size of graph. In the graph size reduction processes for MCP, we reduce the graph size by dropping vertices that do not consist of a clique larger than the best clique C_{best}. In other words, we drop the vertex v from graph G if the upper bound on the clique size that contains the vertex v is not larger than the best clique size $|C_{best}|$. The pseudo code of Reduction for MCP is shown in Fig. 3.

At first, we initialize the subgraph G', the set of vertices V' in G', the set of edges E' in G', and the check vertex set R (Line 1). We select one vertex $v \in CV$ randomly (Line 3). If the upper bound of the vertex v is lager then the best clique size $|C_{best}|$, we drop the vertex v and the edges adjacent to the vertex v from graph G', and add adjacent vertices $v_\gamma \in \Gamma_{(G')}(v)$ to CV (Lines 4–10). This process is repeated until the check vertex set CV is empty (Lines 2–11).

We here show upper bounds on the clique size that contains the vertex. There are three well-known types of upper bounds for maximum clique: degree based upper bound, core number based upper bound, and graph coloring based upper bound [5].

Let $\Gamma_G(v) = (i, (i, v) \in E)$ be the set of adjacent vertices to vertex v, and the degree based upper bound of vertex v is 1 plus size of adjacent vertices $|\Gamma_G(v)|$ that is degree $deg_G(v)$. The degree based upper bound is as follows:

$$UpperBound(G, v) = |\Gamma_G(v)| + |\{v\}| = deg_G(v) + 1 \tag{1}$$

The next upper bound is core number based upper bound. Giver a graph G, the core number $core(G, v)$ is a size of maximal subgraph $G(S)$ in which all vertices have degree at least $|S| - 1$. The core number based upper bound is as follows:

$$UpperBound(G, v) = core(G(\Gamma_G(v)), v) + |\{v\}| = core(G(\Gamma_G(v)), v) + 1 \tag{2}$$

The Graph coloring is often used to obtain an upper bound of clique [19]. The Graph coloring is called vertex coloring problem (VCP) or graph coloring problem, is known to be NP-hard. A vertex coloring is an assignment of colors to the vertices in V such that no two adjacent vertices share the same color. A stable set is a set of pairwise non-adjacent vertices. Hence, a vertex coloring of G is a partition of its vertex set into k

```
k-opt-Local-Search(CC, PA, OM)
begin
1    repeat
2        CC_prev := CC; D := CC_prev; P := V; g := 0; g_max := 0;
3        repeat
4            if |PA ∩ P| > 0 then    // Add Phase
5                select one vertex v ∈ PA ∩ P randomly;
6                CC := CC ∪ {v}; g := g + 1; P := P\{v}
7                if g > g_max then g_max := g; CC_best := CC;
8            else       //Drop Phase (if {PA ∩ P} = ∅)
9                find a vertex v ∈ {CC ∩ P} such that the resulting |PA ∩ P| is maximized;
10               if multiple vertices with the same size of the resulting |PA ∩ P| are found
                 then select one vertex v of D preferentially;
11               CC := CC\{v}; g := g − 1; P := P\{v};
12               if v is contained in CC_prev then D := D\{v};
13           endif
14           update PA, OM;
15       until D = ∅;
16       if g_max > 0 then CC := CC_best else CC := CC_prev;
17   until g_max ≤ 0;
18   return CC;
end
```

Fig. 2. The pseudo code of k-opt-Local-Search used in this paper

```
Reduction(G, C_best)
begin
1    G' := G; V' := V(G'); E' := E(G'); CV := V';
2    repeat
3        select one vertex v ∈ CV randomly;
4        if UpperBound(v) ≤ |C_best| then
5            for each v_γ ∈ Γ_(G')(v) do
6                E' := E'\{(v, v_γ)};
7                if v_γ ∉ CV then CV := CV ∪ {v_γ};
8            endfor
9            V' := V'\{v}; G' := (V', E')
10       endif
11   until CV = ∅;
12   return (V', E');
end
```

Fig. 3. The pseudo code of graph reduction for MCP

stable sets called the colored vertex set. The objective of the VCP is to find a solution s that minimizes the number k of colored vertex set. Simple heuristic algorithms such as order-based coloring, DSATUR, and RLF are frequently applied to obtain the upper bound of clique. The graph coloring based upper bound is as follows:

$$UpperBound(G, v) = color(G(\Gamma_G(v)), v) + |\{v\}| = color(G(\Gamma_G(v)), v) + 1 \quad (3)$$

3.3 Process of Generating Initial Solution

In every time of main loop of MKLS-R, we perform the process of generating an initial solution (line 3 in Fig. 1).

Although one of the simple ways is, as shown in [12], to select a single vertex $v \in V \setminus \{CC_{best}\}$ in fixed G at random, MKLR-R randomly selects a single vertex v contained in current graph G. Note that the graph G is reduced since the best clique C_{best} is found as shown in Sect. 3.2.

Since KLS randomly adds a vertex contained in PA, a resulting solution (clique) obtained by KLS is not expected as the same clique even if the same initial solution is given. Therefore in this paper, we use the most simplest process of generating initial solution of randomly selecting a single vertex v contained in current graph G that is reduced by the graph reduction process.

4 Experimental Results

To evaluate the effectiveness of the graph reduction process for MKLS on massive graphs, we performed computational experiments on the 112 instances of Network-Repository benchmark graphs [1], that have more than 10000 vertices. The proposed algorithm was implemented in C++. All experiments were performed on a machine with a 3.6 GHz Intel Xeon E-2176G CPU and 32GB RAM under Ubuntu 20.04, using the g++ compiler 9.3.0 with '-O3' option. To execute the DIMACS Machine Benchmark, this machine required 0.13 [s] CPU seconds for `r300.5`, 0.82 [s] for `r400.5` and 3.13 [s] for `r500.5`.

In the experiments, the performance of the MKLS with graph Reduction process (MKLS-R) shown in the Sect. 3.2 are compared with that of MKLS (without graph reduction process). Each algorithm was run independently for 10 times on each benchmark graph. We stop the MKLSs when the maximum execution number of local searches is equal to 1000. The results of algorithms is shown in Table 1. The first three columns of the tables contain the instance names, number of vertices $|V|$, and graph density ρ, respectively. We show for each algorithm, the best found clique size "Best" with the number of times in which the best found cliques could be found by the algorithm "(#B)", the average clique size "Avg", the average running time "Time[s]" in seconds in case the algorithm could find the best found cliques. The bold faced data indicate the best results in each of the columns "Best", "Avg", and "Time[s]" in both algorithms for each graph.

We can notice that the average results "Avg" of MKLS with GRP were at least equal to, or better than those of MKLS except for two graphs `tech-p2p-gnutella` and `socfb-BU10`. As a graph showing more specific results, MKLS-R is about 122.6 times faster than the simple MKLS in graph of `ca-coauthors-dblp` that has the massive number of vertices 540486. For the graph of `ca-hollywood-2009` with 1069126 vertices, MKLS-R is about 6.98 times faster than the simple MKLS. We reveal that MKLS-R is effective to focus on promising search space more than MKLS and contributes the improvement of the search performance especially for sparse massive graphs.

Table 1. Results of MKLS-R and MKLS on the network repository graphs.

Instance			MKLS-R			MKLS		
Name	$|V|$	ρ	Best (#B)	Avg	Time[s]	Best (#B)	Avg	Time[s]
ca-HepPh	11204	1.87×10^{-3}	**239(10)**	**239.00**	**0.035**	**239(10)**	**239.00**	0.102
ca-AstroPh	17903	1.23×10^{-3}	**57(10)**	**57.00**	**0.065**	**57(10)**	**57.00**	2.892
ca-CondMat	21363	4.00×10^{-4}	**26(10)**	**26.00**	**0.032**	**26(10)**	**26.00**	2.390
ca-dblp-2010	226413	2.80×10^{-5}	**75(10)**	**75.00**	**0.312**	75(4)	61.70	20.097
ca-citeseer	227320	3.15×10^{-5}	**87(10)**	**87.00**	**0.349**	87(7)	81.90	28.557
ca-dblp-2012	317080	2.09×10^{-5}	**114(10)**	**114.00**	**0.615**	114(7)	98.70	63.542
ca-MathSciNet	332689	1.48×10^{-5}	**25(10)**	**25.00**	**0.610**	25(1)	21.00	91.446
ca-coauthors-dblp	540486	1.04×10^{-4}	**337(10)**	**337.00**	**4.616**	**337(10)**	**337.00**	565.950
ca-hollywood-2009	1069126	9.85×10^{-5}	**2209(10)**	**2209.00**	24.559	**2209(10)**	**2209.00**	171.582
ia-email-EU	32430	1.03×10^{-4}	**12(10)**	**12.00**	**0.028**	**12(10)**	**12.00**	0.085
ia-enron-large	33696	3.19×10^{-4}	**20(10)**	**20.00**	**0.111**	**20(10)**	**20.00**	0.546
ia-wiki-Talk	92117	8.50×10^{-5}	**15(10)**	**15.00**	**0.367**	**15(10)**	**15.00**	2.313
inf-roadNet-PA	1087562	2.61×10^{-6}	**4(10)**	**4.00**	1.873	3(10)	3.00	**1.652**
inf-roadNet-CA	1957027	1.44×10^{-6}	**4(10)**	**4.00**	**1.958**	4(1)	3.10	394.839
rec-amazon	91813	2.98×10^{-5}	**5(10)**	**5.00**	**0.047**	**5(10)**	**5.00**	2.263
sc-nasasrb	54870	8.71×10^{-4}	**24(10)**	**24.00**	**0.165**	**24(10)**	**24.00**	0.246
sc-pkustk11	87804	6.65×10^{-4}	**36(10)**	**36.00**	8.091	**36(10)**	**36.00**	17.036
sc-pkustk13	94893	7.24×10^{-4}	**36(10)**	**36.00**	0.325	**36(10)**	**36.00**	**0.269**
sc-shipsec1	140385	1.73×10^{-4}	**24(10)**	**24.00**	**1.107**	24(5)	22.00	17.627
sc-shipsec5	179104	1.37×10^{-4}	**24(10)**	**24.00**	0.797	**24(10)**	**24.00**	**0.762**
sc-pwtk	217891	2.38×10^{-4}	**24(10)**	**24.00**	**0.063**	**24(10)**	**24.00**	0.066
sc-msdoor	415863	1.08×10^{-4}	**21(10)**	**21.00**	**0.202**	**21(10)**	**21.00**	0.332
sc-ldoor	952203	4.58×10^{-5}	**21(10)**	**21.00**	**0.607**	**21(10)**	**21.00**	0.806
soc-epinions	26588	2.83×10^{-4}	**16(10)**	**16.00**	**0.052**	**16(10)**	**16.00**	1.014
soc-brightkite	56739	1.32×10^{-4}	**37(10)**	**37.00**	**0.105**	**37(10)**	**37.00**	2.743
soc-slashdot	70068	1.46×10^{-4}	**26(10)**	**26.00**	**0.124**	**26(10)**	**26.00**	0.632
soc-BlogCatalog	88784	5.31×10^{-4}	**45(10)**	**45.00**	**0.836**	**45(10)**	**45.00**	4.201
soc-buzznet	101163	5.40×10^{-4}	**31(10)**	**31.00**	**1.362**	**31(10)**	**31.00**	2.132
soc-LiveMocha	104103	4.05×10^{-4}	**15(10)**	**15.00**	0.808	**15(10)**	**15.00**	1.440
soc-douban	154908	2.73×10^{-5}	**11(10)**	**11.00**	**1.372**	11(9)	10.50	18.083
soc-gowalla	196591	4.92×10^{-5}	**29(9)**	**28.20**	**2.748**	**29(5)**	25.00	63.059
soc-twitter-follows	404719	8.71×10^{-6}	**6(9)**	**5.90**	**9.372**	**6(3)**	5.10	38.254
soc-youtube	495957	1.57×10^{-5}	**16(10)**	**16.00**	**1.343**	**16(10)**	**16.00**	6.678
soc-flickr	513969	2.42×10^{-5}	**58(10)**	**58.00**	**1.641**	**58(10)**	**58.00**	33.979
soc-delicious	536108	9.51×10^{-6}	**21(10)**	**21.00**	**2.206**	21(5)	18.80	92.716
soc-FourSquare	639014	1.57×10^{-5}	**30(8)**	**29.80**	20.472	**30(8)**	**29.80**	323.468
soc-digg	770799	1.99×10^{-5}	**50(10)**	**50.00**	**7.041**	50(8)	48.20	233.917
soc-youtube-snap	1134890	4.64×10^{-6}	**17(10)**	**17.00**	**2.322**	**17(10)**	**17.00**	55.621
soc-lastfm	1191805	6.36×10^{-6}	**14(10)**	**14.00**	**4.712**	**14(10)**	**14.00**	43.560
soc-pokec	1632803	1.67×10^{-5}	**29(1)**	**22.00**	101.473	**29(1)**	20.40	943.302
soc-flixster	2523386	2.49×10^{-6}	**31(10)**	**31.00**	**8.136**	**31(10)**	**31.00**	113.475
soc-livejournal	4033137	3.43×10^{-6}	**214(10)**	**214.00**	**27.535**	214(1)	104.90	1858.750
tech-as-caida2007	26475	1.52×10^{-4}	**16(10)**	**16.00**	**0.020**	**16(10)**	**16.00**	0.041
tech-internet-as	40164	1.06×10^{-4}	**16(10)**	**16.00**	**0.034**	**16(10)**	**16.00**	0.085
tech-p2p-gnutella	62561	7.56×10^{-5}	4(6)	3.60	3.354	**4(10)**	**4.00**	3.879
tech-RL-caida	190914	3.33×10^{-5}	**17(10)**	**17.00**	**0.603**	17(9)	16.90	21.800
tech-as-skitter	1694616	7.73×10^{-6}	**67(10)**	**67.00**	**7.369**	**67(10)**	**67.00**	265.017

(continued)

Table 1. (*continued*)

Instance			MKLS-R			MKLS		
Name	$|V|$	ρ	Best (#B)	Avg	Time[s]	Best (#B)	Avg	Time[s]
web-indochina-2004	11358	7.38×10^{-4}	**50(10)**	**50.00**	**0.010**	**50(10)**	**50.00**	1.097
web-BerkStan	12305	2.58×10^{-4}	**29(10)**	**29.00**	**0.006**	**29(10)**	**29.00**	0.359
web-sk-2005	121422	4.54×10^{-5}	**82(10)**	**82.00**	**0.134**	82(9)	81.20	14.993
web-uk-2005	129632	1.40×10^{-3}	**500(10)**	**500.00**	**1.898**	**500(10)**	**500.00**	86.146
web-arabic-2005	163598	1.31×10^{-4}	**102(10)**	**102.00**	**0.439**	**102(10)**	**102.00**	24.253
web-Stanford	281903	5.01×10^{-5}	**61(10)**	**61.00**	**1.292**	61(3)	56.30	37.570
web-it-2004	509338	5.53×10^{-5}	**432(10)**	**432.00**	**1.827**	432(7)	431.10	242.013
socfb-Bingham82	10004	7.25×10^{-3}	**42(10)**	**42.00**	1.048	**42(10)**	**42.00**	1.061
socfb-Mississippi66	10521	1.10×10^{-2}	**48(10)**	**48.00**	**0.744**	**48(10)**	**48.00**	0.802
socfb-Northwestern25	10567	8.75×10^{-3}	**40(10)**	**40.00**	1.765	**40(10)**	**40.00**	**1.535**
socfb-Cal65	11247	5.56×10^{-3}	**50(10)**	**50.00**	**0.357**	**50(10)**	**50.00**	0.619
socfb-BC17	11509	7.35×10^{-3}	**35(10)**	**35.00**	1.275	**35(10)**	**35.00**	**0.998**
socfb-Stanford3	11586	8.47×10^{-3}	**51(10)**	**51.00**	**0.882**	**51(10)**	**51.00**	0.919
socfb-Columbia2	11770	6.42×10^{-3}	**31(10)**	**31.00**	**2.265**	**31(10)**	**31.00**	2.335
socfb-NotreDame57	12155	7.33×10^{-3}	**25(10)**	**25.00**	**0.386**	**25(10)**	**25.00**	1.303
socfb-GWU54	12193	6.32×10^{-3}	**43(10)**	**43.00**	1.110	**43(10)**	**43.00**	**0.889**
socfb-Baylor93	12803	8.30×10^{-3}	**54(10)**	**54.00**	**0.678**	**54(10)**	**54.00**	1.377
socfb-USF51	13377	3.59×10^{-3}	**44(10)**	**44.00**	**0.469**	**44(10)**	**44.00**	0.590
socfb-Syracuse56	13653	5.84×10^{-3}	**47(10)**	**47.00**	1.550	**47(10)**	**47.00**	**1.212**
socfb-Temple83	13686	3.85×10^{-3}	**35(10)**	**35.00**	0.456	**35(10)**	**35.00**	**0.450**
socfb-UC61	13746	4.68×10^{-3}	**48(10)**	**48.00**	**0.873**	**48(10)**	**48.00**	2.502
socfb-Northeastern19	13882	3.96×10^{-3}	**35(10)**	**35.00**	**0.637**	**35(10)**	**35.00**	1.349
socfb-JMU79	14070	4.91×10^{-3}	**39(10)**	**39.00**	**0.856**	**39(10)**	**39.00**	1.070
socfb-UPenn7	14916	6.17×10^{-3}	**42(10)**	**42.00**	**0.970**	**42(10)**	**42.00**	2.111
socfb-UCSB37	14917	4.33×10^{-3}	**53(10)**	**53.00**	**0.464**	**53(10)**	**53.00**	1.955
socfb-UCF52	14940	3.84×10^{-3}	**61(10)**	**61.00**	**0.315**	**61(10)**	**61.00**	2.426
socfb-UCSD34	14948	3.97×10^{-3}	**43(10)**	**43.00**	**1.195**	**43(10)**	**43.00**	2.700
socfb-Harvard1	15126	7.21×10^{-3}	39(7)	38.70	6.340	39(9)	38.90	**4.611**
socfb-MU78	15436	5.45×10^{-3}	**49(10)**	**49.00**	**1.627**	**49(10)**	**49.00**	1.697
socfb-UMass92	16516	3.81×10^{-3}	**35(10)**	**35.00**	**1.311**	**35(10)**	**35.00**	2.385
socfb-UC33	16808	3.70×10^{-3}	**42(10)**	**42.00**	**1.587**	**42(10)**	**42.00**	2.182
socfb-Tennessee95	16979	5.35×10^{-3}	**58(10)**	**58.00**	**1.267**	**58(10)**	**58.00**	1.603
socfb-UVA16	17196	5.34×10^{-3}	**42(10)**	**42.00**	1.570	**42(10)**	**42.00**	**1.548**
socfb-UConn	17206	4.09×10^{-3}	**50(10)**	**50.00**	**0.642**	**50(10)**	**50.00**	1.262
socfb-UConn91	17212	4.08×10^{-3}	**50(10)**	**50.00**	**0.905**	**50(10)**	**50.00**	1.736
socfb-Oklahoma97	17425	5.88×10^{-3}	**58(10)**	**58.00**	**0.959**	**58(10)**	**58.00**	3.072
socfb-USC35	17444	5.27×10^{-3}	**61(10)**	**61.00**	**1.084**	**61(10)**	**61.00**	1.324
socfb-UNC28	18163	4.65×10^{-3}	**47(10)**	**47.00**	**1.837**	**47(10)**	**47.00**	2.762
socfb-Auburn71	18448	5.72×10^{-3}	**57(10)**	**57.00**	**5.723**	57(9)	56.90	6.566
socfb-Cornell15	18660	4.54×10^{-3}	**40(10)**	**40.00**	**1.192**	**40(10)**	**40.00**	2.442
socfb-BU10	19700	3.29×10^{-3}	38(9)	37.90	**4.964**	**38(10)**	**38.00**	6.636
socfb-UCLA	20453	3.57×10^{-3}	**51(10)**	**51.00**	**0.736**	**51(10)**	**51.00**	3.195
socfb-UCLA26	20467	3.57×10^{-3}	**51(10)**	**51.00**	2.266	**51(10)**	**51.00**	**2.424**
socfb-Maryland58	20871	3.42×10^{-3}	**54(10)**	**54.00**	**2.859**	**54(10)**	**54.00**	6.170
socfb-Virginia63	21325	3.07×10^{-3}	**51(10)**	**51.00**	**1.078**	**51(10)**	**51.00**	2.717

(*continued*)

Table 1. (*continued*)

Instance			MKLS-R			MKLS		
Name	$|V|$	ρ	Best (#B)	Avg	Time[s]	Best (#B)	Avg	Time[s]
socfb-NYU9	21679	3.05×10^{-3}	**37(10)**	**37.00**	**1.447**	37(10)	**37.00**	1.596
socfb-Berkeley13	22900	3.25×10^{-3}	**42(10)**	**42.00**	**1.276**	42(10)	**42.00**	1.892
socfb-Wisconsin87	23831	2.94×10^{-3}	**37(10)**	**37.00**	**2.718**	37(10)	**37.00**	2.867
socfb-UGA50	24389	3.95×10^{-3}	**52(10)**	**52.00**	**3.700**	52(10)	**52.00**	7.866
socfb-Rutgers89	24580	2.60×10^{-3}	**46(10)**	**46.00**	**2.316**	46(10)	**46.00**	2.717
socfb-FSU53	27737	2.69×10^{-3}	**56(10)**	**56.00**	**1.084**	56(10)	**56.00**	2.066
socfb-Indiana	29732	2.95×10^{-3}	**48(10)**	**48.00**	**7.832**	48(10)	**48.00**	10.166
socfb-Indiana69	29747	2.95×10^{-3}	**48(10)**	**48.00**	**2.712**	48(10)	**48.00**	5.294
socfb-Michigan23	30147	2.59×10^{-3}	**44(10)**	**44.00**	**2.793**	44(9)	43.80	7.812
socfb-UIllinois	30795	2.67×10^{-3}	**57(10)**	**57.00**	**3.307**	57(10)	**57.00**	5.779
socfb-UIllinois20	30809	2.66×10^{-3}	**57(10)**	**57.00**	**3.138**	57(10)	**57.00**	5.645
socfb-Texas80	31560	2.45×10^{-3}	**59(10)**	**59.00**	**3.422**	59(9)	58.90	11.520
socfb-MSU24	32375	2.13×10^{-3}	**46(10)**	**46.00**	4.700	46(10)	**46.00**	**4.600**
socfb-UF	35111	2.38×10^{-3}	**55(10)**	**55.00**	**2.786**	55(10)	**55.00**	6.149
socfb-UF21	35123	2.38×10^{-3}	**55(10)**	**55.00**	**1.934**	55(10)	**55.00**	3.773
socfb-Texas84	36364	2.41×10^{-3}	**51(10)**	**51.00**	**2.372**	51(10)	**51.00**	8.719
socfb-Penn94	41536	1.58×10^{-3}	**44(10)**	**44.00**	**5.750**	44(8)	43.70	9.594
socfb-OR	63392	4.07×10^{-4}	**30(10)**	**30.00**	**2.142**	30(6)	29.30	20.462
socfb-B-anon	2937612	4.86×10^{-6}	**24(6)**	**23.10**	**192.025**	24(1)	21.20	325.803
socfb-A-anon	3097165	4.93×10^{-6}	**25(4)**	**23.80**	**189.687**	23(7)	22.10	658.927

5   Conclusion

We proposed a Multi start Local Search incorporating k-opt Local Search and Graph Reduction processes, called Multi-start k-opt Local Search with graph Reduction, for solving the MCP on massive graphs. The computational results showed that the Graph Reduction process is effective in that the performance of MKLS can be improved for massive graphs particularly. Future work is to compare the performance of our MKLS with graph Reduction based approach with those of other state-of-the-art metaheuristics on other benchmark graphs such as the DIMACS10 and real world data.

Acknowledgments. This work was supported in part by JSPS KAKENHI Grant Number JP19K12166.

References

1. Network Repository. https://networkrepository.com/index.php
2. Battiti, R., Protasi, M.: Reactive local search for the maximum clique problem. Algorithmica **29**(4), 610–637 (2001). https://doi.org/10.1007/s004530010074
3. Bomze, I.M., Budinich, M., Pardalos, P.M., Pelillo, M.: The maximum clique problem. In: Du, D.-Z., Pardalos, P.M. (eds.) Handbook of Combinatorial Optimization (suppl. Vol. A), pp. 1–74. Kluwer (1999)

4. Cai, S., Lin, J., Wang, Y., Strash, D.: A semi-exact algorithm for quickly computing a maximum weight clique in large sparse graphs. J. Artif. Intell. Res. **72**, 39–67 (2021)
5. Can, L., Xu, Y.J., Hao, W., Yikai, Z.: Finding the maximum clique in massive graphs. Proc. VLDB Endow. **10**(11), 1538–1549 (2017)
6. Chang, L.: Efficient maximum clique computation over large sparse graphs. In: Proceedings of the 25th ACM SIGKDD International Conference on Knowledge Discovery & Data Mining, pp. 529–538 (2019)
7. Cheng, J., Ke, Y., Fu, A.W.C., Yu, J.X., Zhu, L.: Finding maximal cliques in massive networks by h*-graph. In: Proceedings of the 2010 ACM SIGMOD International Conference on Management of Data, pp. 447–458 (2010)
8. Eppstein, D., Löffler, M., Strash, D.: Listing all maximal cliques in large sparse real-world graphs. ACM J. Exp. Algorithmics **18**, 3.1-3.21 (2013)
9. Garey, M.R., Johnson, D.S.: Computers and Intractability: A Guide to the Theory of NP-Completeness. Freeman, New York (1979)
10. Håstad, J.: Clique is hard to approximate within $n^{1-\varepsilon}$. Acta Math. **182**, 105–142 (1999)
11. Johnson, D.S., Trick, M.A.: Cliques, coloring, and satisfiability. In: Second DIMACS Implementation Challenge. DIMACS Series in Discrete Mathematics and Theoretical Computer Science. American Mathematical Society (1996)
12. Katayama, K., Hamamoto, A., Narihisa, H.: An effective local search for the maximum clique problem. Inf. Process. Lett. **95**(5), 503–511 (2005)
13. Kernighan, B.W., Lin, S.: An efficient heuristic procedure for partitioning graphs. Bell Syst. Tech. J. **49**, 291–307 (1970)
14. Khot, S.: Improved inapproximability results for maxclique, chromatic number and approximate graph coloring. In: Proceedings of the 42nd IEEE Symposium on Foundations of Computer Science, pp. 600–609 (2001)
15. Lin, S., Kernighan, B.W.: An effective heuristic algorithm for the traveling salesman problem. Oper. Res. **21**, 498–516 (1973)
16. Modani, N., Dey, K.: Large maximal cliques enumeration in sparse graphs. In: Proceedings of the 17th ACM Conference on Information and Knowledge Management, pp. 1377–1378 (2008)
17. Rossi, R.A., Ahmed, N.K.: The network data repository with interactive graph analytics and visualization. In: AAAI (2015)
18. Tomita, E.: Efficient algorithms for finding maximum and maximal cliques and their applications. In: Poon, S.-H., Rahman, M.S., Yen, H.-C. (eds.) WALCOM 2017. LNCS, vol. 10167, pp. 3–15. Springer, Cham (2017). https://doi.org/10.1007/978-3-319-53925-6_1
19. Tomita, E., et al.: An improved branch-and-bound MCT algorithm for finding a maximum clique. Springer (to appear)
20. Wu, Q., Hao, J.-K.: A review on algorithms for maximum clique problems. Eur. J. Oper. Res. **242**(3), 693–709 (2015)

A Method for Reducing Number of Parameters of Octave Convolution in Convolutional Neural Networks

Yusuke Gotoh[✉] and Yu Inoue

Graduate School of Natural Science and Technology, Okayama University, Okayama, Japan
y-gotoh@okayama-u.ac.jp

Abstract. Computational machine learning has attracted a great deal of attention for its ability to analyze large-scale data. In particular, convolutional neural networks (CNNs) have been proposed in the fields of image recognition and object detection in efforts to develop models with improved accuracy as well as more lightweight models that require a smaller number of parameters and a lower computational cost. Octave Convolution (OctConv) is a method used to reduce the memory and computational cost of a model while also improving its accuracy by replacing the conventional convolutional layer with an OctConv layer. However, the number of parameters used in OctConv is almost the same as that in the case of conventional convolutional processing. In this paper, we propose the Pointwise Octave Convolution (Pointwise OctConv) method, which combines the Pointwise Convolution (Pointwise Conv) method with OctConv to reduce the number of parameters used in OctConv and thus create a lighter model. In the proposed method, the number of parameters is reduced by performing Pointwise Conv before and after the convolution process for each path in the OctConv layer. In an evaluation using ResNet-56, the proposed method reduces the number of parameters by about 63.8% with a loss of classification accuracy of 3.04% when $\alpha = 0.75$.

1 Introduction

Recently, machine learning by computer has attracted a great deal of attention for its ability to analyze large-scale data. In particular, convolutional neural networks (CNNs) have been widely used in the fields of image recognition and object detection [1]. In order to improve the accuracy of CNNs, models such as GoogLeNet [2], VGG [3], ResNet [4], and DenseNet [5] have been proposed. In addition, to reduce the number of parameters and computational cost, lightweight models such as SqueezeNet [6], MobileNets [7], ShuffleNet [8], and MobileNetV2 [9] have been proposed.

Octave Convolution (OctConv) [10] can reduce computational and memory costs while improving the accuracy of CNNs by replacing the conventional convolutional layer with an OctConv layer. However, the number of parameters used in OctConv is almost the same as that in the case of conventional convolutional processing.

In this paper, we propose the Pointwise Octave Convolution (Pointwise OctConv) method, which combines the Pointwise convolution (Pointwise Conv) method with OctConv to reduce the number of parameters used in OctConv and thus create a lighter

© The Author(s), under exclusive license to Springer Nature Switzerland AG 2022
L. Barolli et al. (Eds.): EIDWT 2022, LNDECT 118, pp. 212–222, 2022.
https://doi.org/10.1007/978-3-030-95903-6_23

model. In the proposed method, the number of parameters is reduced by performing Pointwise Conv before and after the convolution process for each path in the OctConv layer.

The remainder of the paper is organized as follows. Convolutional neural networks are introduced in Sect. 2. In Sect. 3, we give an overview of the conventional methods developed for reducing the number of parameters in the convolutional layer. Next, Oct-Conv is introduced and explained in Sect. 4. In Sect. 5, we present the details of the proposed method, and then we evaluate its performance in Sect. 6. Finally, we conclude the paper in Sect. 7.

2   Convolutional Neural Networks

Convolutional neural networks (CNNs) are composed of multiple layers, and they offer high performance in image recognition. Here, we describe the convolutional and pooling layers that make up CNNs.

2.1   Convolutional Layer

In the convolutional layer, feature points are extracted from the original image using a set of small feature detectors called kernels. Each kernel can extract the characteristic structure of the original image by sliding pixel by pixel over a fixed area in the image and calculating the weighted sum in each area.

By iteratively performing this calculation, we can output a feature map with a characteristic structure by the kernel. By running through kernels in local regions, the convolutional computation can extract features at all locations in the image.

2.2   Pooling Layer

The pooling layer compresses the original image without missing any important feature information. By compressing the image in the pooling layer, the effects of positional changes, such as movement and rotation of the image, can be reduced.

3   Conventional Methods for Reducing Number of Parameters in Convolutional Layer

3.1   Pruning

The pruning method in CNNs reduces the number of parameters and computational cost by setting the weights of the convolutional layer and all of the combined layers to zero. In addition, pruning is robust against noise [11]. CNNs typically alternate between training the network and pruning the neural networks.

There are two types of pruning methods: Unstructured Pruning [12, 13] and Structured Pruning [14]. Unstructured Pruning reduces the weights based on the importance of each element in the kernel. By reducing the number of unimportant neurons scattered through the model, Unstructured Pruning can reduce the number of neurons at a high

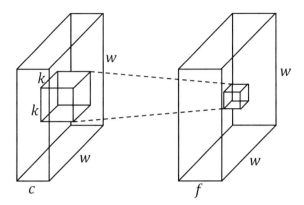

Fig. 1. Conventional convolutional method

rate while maintaining the classification accuracy. On the other hand, unstructured pruning has difficulty in improving the computational efficiency. Structured Pruning reduces the weights based on the importance of each kernel or channel. Since the kernel can be reconstructed for a network with a reduced number of channels, the inference speed of Structured Pruning can be accelerated.

3.2   Decomposition of Convolution

As a means to reduce the number of parameters in the convolutional layer, methods such as Pointwise convolution (Pointwise Conv) and Depthwise convolution (Depthwise Conv) have been proposed. In this section, we explain the number of parameters and computational cost required for the conventional convolutional method, the Pointwise Conv method, and the Depthwise Conv method.

Conventional Convolutional Method

The structure of a conventional convolutional method using a convolutional layer is shown in Fig. 1. The size of the feature map in the input layer is $w \times w$, the number of channels in the input layer is c, the kernel size is $k \times k$, and the number of channels in the output layer is f. The computational cost of convolution on the feature map in the input layer is k^2c per location. By applying this amount of computation to w^2 locations in the feature map of the input layer, a feature map of the output layer with a single channel is generated. Therefore, the computational cost for f feature maps in the output layer is k^2cw^2f. Furthermore, since the number of parameters in the feature map of one channel in the output layer is k^2c, the number of parameters for f feature maps in the output layer is k^2cf. Pointwise Conv is a method using a conventional convolutional layer at $k = 1$. In this case, the computational cost is w^2cf and the number of parameters is cf.

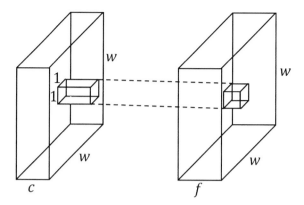

Fig. 2. Pointwise convolution method

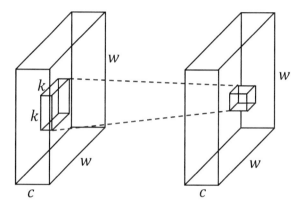

Fig. 3. Depthwise convolution

Pointwise Convolution Method

The pointwise convolution (Pointwise Conv) method [7], shown in Fig. 2, is a convolution with a 1×1 kernel used for the Inception module in GoogLeNet [2] and for the bottleneck structure in ResNet [4]. In Pointwise Conv, the number of parameters can be reduced by performing convolution on the feature map in the channel direction instead of in the spatial direction. This method can also be used to vary the number of channels in a feature map.

Depthwise Convolution Method

As shown in Fig. 3, depthwise convolution [7] performs convolution in the spatial direction for each channel on the feature map. Since depthwise convolution (Depthwise Conv) does not perform convolution in the channel direction, the computational cost for one convolution is k^2. Moreover, the number of input channels is equal to the number of output channels, and thus $c = f$. Therefore, the computational cost of Depthwise Conv is w^2ck^2, and the number of parameters is k^2c.

Neural network models such as MobileNets [7], MobileNetV2 [9], SliceNet [15], and Xception [16] combine Pointwise Conv and Depthwise Conv to perform convolution. These models reduce the number of parameters and computational cost compared to conventional convolutional layers that convolute simultaneously in the spatial and channel directions.

4   Octave Convolution

4.1   Constitution of Octave Convolution

Octave convolution (OctConv) [10] is a method designed to reduce the computational cost of CNNs by decomposing the feature map into high-frequency and low-frequency components and performing convolution on each component. Let X be the input tensor and Y be the output tensor satisfying $X, Y \in \mathbb{R}^{c \times h \times w}$. Next, decompose the input X into its high-frequency and low-frequency components and set them as $X^H \in \mathbb{R}^{(1-\alpha)c \times h \times w}$ and $X^L \in \mathbb{R}^{\alpha c \times \frac{h}{2} \times \frac{w}{2}}$, respectively. Here, h and w are spatial dimensions, c is the number of channels, and $\alpha \in [0, 1]$ is the ratio of channels assigned to the low-frequency component.

Similarly, let $Y^H = Y^{H \to H} + Y^{L \to H}$ denote the high-frequency component and $Y^L = Y^{L \to L} + Y^{H \to L}$ denote the low-frequency component of the output Y. Here, $Y^{A \to B}$ represents the update of the convolution from group A to group B on the feature map. Accordingly, $Y^{H \to H}$ and $Y^{L \to L}$ update the information at the same frequency, while $Y^{H \to L}$ and $Y^{L \to H}$ update information at different frequencies.

Let $W \in \mathbb{R}^{c \times k \times k}$ be the convolution kernel of $k \times k$. This kernel performs the convolution of X^H and X^L by partitioning the components into $W = [W^H, W^L]$. Each component can also be partitioned to separate the same and different frequencies as $W^H = [W^{H \to H}, W^{L \to H}]$ and $W^L = [W^{L \to L}, W^{H \to L}]$.

The design of OctConv is shown in Fig. 4. OctConv consists of four different computational paths, and the output $Y = \{Y^H, Y^L\}$ can be expressed by the following equations.

$$Y^H = f(X^H; W^{H \to H}) + \text{upsample}(f(X^L; W^{L \to H}), 2), \tag{1}$$

$$Y^L = f(X^L; W^{L \to L}) + f(\text{pool}(X^H, 2); W^{H \to L}), \tag{2}$$

where $\text{pool}(X; W)$ is the convolutional computation with parameter W, $\text{pool}(X; k)$ is the Average Pooling when the kernel size is $k \times k$ and the stride is k, and $\text{upsample}(X; k)$ is the UpSampling of the coefficient k in the Nearest-neighbor interpolation. The two green arrows in Fig. 4 indicate the update of the feature map within the high-frequency component and within the low-frequency component. The two red arrows indicate the information exchange between the two frequencies.

OctConv can improve the accuracy of image recognition by dividing the feature map into high and low frequency components in space and exchanging their information. In addition, OctConv can reduce the computational cost and memory usage by reducing the resolution of low-frequency components.

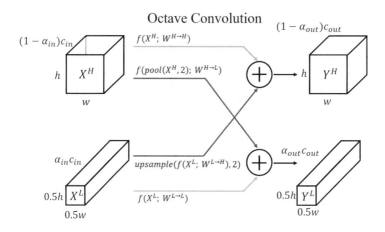

Fig. 4. Constitution of OctConv

4.2 Computational Cost and Number of Parameters in Octave Convolution

Next, we describe the computational cost and number of parameters in OctConv. Oct-Conv performs convolution calculations on four different paths: $H \rightarrow H, H \rightarrow L, L \rightarrow H$, and $L \rightarrow L$. When $c_{in} = c_{out} = c$ and $\alpha_{in} = \alpha_{out} = \alpha$, the computational cost for each path is shown below.

$$FLOPS(Y^{H \rightarrow H}) = h \times w \times k^2 \times (1 - \alpha)^2 \times c^2, \tag{3}$$

$$FLOPS(Y^{H \rightarrow L}) = \frac{h}{2} \times \frac{w}{2} \times k^2 \times \alpha \times (1 - \alpha) \times c^2, \tag{4}$$

$$FLOPS(Y^{L \rightarrow H}) = \frac{h}{2} \times \frac{w}{2} \times k^2 \times (1 - \alpha) \times \alpha \times c^2, \tag{5}$$

$$FLOPS(Y^{L \rightarrow L}) = \frac{h}{2} \times \frac{w}{2} \times k^2 \times \alpha^2 \times c^2. \tag{6}$$

Therefore, the computational cost of OctConv can be expressed by summing the values in Eqs. (3) to (6) as follows.

$$FLOPS([Y^H, Y^L]) = (1 - \frac{3}{4}\alpha(2 - \alpha)) \times h \times w \times k^2 \times c^2. \tag{7}$$

The ratio of the computational cost between OctConv and conventional convolution is $1 - \frac{3}{4}\alpha(2 - \alpha)$ for the same number of input and output channels and kernel size. This ratio increases as the value of α approaches one, which reduces the computational cost of OctConv. The number of parameters for OctConv is shown in the following equation, which is the same as the number of parameters for general convolution.

$$k^2(1 - \alpha)^2 c^2 + k^2 \alpha(1 - \alpha)c^2 + k^2(1 - \alpha)\alpha c^2 + k^2 \alpha^2 c^2 = k^2 c^2. \tag{8}$$

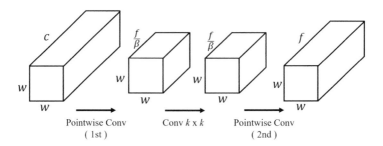

Fig. 5. Bottleneck structure

5 Proposed Method

5.1 Bottleneck Structure

We propose the Pointwise Octave Convolution (Pointwise OctConv) method, which combines OctConv and Pointwise Conv to reduce the number of parameters. In the proposed method, we introduce the bottleneck structure [17] in the Residual Units of ResNet [4]. In the bottleneck structure, the number of parameters is reduced by sandwiching the convolutional layers with Pointwise Conv.

As shown in Fig. 5, the input feature map is $w \times w$, the number of input channels is c, the kernel size is $k \times k$, the number of output channels is f, and the channel compression ratio is β. By inserting the first Pointwise Conv before Conv $k \times k$, the number of channels is compressed from c to f/β. In this case, the number of input channels at Conv $k \times k$ is reduced from c to f/β. Next, by inserting a second Pointwise Conv after Conv $k \times k$, the number of channels increases from f/β to f. In this case, the number of output channels in Conv $k \times k$ is reduced from c to f/β.

The number of parameters in the convolutional layer is ck^2f, which is proportional to the number of input channels c and the number of output channels f. Therefore, by sandwiching the bottleneck convolutional layer with Pointwise Conv, the number of input and output channels is reduced, thus reducing the number of parameters.

5.2 Structure of Proposed Method

In the proposed method, the convolutional layer of each path is sandwiched by Pointwise Conv. For the two paths $H \rightarrow H$ and $L \rightarrow L$, the proposed method simply sandwiches the convolution layer with Pointwise Conv as shown in Fig. 6(a). For the $H \rightarrow L$ path, the proposed method sandwiches the convolutional layer with Pointwise Conv after performing Average Pooling, as shown in Fig. 6(b). In the $L \rightarrow H$ path, UpSampling is performed after the convolutional layer is sandwiched by Pointwise Conv, as shown in Fig. 6(c). Finally, we evaluate each of these four paths and select the one that has the greatest effect on reducing the number of parameters while minimizing the degradation of classification accuracy.

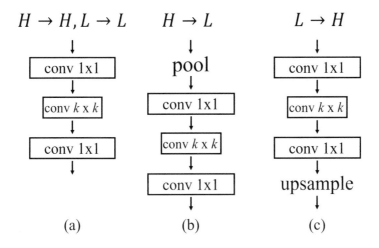

Fig. 6. Structure of proposed method

5.3 Computational Cost

In Fig. 5, the computational cost for the first Pointwise Conv is $w^2 c \frac{f}{\beta}$. Since the computational cost for the Conv $k \times k$ sandwiched between the previous and next Pointwise Conv is $k^2 w^2 \left(\frac{f}{\beta}\right)^2$, the computational cost for the second Pointwise Conv is $w^2 \frac{f^2}{\beta}$. Therefore, the overall computational cost is calculated by the following equation.

$$w^2 c \frac{f}{\beta} + k^2 w^2 \left(\frac{f}{\beta}\right)^2 + w^2 \frac{f^2}{\beta} = \frac{w^2 f}{\beta^2}(k^2 f + \beta(c + f)). \tag{9}$$

In this case, when $w = 32$, $c = f = 16$, $k = 3$, and $\beta = 2$, the computational cost decreases by about 63.9%. Finally, based on Eq. (9), the computational cost of using the bottleneck structure for each path in OctConv is calculated by the following formula.

$$FLOPS(Y^{H \to H}) = \frac{(1 - \alpha)^2}{\beta^2} \cdot hwc^2(k^2 + 2\beta), \tag{10}$$

$$FLOPS(Y^{H \to L}) = \frac{\alpha}{4\beta^2} \cdot hwc^2(\alpha k^2 + \beta), \tag{11}$$

$$FLOPS(Y^{L \to H}) = \frac{(1 - \alpha)}{4\beta^2} \cdot hwc^2((1 - \alpha)k^2 + \beta), \tag{12}$$

$$FLOPS(Y^{L \to L}) = \frac{\alpha^2}{4\beta^2} \cdot hwc^2(k^2 + 2\beta). \tag{13}$$

5.4 Number of Parameters

The number of parameters for the proposed method is calculated by the following equation.

$$c \cdot \frac{f}{\beta} + k^2 \cdot \frac{f}{\beta} \cdot \frac{f}{\beta} + \frac{f}{\beta} \cdot f = \frac{f}{\beta^2}(k^2 f + \beta(c + f)). \tag{14}$$

Table 1. Classification accuracy, number of parameters, and computational complexity for each method

Model	α	Classification accuracy [%]	Number of parameters	Computational complexity
ResNet-56	–	92.58	0.86×10^6	1.25×10^8
Oct-ResNet-56	0.25	93.13	0.86×10^6	0.85×10^8
Pwise-Oct-ResNet-56		90.31	0.31×10^6	0.36×10^8
Oct-ResNet-56	0.75	91.62	0.86×10^6	0.38×10^8
Pwise-Oct-ResNet-56		88.58	0.31×10^6	0.18×10^8

Similar to Eq. (9) in Subsect. 5.3, when $w = 32$, $c = f = 16$, $k = 3$, and $\beta = 2$, the number of parameters is reduced by about 63.9%.

6   Evaluation

6.1   Constitution of CNNs

In the evaluation, we compared the proposed method with the conventional methods using the dataset CIFAR-10 [18]. We use ResNet [4] as the neural network model. The CNN model that applies the existing OctConv method [10] to ResNet is called Oct-ResNet, and the CNN model that applies the proposed Pointwise OctConv method to ResNet is called Pwise-Oct-ResNet. In Oct-ResNet, all convolutional layers except for the first one are replaced by OctConv layers. For all layers except the first and last OctConv layers, we set $\alpha_{in} = \alpha_{out} = \alpha$. In the first OctConv layer, we set $\alpha_{in} = 0$ and $\alpha_{out} = \alpha$. In the last OctConv layer, $\alpha_{in} = 0$ and $\alpha_{out} = \alpha$ are set.

In PwiseOct-ResNet, the convolutional layers of each path are sandwiched between Pointwise Conv layers with the same settings as in Oct-ResNet. All models use momentum as an optimizer to represent the optimization algorithm used in the CNNs.

6.2   Computer Environment

The OS is Ubuntu 18.04 LTS, the CPU is an Intel Core 17-8700K (3.70 GHz), and the GPU is a GeForce RTX 2080. We used Keras as the neural network library for evaluation.

6.3   Evaluation Result

Table 1 shows the classification accuracies of ResNet-56, Oct-ResNet-56, and Pwide-Oct-ResNet-56 when Pointwise Conv is applied to all paths. Here, $\beta = 2$, and the computational complexity was calculated using pseudocode.

As shown in the table, the proposed method reduced the number of parameters by about 63.8% compared to Oct-ResNet-56 in the case of $\alpha = 0.75$, where the ratio of low-frequency channels was large. On the other hand, the classification accuracy of the proposed method was reduced by about 3.04% compared to Oct-ResNet-56. In the proposed method, the number of parameters was greatly reduced by applying Pointwise Conv to all of the paths, which has a large impact on the classification accuracy.

7 Conclusion

In this paper, as a way to reduce the number of parameters used in OctConv and thus create a lighter model, we proposed the Pointwise OctConv method, which applies the conventional method of pointwise convolution to OctConv. The proposed method reduces the number of parameters by performing pointwise convolution before and after the convolution process of each path in the OctConv layer. In our evaluation, we confirmed that the proposed method can reduce the number of parameters more than the conventional method when the ratio of low-frequency channels is large.

In the future, we will evaluate our method using models other than ResNet as well as larger datasets.

Acknowledgement. This work was supported by JSPS KAKENHI Grant Number 18K11265 and 21H03429, and JGC-S Scholarship Foundation.

References

1. Krizhevsky, A., Sutskever, I., Hinton, G.E.: ImageNet classification with deep convolutional neural networks. In: Advances in Neural Information Processing Systems, pp. 1097–1105 (2012)
2. Szegedy, C., et al.: Going deeper with convolutions. In: Proceedings of the IEEE Conference on Computer Vision and Pattern Recognition, pp. 1–9 (2015)
3. Simonyan, K., Zisserman, A.: Very deep convolutional networks for large-scale image recognition. arXiv (online). https://arxiv.org/pdf/1409.1556.pdf. Accessed 15 Nov 2021
4. He, K., Zhang, X., Ren, S., Sun, J.: Deep residual learning for image recognition, arXiv (online). https://arxiv.org/pdf/1512.03385.pdf. Accessed 15 Nov 2021
5. Huang, G., Liu, Z., Maaten, L., Weinberger, K.Q.: Densely connected convolutional networks, arXiv (online). https://arxiv.org/pdf/1608.06993.pdf. Accessed 15 Nov 2021
6. Iandola, F.N., Han, S., Moskewicz, M.W., Ashraf, K., Dally, W.J., Keutzer, K.: SqueezeNet: AlexNet-level accuracy with 50x fewer parameters and < 0.5 MB model size, arXiv (online). https://arxiv.org/pdf/1602.07360.pdf. Accessed 15 Nov 2021
7. Howard, A.G., et al.: MobileNets: efficient convolutional neural networks for mobile vision applications, arXiv (online). https://arxiv.org/pdf/1704.04861.pdf. Accessed 15 Nov 2021
8. Zhang, X., Zhou, X., Lin, M., Sun, J.: ShuffleNet: an extremely efficient convolutional neural networks for mobile devices, arXiv (online). https://arxiv.org/pdf/1707.01083.pdf. Accessed 15 Nov 2021
9. Sandler, M., Howard, A., Zhu, M., Zhmoginov, A., Chen, L.C.: MobileNetV2: inverted residuals and linear bottlenecks. In: Proceedings of the IEEE Conference on Computer Vision and Pattern Recognition, pp. 4510–4520 (2018)
10. Chen, Y., et al.: Drop an octave: reducing spatial redundancy in convolutional neural networks with octave convolution, arXiv (online). https://arxiv.org/pdf/1904.05049.pdf. Accessed 15 Nov 2021

11. Ahmad, S., Scheinkman, L.: How can we be so dense? The benefits of using highly sparse representations, arXiv (online). https://arxiv.org/pdf/1903.11257.pdf. Accessed 15 Nov 2021
12. Han, S., Pool, J., Tran, J., Dally, W.J.: Learning both weights and connections for efficient neural networks, arXiv (online). https://arxiv.org/pdf/1506.02626v3.pdf. Accessed 15 Nov 2021
13. Han, S., Mao, H., Dally, W.J.: Deep compression: compressing deep neural networks with pruning, trained quantization and Huffman coding, arXiv (online). https://arxiv.org/pdf/1510.00149.pdf. Accessed 15 Nov 2021
14. Li, H., Kadav, A., Durdanovic, I., Samet, H., Graf, H.P.: Pruning filters for efficient ConvNets, arXiv (online). https://arxiv.org/pdf/1608.08710.pdf. Accessed 15 Nov 2021
15. Kaiser, L., Gomez, A.N., Chollet, F.: Depthwise separable convolutions for neural machine translation, arXiv (online). https://arxiv.org/pdf/1706.03059.pdf. Accessed 15 Nov 2021
16. Chollet, F.: Xception: deep learning with depthwise separable convolutions, arXiv (online). https://arxiv.org/pdf/1610.02357.pdf. Accessed 15 Nov 2021
17. He, K., Zhang, X., Ren, S., Sun, J.: Identity mappings in deep residual networks, arXiv (online). https://arxiv.org/pdf/1603.05027.pdf. Accessed 15 Nov 2021
18. The CIFAR-10 and CIFAR-100 datasets (online). https://www.cs.toronto.edu/~kriz/cifar.html. Accessed 15 Nov 2021

A Comparison Study of RIWM with RDVM and CM Router Replacement Methods for WMNs Considering Boulevard Distribution of Mesh Clients

Admir Barolli[1(✉)], Phudit Ampririt[2], Shinji Sakamoto[3], Elis Kulla[4], and Leonard Barolli[5]

[1] Department of Information Technology, Aleksander Moisiu University of Durres, L.1, Rruga e Currilave, Durres, Albania
[2] Graduate School of Engineering, Fukuoka Institute of Technology, 3-30-1 Wajiro-Higashi, Higashi-Ku, Fukuoka 811-0295, Japan
bd21201@bene.fit.ac.jp
[3] Department of Information and Computer Science, College of Engineering, Kanazawa Institute of Technology (KIT), 7-1 Ohgigaoka, Nonoichi, Ishikawa 921-8501, Japan
shinji.sakamoto@ieee.org
[4] Department of Information and Computer Engineering, Okayama University of Science (OUS), 1-1 Ridaicho, Kita-Ku, Okayama 700-0005, Japan
kulla@ice.ous.ac.jp
[5] Department of Information and Communication Engineering, Fukuoka Institute of Technology, 3-30-1 Wajiro-Higashi, Higashi-Ku, Fukuoka 811-0295, Japan
barolli@fit.ac.jp

Abstract. The Wireless Mesh Networks (WMNs) have attracted attention for different applications. They are an important networking infrastructure and they have many advantages such as low cost and high-speed wireless Internet connectivity. However, they have some problems such as router placement, covering of mesh clients and load balancing. To deal with these problems, in our previous work, we implemented a hybrid simulation system based on Particle Swarm Optimization (PSO) and Distributed Genetic Algorithm (DGA) called WMN-PSODGA. Moreover, we added in the fitness function a new parameter for the load balancing of the mesh routers called NCMCpR (Number of Covered Mesh Clients per Router). In this paper, we consider Boulevard distribution of mesh clients and three router replacement methods: Random Inertia Weight Method (RIWM), Rational Decrement of Vmax Method (RDVM) and Constriction Method (CM). We carry out simulations using WMN-PSODGA hybrid simulation system and compare the performance of RIWM with RDVM and CM. The simulation results show that RIWM has better loading balancing than RDVM and CM.

L. Barolli et al. (Eds.): EIDWT 2022, LNDECT 118, pp. 223–235, 2022.
https://doi.org/10.1007/978-3-030-95903-6_24

1   Introduction

The wireless networks and devices can provide users access to information and communication anytime and anywhere [3, 8–11, 14, 20, 26, 27, 29, 33]. The Wireless Mesh Networks (WMNs) are gaining a lot of attention because of their low-cost that makes them attractive for providing wireless Internet connectivity. A WMN is dynamically self-organized and self-configured. The nodes in the network automatically establish and maintain the mesh connectivity among itself by creating an ad-hoc network. This feature brings many advantages to WMN such as easy network maintenance, robustness and reliable service coverage [1]. Moreover, such infrastructure can be deployed in community networks, metropolitan area networks, municipal networks to support applications for urban areas, medical, transport and surveillance systems.

Mesh node placement in WMNs can be seen as a family of problems, which is shown (through graph theoretic approaches or placement problems, e.g. [6, 15]) to be computationally hard to solve for most of the formulations [37].

In this work, we consider the version of mesh router nodes placement problem in which we are given a grid area. We consider where to deploy a number of mesh router nodes and a number of mesh client nodes of fixed positions (of an arbitrary distribution) in the grid area. The objective is to find a location assignment for the mesh routers to the cells of the grid area that maximizes the network connectivity, client coverage and consider load balancing for each router. Network connectivity is measured by Size of Giant Component (SGC) of the resulting WMN graph, while the user coverage is simply the number of mesh client nodes that fall within the radio coverage of at least one mesh router node and is measured by Number of Covered Mesh Clients (NCMC). For load balancing, we added in the fitness function a new parameter called NCMCpR (Number of Covered Mesh Clients per Router).

Node placement problems are known to be computationally hard to solve [12, 13, 38]. In previous works, some intelligent algorithms have been recently investigated for node placement problem [4, 7, 16, 18, 21–23, 31, 32].

In [24], we implemented a Particle Swarm Optimization (PSO) based simulation system called WMN-PSO and another simulation system based on Genetic Algorithm (GA) called WMN-GA [19], for solving node placement problem in WMNs. Then, we designed and implemented a hybrid simulation system based on PSO and Distributed GA (DGA). We call this system WMN-PSODGA.

In this paper, we present the performance analysis of WMNs using WMN-PSODGA system considering Boulevard distribution of mesh clients and three router replacement methods: Random Inertia Weight Method (RIWM), Rational Decrement of Vmax Method (RDVM) and Constriction Method (CM). We compare the performance of RIWM with RDVM and CM.

The rest of the paper is organized as follows. We present our designed and implemented hybrid simulation system in Sect. 2. The simulation results are given in Sect. 3. Finally, we give conclusions and future work in Sect. 4.

2   Proposed and Implemented Simulation System

2.1   Particle Swarm Optimization

In PSO, a number of simple entities (the particles) are placed in the search space of some problems or functions and each evaluates the objective function at its current location. The objective function is often minimized and the exploration of the search space is not through evolution [17].

Each particle then determines its movement through the search space by combining some aspect of the history of its own current and best (best-fitness) locations with those of one or more members of the swarm, with some random perturbations. The next iteration takes place after all particles have been moved. Eventually the swarm as a whole, like a flock of birds collectively foraging for food, is likely to move close to an optimum of the fitness function.

Each individual in the particle swarm is composed of three $\mathcal{D}$-dimensional vectors, where $\mathcal{D}$ is the dimensionality of the search space. These are the current position $\vec{x}_i$, the previous best position $\vec{p}_i$ and the velocity $\vec{v}_i$.

The particle swarm is more than just a collection of particles. A particle by itself has almost no power to solve any problem; progress occurs only when the particles interact. Problem solving is a population-wide phenomenon, emerging from the individual behaviors of the particles through their interactions. In any case, populations are organized according to some sort of communication structure or topology, often thought of as a social network. The topology typically consists of bidirectional edges connecting pairs of particles, so that if j is in i's neighborhood, i is also in j's. Each particle communicates with some other particles and is affected by the best point found by any member of its topological neighborhood. This is just the vector $\vec{p}_i$ for that best neighbor, which we will denote with $\vec{p}_g$. The potential kinds of population "social networks" are hugely varied, but in practice certain types have been used more frequently. We show the pseudo code of PSO in Algorithm 1.

In the PSO process, the velocity of each particle is iteratively adjusted so that the particle stochastically oscillates around $\vec{p}_i$ and $\vec{p}_g$ locations.

2.2   Distributed Genetic Algorithm

The Distributed Genetic Algorithm (DGA) has been used in various fields of science. The DGA has shown their usefulness for the resolution of many computationally hard combinatorial optimization problems. We show the pseudo code of DGA in Algorithm 2.

Population of individuals: Unlike local search techniques that construct a path in the solution space jumping from one solution to another one through local perturbations, DGA use a population of individuals giving thus the search a larger scope and chances to find better solutions. This feature is also known as "exploration" process in difference to "exploitation" process of local search methods.

Algorithm 1. Pseudo code of PSO.

/* Initialize all parameters for PSO */
Computation maxtime:= Tp_{max}, $t := 0$;
Number of particle-patterns:= m, $2 \leq m \in \mathbf{N}^1$;
Particle-patterns initial solution:= $\mathbf{P}_i^0$;
Particle-patterns initial position:= $\mathbf{x}_{ij}^0$;
Particles initial velocity:= $\mathbf{v}_{ij}^0$;
PSO parameter:= ω, $0 < \omega \in \mathbf{R}^1$;
PSO parameter:= C_1, $0 < C_1 \in \mathbf{R}^1$;
PSO parameter:= C_2, $0 < C_2 \in \mathbf{R}^1$;
/* Start PSO */
Evaluate($\mathbf{G}^0, \mathbf{P}^0$);
while $t < Tp_{max}$ **do**
    /* Update velocities and positions */
    $v_{ij}^{t+1} = \omega \cdot v_{ij}^t$
        $+C_1 \cdot \mathrm{rand}() \cdot (best(P_{ij}^t) - x_{ij}^t)$
        $+C_2 \cdot \mathrm{rand}() \cdot (best(G^t) - x_{ij}^t)$;
    $x_{ij}^{t+1} = x_{ij}^t + v_{ij}^{t+1}$;
    /* if fitness value is increased, a new solution will be accepted. */
    Update_Solutions($\mathbf{G}^t, \mathbf{P}^t$);
    $t = t + 1$;
end while
Update_Solutions($\mathbf{G}^t, \mathbf{P}^t$);
return Best found pattern of particles as solution;

Fitness: The determination of an appropriate fitness function, together with the chromosome encoding are crucial to the performance of DGA. Ideally we would construct objective functions with "certain regularities", i.e. objective functions that verify that for any two individuals which are close in the search space, their respective values in the objective functions are similar.

Selection: The selection of individuals to be crossed is another important aspect in DGA as it impacts on the convergence of the algorithm. Several selection schemes have been proposed in the literature for selection operators trying to cope with premature convergence of DGA. There are many selection methods in GA. In our system, we implement 2 selection methods: Random method and Roulette wheel method.

Crossover operators: Use of crossover operators is one of the most important characteristics. Crossover operator is the means of DGA to transmit best genetic features of parents to offsprings during generations of the evolution process. Many methods for crossover operators have been proposed such as Blend Crossover (BLX-α), Unimodal Normal Distribution Crossover (UNDX), Simplex Crossover (SPX).

Mutation operators: These operators intend to improve the individuals of a population by small local perturbations. They aim to provide a component of randomness in the neighborhood of the individuals of the population. In our system, we implemented two mutation methods: uniformly random mutation and boundary mutation.

Escaping from local optima: GA itself has the ability to avoid falling prematurely into local optima and can eventually escape from them during the search process. DGA has one more mechanism to escape from local optima by considering some islands. Each island computes GA for optimizing and they migrate its gene to provide the ability to avoid from local optima (see Fig. 1).

Convergence: The convergence of the algorithm is the mechanism of DGA to reach to good solutions. A premature convergence of the algorithm would cause that all individuals of the population be similar in their genetic features and thus the search would result ineffective and the algorithm getting stuck into local optima. Maintaining the diversity of the population is therefore very important to this family of evolutionary algorithms.

Algorithm 2. Pseudo code of DGA.

/* Initialize all parameters for DGA */
Computation maxtime:= Tg_{max}, $t := 0$;
Number of islands:= n, $1 \leq n \in N^1$;
initial solution:= $\boldsymbol{P}_i^0$;
/* Start DGA */
Evaluate($\boldsymbol{G}^0, \boldsymbol{P}^0$);
while $t < Tg_{max}$ **do**
   for all islands **do**
      Selection();
      Crossover();
      Mutation();
   end for
   $t = t + 1$;
end while
Update_Solutions($\boldsymbol{G}^t, \boldsymbol{P}^t$);
return Best found pattern of particles as solution;

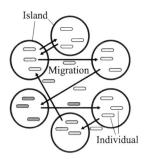

Fig. 1. Model of migration in DGA.

2.3   WMN-PSODGA Hybrid Simulation System

In this subsection, we present the initialization, particle-pattern, gene coding, fitness function and replacement methods. The pseudo code of our implemented system is shown in Algorithm 3. Also, our implemented simulation system uses Migration function as shown in Fig. 2. The Migration function swaps solutions among lands included in PSO part.

Algorithm 3. Pseudo code of WMN-PSODGA system.

Computation maxtime:= T_{max}, $t := 0$;
Initial solutions: $\boldsymbol{P}$.
Initial global solutions: $\boldsymbol{G}$.
/* Start PSODGA */
while $t < T_{max}$ **do**
   Subprocess(PSO);
   Subprocess(DGA);
   WaitSubprocesses();
   Evaluate($\boldsymbol{G^t}, \boldsymbol{P^t}$)
   /* Migration() swaps solutions (see Fig. 2). */
   Migration();
   $t = t + 1$;
end while
Update_Solutions($\boldsymbol{G^t}, \boldsymbol{P^t}$);
return Best found pattern of particles as solution;

Initialization

We decide the velocity of particles by a random process considering the area size. For instance, when the area size is $W \times H$, the velocity is decided randomly from $-\sqrt{W^2 + H^2}$ to $\sqrt{W^2 + H^2}$.

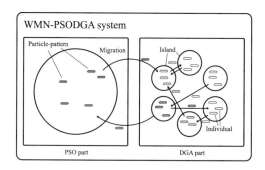

Fig. 2. Model of WMN-PSODGA migration.

Particle-Pattern

A particle is a mesh router. A fitness value of a particle-pattern is computed by combination of mesh routers and mesh clients positions. In other words, each particle-pattern is a solution as shown is Fig. 3.

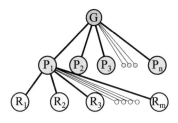

G: Global Solution
P: Particle-pattern
R: Mesh Router
n: Number of Particle-patterns
m: Number of Mesh Routers

Fig. 3. Relationship among global solution, particle-patterns, and mesh routers in PSO part.

Gene Coding

A gene describes a WMN. Each individual has its own combination of mesh nodes. In other words, each individual has a fitness value. Therefore, the combination of mesh nodes is a solution.

Fitness Function

The fitness function of WMN-PSODGA is used to evaluate the temporary solution of the router's placements. The fitness function is defined as:

$$Fitness = \alpha \times NCMC(\boldsymbol{x}_{ij}, \boldsymbol{y}_{ij}) + \beta \times SGC(\boldsymbol{x}_{ij}, \boldsymbol{y}_{ij}) + \gamma \times NCMCpR(\boldsymbol{x}_{ij}, \boldsymbol{y}_{ij}).$$

This function uses the following indicators.

- NCMC (Number of Covered Mesh Clients)
  The NCMC is the number of clients covered by routers.
- SGC (Size of Giant Component)
  The SGC is the maximum number of connected routers.
- NCMCpR (Number of Covered Mesh Clients per Router)
  The NCMCpR is the number of clients covered by each router. The NCMCpR indicator is used for load balancing.

WMN-PSODGA aims to maximize the value of fitness function in order to optimize the placements of routers using the above three indicators. Weight-coefficients of the fitness function are α, β, and γ for NCMC, SGC, and NCMCpR, respectively. Moreover, the weight-coefficients are implemented as $\alpha + \beta + \gamma = 1$.

Router Replacement Methods
A mesh router has x, y positions, and velocity. Mesh routers are moved based on velocities. There are many router replacement methods as shown in following. In this paper, we consider RIWM, RDVM and CM.

Constriction Method (CM)
    CM is a method which PSO parameters are set to a week stable region ($\omega = 0.729, C_1 = C2 = 1.4955$) based on analysis of PSO by M. Clerc et al. [2,5,35].
Random Inertia Weight Method (RIWM)
    In RIWM, the ω parameter is changing ramdomly from 0.5 to 1.0. The C_1 and C_2 are kept 2.0. The ω can be estimated by the week stable region. The average of ω is 0.75 [28,35].
Linearly Decreasing Inertia Weight Method (LDIWM)
    In LDIWM, C_1 and C_2 are set to 2.0, constantly. On the other hand, the ω parameter is changed linearly from unstable region ($\omega = 0.9$) to stable region ($\omega = 0.4$) with increasing of iterations of computations [35,36].
Linearly Decreasing Vmax Method (LDVM)
    In LDVM, PSO parameters are set to unstable region ($\omega = 0.9, C_1 = C_2 = 2.0$). A value of V_{max} which is maximum velocity of particles is considered. With increasing of iteration of computations, the V_{max} is kept decreasing linearly [30,34].
Rational Decrement of Vmax Method (RDVM)
    In RDVM, PSO parameters are set to unstable region ($\omega = 0.9, C_1 = C_2 = 2.0$). The V_{max} is kept decreasing with the increasing of iterations as

$$V_{max}(x) = \sqrt{W^2 + H^2} \times \frac{T - x}{x}.$$

Table 1. The common parameters for each simulation.

Parameters	Values
Distribution of Mesh Clients	Boulevard Distribution
Number of Mesh Clients	48
Number of Mesh Routers	16
Radius of a Mesh Router	3.0 - 3.5
Number of GA Islands	16
Number of Migrations	200
Evolution Steps	9
Crossover Rate	0.8
Mutation Rate	0.2
Replacement Methods	RDVM, RIWM, CM
Area Size	32.0×32.0

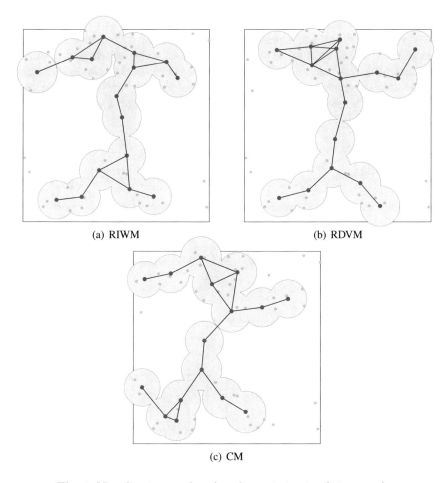

(a) RIWM

(b) RDVM

(c) CM

Fig. 4. Visualization results after the optimization (16 routers).

Where, W and H are the width and the height of the considered area, respectively. Also, T and x are the total number of iterations and a current number of iteration, respectively [25].

3   Simulation Results

In this section, we present the simulation results. Table 1 shows the common parameters for each simulation. In Fig. 4 are shown the visualization results after the optimization for the case of 16 mesh routers. For this scenario, the SGC is maximized because all routers are connected. In Fig. 5 are shown the transition of standard deviations. The standard deviation is related to load balancing. When the standard deviation is increased, the number of mesh clients for each router tends to be different. On the other hand, when the standard deviation is decreased, the number of mesh clients for each router tends to go close to each other. The value of r in Fig. 5 means the correlation coefficient. Comparing Fig. 5(a) with Fig. 5(b) and 5(c), we see that the standard deviation of three router replacement methods is decreased. But, the correlation coefficient of RIWM is lower than RDVM and CM. So, the load balancing of RIWM is better than RDVM and CM.

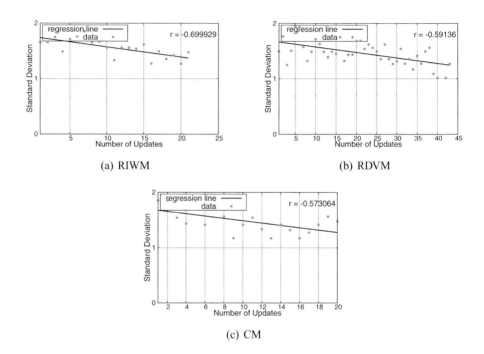

Fig. 5. Standard deviation for RDVM, RIWM and CM (16 routers).

4   Conclusions

In this work, we evaluated the performance of WMNs using a hybrid simulation system based on PSO and DGA (called WMN-PSODGA). We considered Boulevard distribution of mesh clients and three router replacement methods. We compared the performance of RIWM with RDVM and CM. The simulation results show that RIWM has better load balancing than RDVM and CM. In future work, we will consider other distributions of mesh clients.

References

1. Akyildiz, I.F., Wang, X., Wang, W.: Wireless mesh networks: a survey. Comput. Netw. **47**(4), 445–487 (2005)

2. Barolli, A., Sakamoto, S., Ozera, K., Ikeda, M., Barolli, L., Takizawa, M.: Performance evaluation of WMNs by WMN-PSOSA simulation system considering constriction and linearly decreasing Vmax methods. In: Xhafa, F., Caballé, S., Barolli, L. (eds.) 3PGCIC 2017. LNDECT, vol. 13, pp. 111–121. Springer, Cham (2018). https://doi.org/10.1007/978-3-319-69835-9_10

3. Barolli, A., Sakamoto, S., Barolli, L., Takizawa, M.: Performance analysis of simulation system based on particle swarm optimization and distributed genetic algorithm for WMNs considering different distributions of Mesh clients. In: Barolli, L., Xhafa, F., Javaid, N., Enokido, T. (eds.) IMIS 2018. AISC, vol. 773, pp. 32–45. Springer, Cham (2019). https://doi.org/10.1007/978-3-319-93554-6_3

4. Barolli, A., Sakamoto, S., Ozera, K., Barolli, L., Kulla, E., Takizawa, M.: Design and implementation of a hybrid intelligent system based on particle swarm optimization and distributed genetic algorithm. In: Barolli, L., Xhafa, F., Javaid, N., Spaho, E., Kolici, V. (eds.) EIDWT 2018. LNDECT, vol. 17, pp. 79–93. Springer, Cham (2018). https://doi.org/10.1007/978-3-319-75928-9_7

5. Clerc, M., Kennedy, J.: The particle swarm-explosion, stability, and convergence in a multidimensional complex space. IEEE Trans. Evol. Comput. **6**(1), 58–73 (2002)

6. Franklin, A.A., Murthy, C.S.R.: Node placement algorithm for deployment of two-tier wireless Mesh networks. In: Proceedings of Global Telecommunications Conference, pp. 4823–4827 (2007)

7. Girgis, M.R., Mahmoud, T.M., Abdullatif, B.A., Rabie, A.M.: Solving the wireless Mesh network design problem using genetic algorithm and simulated annealing optimization methods. Int. J. Comput. Appl. **96**(11), 1–10 (2014)

8. Goto, K., Sasaki, Y., Hara, T., Nishio, S.: Data gathering using mobile agents for reducing traffic in dense mobile wireless sensor networks. Mob. Inf. Syst. **9**(4), 295–314 (2013)

9. Inaba, T., Elmazi, D., Sakamoto, S., Oda, T., Ikeda, M., Barolli, L.: A secure-aware call admission control scheme for wireless cellular networks using fuzzy logic and its performance evaluation. J. Mobile Multimed. **11**(3&4), 213–222 (2015)

10. Inaba, T., Obukata, R., Sakamoto, S., Oda, T., Ikeda, M., Barolli, L.: Performance evaluation of a QoS-aware fuzzy-based CAC for LAN access. Int. J. Space-Based Situated Comput. **6**(4), 228–238 (2016)

11. Inaba, T., Sakamoto, S., Oda, T., Ikeda, M., Barolli, L.: A testbed for admission control in WLAN: a fuzzy approach and its performance evaluation. In: BWCCA 2016. LNDECT, vol. 2, pp. 559–571. Springer, Cham (2017). https://doi.org/10.1007/978-3-319-49106-6_55

12. Lim, A., Rodrigues, B., Wang, F., Xu, Z.: k-center problems with minimum coverage. Theoret. Comput. Sci. **332**(1–3), 1–17 (2005)

13. Maolin, T., et al.: Gateways placement in backbone wireless Mesh networks. Int. J. Commun. Netw. Syst. Sci. **2**(1), 44–50 (2009)

14. Matsuo, K., Sakamoto, S., Oda, T., Barolli, A., Ikeda, M., Barolli, L.: Performance analysis of WMNs by WMN-GA simulation system for two WMN architectures and different TCP congestion-avoidance algorithms and client distributions. Int. J. Commun. Netw. Distrib. Syst. **20**(3), 335–351 (2018)

15. Muthaiah, S.N., Rosenberg, C.P.: Single gateway placement in wireless Mesh networks. In: Proceedings of 8th International IEEE Symposium on Computer Networks, pp. 4754–4759 (2008)

16. Naka, S., Genji, T., Yura, T., Fukuyama, Y.: A hybrid particle swarm optimization for distribution state estimation. IEEE Trans. Power Syst. **18**(1), 60–68 (2003)

17. Poli, R., Kennedy, J., Blackwell, T.: Particle swarm optimization. Swarm Intell. **1**(1), 33–57 (2007). https://doi.org/10.1007/s11721-007-0002-0

18. Sakamoto, S., Kulla, E., Oda, T., Ikeda, M., Barolli, L., Xhafa, F.: A comparison study of simulated annealing and genetic algorithm for node placement problem in wireless Mesh networks. J. Mobile Multimed. **9**(1–2), 101–110 (2013)

19. Sakamoto, S., Kulla, E., Oda, T., Ikeda, M., Barolli, L., Xhafa, F.: A comparison study of hill climbing, simulated annealing and genetic algorithm for node placement problem in WMNs. J. High Speed Netw. **20**(1), 55–66 (2014)

20. Sakamoto, S., Kulla, E., Oda, T., Ikeda, M., Barolli, L., Xhafa, F.: A simulation system for WMN based on SA: performance evaluation for different instances and starting temperature values. Int. J. Space-Based Situated Comput. **4**(3–4), 209–216 (2014)

21. Sakamoto, S., Kulla, E., Oda, T., Ikeda, M., Barolli, L., Xhafa, F.: Performance evaluation considering iterations per phase and SA temperature in WMN-SA system. Mob. Inf. Syst. **10**(3), 321–330 (2014)

22. Sakamoto, S., Lala, A., Oda, T., Kolici, V., Barolli, L., Xhafa, F.: Application of WMN-SA simulation system for node placement in wireless mesh networks: a case study for a realistic scenario. Int. J. Mobile Comput. Multimed. Commun. (IJMCMC) **6**(2), 13–21 (2014)

23. Sakamoto, S., Oda, T., Ikeda, M., Barolli, L., Xhafa, F.: An integrated simulation system considering WMN-PSO simulation system and network simulator 3. In: BWCCA 2016. LNDECT, vol. 2, pp. 187–198. Springer, Cham (2017). https://doi.org/10.1007/978-3-319-49106-6_17

24. Sakamoto, S., Oda, T., Ikeda, M., Barolli, L., Xhafa, F.: Implementation and evaluation of a simulation system based on particle swarm optimisation for node placement problem in wireless Mesh networks. Int. J. Commun. Netw. Distrib. Syst. **17**(1), 1–13 (2016)

25. Sakamoto, S., Oda, T., Ikeda, M., Barolli, L., Xhafa, F.: Implementation of a new replacement method in WMN-PSO simulation system and its performance evaluation. In: The 30th IEEE International Conference on Advanced Information Networking and Applications (AINA-2016), pp. 206–211 (2016)

26. Sakamoto, S., Obukata, R., Oda, T., Barolli, L., Ikeda, M., Barolli, A.: Performance analysis of two wireless mesh network architectures by WMN-SA and WMN-TS simulation systems. J. High Speed Netw. **23**(4), 311–322 (2017)

27. Sakamoto, S., Ozera, K., Barolli, A., Ikeda, M., Barolli, L., Takizawa, M.: Implementation of an intelligent hybrid simulation systems for WMNs based on particle swarm optimization and simulated annealing: performance evaluation for different replacement methods. Soft. Comput. **23**(9), 3029–3035 (2017)

28. Sakamoto, S., Ozera, K., Barolli, A., Ikeda, M., Barolli, L., Takizawa, M.: Performance evaluation of WMNs by WMN-PSOSA simulation system considering random inertia weight method and linearly decreasing Vmax method. In: Barolli, L., Xhafa, F., Conesa, J. (eds.) BWCCA 2017. LNDECT, vol. 12, pp. 114–124. Springer, Cham (2018). https://doi.org/10.1007/978-3-319-69811-3_10

29. Sakamoto, S., Ozera, K., Ikeda, M., Barolli, L.: Implementation of intelligent hybrid systems for node placement problem in WMNs considering particle swarm optimization, hill climbing and simulated annealing. Mobile Netw. Appl. **23**(1), 27–33 (2017). https://doi.org/10.1007/s11036-017-0897-7

30. Sakamoto, S., Ozera, K., Ikeda, M., Barolli, L.: Performance evaluation of WMNs by WMN-PSOSA simulation system considering constriction and linearly decreasing inertia weight methods. In: Barolli, L., Enokido, T., Takizawa, M. (eds.) NBiS 2017. LNDECT, vol. 7, pp. 3–13. Springer, Cham (2018). https://doi.org/10.1007/978-3-319-65521-5_1

31. Sakamoto, S., Ozera, K., Oda, T., Ikeda, M., Barolli, L.: Performance evaluation of intelligent hybrid systems for node placement in wireless Mesh networks: a comparison study of WMN-PSOHC and WMN-PSOSA. In: Barolli, L., Enokido, T. (eds.) IMIS 2017. AISC, vol. 612, pp. 16–26. Springer, Cham (2018). https://doi.org/10.1007/978-3-319-61542-4_2

32. Sakamoto, S., Ozera, K., Oda, T., Ikeda, M., Barolli, L.: Performance evaluation of WMN-PSOHC and WMN-PSO simulation systems for node placement in wireless mesh networks: a comparison study. In: Barolli, L., Zhang, M., Wang, X.A. (eds.) EIDWT 2017. LNDECT, vol. 6, pp. 64–74. Springer, Cham (2018). https://doi.org/10.1007/978-3-319-59463-7_7

33. Sakamoto, S., Ozera, K., Barolli, A., Barolli, L., Kolici, V., Takizawa, M.: Performance evaluation of WMN-PSOSA considering four different replacement methods. In: Barolli, L., Xhafa, F., Javaid, N., Spaho, E., Kolici, V. (eds.) EIDWT 2018. LNDECT, vol. 17, pp. 51–64. Springer, Cham (2018). https://doi.org/10.1007/978-3-319-75928-9_5

34. Schutte, J.F., Groenwold, A.A.: A study of global optimization using particle swarms. J. Global Optim. **31**(1), 93–108 (2005)

35. Shi, Y.: Particle swarm optimization. IEEE Connect. **2**(1), 8–13 (2004)

36. Shi, Y., Eberhart, R.C.: Parameter selection in particle swarm optimization. In: Evolutionary Programming VII, pp. 591–600 (1998)

37. Vanhatupa, T., Hannikainen, M., Hamalainen, T.: Genetic algorithm to optimize node placement and configuration for WLAN planning. In: Proceedings of The 4th IEEE International Symposium on Wireless Communication Systems, pp. 612–616 (2007)

38. Wang, J., Xie, B., Cai, K., Agrawal, D.P.: Efficient mesh router placement in wireless mesh networks. In: Proceedings of IEEE International Conference on Mobile Adhoc and Sensor Systems (MASS-2007), pp. 1–9 (2007)

A Fuzzy-Based System for Safe Driving in VANETs Considering Impact of Driver Impatience on Stress Feeling Level

Kevin Bylykbashi[1]([✉]), Ermioni Qafzezi[2], Phudit Ampririt[2], Makoto Ikeda[1], Keita Matsuo[1], and Leonard Barolli[1]

[1] Department of Information and Communication Engineering, Fukuoka Institute of Technology (FIT), 3-30-1 Wajiro-Higashi, Higashi-Ku, Fukuoka 811-0295, Japan
kevin@bene.fit.ac.jp, makoto.ikd@acm.org, {kt-matsuo,barolli}@fit.ac.jp
[2] Graduate School of Engineering, Fukuoka Institute of Technology (FIT), 3-30-1 Wajiro-Higashi, Higashi-Ku, Fukuoka 811-0295, Japan
{bd20101,bd21201}@bene.fit.ac.jp

Abstract. In this paper, we propose and implement an intelligent system based on Fuzzy Logic (FL) for determining the driver's stress in Vehicular Ad hoc Networks (VANETs) considering driver's impatience, the behavior of other drivers, and the traffic condition as input parameters. The proposed system, called Fuzzy-based System for Determining the Stress Feeling Level (FSDSFL), can invoke a certain action that improves the driver's mood by providing the appropriate driving support. We show through simulations the effect of the considered parameters on the determination of the stress feeling level and demonstrate some actions that can be performed accordingly.

1 Introduction

Road traffic accidents claim approximately 1.35 million lives each year and cause up to 50 million non-fatal injuries, with many of those injured people incurring a disability as a result of those injuries. The fact is that each of those deaths and injuries is totally preventable [11]. In this regard, industry, governmental institutions, and academic researchers are conducting substantial research to provide proper systems and infrastructure for car accident prevention. The initiatives of many governments for a collaboration of such researchers have concluded to the establishment of Intelligent Transport Systems (ITSs). ITSs focus on the deployment of intelligent transportation technologies by combining cutting-edge information, communication, and control technologies to design sustainable information networks based on people, vehicles, and roads.

As a key component of ITS, Vehicular Ad hoc Networks (VANETs) aim not only at saving lives but also improving traffic mobility, increasing efficiency,

L. Barolli et al. (Eds.): EIDWT 2022, LNDECT 118, pp. 236–244, 2022.
https://doi.org/10.1007/978-3-030-95903-6_25

and promoting travel convenience of drivers and passengers. In VANETs, network nodes (vehicles) are equipped with networking functions to exchange essential information such as safety messages and traffic/road information with one another via vehicle-to-vehicle (V2V) and with roadside units (RSUs) through vehicle-to-infrastructure (V2I) communications. Although VANETs are already implemented in reality introducing several applications, the current architectures face numerous challenges.

To overcome the encountered challenges, the integration of many emerging technologies—Software Defined Networking (SDN), Cloud Computing, Edge/Fog Computing, 5G, Information-Centric Networking (ICN), Blockchain, and so forth—within current VANETs is actively being proposed. Although these emerging technologies promise to solve several issues, other approaches and technologies—Wireless Sensor Networks (WSNs), Internet of Things(IoT)—that have been around us for years can be used as an effective complement to alleviate various limitations.

Alongside these technologies, various artificial intelligence approaches, including Fuzzy Logic (FL) and Machine Learning (ML), are paving the way not only for a complete deployment of VANETs but also for reaching a bigger goal, that of putting fully autonomous vehicles on the roads. Nevertheless, fully driverless cars still have a long way to go, and the current advances fall only between the Level 2 and 3 of the Society of Automotive Engineers (SAE) levels [9]. Until we have self-driving vehicles, many automotive companies and academic researchers will continue working on Driver-Assistance Systems (DASs) as a principal safety feature to enhance driving safety in non-automated vehicles. These intelligent systems reside inside the vehicle and rely on the measurement and perception of the surrounding environment, and based on the acquired information, a variety of actions is taken to ease the driving operation.

Aside from the external factors, there are other determinants that affect the driving operation, such as the drivers and their behaviors. In fact, according to some traffic safety facts provided by a survey of the U.S. Department of Transportation [10], the drivers are the immediate reason for more than 94% of the investigated car crashes. While most of the errors committed by drivers are considered involuntary, other errors often come as a result of their behavior, and this must be utterly preventable. These errors are, in most cases, associated with the stress that drivers feel when they are behind the wheel. Determining the factors that cause the stress is consequently a need that requires careful and immediate work.

In this paper, we consider FL and propose an approach that determines the driver's stress in real-time based on factors such as the driver's impatience, traffic condition, and the behavior of other drivers. We implement an intelligent system considering these three factors as input parameters and determine the stress feeling level, which is the output of the proposed system. We evaluate the proposed system by computer simulations. Based on the output value, the proposed system can determine whether the driver poses a risk for himself and other road users.

The structure of the paper is as follows. Section 2 presents a brief overview of VANETs. Section 3 describes the proposed fuzzy-based simulation system and its implementation. Section 4 discusses the simulation results. Finally, conclusions and future work are given in Sect. 5.

2   Overview of VANETs

VANETs are a special case of Mobile Ad hoc Networks (MANETs) in which the mobile nodes are vehicles. In VANETs, nodes (vehicles) have high mobility and tend to follow organized routes instead of moving randomly. Moreover, vehicles offer attractive features such as higher computational capability and localization through GPS.

VANETs have huge potential to enable applications ranging from road safety, traffic optimization, infotainment, commercial to rural and disaster scenario connectivity. Among these, road safety and traffic optimization are considered the most important ones as they have the goal to reduce drastically the high number of accidents, guarantee road safety, make traffic management, and create new forms of inter-vehicle communications in ITSs. The ITSs manage the vehicle traffic, support drivers with safety and other information, and provide some services such as automated toll collection and driver-assist systems [4].

Despite the attractive features, VANETs are characterized by very large and dynamic topologies, variable capacity wireless links, bandwidth and hard delay constraints, and by short contact durations which are caused by the high mobility, high speed, and low density of vehicles. In addition, limited transmission ranges, physical obstacles, and interferences make these networks characterized by disruptive and intermittent connectivity.

To make VANETs applications possible, it is necessary to design proper networking mechanisms that can overcome relevant problems that arise from vehicular environments.

3   Proposed Fuzzy-Based System

The highly competitive and rapidly advancing autonomous vehicle race has been on for several years now, and it is a matter of time until we have these vehicles on the roads. However, even if the automotive companies do all it takes to create fully automated cars, there will still be one big obstacle, the infrastructure. In addition, this could take decades, even in the most developed countries. Moreover, 93% of the world's fatalities on the roads occur in low- and middle-income countries [11] and considering all these facts, DASs should remain the focus of interest for the foreseeable future.

DASs can be very helpful in many situations as they do not depend on the infrastructure as much as driverless vehicles do. Furthermore, DASs can provide driving support with very little cost, thus help the low- and middle-income

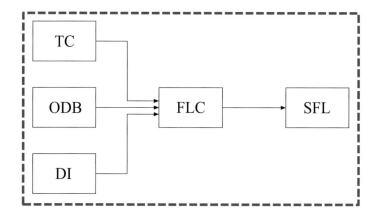

Fig. 1. A diagram of the proposed fuzzy-based system.

countries in the long battle against car accidents. Most of these systems focus on the drivers' situations and intervene when they seem incapable of driving safely.

Our research work focuses on developing an intelligent non-complex driving support system that determines the driving risk level in real-time by considering different types of parameters. In previous works, we have considered different parameters, including in-car environment parameters such as the ambient temperature and noise, and driver's vital signs, i.e., heart and respiratory rate, for which we implemented a testbed and conducted experiments in a real scenario [3]. The considered parameters include environmental factors and driver's health condition, as these parameters affect the driver's capability and vehicle performance. In [1], we presented an integrated fuzzy-based system, which in addition to those parameters, considers the following inputs: vehicle speed, weather and road condition, driver's body temperature, and vehicle interior relative humidity. The inputs were categorized based on the way they affect the driving operation. In a more recent work [2], we proposed a system that decides the driver's impatience since the impatient drivers are often an immediate cause of many road accidents. In this work, we propose a system that can determine the driver's stress feeling level based on the degree of impatience, traffic condition, and other driver's behavior.

We use FL to implement the proposed system as it can make a real-time decision based on the uncertainty and vagueness of the provided information [5–8,12,13]. The proposed system, called Fuzzy-based System for Determining Stress Feeling Level (FSDSFL), is shown in Fig. 1. FSDSFL has the following inputs: Traffic Condition (TC), Other Driver's Behavior (ODB), and Driver's Impatience (DI). The output of the system is Stress Feeling Level (SFL). The term sets for the system parameters are given in Table 1.

Based on the linguistic description of input and output parameters, the Fuzzy Rule Base (FRB) of the proposed system forms a fuzzy set of dimensions $| T(x_1) | \times | T(x_2) | \times \cdots \times | T(x_n) |$, where $| T(x_i) |$ is the number of terms

Table 1. Parameters and their term sets for FSDSFL.

Parameters	Term set
Traffic Condition (TC)	Light (Li), Moderate (Mo), Heavy (He)
Other Drivers' Behavior (ODB)	Very Bad (VB), Bad (Ba), Good (Go)
Driver's Impatience (DI)	Low (L), Moderate (M), High (H)
Stress Feeling Level (SFL)	Low (Lw), Moderate (Md), High (Hg), Very High (VH), Extremely High (EH)

Table 2. FRB of FSDSFL.

No	TC	ODB	DI	SFL
1	Li	VB	L	Md
2	Li	VB	M	Hg
3	Li	VB	H	VH
4	Li	Bd	L	Lw
5	Li	Bd	M	Md
6	Li	Bd	H	Hg
7	Li	Go	L	Lw
8	Li	Go	M	Lw
9	Li	Go	H	Md
10	Mo	VB	L	Hg
11	Mo	VB	M	VH
12	Mo	VB	H	EH
13	Mo	Bd	L	Md
14	Mo	Bd	M	Hg
15	Mo	Bd	H	VH
16	Mo	Go	L	Lw
17	Mo	Go	M	Md
18	Mo	Go	H	Hg
19	He	VB	L	VH
20	He	VB	M	EH
21	He	VB	H	EH
22	He	Bd	L	Hg
23	He	Bd	M	VH
24	He	Bd	H	EH
25	He	Go	L	Md
26	He	Go	M	Hg
27	He	Go	H	VH

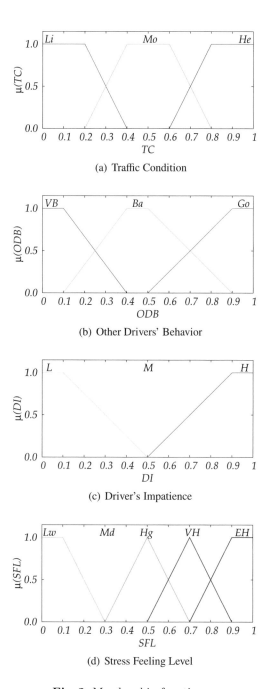

(a) Traffic Condition

(b) Other Drivers' Behavior

(c) Driver's Impatience

(d) Stress Feeling Level

Fig. 2. Membership functions.

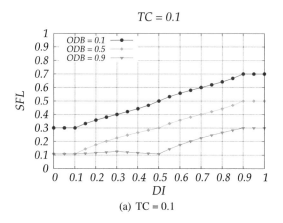

(a) TC = 0.1

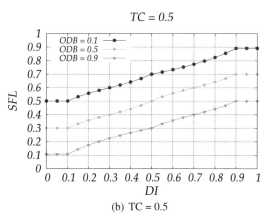

(b) TC = 0.5

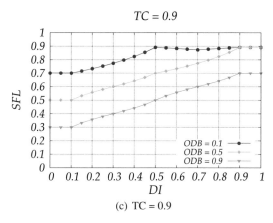

(c) TC = 0.9

Fig. 3. Simulation results.

on $T(x_i)$ and n is the number of input parameters. FSDSFL has three input parameters with three linguistic terms each, therefore there are 27 rules in the FRB, which are shown in Table 2. The control rules of FRB have the form: IF "conditions" THEN "control action". The membership functions are shown in Fig. 2. We use triangular and trapezoidal membership functions because these types of functions are more suitable for real-time operation.

4 Simulation Results

In this section, we present the simulation results for our proposed system. The simulation results are presented in Fig. 3. We consider TC and ODB as constant parameters. We show the relation between SFL and DI for different ODB values. The TC values of 0.1, 0.5, and 0.9, simulate light, moderate, and heavy traffic conditions, respectively.

In Fig. 3(a), we show the scenarios where there is no or very little traffic congestion. We can see that there is no situation where the driver is considered to experience severe stress and only one where he feels very stressed. The latter scenario happens only when he seems to be too impatient and when the behavior of other drivers is very bad. Other than in these extreme scenarios, the stress feeling level is determined as moderate and below for most scenarios.

However, when the traffic congestion increases, we can see that for all scenarios the driver is under more stress. The impact of traffic condition can be seen in Fig. 3(b) and Fig. 3(c). For TC = 0.5, the only situation where the driver is determined to have no stress is when he is patient and the other drivers are not violating traffic rules. On the other hand, when the traffic is heavy the driver is always under some stress, regardless of the behavior of other drivers or the patience showed.

In these cases, when the driver is considered to be under too much stress, the system can take several actions that can improve the driving situations and therefore reduce the risks of accidents. For example, the system can suggest the use of an alternative route that has less congestion, or it can adjust the interior environment of the car to help the driver manage the driving situations patiently.

5 Conclusions

In this work, we presented the implementation of an FL approach that determines the driver's stress feeling level in real-time by considering the driver's impatience, the traffic conditions, and the behavior of other drivers. We showed through simulations the effect of the considered parameters on the determination of the stress feeling level. The simulations show that when the traffic is heavier than usual, drivers tend to show an increased level of stress. The stress is even higher when other drivers violate the traffic rules. However, when the drivers have much patience, they can manage to drive smoothly in almost every driving situation.

In the future, we would like to make extensive simulations and experiments to evaluate the proposed system and compare the performance with other systems.

References

1. Bylykbashi, K., Qafzezi, E., Ampririt, P., Ikeda, M., Matsuo, K., Barolli, L.: Performance evaluation of an integrated fuzzy-based driving-support system for real-time risk management in VANETs. Sensors **20**(22), 6537 (2020). https://doi.org/10.3390/s20226537
2. Bylykbashi, K., Qafzezi, E., Ampririt, P., Ikeda, M., Matsuo, K., Barolli, L.: A fuzzy-based system for deciding driver impatience in VANETs. In: Barolli, L. (ed.) 3PGCIC 2021. LNNS, vol. 343, pp. 129–137. Springer, Cham (2022). https://doi.org/10.1007/978-3-030-89899-1_13
3. Bylykbashi, K., Qafzezi, E., Ikeda, M., Matsuo, K., Barolli, L.: Fuzzy-based driver monitoring system (FDMS): implementation of two intelligent FDMSs and a testbed for safe driving in VANETs. Futur. Gener. Comput. Syst. **105**, 665–674 (2020). https://doi.org/10.1016/j.future.2019.12.030
4. Hartenstein, H., Laberteaux, L.: A tutorial survey on vehicular ad hoc networks. IEEE Commun. Mag. **46**(6), 164–171 (2008)
5. Kandel, A.: Fuzzy Expert Systems. CRC Press, Boco Raton (1991)
6. Klir, G.J., Folger, T.A.: Fuzzy Sets, Uncertainty, and Information. Prentice Hall Inc., Upper Saddle River (1987)
7. McNeill, F.M., Thro, E.: Fuzzy Logic: A Practical Approach. Academic Press, Cambridge (1994)
8. Munakata, T., Jani, Y.: Fuzzy systems: an overview. Commun. ACM **37**(3), 69–77 (1994). https://doi.org/10.1145/175247.175254
9. SAE On-Road Automated Driving (ORAD) committee: Taxonomy and definitions for terms related to driving automation systems for on-road motor vehicles. Technical report, Society of Automotive Engineers (SAE) (2018). https://doi.org/10.4271/J3016_201806
10. Singh, S.: Critical reasons for crashes investigated in the national motor vehicle crash causation survey. Technical report (2015)
11. World Health Organization: Global status report on road safety 2018: summary. World Health Organization, Geneva, Switzerland (2018). (WHO/NMH/NVI/18.20). Licence: CC BY-NC-SA 3.0 IGO)
12. Zadeh, L.A., Kacprzyk, J.: Fuzzy Logic for the Management of Uncertainty. Wiley, New York (1992)
13. Zimmermann, H.J.: Fuzzy Set Theory and Its Applications. Springer, New York (1996). https://doi.org/10.1007/978-94-015-8702-0

Mobility-Aware Narrow Routing Protocol for Underwater Wireless Sensor Networks

Elis Kulla[1]([☒])[ID], Kuya Shintani[1], and Keita Matsuo[2][ID]

[1] Okayama University of Science (OUS),
1-1 Ridaicho, Kita-ku, Okayama 700-0005, Japan
kulla@ice.ous.ac.jp, t18j041sk@ous.jp
[2] Fukuoka Institute of Technology (FIT),
3-30-1 Wajiro-Higashi, Higashi-Ku, Fukuoka 811-0295, Japan
kt-matsuo@fit.ac.jp

Abstract. Underwater Wireless Sensor Networks (UWSN) applications should consider delays, disruptions and disconnections between nodes, because underwater environment is not appropriate for real time communications. Instead of electromagnetic radio waves, acoustic waves have a better propagation in underwater environment. In these conditions, the usage of Delay Tolerant Networks (DTN) in UWSN seems reasonable and beneficial. One characteristic of routing protocols for DTN is flooding of messages to increase the delivery probability. For instance, Epidemic Routing (ER) protocol floods the network with copies of generated messages. This creates a lot overhead in each node's buffer, and uses a lot of valuable energy from the relay nodes. In this paper, we present a routing protocol for Underwater Wireless Sensor Networks, which is based on the existing Focused Beam Routing Protocol. Mobility-Aware Narrow Routing (MANR) enables the forwarding nodes to take into consideration the next destination of the receiving nodes before making a forwarding decision. MANR considers that each node knows its next destination approximately in 2D coordinates. We evaluate MANR using The ONE simulator.

Keywords: Underwater Wireless Sensor Networks · UWSN · Mobility-Aware Narrow Routing · MANR · ONE simulator

1 Introduction

Underwater communications are becoming popular in commercial applications, such as disaster detection, environment monitoring, biological research on marine animals, ocean floor mapping, robot coordination, and so on [1,2]. An example of UWSN, includes several Autonomous Underwater Vehicles (AUVs) with various sensors (environmental, chemical, navigation and so on). They use their sensors to collect data and submit the sensed data up to the sea surface, where air-water interface ships, boats or buoys are located (see Fig. 1). Furthermore, these air-water interfaces might also use electromagnetic waves to send the aggregated data back to monitoring centers or data processing centers.

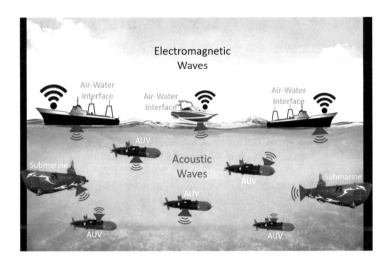

Fig. 1. Underwater wireless sensor network application.

The preferred communication media for underwater communications is acoustic waves in contrast to electromagnetic radio waves, which are widely used in the air. In fact, low frequency electromagnetic radio waves (30 Hz–300 Hz) can propagate for longer distances, but they require large antennae and high transmission power. Optical waves have better propagation, but they require directional coordination, which is almost impossible in underwater applications, where the devices are in constant movement [2].

Most of the research regarding the physical layer in underwater communications uses Phase Shift Keying (PSK) and Quadrature Amplitude Modulation (QAM) modulation techniques. With the increasing of processing capabilities of small devices, Orthogonal Frequency Division Modulation (OFDM) is also considered.

At the Medium Access Control (MAC) layer, Frequency Division Multiple Access (FDMA) does not work well with limited-band acoustic signals because they are affected by fading and multi-path [2]. Moreover, in acoustic communications there is a difference in delays between consecutive packets (jitter), which makes it very difficult to implement Time Division Multiple Access (TDMA) techniques. CDMA is so far the best solution, but newly designed techniques are yet to be developed, tested and implemented [3,4].

Because it is costly to implement network infrastructure in underwater environment, the network layer is mostly oriented towards adhoc architecture, where all participating nodes forward packets to other nodes until the packets reach the intended destination. In fact, depending on the application, the adhoc architecture may consist of:

Real-time data transmission, where participating nodes, either already know where to forward the received data, or the can find out in a very short amount of time. Such amount of time is usually considered as "the real time".

Delay-tolerant data transmission, where different techniques are used to increase packet delivery ratio (PDR) even when there are no available routes to the intended destination. Delay-tolerant data transmissions, us the store-carry-forward paradigm, to transmit data from one node to another, until the data reach the destination.

Thus, existing adhoc and Delay Tolerant Network (DTN) Routing Protocols need to be redesigned, in order to deal with unstable links and high delay variance in underwater environment.

DTNs are enabling various networking applications in different fields. Work have been done towards using DTN in order to enable vehicular communication in VANET [5–7]. Other applications of DTN consider underwater environment. In [8], the authors present an underwater framework, which uses DTN to enable communication between AUVs. They also discuss various complex application scenarios for control operations involving single and multiple AUVs. Other works focus on acoustic signal generation in underwater environment.

In this paper, we introduce our proposed routing scheme, show its implementation in The One simulator and analyze its performance in terms of different evaluation parameters.

The structure of the paper is explained in the following. In Sect. 2, we discuss DTN routing protocols and their usage in UWSN. In Sect. 3, we proposed MANR and its implementation in The ONE simulator. We describe our simulation settings in underwater scenario in Sect. 4. In Sect. 5, we show simulation results and discuss the findings. Finally, we draw conclusions in Sect. 6.

2   Focused Beam Routing

A more detailed description of FBR, can be found in [9]. Here we will shortly explain FBR in contrast to other routing paradigms in UWSN.

In UWSNs, the store-carry-forward paradigm is popular among routing protocols, hence Epidemic Routing (ER), which is visualized in Fig. 2(a), is the main forwarding technique. Here, the forwarding area is defined from the communicating distance.

Since the objective of routing protocols for UWSN in the majority of applications is to collect data to the modems in the water surface, and in order to decrease the number of copied packets, forcing nodes to strictly forward packets only to nodes that are closer to the surface has proven beneficial in terms of performance for these applications. This category of routing protocols is called depth-based routing (DBR) and is simplified in Fig. 2(b).

In order to decrease the overhead and energy consumption, some routing protocols, focus their forwarding area towards the surface data collector, and define it by a relatively small angle, as shown in Fig. 2(c). We implemented FBR as explained in the following.

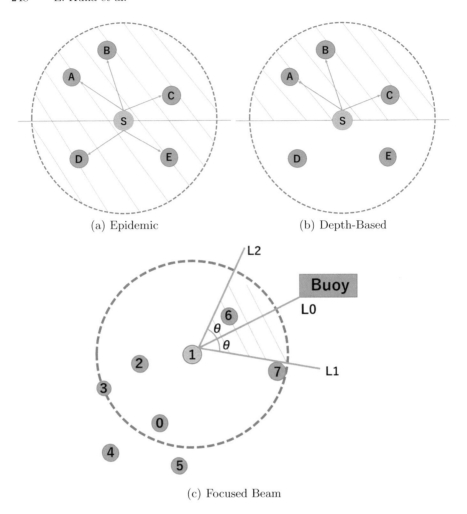

(a) Epidemic   (b) Depth-Based

(c) Focused Beam

Fig. 2. UWSN routing protocol paradigms

3   Implementation of FBR in the ONE Simulator

While there are a lot of routing protocol proposed and implemented around the world, we could not find an open implementation of FBR that could be easily editable and verifiable in different scenarios. Thus we implemented FBR in the well-known The One Simulator. The following assumptions are made, in order to simplify the implementation:

– We consider a two dimensional environment: width on the horizontal axis and depth in the vertical axis.
– We assume that every nodes' location is known globally in every moment, which in fact is very difficult to achieve in reality without a well-established infrastructure.

Table 1. Simulation parameters' settings

Parameters	Value
Movement Model	Random Waypoint
Simulation Time	5400 s (1.5 h)
Simulation Area	500m × 500 m
Message Size	Small: 180 kB–220 kB Large: 1300 kB–1700 kB
FBR's Angle (θ)	30°, 45°, 90°

Table 2. Participating nodes in simulations

Parameters	Collector node	Sensor nodes	Relay nodes (AUVs)
Number of Nodes	1	50	10
Mobility	Static at (250, 0)	Random Waypoint Mobility Model	
Moving Speed	0	0.5–1.5 m/s	2–5 m/s
Buffer Size	Unlimited	10 MB	300 MB
Interface	High Speed Interface HSI	Low Speed Interface LSI	HSI and LSI
Transmission Speed	10 Mbps	250 kbps	
Transmission Distance	100 m	50 m	

- Every participating node is mobile and moves based on the Random Waypoint Mobility Model, as implemented in The ONE simulator.
- All transmissions are directed towards a buoy sink, which is located in the top-center of our simple 2D environment.

Then, while referring to Fig. 2(c), whenever a participating node (node 1) receives a new message or contacts a new node, the following happens.

- First, forwarding nodes calculate the angle in degrees of line L_0 directed towards the buoy sink.
- Then based on the angle of transmission θ, we define the transmission area on both sides of line L_0. FBR transmission area is confined by lines L_1 and L_2.
- Finally, the transmission area (shaded area) is defined by the communication distance.

4   Simulation Settings and Results

In order to analyse the proposed MANR protocol and compare its performance to that of FBR protocol, we conducted simulations in different scenarios, and set the environment as shown in Table 1. We used The ONE Simulator [10], which has already implemented the ER protocol and the store-carry-forward mechanism.

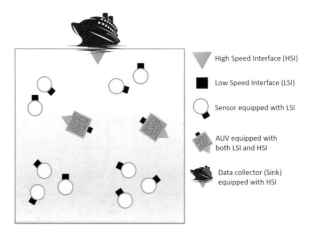

Fig. 3. Simulation scenario

In Table 2, the settings of each type of nodes used in our simulations is summarized. As shown in Fig. 3, a single data collector is located at the top-center (250 m, 0 m) of our simulation area (500 m × 500 m). 40 underwater sensor nodes, which are usually small devices attached to marine animals or floating in the water, generate data in two different data patterns: Small message and Large message. They are equipped with Low Speed Interface (LSI), in order to forward their sensed data to each-other and the relay nodes. Relay nodes are more powerful devices equipped with extra resources and they act as data mules for underwater sensors. In our case, they are equipped with 2 interfaces; one LSI interface in order to get data from senors and a High Speed Interface (HSI) in order to send data to the data collector. The data collector node receives messages only from relay nodes, by using its HSI.

5 Simulation Results

We present our simulation results for Overhead Ratio (OR) and Delivery Probability (DP), in Figs. 4 and 5, respectively. We note that, when transmission angle is 180°, both protocols act as ER protocol. This is confirmed by showing exactly the same performance when transmission angle is 180°.

In Fig. 4, we notice that when generated messages are small (200 kB), OR is smaller for all communication angles and both protocols. We assume that smaller messages can be forwarded faster and stored easier in node's buffers. Moreover, it can be seen that MANR has lower OR for smaller communication angles (30° and 45°).

By achieving lower overhead for each communication, we claim that our proposed MANR shows better performance than FBR, mainly for smaller communication angles. From Fig. 5, we can observe that DP for MANR is slightly smaller, but the improvement in OR is more significant.

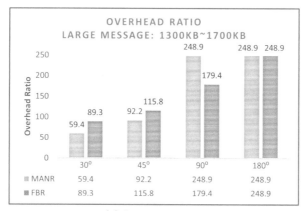

(a) Large message.

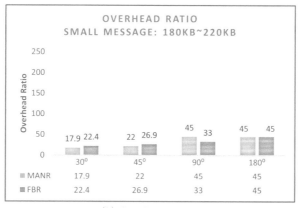

(b) Small message.

Fig. 4. Overhead ratio for 3 different message sizes, comparing FBR and MANR for different communication angles.

MANR shows great performance in terms of decreasing resource usage, with slightly lower delivery ratio. Both protocols outperform ER protocol for OR, ANH and DP. When sensors send smaller (but frequent) messages in the network, all protocols show lower DPs, but better OR and ANH.

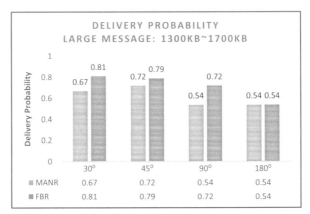

(a) Large message.

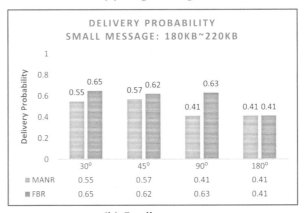

(b) Small message.

Fig. 5. Delivery probability for 3 different message sizes, comparing FBR and MANR for different communication angles.

6   Conclusions and Future Works

In this paper, we proposed a Mobility-Aware Narrow Routing Protocol (MANR) for Underwater Wireless Sensor Networks, which makes forwarding decisions based on the next destination of each neighbor. We implemented MANR in The ONE Simulator, and compared its performances with FBR, two well-known routing protocols for Delay Tolerant Networks. From the simulation results, we draw the following conclusions.

MANR outperforms FBR for small communication angles.
When sensors send smaller (but frequent) messages, all protocols show lower DPs, but better OR.

We believe that, MANR needs further improvements, in order to increase its predictability. We are continuously improving and testing MANR, in order to furthermore decrease resource usage (i.e. OR), while keeping an acceptable performance in DP.

References

1. Potter, J., Alves, J., Green, D., Zappa, G., Nissen, I., McCoy, K.: The JANUS underwater communications standard. In: 2014 Underwater Communications and Networking (UComms), Sestri Levante, pp. 1–4 (2014)
2. Stojanovic, M.: Underwater acoustic communication. In: Wiley Encyclopedia of Electrical and Electronics Engineering (1999). https://doi.org/10.1002/047134608X.W5411
3. Akyildiz, I.F., Pompili, D., Melodia, T.: Underwater acoustic sensor networks: research challenges. Ad Hoc Netw. **3**(3), 257–279 (2005). https://doi.org/10.1016/j.adhoc.2005.01.004
4. Yin, J., Du, P., Yang, G., Zhou, H.: Space-division multiple access for CDMA multiuser underwater acoustic communications. J. Syst. Eng. Electron. **26**(6), 1184–1190 (2015)
5. Magdum, S.S., Sharma, M., Kala, S.M., Franklin, A.A., Tamma, B.R.: Evaluating DTN routing schemes for application in vehicular networks. In: 2019 11th International Conference on Communication Systems & Networks (COMSNETS), pp. 771–776 (2019)
6. Bylykbashi, K., Spaho, E., Barolli, L., Xhafa, F.: Routing in a many-to-one communication scenario in a realistic VDTN. Int. J. High Speed Netw. **24**(2), 107–118 (2018)
7. Cuka, M., Shinko, I., Spaho, E., Oda, T., Ikeda, M., Barolli, L.: A simulation system based on ONE and SUMO simulators: performance evaluation of different vehicular DTN routing protocols. Int. J. High Speed Netw. **23**(1), 59–66 (2017)
8. Marques, E.R.B., et al.: AUV control and communication using underwater acoustic networks. In: OCEANS 2007 - Europe, pp. 1–6 (2007). https://doi.org/10.1109/OCEANSE.2007.4302469
9. Jornet, J.M., Stojanovic, M., Zorzi, M.: Focused beam routing protocol for underwater acoustic networks. In: Proceedings of the third ACM international workshop on Underwater Networks (WuWNeT '08), pp. 75–82. Association for Computing Machinery New York (2008)
10. Keränen, A., Ott, J., Kärkkäinen, T.: The ONE simulator for DTN protocol evaluation. In: Proceedings of the 2nd International Conference on Simulation Tools and Techniques, Rome, Italy (2009)

Design and Implementation of a Testbed for Delay Tolerant Networks: Work in Progress

Kuya Shintani[1], Elis Kulla[1(✉)], Makoto Ikeda[2], Leonard Barolli[2], and Evjola Spaho[3]

[1] Okayama University of Science (OUS), 1-1 Ridaicho, Kita-ku, Okayama 700-0005, Japan
`t18j041sk@ous.jp, kulla@ice.ous.ac.jp`
[2] Fukuoka Institute of Technology (FIT), 3-30-1 Wajiro-Higashi, Higashi-Ku, Fukuoka 811-0295, Japan
`makoto.ikd@acm.org, barolli@fit.ac.jp`
[3] Polytechnic University of Tirana (UPT), Mother Teresa Square, No. 4, Tirana, Albania

Abstract. One characteristic of routing protocols for Delay Tolerant Networks (DTNs) is flooding of messages to increase the delivery probability. For instance, Epidemic Routing (ER) protocol floods the network with copies of generated messages. This creates a lot overhead in each node's buffer, and uses a lot of valuable energy from the relay nodes. In order to minimize resource usage, designing energy-aware protocols is a challenge. These protocols need to be tested by simulation and experiments. While simulations are easy to repeatedly perform, experiments require a lot of time and human effort. In this paper, we present our implementation of a DTN testbed, by using Raspberry Pi as a DTN node. Each DTN node is equipped with different sensors and interfaces, and we have developed appropriate software modules to manage these sensors and interfaces.

Keywords: Delay Tolerant Networks · DTN · Experimental environment · Testbed · Bundle protocol · Epidemic routing

1 Introduction

Delay Tolerant Networks (DTNs) are characterised by the absence of a continuous path between the source and destination [1]. DTNs nodes use store-carry-forward approach. They store message copies and carry them until they encounter other nodes to which they forward these copies. This repeated process improves data delivery and reduces average delay, but the overhead in the network will be high.

Every mobile node equipped with a wireless interface and a buffer, can be used as a DTN node to store-carry-forward messages. However, storing, sending and receiving messages are processes that consumes energy. We believe, it is

L. Barolli et al. (Eds.): EIDWT 2022, LNDECT 118, pp. 254–262, 2022.
https://doi.org/10.1007/978-3-030-95903-6_27

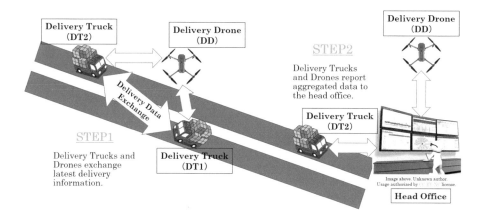

Fig. 1. Typical scenario of delivery units using DTN to exchange delivery data.

important to design the appropriate routing protocols for DTN, that consumes less resources (energy and buffer) and have a high delivery probability.

DTNs are first designed for the Interplanetary communication where opportunistic contacts and store-carry-forward approach are used. However DTN communication architectures can be used also in applications for remote regions and developing areas [2–4].

DTNs are a type of adhoc networks and perform well in real-time applications if the network is dense. But their power lies in delivering data even when a route to the destination does not exist.

DTNs are enabling various networking applications in different fields. Work have been done towards using DTN in order to enable vehicular communication in VANET [5–7]. Many wireless adhoc geographical routing protocols use positioning and other geographical data to make forwarding decisions. While it is easy to get positioning data from gnss satellite system, in underwater environment, global positioning knowledge is not an easy problem. Extensive research have been done in localization and tracking in underwater environment. In [8], the authors propose a routing method that utilizes number of interactions and contact duration between nodes in order to make routing decisions. The authors claim that the proposed method improves overhead cost and achieves shorter delivery delay while not decreasing its performance.

A typical commercial scenario of DTN is a proprietary network of delivery trucks and drones (delivery units), as shown in Fig. 1. While on delivery service, trucks and drones can exchange delivery data with each-other, then forward those data to other delivery units or to head office data centers, when available. Delivery units are equipped with wireless interfaces, which can be configured to work as DTN nodes. Apart from delivery information, each unit can collect data about their technical condition, road traffic, environment conditions and so on.

In this paper, we introduce our testbed for DTNs, implemented in our laboratory in Okayama University of Science, in collaboration with Fukuoka Institute of Technology. This implementation enables testing and verification of different protocols for routing and data management.

The structure of the paper is explained in the following. In Sect. 2, we give a short overview of DTNs and its main routing protocols. In Sect. 3, we describe the design and implementation of our testbed for DTNs. We conclude the paper in Sect. 4, with conclusions and future works.

2 Delay Tolerant Networks (DTN)

2.1 DTN Overview

Delay Tolerant Networks (DTNs) are occasionally connected networks, characterized by the absence of a continuous path between the source and destination [1]. DTNs enable communication in sparse areas where end-to-end connectivity is nonexistent by using store-carry-forward approach. A DTN node stores messages and carries them until it encounters other nodes, to which it forwards copies of these messages. The receiving node will repeat forwarding copies until the destination node is encountered or the message time to live (TTL) is expired. This process is called message switching. Waiting for a link to establish may take a long time and this will cause delay and effect also to the asymmetric data rate.

Different copies of the same message will be routed independently in the network. This approach will improve the data delivery and reduce the average delay. However, the drawback of this approach is waste of network bandwidth and nodes storage, especially in dense networks, where the same copy of a message will be forwarded and stored by several nodes.

In order to realize the above process, bundle protocol has been designed as an implementation of the DTN architecture. A bundle is a basic data unit of the DTN bundle protocol. Each bundle comprises a sequence of two or more blocks of protocol data, which serve for various purposes. The Bundle protocol is specified in RFC 5050 [9]. It is responsible for accepting messages from the application and sending them as one or more bundles via store-carry-forward operations to the destination DTN node. The Bundle protocol runs above the TCP/IP level.

2.2 DTN Routing Protocols

Various routing protocols for DTNs are proposed and implemented. DTN routing protocols can be classified in single-copy and multiple-copy protocols or forwarding and replicating protocols. In single-copy routing protocols or forwarding protocols, for each message only one copy exists in the network. In multiple-copy routing protocols or replicating protocols, there are two subgroups: N-copy and unlimited copy routing protocols. N-copy protocols limits the number of message copies that will be created to a configurable maximum by using probabilistic approaches and distribute these copies to contacts until the number of copies is reached. Unlimited copy routing protocols perform variants of flooding.

Epidemic Routing Protocol. Epidemic [10] is a routing protocol that uses the flooding mechanism 2. Each message is spread in the network with no priority and no limit (Fig. 2). In the first phase, when two nodes are within the communication range of each other will exchange the list of message IDs and compare them to find the messages that are not already in the storage of the other node. The next phase is a check of available buffer storage space. This protocol can be used in scenarios where there is no information for the network topology and the mobility of nodes. In these scenarios epidemic protocol tends to maximize the delivery probability and to minimize the latency.

Spray and Wait Routing Protocol. Spray and Wait [11] is composed of the spray phase and the wait phase. Any time a new message is created in the network, a maximum of L number of copies of the message is created in the network. In the spray phase, the source of the message will spray one copy of this message to L different "relays". When a relay receives the copy, it enters the wait phase, where it will simply hold that message until the destination is encountered directly. This routing protocol benefits from the replication-based routing to achieve high delivery ratio and from forwarding-based routing to achieve low resource utilization.

Maxprop Routing Protocol. Maxprop [12] uses priority for the schedule of packets transmitted to other nodes and the schedule of packets to be dropped. These priorities are based on historical data and some complementary mechanisms. Three used mechanisms are: head start for new bundles, lists of previous intermediaries, and system-wide acknowledgements. In order to guarantee that all bundles have a chance of being propagated in the network, head start and priority is given to new bundles to be transmitted. Lists of previous intermediaries are maintained to prevent bundles of being sent to the same node again. System-wide acks are propagated through the network in order to notify nodes to eliminate redundant copies of the bundles that have already been delivered to their destination.

Prophet Routing Protocol. Prophet (Probabilistic Routing Protocol using History of Encounters and Transitivity) [13] is a variant of the epidemic routing protocol. It operates by pruning the epidemic distribution tree to minimize resource usage while still attempting to achieve the best case routing capabilities of epidemic routing. This protocol uses the delivery predictability, and attempts to estimate based on node encounter history, which node has the higher probability of successful delivery of a message to the final destination. When two nodes are in communication range, a new message copy is transferred only if the other node has a better probability of delivering it to the destination.

First Contact Routing Protocol. In First Contact [14], the nodes forward bundles to the first node they encounter. This results in a random search for the

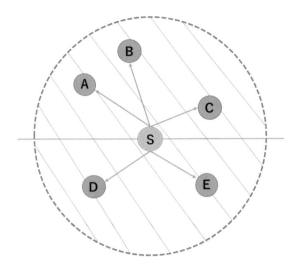

Fig. 2. Epidemic Routing allows each node to send copies of saved messages to each node in its communication range.

destination node. If any node comes first in to the radio range of the source node will be given the message. It doesn't determine the next best hop moving to the destination. The message is forwarded randomly when two or more nodes come in contact with the source node at the same time. Local copy of the message is eliminated after successful transfer from one node to another node. Thus, a single copy of the message flows in the networks.

3   Implementation of DTN Testbed

In this work, we present the implementation of our testbed for DTN, and its potential applications. We have developed an experimental environment for DTNs consisting of 5 Raspberry PIs (DTN Node). The DTN Node is equipped with several sensors and interfaces, and runs various software modules, as shown in Fig. 3. They can be static nodes and attached to AC power around our campus, or mobile and connected to a mobile power source (battery). In order to implement mobility and dynamic topology, we have developed an iptables-based MAC-filtering technique to add nodes to the network and remove them at different points in time. Our system also enables data logging for further analysis and improvement.

3.1   Data Sensing and Generation

For sensing environmental and gps data, we use Sony's Spresense main board, which is equipped with a GPS module. We also added a BME280 sensor module to sense environmental data such as temperature, humidity and atmospheric

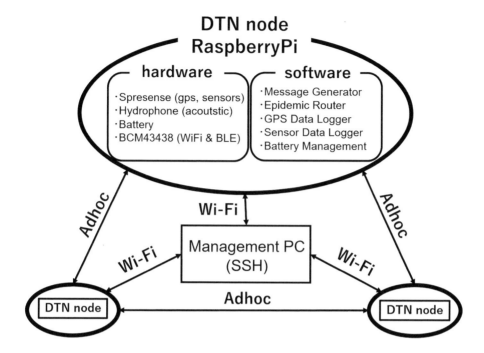

Fig. 3. Experimental system (testbed) overview.

pressure. For both hardware modules, we have developed a software module to read data from the sensors. Spresense is a low-power board computer for the IoT that is equipped with a GPS receiver and allows for IoT versatility and can be developed for a vast range of applications [15].

DTN Node is also equipped with a data generation software module that generates different patterns of data, in order to simulate experimental conditions.

3.2 Network Settings and Management

Each of 5 DTN nodes (RaspberryPI 3 B+) is equipped with a network module, which supports 2.4 GHz and 5 GHz IEEE 802.11.b/g/n/ac wireless LAN, Bluetooth 4.2, BLE and wired ethernet. We use its wireless LAN interface to configure an adhoc network among DTN nodes. This is where we run our DTN routing protocol software and data transmission.

In order to manage each DTN node around campus, we have added a Raspberry Pi USB Wifi dongle and set it up to connect to our on campus network. By doing this, we can remotely control each node from our laboratory's PC using SSH protocol. This introduces some overhead in DTN communications, but we try to avoid SSH connections when experiments are running.

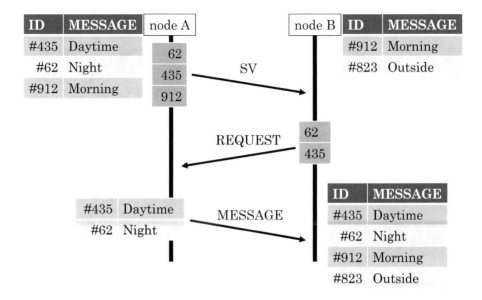

Fig. 4. Bundle protocol implementation example.

3.3 Data Forwarding

In order to enable data forwarding among DTN nodes, we have implemented a simplified version of the bundle protocol, as described in RFC 5050 [9]. As shown in Fig. 4, when node A contacts node B, it sends an SV message, which contains the hashed ids of all messages saved in its buffer (62, 435, 912). Then, node B compares message ids from the received SV message and its own buffer, in order to identify messages that are not present in its buffer. Then it sends a Request message towards node A, to request the messages, which ids were not present in its buffer (62, 435). Finally, node A sends the message bundle (62: "Night", 435: "Daytime"). This process requires some time depending on the number of messages and the message size. For DTNs, where links are unstable and unreliable, the transmission of message bundles might be interrupted.

When DTN nodes repeat the bundle process for every encountering node, they are using Epidemic Routing and multiple copies of the same message are flooding around the network. Thus, the overhead in the network is relatively high, especially when the number of nodes increases and the message size is big. In order to avoid this we are implementing different routing approaches, including traditional Spray-and-Wait routing and newly designed protocols. One tries to minimize forwarded messages based on node's direction, which we get from GPS logger module and include in SV messages. Another approach, aims to increase network lifetime, by excluding nodes with low residual battery levels. Both approaches are work in progress and will be published in the near future, after extensive testing and fine-tuning.

4   Conclusions and Future Works

In this paper we introduced our implemented testbed for DTNs. Raspberry Pi 3 B+ is used as a DTN Node, and equipped with additional hardware (sensors, battery) and software (data generator, bundle protocol). From our experience with this implementation we learned the following:

- DTNs fit in various applications, including real-time communications (when adequate relay nodes are present) and store-carry-forward communications for sparser networks.
- Epidemic Routing as implemented by bundle protocol, increases the number of copies in the network. In sparse networks, this is required, but for denser networks it introduces a high overhead. Thus, in order to minimize the overhead, newly designed routing approaches are important.
- In order to evaluate DTN protocols, experiments are time-consuming, because they need to verify delay-tolerant aspects of these protocols.
- Implementing a real experiment environment, requires technical skills in various fields, such as operating system knowledge, thread programming, hardware connectivity, interface adaptability, and so on.

In the future, we would like to follow up with the following objectives:

- Enrich our testbed with more functionalities, including automatic mobility and indoor localization.
- Continue the implementation of routing approached to DTN, in order to minimize routing overhead.
- Implement our DTN testbed in off-campus environment, in collaboration with marine life research from our university.

References

1. Fall, K.: A delay-tolerant network architecture for challenged internets. In: SIGCOMM-2003, pp. 27–34 (2003)
2. Demmer, M.J., Fall, K.R.: DTLSR: delay tolerant routing for developing regions. In: Proceedings of the 2007 Workshop on Networked Systems for Developing Regions, NSDR 2007, Kyoto, Japan, 5 p. (2007)
3. Pentland, A., Fletcher, R., Hasson, A.: DakNet: rethinking connectivity in developing nations. IEEE Comput. **37**(1), 78–83 (2004)
4. Guo, S., et al.: Design and implementation of the KioskNet system. Comput. Netw. **55**(1), 264–281 (2011)
5. Magdum, S.S., Sharma, M., Kala, S.M., Franklin, A.A., Tamma, B.R.: Evaluating DTN routing schemes for application in vehicular networks. In: 2019 11th International Conference on Communication Systems & Networks (COMSNETS), pp. 771–776 (2019)
6. Bylykbashi, K., Spaho, E., Barolli, L., Xhafa, F.: Routing in a many-to-one communication scenario in a realistic VDTN. Int. J. High Speed Netw. **24**(2), 107–118 (2018)

7. Cuka, M., Shinko, I., Spaho, E., Oda, T., Ikeda, M., Barolli, L.: A simulation system based on ONE and SUMO simulators: performance evaluation of different vehicular DTN routing protocols. Int. J. High Speed Netw. **23**(1), 59–66 (2017)
8. Yoon, J., Kim, S., Lee, J., Jang, K.: An enhanced friendship-based routing scheme exploiting regularity in an opportunistic network. In: 2016 IEEE International Conference on Internet of Things (iThings) and IEEE Green Computing and Communications (GreenCom) and IEEE Cyber, Physical and Social Computing (CPSCom) and IEEE Smart Data (SmartData), pp. 51–57 (2016)
9. Scott, K., Burleigh, S.: Bundle protocol specification. In: IETF RFC 5050 (Experimental) (2007)
10. Vahdat, A., Becker, D.: Epidemic routing for partially-connected adhoc networks. Handbook of Systemic Autoimmune Diseases, Technical report (2000)
11. Spyropoulos, T., Psounis, K., Raghavendra, C.: Spray and wait: an efficient routing scheme for intermittently connected mobile networks. In: Proceedings of ACM SIGCOMM-2005 Workshop on Delay-Tolerant Networking, pp. 252–259 (2005)
12. Burgess, J., Gallagher, B., Jensen, D., Levine, B.N.: Maxprop: routing for vehicle-based disruption-tolerant networks. In: Proceedings of IEEE INFOCOM 2006, pp. 1688–1698 (2006)
13. Lindgren, A., Doria, A., Davies, E.B., Grasic, S.: Probabilistic routing protocol for intermittently connected networks. ACM SIGMOBILE Mobile Comput. Commun. Rev. **7**(3), 19–20 (2003)
14. Jain, S., Fall, K., Patra, R.: Routing in a delay tolerant network. In: Proceedings of ACM SIGCOMM-04, pp. 145–158 (2004)
15. Sony Spresense Board. https://www.sony-semicon.co.jp/e/products/smart-sensing/spresense/

Evaluation of Focused Beam Routing Protocol on Delay Tolerant Network for Underwater Optical Wireless Communication

Keita Matsuo[1(✉)], Elis Kulla[2], and Leonard Barolli[1]

[1] Department of Information and Communication Engineering,
Fukuoka Institute of Technology (FIT), 3-30-1 Wajiro-Higashi, Higashi-Ku,
811-0295 Fukuoka, Japan
{kt-matsuo,barolli}@fit.ac.jp
[2] Department of Information and Computer Engineering,
Okayama University of Science (OUS), 1-1 Ridaicho, Kita-Ku,
700-0005 Okayama, Japan
kulla@ice.ous.ac.jp

Abstract. Recently, underwater communication has been developing in many ways, such as wired communication, underwater acoustic communication (UAC), underwater radio wave wireless communication (URWC), underwater optical wireless communication (UOWC). The main issue in underwater communication is the communication interruption because signals are affected by various factors in underwater environment. Consequently, communication links are unstable and real time communication is almost impossible. Therefore, we considered combining both underwater communication and delay tolerant network technologies. In this paper, we present the evaluation of focused beam routing (FBR) protocol for UOWC in The ONE simulator for different FBR angles. The results show that the highest delivery probability is achieved for $15°$, while for angles over $30°$, the values drastically decreased, because of the short communication ranges. Furthermore, when we used narrow FBR angles, both latency and overhead decreased. From these results, we can claim that FBR protocol can be used for effective communication for UOWC.

1 Introduction

Recently, underwater communication has been developing in many ways, such as wired communication (WC), underwater acoustic communication (UAC), underwater radio wave wireless communication (URWC), underwater optical wireless communication (UOWC). So far, underwater communication remains realized until nowadays via communication cables due to the limited development of underwater wireless communications [4], and the high cost of hydrophones and other equipment.

L. Barolli et al. (Eds.): EIDWT 2022, LNDECT 118, pp. 263–271, 2022.
https://doi.org/10.1007/978-3-030-95903-6_28

UAC is a more popular technique which uses sound signals in water. This way, it is able to transmit signals emitted from sensors, robots or submarines to longer distances. However, the bandwidth is narrow and transmission speed is relatively low. Moreover, the speed of sound is affected by temperature, depth and salinity of underwater environment. These factors produce variations in speed of sound in underwater environment [2].

In URWC, radio waves are used for communication. The radio wave employed for terrestrial wireless communication are also utilized for underwater communication, it achieves high data rate for short communication range and suffers from Doppler effect [4]. There are some former research for absorption losses in water. According to [7], the absorption losses are about 9 dB to 19 dB in freshwater at 2.4 GHz, the absorption losses are about 19 dB to 24 dB in river water at 2.4 GHz, and absorption losses are about 25 dB to 30 dB in seawater at 2.4 GHz. If we use URWC in the sea environment, the losses are bigger than other kind of water. For this reason, it is very difficult to use high frequency radio waves in underwater environment. For instance, if we use radio waves in water environment, the transmission distance would be few centimeters.

UOWC uses visible light for communication, which allows longer communication distances in the water. According to [8], the communication distance in the water by using visible light is over 100 m at 20 Mbps. Recently, laser diodes (LDs) and light emitting diode (LEDs) have developed quickly, therefore many lighting devices are shifting from conventional light to diodes as a LEDs light. The main features of LEDs are high energy efficiency and quick response which is sufficient to achieve high speed communication by using UOWC, it is possible to implement broadband underwater network.

As we describe above, underwater communication links are unstable because they are affected by various factors in underwater environment. Therefore, we considered a technology such as a delay tolerant network (DTN) is a suitable paradigm for underwater environment.

In Table 1 is shown comparison of underwater communications.

These communication ways have characteristics strengths and weaknesses. In order to use high speed wireless network in under water environment, we need to choose both URWC or UOWC. Especially, in URWC that can extendet the communication distance by decreasing the radio frequency, but in this case the bandwidth would become narrower, causing lower communication speed. So, we have focused on UOWC by using DTNs technologies to make good communication environment in underwater.

In this paper, we present the evaluation of focused beam routing (FBR) protocol for underwater optical wireless network. The evaluation results show that the FBR protocol was able to carry the data to the surface station.

The rest of this paper is structured as follows. In Sect. 2, we introduce the related work. In Sect. 3, we present the proposed FBR protocol with DTNs for UOWC. In Sect. 4, we describe implementation of FBR in The ONE Simulator. In Sect. 5, we show the simulation results. Finally, conclusions and future work are given in Sect. 6.

Table 1. Comparison of underwater communications.

Comunication method	Communication distance	Transmission speed
WC	Depend on cable length	High
UAC	Long	Low
URWC	Short	High
UOWC	Middle	High

2   Related Work

So far, many studies of UOWC have published in several years. The UOWC reduced absorption window for blue-green light. Due to its higher bandwidth, underwater optical wireless communications can support higher data rates at low latency levels compared to acoustic and RF counterparts [9]. Moreover some commercial UOWC modems were released. One of the research teams started developing UOWC, their try was achieved communication with over 100 m range at 20 Mbps in bi-directional with prototype UOWC modem [8]. The research of UOWC is not new that was started from 1960s to make high speed communication techniques in the lake or sea. However, it has accelerated after invented of blue LEDs. One of the interest research regions is compared LEDs colors that compared to the open ocean where blue LEDs perform well for data communications, in coastal and harbor environments optical transmission becomes worse and the color of lowest attenuation shifts to green. Another problem concerns the "green-yellow gap" of LEDs, as the quantum efficiency of current commercially available green LEDs is poor [11]. The direction of UOWS signals is narrow. If transmitter makes the extremely narrow range visible light signals as laser's light by using LEDs, it will achieve long distance communication, while it makes the wide range visible light signal, the receiver becomes easier to find the signals instead of long communication distance.

Delay tolerant networks (DTNs) are intermittently connected communications that would be helping from repeatedly and long time disconnection because of various reasons such as out of communication range, short of battery, and so on. The DTNs include wireless sensor networks using scheduled intermittent connectivity, mobile ad hoc networks, satellite networks with periodic connectivity, village networks, wildlife tracking networks, and pocket switched networks, etc. Due to the broad application prospect, delay tolerant networks attract much attention [14]. The DTNs have been utilized in various communication paradigms such as vehicular delay tolerant networks (VDTNs) [1,3,10], and DTNs system in a communication network on a railway line [12,13].

FBR is a well-known routing protocol for under water sensor network (UWSN), which considers as forward candidates, only active neighbors which location is inside a region specified by the direction from the forwarding node to the destination by considering an arbitrary angle and the communication distance [6].

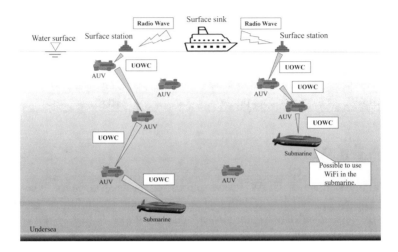

Fig. 1. Image of UOWC by using DTNs.

3   Proposed FBR Protocol with DTNs for UOWC

In this section, we describe the FBR protocol with DTNs for UOWC. In Fig. 1 is shown image of UOWC by using DTNs, autonomous underwater vehicles (AUVs) can communicate using DTNs' store-carry-and-forward paradigm. Surface station could receive data from AUVs then the station send the data to the surface sink or monitoring center in ground.

Conventional marine research using submarine could not exchange data with ground unless WC was used. Therefore, it was necessary to wait for the submarine to return in order to get the data. However, using the method that we propose, it will be possible to send the data from the underwater to the sea surface. Also, it can send large size data such as photos, sounds and videos because of UOWN links. Moreover, by using UOWC various protocols and real time applications can be implemented.

If the communication are become intermittently connected between nodes (AUVs, submarine, surface station), they can send the data by DTNs with FBR. If the emitting visible lights of UOWN (signals) is narrow, the communication distance will be longer, whereas when the emitting visible lights has wide range, the distance will be short. We show the image of FBR protocol for UOWC in Fig. 2. When we use FBR with 1°, it will transmit the signals to longer distance, while if we use 30°, the distance will be short. However, it can transmit the signals to more nodes. Also, if it uses 180° the sender node emits the signal to omnidirectional like epidemic, though sender node sends many receiver nodes the signals, the transmit distance will be minimum.

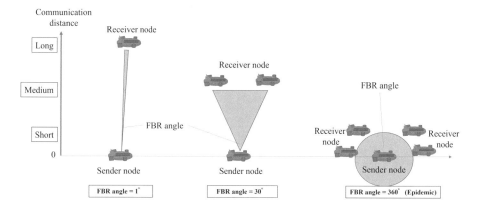

Fig. 2. Image of FBR protocol for UOWC.

4   Implementation of FBR in the ONE Simulator

The Opportunistic Networking Environment (ONE) simulator specifically designed for evaluating DTNs routing and application protocols. It allows users to create scenarios based upon different synthetic movement models and real-world traces and offers a framework for implementing routing and application protocols [5]. Thus, we implemented FBR protocol in The ONE Simulator. The following assumptions are made, in order to simplify the implementation:

- The environment is considered with only 2 dimensions: the width on the horizontal axis and the depth in the vertical axis.
- Every node knows self location and destination.
- Every participating node is mobile and moves based on the Random Waypoint mobility model, as implemented in The ONE Simulator.
- There is only one surface station, and all transmissions are directed towards this surface station, which is located in the middle top of the simplified 2D environment.

5   Simulations

In order to investigate for FBR protocol on UOWC, we used The ONE Simulator. We implemented FBR protocol in the simulator. We assumed the communication ranges of UOWC as shown in Table 2. In Table 3 is shown simulation parameters' settings. The FBR angle is always directed towards surface station.

Table 2. FBR angles and communication ranges.

FBR angles [degree]	Communication Ranges [m]
1	100
15	50
30	30
60	15
90	10
180	5

Table 3. Simulation parameters' settings.

Parameters	Value
Transmit speed	20 [Mbps]
Data size	800–1200 [kB]
Ivent interval	8–12 [s]
Number of Surface stations	1 Static (middle of top)
Noumber of nodes	16
Movement Model	Random Waypoint
Simulation Time	3600 [s]
Simulation area	400 × 400 [m]
Buffer size	300 MBytes

For different angles and different communication range, we show the simulation results for delivery probability, latency average, hop count average and overhead ratio in Figs. 3, 4, 5 and 6, respectively. We simulated each FBR angle for 3600 s.

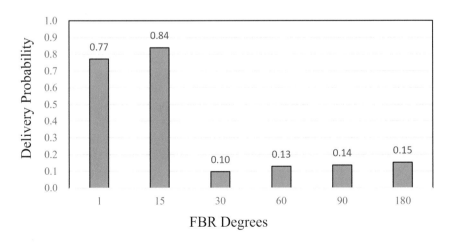

Fig. 3. Delivery probability of FBR protocol for UOWC.

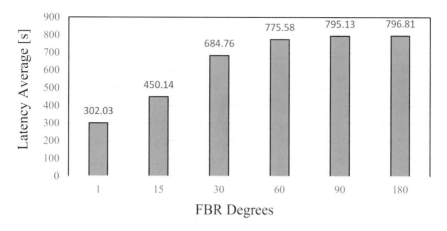

Fig. 4. Latency average of FBR protocol for UOWC.

From the result, the highest delivery probability is when FBR angle is 15°, while when FBR angle is over 30°, the results are similar. Latency average increased when FBR angles increased. The hop count average is steadily growing after 60°. When the communication range is long, the overhead ratio decreases as show in Fig. 6.

From simulation results, we can conclude that the ideal FBR angles are between 15 and within 30°, because the value of delivery probability was better and latency, hop count were lower. Moreover, the overhead ratio was good.

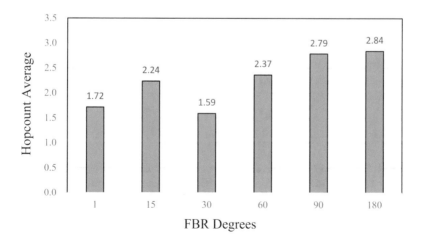

Fig. 5. Hopcount average of FBR protocol for UOWC.

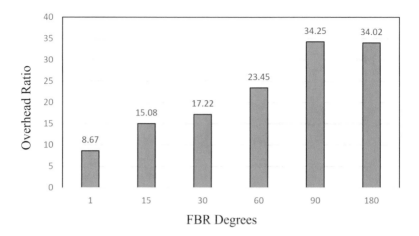

Fig. 6. Overhead ratio of FBR protocol for UOWC.

6   Conclusions and Future Work

In this paper, we introduced some of underwater communication ways and compared them. Also, we explained UOWC in detail. UOWC has high potential in the future underwater communication. We implemented FBR protocol in The ONE Simulator. In the simulation, we defined six FBR angles, then we evaluated the performance of FBR protocol for UOWC. The results show the highest delivery probability was between 15 and within 30°, while over 30° the values drastically decrease, because of the short communication ranges. Also, when we use narrow FBR degrees, both latency and overhead decreased. From these results we can assume that FBR protocol can be used for effective communication for UOWC.

In the future, we would like to add other protocols and functions to the simulation system and carry out extensive simulations to evaluate various underwater environments.

References

1. Ahmed, S.H., Kang, H., Kim, D.: Vehicular delay tolerant network (VDTN): Routing perspectives. In: 2015 12th Annual IEEE Consumer Communications and Networking Conference (CCNC), pp. 898–903. IEEE (2015)
2. Awan, K.M., Shah, P.A., Iqbal, K., Gillani, S., Ahmad, W., Nam, Y.: Underwater wireless sensor networks: a review of recent issues and challenges. Wireless Communications and Mobile Computing 2019 (2019)
3. Azuma, M., Uchimura, S., Tada, Y., Ikeda, M., Barolli, L.: A hybrid message delivery method for vehicular DTN considering impact of shuttle buses and roadside units. In: Barolli, L., Woungang, I., Enokido, T. (eds.) AINA 2021. LNNS, vol. 227, pp. 211–218. Springer, Cham (2021). https://doi.org/10.1007/978-3-030-75078-7_22

4. Jouhari, M., Ibrahimi, K., Tembine, H., Ben-Othman, J.: Underwater wireless sensor networks: a survey on enabling technologies, localization protocols, and internet of underwater things. IEEE Access **7**, 96879–96899 (2019)
5. Keränen, A., Ott, J., Kärkkäinen, T.: The one simulator for dtn protocol evaluation. In: Proceedings of the 2nd International Conference on Simulation Tools and Techniques, pp. 1–10 (2009)
6. Kulla, E., Katayama, K., Matsuo, K., Barolli, L.: Enhanced focused beam routing in underwater wireless sensor networks. In: Barolli, L., Natwichai, J., Enokido, T. (eds.) EIDWT 2021. LNDECT, vol. 65, pp. 1–9. Springer, Cham (2021). https://doi.org/10.1007/978-3-030-70639-5_1
7. Qureshi, U.M., et al.: RF path and absorption loss estimation for underwater wireless sensor networks in different water environments. Sensors **16**(6), 890 (2016)
8. Sawa, T., Nishimura, N., Tojo, K., Ito, S.: Practical performance and prospect of underwater optical wireless communication: -results of optical characteristic measurement at visible light band under water and communication tests with the prototype modem in the sea-. IEICE Trans. Fundam. Electron. Commun. Comput. Sci. **102**(1), 156–167 (2019)
9. Schirripa Spagnolo, G., Cozzella, L., Leccese, F.: Underwater optical wireless communications: overview. Sensors **20**(8), 2261 (2020)
10. Spaho, E., Barolli, L., Kolici, V., Lala, A.: Performance evaluation of different routing protocols in a vehicular delay tolerant network. In: 2015 10th International Conference on Broadband and Wireless Computing, Communication and Applications (BWCCA), pp. 157–162. IEEE (2015)
11. Sticklus, J., Hoeher, P.A., Röttgers, R.: Optical underwater communication: the potential of using converted green LEDs in coastal waters. IEEE J. Ocean. Eng. **44**(2), 535–547 (2018)
12. Tikhonov, E., Schneps-Schneppe, D., Namiot, D.: Delay tolerant network potential in a railway network. In: 2020 26th Conference of Open Innovations Association (FRUCT), pp. 438–448. IEEE (2020)
13. Tikhonov, E., Schneps-Schneppe, D., Namiot, D.: Delay tolerant network protocols for an expanding network on a railway. In: 2020 International Conference on Innovation and Intelligence for Informatics, Computing and Technologies (3ICT), pp. 1–6. IEEE (2020)
14. Xiao, M., Huang, L.: Delay-tolerant network routing algorithm. J. Comput. Res. Dev. **46**(7), 1065 (2009)

A Fuzzy-Based System for Slice Service Level Agreement in 5G Wireless Networks: Effect of Traffic Load Parameter

Phudit Ampririt[1(✉)], Ermioni Qafzezi[1], Kevin Bylykbashi[2], Makoto Ikeda[2], Keita Matsuo[2], and Leonard Barolli[2]

[1] Graduate School of Engineering, Fukuoka Institute of Technology, 3-30-1 Wajiro-Higashi, Higashi-Ku, Fukuoka 811-0295, Japan
{bd21201,bd20101}@bene.fit.ac.jp

[2] Department of Information and Communication Engineering, Fukuoka Institute of Technology, 3-30-1 Wajiro-Higashi, Higashi-Ku, Fukuoka 811-0295, Japan
kevin@bene.fit.ac.jp, makoto.ikd@acm.org,
{kt-matsuo,barolli}@fit.ac.jp

Abstract. The Fifth Generation (5G) wireless network is expected to be flexible to satisfy user requirements and the Software-Defined Network (SDN) with Network Slicing will be a good approach for admission control. In this paper, we propose a fuzzy-based system for user Service Level Agreement considering four parameters: Reliability (Re), Availability (Av), Latency (La) and Traffic Load (Tl) as a new parameter. From simulation results, we conclude that the considered parameters have different effects on the SLA. When Re and Av are increasing, the SLA parameter is increased but when La and Tl are increasing, the SLA parameter is decreased.

1 Introduction

Recently, the growth of wireless technologies and user's demand of services are increasing rapidly. Especially in 5G wireless networks, there will be billions of new devices with unpredictable traffic pattern which provide high data rates. With the appearance of Internet of Things (IoT), these devices will generate Big Data to the Internet, which will cause the congestion and deterioration of QoS [1].

The 5G wireless network will provide users with new experiences such as Ultra High Definition Television (UHDT) on Internet and support a lot of IoT devices with long battery life and high data rate on hotspot areas with high user density. In the 5G technology, the routing and switching technologies aren't important anymore or coverage area is shorter than 4G because it uses high frequency for facing higher device's volume and high user density [2–4].

There are many research work that try to build systems which are suitable for 5G era. The SDN is one of them [5]. For example, the mobile handover mechanism with SDN is used for reducing the delay in handover processing. Also, the QoS can be improved by applying Fuzzy Logic (FL) on SDN controller [6–8].

L. Barolli et al. (Eds.): EIDWT 2022, LNDECT 118, pp. 272–282, 2022.
https://doi.org/10.1007/978-3-030-95903-6_29

In our previous work [9–12], we presented a Fuzzy-based system for admission decision considering four input parameters: Quality of service (QoS), Slice Priority (SP), Service Level Agreement (SLA) and Slice Overloading Cost (SOC). The output parameter was Admission Decision (AD). In this paper, we propose a Fuzzy-based system for user Service Level Agreement considering four parameters: Reliability (Re), Availability (Av), Latency (La) and Traffic Load (Tl) as a new paramter.

The rest of the paper is organized as follows. In Sect. 2 is presented an overview of SDN. In Sect. 3, we present application of Fuzzy Logic for admission control. In Sect. 4, we describe the proposed Fuzzy-based system and its implementation. In Sect. 5, we discuss the simulation results. Finally, conclusions and future work are presented in Sect. 6.

2 Software-Defined Networks (SDNs)

The SDN is a new networking paradigm that decouples the data plane from control plane in the network. By SDN is easy to manage and provide network software based services from a centralised control plane. The SDN control plane is managed by SDN controller or cooperating group of SDN controllers. The SDN structure is shown in Fig. 1 [13, 14].

- **Application Layer** builds an abstracted view of the network by collecting information from the controller for decision-making purposes. The types of applications are related to: network configuration and management, network monitoring, network troubleshooting, network policies and security.
- **Control Layer** receives instructions or requirements from the Application Layer and control the Infrastructure Layer by using intelligent logic.
- **Infrastructure Layer** receives orders from SDN controller and sends data among them.

The SDN can manage network systems while enabling new services. In congestion traffic situation, management system can be flexible, allowing users to easily control and adapt resources appropriately throughout the control plane. Mobility management is easier and quicker in forwarding across different wireless technologies (e.g. 5G, 4G, Wifi and Wimax). Also, the handover procedure is simple and the delay can be decreased.

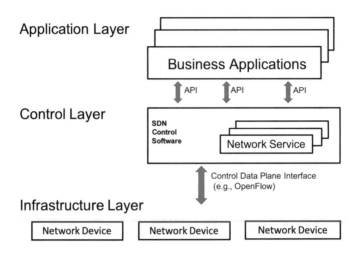

Fig. 1. Structure of SDN.

3 Outline of Fuzzy Logic

A FL system is a nonlinear mapping of an input data vector into a scalar output, which is able to simultaneously handle numerical data and linguistic knowledge. The FL can deal with statements which may be true, false or intermediate truth-value. These statements are impossible to quantify using traditional mathematics. The FL system is used in many controlling applications such as aircraft control (Rockwell Corp.), Sendai subway operation (Hitachi), and TV picture adjustment (Sony) [15–17].

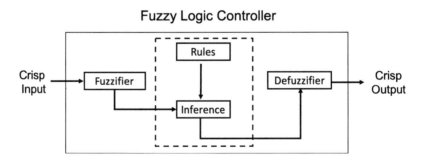

Fig. 2. FLC structure.

In Fig. 2 is shown Fuzzy Logic Controller (FLC) structure, which contains four components: fuzzifier, inference engine, fuzzy rule base and defuzzifier.

- **Fuzzifier** is needed for combining the crisp values with rules which are linguistic variables and have fuzzy sets asSLAiated with them.

- **The Rules** may be provided by expert or can be extracted from numerical data. In engineering case, the rules are expressed as a collection of IF-THEN statements.
- **The Inference Engine** infers fuzzy output by considering fuzzified input values and fuzzy rules.
- **The Defuzzifier** maps output set into crisp numbers.

3.1 Linguistic Variables

A concept that plays a central role in the application of FL is that of a linguistic variable. The linguistic variables may be viewed as a form of data compression. One linguistic variable may represent many numerical variables. It is suggestive to refer to this form of data compression as granulation.

The same effect can be achieved by conventional quantization, but in the case of quantization, the values are intervals, whereas in the case of granulation the values are overlapping fuzzy sets. The advantages of granulation over quantization are as follows:

- it is more general;
- it mimics the way in which humans interpret linguistic values;
- the transition from one linguistic value to a contiguous linguistic value is gradual rather than abrupt, resulting in continuity and robustness.

For example, let Temperature (T) be interpreted as a linguistic variable. It can be decomposed into a set of Terms: T (Temperature) = {Freezing, Cold, Warm, Hot, Blazing}. Each term is characterised by fuzzy sets which can be interpreted, for instance, "Freezing" as a temparature below $0\,^{\circ}$C, "Cold" as a temparature close to $10\,^{\circ}$C.

3.2 Fuzzy Control Rules

Rules are usually written in the form "IF x is S THEN y is T" where x and y are linguistic variables that are expressed by S and T, which are fuzzy sets. The x is a control (input) variable and y is the solution (output) variable. This rule is called Fuzzy control rule. The form "IF ... THEN" is called a conditional sentence. It consists of "IF" which is called the antecedent and "THEN" is called the consequent.

3.3 Defuzzificaion Method

There are many defuzzification methods, which are showing in following:

- The Centroid Method;
- Tsukamoto's Defuzzification Method;
- The Center of Are (COA) Method;
- The Mean of Maximum (MOM) Method;
- Defuzzification when Output of Rules are Function of Their Inputs.

4   Proposed Fuzzy-based System

In this work, we use FL to implement the proposed system. In Fig. 3, we show the overview of our proposed system. Each evolve Base Station (eBS) will receive controlling order from SDN controller and they can communicate and send data with User Equipment (UE). On the other hand, the SDN controller will collect all the data about network traffic status and controlling eBS by using the proposed Fuzzy-based system. The SDN controller will be a communicating bridge between eBS and 5G core network. The proposed system is called Integrated Fuzzy-based Admission Control System (IFACS) in 5G wireless networks. The structure of IFACS is shown in Fig. 4. For the implementation of our system, we consider four input parameters: Quality of Service (QoS), Slice Priority (SP), Slice Overloading Cost (SOC), Service Level Agreement (SLA) and the output parameter is Admission Decision (AD). We applied FL to evaluate QoS, SP, SOC and SLA. The QoS is considering four parameters: Slice Throughput (ST), Slice Delay (SD), Slice Loss (SL) and Slice Reliability (SR). The SP is considering three parameters: Slice Traffic Volume (STV), Slice Interference from Other Slices (SIOS) and Slice Connectivity (SC). The SOC is considering three parameters: Virtual Machine Overloading Cost (VMOC), Link Overloading Cost (LOC) and Switches Overloading Cost (SWOC). In this paper, we apply FL to evaluate the SLA by considering 4 parameters: Reliability (Re), Availability (Av), Latency (La) and Traffic Load (Tl) as a new parameter. The output parameter is SLA.

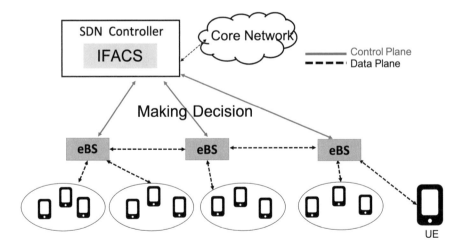

Fig. 3. Proposed system overview.

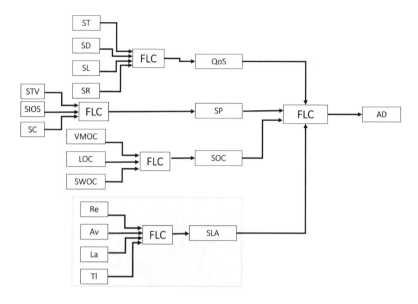

Fig. 4. Proposed system structure.

Reliability (Re): The Re value is the required user reliability. When Re value is high, the SLA is high.

Availability (Av): The Av is the required user availability. When Av value is high, the SLA is high.

Latency (La): The La is the required user latency. When La value is high, the SLA is low.

Traffic Load (Tl): The Tl is the total traffic of users in a slice. When Tl value is high, the SLA is low.

Service Level Agreement (SLA): The SLA is described based on these four parameters. The user request is characterized by its required SLA.

Table 1. Parameter and their term sets.

Parameters	Term set
Reliability (Re)	Low (Lo), Medium (Me), High (Hi)
Availability (Av)	Low (Lw), Medium (Md), High (Hg)
Latency (La)	Low (L), Medium (M), High (H)
Traffic Load (Tl)	Small (Sm), Intermediate (In), Huge (Hu)
Service Level Agreement (SLA)	SLA1, SLA2, SLA3, SLA4, SLA5, SLA6, SLA7

Table 2. Fuzzy Rule base

Rule	Re	Av	La	Tl	SLA	Rule	Re	Av	La	Tl	SLA
1	Lo	Lw	L	Sm	SLA4	41	Me	Md	M	In	SLA4
2	Lo	Lw	L	In	SLA3	42	Me	Md	M	Hu	SLA3
3	Lo	Lw	L	Hu	SLA2	43	Me	Md	H	Sm	SLA4
4	Lo	Lw	M	Sm	SLA3	44	Me	Md	H	In	SLA4
5	Lo	Lw	M	In	SLA2	45	Me	Md	H	Hu	SLA2
6	Lo	Lw	M	Hu	SLA1	46	Me	Hg	L	Sm	SLA7
7	Lo	Lw	H	Sm	SLA2	47	Me	Hg	L	In	SLA6
8	Lo	Lw	H	In	SLA1	48	Me	Hg	L	Hu	SLA5
9	Lo	Lw	H	Hu	SLA1	49	Me	Hg	M	Sm	SLA6
10	Lo	Md	L	Sm	SLA5	50	Me	Hg	M	In	SLA5
11	Lo	Md	L	In	SLA4	51	Me	Hg	M	Hu	SLA4
12	Lo	Md	L	Hu	SLA3	52	Me	Hg	H	Sm	SLA5
13	Lo	Md	M	Sm	SLA4	53	Me	Hg	H	In	SLA4
14	Lo	Md	M	In	SLA3	54	Me	Hg	H	Hu	SLA3
15	Lo	Md	M	Hu	SLA1	55	Hi	Lw	L	Sm	SLA6
16	Lo	Md	H	Sm	SLA3	56	Hi	Lw	L	In	SLA5
17	Lo	Md	H	In	SLA2	57	Hi	Lw	L	Hu	SLA4
18	Lo	Md	H	Hu	SLA1	58	Hi	Lw	M	Sm	SLA5
19	Lo	Hg	L	Sm	SLA6	59	Hi	Lw	M	In	SLA4
20	Lo	Hg	L	In	SLA5	60	Hi	Lw	M	Hu	SLA3
21	Lo	Hg	L	Hu	SLA4	61	Hi	Lw	H	Sm	SLA4
22	Lo	Hg	M	Sm	SLA5	62	Hi	Lw	H	In	SLA3
23	Lo	Hg	M	In	SLA4	63	Hi	Lw	H	Hu	SLA2
24	Lo	Hg	M	Hu	SLA3	64	Hi	Md	L	Sm	SLA7
25	Lo	Hg	H	Sm	SLA4	65	Hi	Md	L	In	SLA6
26	Lo	Hg	H	In	SLA3	66	Hi	Md	L	Hu	SLA5
27	Lo	Hg	H	Hu	SLA2	67	Hi	Md	M	Sm	SLA6
28	Me	Lw	L	Sm	SLA5	68	Hi	Md	M	In	SLA5
29	Me	Lw	L	In	SLA4	69	Hi	Md	M	Hu	SLA4
30	Me	Lw	L	Hu	SLA3	70	Hi	Md	H	Sm	SLA5
31	Me	Lw	M	Sm	SLA4	71	Hi	Md	H	In	SLA4
32	Me	Lw	M	In	SLA3	72	Hi	Md	H	Hu	SLA3
33	Me	Lw	M	Hu	SLA2	73	Hi	Hg	L	Sm	SLA7
34	Me	Lw	H	Sm	SLA3	74	Hi	Hg	L	In	SLA7
35	Me	Lw	H	In	SLA2	75	Hi	Hg	L	Hu	SLA6
36	Me	Lw	H	Hu	SLA1	76	Hi	Hg	M	Sm	SLA7
37	Me	Md	L	Sm	SLA6	77	Hi	Hg	M	In	SLA6
38	Me	Md	L	In	SLA5	78	Hi	Hg	M	Hu	SLA5
39	Me	Md	L	Hu	SLA4	79	Hi	Hg	H	Sm	SLA6
40	Me	Md	M	Sm	SLA5	80	Hi	Hg	H	In	SLA5
						81	Hi	Hg	H	Hu	SLA4

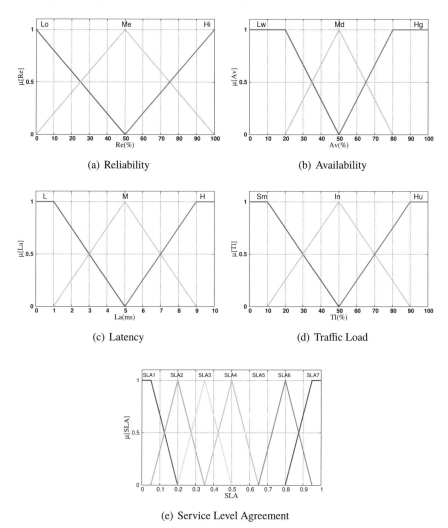

Fig. 5. Membership functions.

The membership functions are shown in Fig. 5. We use triangular and trapezoidal membership functions because they are more suitable for real-time operations [18–21]. We show parameters and their term sets in Table 1. The Fuzzy Rule Base (FRB) is shown in Table 2 and has 81 rules. The control rules have the form: IF "condition" THEN "control action". For example, for Rule 1:"IF Re is Lo, Av is Lw, La is L and Tl is Sm THEN SLA is SLA4".

5 Simulation Results

In this section, we present the simulation result of our proposed system. The simulation results are shown in Fig. 6, Fig. 7 and Fig. 8. They show the relation of SLA with Tl for different La values considering Re and Av as constant parameters.

In Fig. 6 (a), we consider the Re value 10% and the Av value 10%. When Tl is increased, we see that AD is decreased. For Tl 30%, when La is increased from 1 ms to 5 ms and 5 ms to 9 ms, the SLA is decreased by 15% and 11.84%, respectively.

We compare Fig. 6 (a) with Fig. 6 (b) to see how Av has affected SLA. We change the Av value from 10% to 90%. The SLA is increasing by 30% when the Tl value is 50% and the La is 1 ms. In Fig. 6 (b), when we changed the Tl value from 20% to 80%, the AD is decreased 16.62% when the La value is 1 ms.

In Fig. 7, we increase the value of Re to 50%. The SLA values in Fig. 7 are higher than in Fig. 6. When the Av increases from 10% to 90%, the SLA increases 30% when Tl value is 60% and the La value is 1 ms. In Fig. 7 (b), when La is 5 ms, all SLA values are higher than 0.5. This means that users fulfill the required SLA.

In Fig. 8, we increase the value of Re to 90%. We see that the SLA values are increased much more compared with the results of Fig. 6 and Fig. 7. When we changed the Av value from 10% to 90% and La values are 5 ms and 1 ms, all SLA values are higher than 0.5.

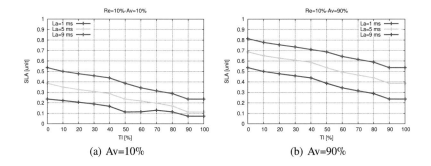

(a) Av=10%   (b) Av=90%

Fig. 6. Simulation results for Re = 10%.

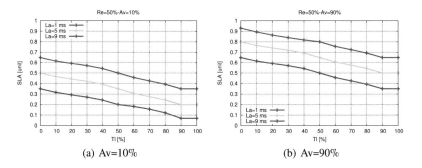

(a) Av=10%   (b) Av=90%

Fig. 7. Simulation results for Re = 50%.

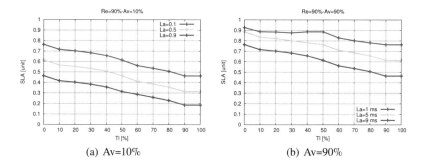

(a) Av=10%   (b) Av=90%

Fig. 8. Simulation results for Re = 90%.

6   Conclusions and Future Work

In this paper, we proposed and implemented a Fuzzy-based system for user SLA. The admission control mechanism will search for a set of slices and try to make connection between a new user and a slice that match the required SLA. So, the SLA parameter will be used as an input parameter for admission control in 5G wireless networks. We evaluated the proposed system by simulations. From the simulation results, we found that four parameters have different effects on the SLA. When Re and Av value are increased, the SLA value is increased but when La and Tl value are increased, the SLA value is decreased.

In the future, we will consider other parameters and make extensive simulations to evaluate the proposed system.

References

1. Navarro-Ortiz, J., Romero-Diaz, P., Sendra, S., Ameigeiras, P., Ramos-Munoz, J.J., Lopez-Soler, J.M.: A survey on 5g usage scenarios and traffic models. IEEE Commun. Surv. Tutor. **22**(2), 905–929 (2020)
2. Hossain, S.: 5G wireless communication systems. Am. J. Eng. Res. (AJER) **2**(10), 344–353 (2013)
3. Giordani, M., Mezzavilla, M., Zorzi, M.: Initial access in 5g mmwave cellular networks. IEEE Commun. Mag. **54**(11), 40–47 (2016)
4. Kamil, I.A., Ogundoyin, S.O.: Lightweight privacy-preserving power injection and communication over vehicular networks and 5g smart grid slice with provable security. Internet Things **8**(100116), 100–116 (2019)
5. Hossain, E., Hasan, M.: 5g cellular: key enabling technologies and research challenges. IEEE Instrum. Meas. Mag. **18**(3), 11–21 (2015)
6. Yao, D., Su, X., Liu, B., Zeng, J.: A mobile handover mechanism based on fuzzy logic and MPTCP protocol under SDN architecture. In: 18th International Symposium on Communications and Information Technologies (ISCIT-2018), pp. 141–146 (September 2018)
7. Lee, J., Yoo, Y.: Handover cell selection using user mobility information in a 5g sdn-based network. In: 2017 Ninth International Conference on Ubiquitous and Future Networks (ICUFN-2017), pp. 697–702 (July 2017)

8. Moravejosharieh, A., Ahmadi, K., Ahmad, S.: A fuzzy logic approach to increase quality of service in software defined networking. In: 2018 International Conference on Advances in Computing,Communication Control and Networking (ICACCCN-2018), pp. 68–73 (October 2018)

9. Ampririt, P., Ohara, S., Qafzezi, E., Ikeda, M., Barolli, L., Takizawa, M.: Integration of software-defined network and fuzzy logic approaches for admission control in 5g wireless networks: a fuzzy-based scheme for qos evaluation. In: Barolli, L., Takizawa, M., Enokido, T., Chen, H.-C., Matsuo, K. (eds.) BWCCA 2020. LNNS, vol. 159, pp. 386–396. Springer, Cham (2021). https://doi.org/10.1007/978-3-030-61108-8_38

10. Ampririt, P., Ohara, S., Qafzezi, E., Ikeda, M., Barolli, L., Takizawa, M.: Effect of slice overloading cost on admission control for 5g wireless networks: a fuzzy-based system and its performance evaluation. In: Barolli, L., Natwichai, J., Enokido, T. (eds.) EIDWT 2021. LNDECT, vol. 65, pp. 24–35. Springer, Cham (2021). https://doi.org/10.1007/978-3-030-70639-5_3

11. Ampririt, P., Qafzezi, E., Bylykbashi, K., Ikeda, M., Matsuo, K., Barolli, L.: A fuzzy-based system for user service level agreement in 5g wireless networks. In: Barolli, L., Chen, H.-C., Miwa, H. (eds.) INCoS 2021. LNNS, vol. 312, pp. 96–106. Springer, Cham (2022). https://doi.org/10.1007/978-3-030-84910-8_10

12. Ampririt, P., Qafzezi, E., Bylykbashi, K., Ikeda, M., Matsuo, K., Barolli, L.: An intelligent system for admission control in 5g wireless networks considering fuzzy logic and sdns: effects of service level agreement on acceptance decision. In: Barolli, L. (ed.) 3PGCIC 2021. LNNS, vol. 343, pp. 185–196. Springer, Cham (2022). https://doi.org/10.1007/978-3-030-89899-1_19

13. Li, L.E., Mao, Z., M., Rexford, J.: Toward software-defined cellular networks. In: 2012 European Workshop on Software Defined Networking, pp. 7–12 (October 2012)

14. Mousa, M., Bahaa-Eldin, A.M., Sobh, M.: Software defined networking concepts and challenges. In: 2016 11th International Conference on Computer Engineering & Systems (ICCES-2016), pp. 79–90. IEEE (2016)

15. Jantzen, J.: Tutorial on fuzzy logic, Technical University of Denmark. Dept. of Automation, Technical report (1998)

16. Mendel, J.M.: Fuzzy logic systems for engineering: a tutorial. Proc. IEEE **83**(3), 345–377 (1995)

17. Zadeh, L.A.: Fuzzy logic. Computer **21**, 83–93 (1988)

18. Norp, T.: 5g requirements and key performance indicators. J. ICT Stand. **6**(1), 15–30 (2018)

19. Parvez, I., Rahmati, A., Guvenc, I., Sarwat, A.I., Dai, H.: A survey on low latency towards 5g: Ran, core network and caching solutions. IEEE Commun. Surv. Tutor. **20**(4), 3098–3130 (2018)

20. Kim, Y., Park, J., Kwon, D.H., Lim, H.: Buffer management of virtualized network slices for quality-of-service satisfaction. In: 2018 IEEE Conference on Network Function Virtualization and Software Defined Networks (NFV-SDN-2018), pp. 1–4 (2018)

21. Barolli, L., Koyama, A., Yamada, T., Yokoyama, S.: An integrated cac and routing strategy for high-speed large-scale networks using cooperative agents. IPSJ J. **42**(2), 222–233 (2001)

A River Monitoring and Predicting System Considering a Wireless Sensor Fusion Network and LSTM

Yuki Nagai[1], Tetsuya Oda[1(✉)], Tomoya Yasunaga[1], Chihiro Yukawa[1], Aoto Hirata[2], Nobuki Saito[2], and Leonard Barolli[3]

[1] Department of Information and Computer Engineering, Okayama University of Science (OUS), 1-1 Ridaicho, Kita-ku, Okayama 700–0005, Japan
{t18j057ny,t18j091yt,t18j097yc}@ous.jp, oda@ice.ous.ac.jp
[2] Graduate of Engineering, Okayama University of Science (OUS), 1-1 Ridaicho, Kita-ku, Okayama 700–0005, Japan
{t21jm02zr,t21jm01md}@ous.jp
[3] Department of Information and Communication Engineering, Fukuoka Insitute of Technology, 3-30-1 Wajiro-Higashi-ku, Fukuoka 811-0295, Japan
barolli@fit.ac.jp

Abstract. Flooding caused by bad weather conditions, such as flooding caused by heavy rains, can cause great damage and put many people at risk. In order to avoid danger and protect people, the evacuation action should start before the flood occurs. Therefore, it is necessary to take measures such as predicting flood damage by monitoring rivers and understanding the weather conditions. However, it is difficult to monitor the entire river over a long distance. Wireless sensor fusion networks have the advantage of being able to collect and analyze a variety of information from a wide range of sources. They can be effective in monitoring rivers and embankments by predicting and preventing flood damage in rivers caused by various factors. In this paper, we developed some sensor devices and propose a wireless sensor fusion network for monitoring rivers and levees. By experiments, we confirmed the accuracy of sensor devices in a real environment we collected sensor data by wireless sensor fusion network and used LSTM to predict the river conditions.

1 Introduction

Freakish weather is occurring frequently in many countries and the resulting natural disasters have become a major problem. According to the World Disasters Report 2020 [1], the number of natural disasters is $2,850$ from 2010 to 2019, of which $1,298$, or $45.5\,[\%]$ were reported to be flooding. The number of people affected by the floods is approximately 673 million. In order to escape from floods and disasters, it is necessary that the evacuating be in advance. But, there is a great possibility of encountering floods due to misjudgments of prevention evacuation time. Therefore, there is an urgent need to monitor rivers and predict floods by grasping and analyzing weather conditions and taking measures for disaster prevention and mitigation. However, since rivers extend over long distances and wide areas and condition changes depending on topography and weather

L. Barolli et al. (Eds.): EIDWT 2022, LNDECT 118, pp. 283–290, 2022.
https://doi.org/10.1007/978-3-030-95903-6_30

conditions, it is difficult for humans to monitor the situation, predict disasters and provide information to the areas around rivers. In this paper, we develop a sensing device and propose a wireless sensor fusion network [2–8] to monitor the river. The wireless sensor fusion network collects data as a batch to obtain various information from a wide area as real-time streaming, analyzing and predicting flood damage based on an intelligent system. Therefore, we examine the effectiveness of methods for monitoring rivers and predicting floods caused by various factors and predicting weather conditions that change in real-time. The paper is organized as follows. In Sect. 2, we present evaluation the proposed method. In Sect. 3, we discuss the results. Finally, in Sect. 4, we conclude the paper.

2 Proposed System

Figure 1 shows the proposed system. In the proposed system, water level, rainfall, soil moisture, temperature, humidity, atmospheric pressure, water temperature and the above are measure by the sensing devices describe below. The wireless sensor network installed along the river covers a wide area of the river and collects sensing data, which are used to predict river and weather conditions based on the time-series analysis. The proposed system uses the ZigBee [9–14] communication protocol with wireless Xbee modules for wireless communication between sensing devices and Jetson Nano. The HTTP mapping over QUIC (HTTP/3) [15,16] is considered as the data transmission protocol.

2.1 Sensors for River Monitoring

Figure 2 shows sensing devices used for the measurement. We developed a water level indicator Fig. 2(a), a tipping bucket type rain gauge Fig. 2(b), a soil moisture indicator Fig. 2(c) and a indicator of temperature, humidity and atmospheric pressure Fig. 2(d) as sensing devices for monitoring rivers. We developed the waterproof mechanism and some parts of the sensing devices with a 3D printer in order to have a low cost [17–20]. The water level indicator is calculated from the water pressure w [Mpa] measured by the water pressure gauge in the water and the atmospheric pressure P [hPa] measured by the atmospheric pressure indicator on the ground where the pressure W [Mpa] is $W = w - P$ and the water level h [m] is $h = W \times 0.0098$. The tipping bucket type raingauge is a rainwater accumulator and the tipping bucket tilts and drains every 0.500 [mm]. The measured rainfall Q [mm] is $Q = 0.500 \times N$ based on the number of times the tipping bucket tilts per hour N. The soil moisture indicator measures the relative permittivity of water $(0.000 \leq {}^\circ C \leq 90.000)$ to obtain the soil moisture content $(58.150 \leq \mu \leq 88.150)$.

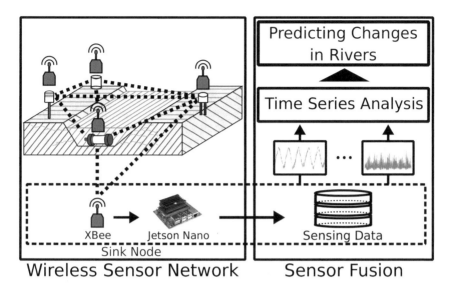

Fig. 1. Proposed system.

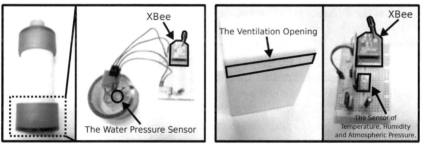

(a) The water level indicator.

(b) Indicator of temperature, humidity and atmospheric pressure.

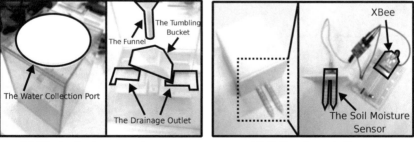

(c) Tipping bucket type raingauge.

(d) Soil moisture indicator.

Fig. 2. Sensing devices for river monitoring.

2.2 Sensor Fusion of Measurement Data

The collect sensing data are stored in the sink node. Since the water level in the river and the soil moisture in the embankment change with time, it is necessary to analyze the time series in order to predict the changes in the river condition [21]. Therefore, we perform sensor fusion [22] with the accumulated sensing data and use Long Short-Term Memory (LSTM) [23–26] to predict the river fluctuation. In order to improve the prediction accuracy, the LSTM is trained in advance by the observed data, taking into account the dataset of temperature, humidity, atmospheric pressure and rainfall of Automated Meteorological Data Acquisition System (AMeDAS) installed near the observation points provided by the Japan Meteorological Agency. Then, the proposed system learns the measured values collected by the sensing devices.

3 Evaluation Results

3.1 Experimental Scenario

Figure 3 shows the experimental environment. The indicator of temperature, humidity and atmospheric pressure is installed near the flume Fig. 3(a) and the soil moisture indicator is installed near the settling tank Fig. 3(b) for measurement. In order to reduce the error due to the difference in the measurement environment of the sensing devices, we consider the standard set by the World Meteorological Organization (WMO) [27]. The sensing device should be install at a distance of at least twice the height of the structure or at a distance of at least 4 [m] from the structure. The indicator of temperature, humidity and atmospheric pressure should be installed inside a box that is 1.5 [m] from the ground, out of direct sunlight and with access to the outside air. A soil moisture indicator measures soil moisture by burying the sensor part in the soil. The Jetson Nano is installed at a distance of 1 [m] from the sensing device. The measurements are always sent to the sink node and the data received at the nearest time is stored every minute. The LSTM is trained with a model that predicts the value one hour later from the input value by the temperature data of AMeDAS near the observation point for 10 years from January 2010.

(a) Flume.   (b) Settling Tank.

Fig. 3. The experimental environment.

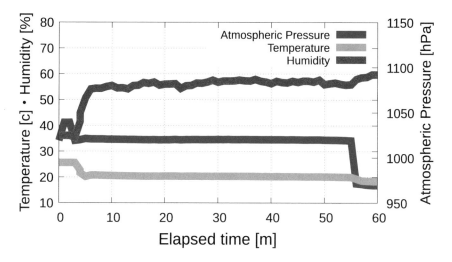

Fig. 4. Measured value of temperature, humidity and atomspheric pressure.

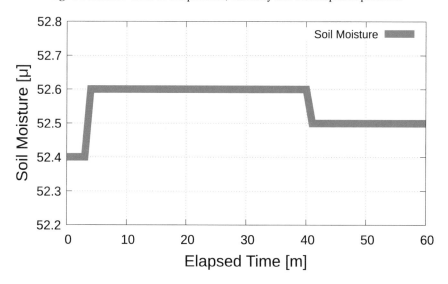

Fig. 5. Measured values of soil moisture.

3.2 Experimental Results of Wireless Sensor Network

Figure 4 and Fig. 5 shows the measured values of the indicator of temperature, humidity and atmospheric pressure and soil moisture indicator, respectively. The sensing data was collected for one hour in the experimental environment. From Fig. 4 and Fig. 5, we confirmed the operation and communication of the developed sensing devices in the real environment. Since the values measured by the sensing device were measured without large errors, we considered that the accuracy was enough to be used for LSTM training.

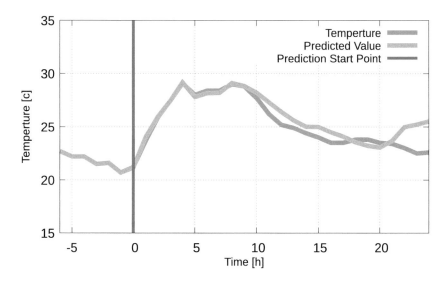

Fig. 6. Results of temperature prediction by LSTM.

3.3   Experimental Results of Temperature Prediction with LSTM

Figure 6 shows the results of the temperature prediction using the pre-trained LSTM. In the experiment, the temperature data of October 15, 2021 at the same location as the observation point of the temperature data with training are used as validation data. Temperature data are used by LSTM to predict the temperature for 24 h. From the temperature prediction results, we confirmed that the Mean Squared Error of the predicted values is 3.016.

4   Conclusions

In this paper, we developed some sensing devicse for river monitoring, collected sensing data by wireless sensor network and confirmed the prediction accuracy by LSTM. In the future, we will consider other analysis methods of the sensing data, such as Echelon analysis [28] with geographic information and reservoir computing.

Acknowledgement. This work was supported by JSPS KAKENHI Grant Number JP20K19793 and Grant for Promotion of OUS Research Project (OUS-RP-20-3).

References

1. International Federation of Red Cross and Red Crescent Societies: World Disasters Report 2020 (2020)
2. Lewi, T., et al.: Aerial sensing system for wildfire detection. In: Proceedings of the 18-th ACM Conference on Embedded Networked Sensor Systems, pp. 595–596 (2020)

3. Mulukutla, G., et al.: Deployment of a large-scale soil monitoring geosensor network. SIGSPATIAL Spec. **7**, 3–13 (2015)
4. Gayathri, M., et al.: A low cost wireless sensor network for water quality monitoring in natural water bodies. In: Proceedings of the IEEE Global Humanitarian Technology Conference, pp. 1–8 (2017)
5. Gellhaar, M., et al.: Design and evaluation of underground wireless sensor networks for reforestation monitoring. In: Proceedings of the 41-st International Conference on Embedded Wireless Systems and Networks, pp. 229–230 (2016)
6. Yu, A., et al.: Research of the factory sewage wireless monitoring system based on data fusion. In: Proceedings of the 3rd International Conference on Computer Science and Application Engineering, no. 65, pp. 1–6 (2019)
7. Suzuki, M., et al.: A high-density earthquake monitoring system using wireless sensor networks. In: Proceedings of the 5-th International Conference on Embedded Networked Sensor Systems, pp. 373–374 (2007)
8. Oda, T., et al.: Design and implementation of a simulation system based on deep q-network for mobile actor node control in wireless sensor and actor networks. In: Proceedings of the IEEE 31st International Conference on Advanced Information Networking and Applications Workshops, pp. 195–200 (2017)
9. Oda, T., et al.: A genetic algorithm-based system for wireless mesh networks: analysis of system data considering different routing protocols and architectures. Soft. Comput. **20**(7), 2627–2640 (2016)
10. Oda, T., et al.: Evaluation of WMN-GA for different mutation operators. Int. J. Space-Based Situated Comput. **2**(3), 149–157 (2012)
11. Oda, T., et al.: WMN-GA: a simulation system for WMNs and its evaluation considering selection operators. J. Ambient. Intell. Humaniz. Comput. **4**(3), 323–330 (2013)
12. Nishikawa, Y., et al.: Design of stable wireless sensor network for slope monitoring. In: Proceedings of the IEEE Topical Conference on Wireless Sensors and Sensor Networks, pp. 8–11 (2018)
13. Hirata, A., et al.: A coverage construction method based hill climbing approach for mesh router placement optimization. In: Proceedings of the 15th International Conference on Broad-Band and Wireless Computing, Communication and Applications, pp. 355–364 (2020)
14. Hirata, A., et al.: A coverage construction and hill climbing approach for mesh router placement optimization: simulation results for different number of mesh routers and instances considering normal distribution of mesh clients. In: Proceedings of the 15th International Conference on Complex, Intelligent and Software Intensive Systems, pp. 161–171 (2021)
15. Roskind, J., et al.: The QUIC transport protocol: design and internet-scale deployment. In: Proceedings of the 50-th ACM Conference on Special Interest Group on Data Communication, pp. 183–196 (2017)
16. Sharma, A., Kamthania, D.: QUIC protocol based monitoring probes for network devices monitor and alerts. In: Singh, U., Abraham, A., Kaklauskas, A., Hong, T.-P. (eds.) Smart Sensor Networks. SBD, vol. 92, pp. 127–150. Springer, Cham (2022). https://doi.org/10.1007/978-3-030-77214-7_6
17. Saito, N., et al.: Design and implementation of a DQN based AAV. In: Proceedings of the 15th International Conference on Broad-Band and Wireless Computing, Communication and Applications, pp. 321-329, 2020
18. Saito, N., et al.: Simulation results of a DQN based AAV testbed in corner environment: a comparison study for normal DQN and TLS-DQN. In: Proceedings of the 15th International Conference on Innovative Mobile and Internet Services in Ubiquitous Computing, pp. 156–167 (2021)

19. Saito, N., et al.: A tabu list strategy based dqn for aav mobility in indoor single-path environment: implementation and performance evaluation. Internet Things **14**, 100394 (2021)

20. Saito, N., et al.: A LiDAR based mobile area decision method for TLS-DQN: improving control for AAV mobility. In: Proceedings of the 16th International Conference on P2P, Parallel, Grid, Cloud and Internet Computing, pp. 30–42 (2021)

21. Sharma, P., et al.: A machine learning approach to flood severity classification and alerting. In: Proceedings of the 4-th ACM SIGSPATIAL International Workshop on Advances in Resilient and Intelligent, pp. 42–47 (2021)

22. Hang, C., et al.: Recursive truth estimation of time-varying sensing data from online open sources. In: Proceedings of the 14-th International Conference on Distributed Computing in Sensor Systems, pp. 25–34 (2018)

23. Hochreiter, S., et al.: Long short-term memory. Neural Comput. **9**, 1735–1780 (1997)

24. Karevan, Z., et al.: Transductive LSTM for time-series prediction: an application to weather forecasting. Neural Netw. **125**, 1–9 (2019)

25. Li, Y., et al.: Hydrological time series prediction model based on attention-LSTM neural network. In: Proceedings of the 2nd International Conference on Machine Learning and Machine, pp. 21–25 (2019)

26. Toyoshima, K., et al.: Proposal of a haptics and LSTM based soldering motion analysis system. In: Proceedings of the IEEE 10th Global Conference on Consumer Electronics, pp. 1–2 (2021)

27. World Meteorological Organization (WMO): Guide to Instruments and Methods of Observation (2018)

28. Myers, W., et al.: Echelon approach to areas of concern in synoptic regional monitoring. Environ. Ecol. Stat. 131–152 (1997)

Social Experiment of Realtime Road State Sensing and Analysis for Autonomous EV Driving in Snow Country

Yositaka Shibata[1]([✉]), Akira Sakuraba[2], Yoshikazu Arai[1], Yoshiya Saito[1], and Noriki Uchida[3]

[1] Iwate Prefectural University, Takizawa, Japan
{shibata,arai,y-saito}@iwate-pu.ac.jp
[2] Tokyo Metropolitan Industrial Technology Research Institute, Tokyo, Japan
sakuraba.akira@iri-tokyo.jp
[3] Fukuoka Institute Technology, Fukuoka, Japan
n-uchida@fit.ac.jp

Abstract. In order to realize autonomous electric vehicle system in snow country, various road state information including dry, wet, slush, snowy, icy states on the road are determined in realtime using various environmental sensors. These state information is not only exchanged directly between EVs through V2X, but also collected into cloud servers on Internet and organized as wide area road state information platform for ordinal users to present as a viewer system of the road state using smartphone and tablet terminals. In this paper, the basic consideration of architecture of autonomous electric vehicle with road sensing road state in snow country and predicted viewer system for safety driving through V2X communication system and social experiment by prototype system are discussed.

1 Introduction

Autonomous driving systems is expected as future safer and effective vehicle and have been investigated and developed in industrial countries and driving the exclusive roads and highway roads at level 3 or 4 [1]. Those roads on which autonomous car run are ideal for autonomous car driving because the driving lanes are clear, the driving direction is the same and opposite driving car lanes are completely separated in these countries. There is also no obstacles and cracks on the roads. However, the road infrastructures in many developing countries are not always well maintained compared with improvement of vehicles. In particular, the road states in developing countries are so bad and dangerous due to luck of regular road maintenance, falling objects from other vehicles or precipice from roadside. Therefore, in order to maintain safe and reliable autonomous driving, the vehicles have to detect those obstacles in advance and avoid when they pass away.

On the other hand, in the cold or snow countries, such as Japan and Northern European countries, most of the road surfaces are occupied with heavy snow and iced surface in winter and many slip accidents occur even though the vehicles attach anti-snow slip tires. In fact almost more than 90% of traffic accidents in northern part of Japan is caused from slipping car on snowy or iced road [2].

L. Barolli et al. (Eds.): EIDWT 2022, LNDECT 118, pp. 291–300, 2022.
https://doi.org/10.1007/978-3-030-95903-6_31

In many snow countries, since the car traffic and the number of the residents are not so many, it is better to apply the limited line or dedicated leased line based autonomous driving where the leased line along the road is dedicated such as street car system and the autonomous vehicle can autonomously run on the leased line while sharing the road by other ordinal cars. One of the typical autonomous vehicles run on the leased line which includes magnetic force line under about 10–20 cm in depth. The various sensors to keep the correct car position, 3D far-infrared camera to observe the other ordinal cars and detect the forward obstacles, GPS and 3D map to observe correct the own positon and autopilot control system to maintain the autonomous driving on the leased line are on-boarded. This leased line based autonomous driving system can be realized by low maintenance cost for rural and snow areas.

In fact, there are more than 18 small local areas to introduce the leased line based autonomous driving on March 2021 in Japan as pilot project by Ministry of Land, Infrastructure, Transport and Tourism. No accident are reported in those local areas so far since the driving vehicle speed is always maintained under 12 km/h for safe driving to prepare for bad road condition. However, there is no road sensor based driving system and communication system to share the road condition. Usually, in those rural areas or snow areas, the information and communication infrastructure is not well developed and even does not work for 3G cellular network functions, so called challenged network environment. Thus, once a traffic accident or disaster information collection, transmission and sharing are delayed or even cannot be made.

In order to resolve those problems, we introduce a new generation wide area road surface state information platform based on crowd sensing and veheV2X technologies. In road sensing, the sensor data from various environmental sensors including accelerator, gyro sensor, infrared temperature sensor, quasi electrical static sensor, camera and GPS attached on vehicle are integrated to precisely determine the various road states and identify the dangerous locations. This road information is transmitted to the neighbor vehicles and road side servers in realtime using V2X communication network. Eventually those road state information at the received vehicle are displayed on the viewer system by alerting the dangerous locations. In this paper, the a new road state information platform based on those IoT road state sensing, V2X communication system and viewing system is introduced to realize EV autonomous EV driving,

In the followings, the related works with road state information by sensor and V2X technologies are explained in Sect. 2. System and architecture of the proposed road state sensing information platform are explained in Sect. 3. Next, EV autonomous driving system is explained in Sect. 4. Preliminary experiment by prototyped platform to evaluate our proposed system is precisely explained in Sect. 5. In final, conclusion and future works are summarized in Sect. 6.

2   Related Works

With the road state sensing method, there are several related works so far. Particularly road surface temperature is essential to know the road state such as snowy or icy in winter season whether the road surface temperature is under minus 4 °C or over.

In the paper [3], the road surface temperature model by taking account of the effects of surrounding road environment to facilitate proper snow and ice control operations is introduced. In this research, the fixed sensor system along road is used to observe the precise temperature using the monitoring system with long-wave radiation. They build the road surface temperature model using heat balance method.

In the paper [4–6], cost effective and simple wide area road surface temperature prediction method while maintaining the prediction accuracy of current model is developed. Using the relation between the air temperature and the meshed road surface temperature, statistical thermal map data are calculated to improve the accuracy of the road surface temperature model. Although the predicted accuracy is high, the difference between the ice and snow states was not clearly resolved.

In the paper [7], a road state data collection system of roughness of urban area roads is introduced. In this system, mobile profilometer using the conventional accelerometers to measure realtime roughness and road state GIS is introduced. This system provides general and wide area road state monitoring facility in urban area, but snow and icy states are note considered.

In the paper [8], a measuring method of road surface longitudinal profile using build-in accelerometer and GPS of smartphone is introduced to easily calculate road flatness and International Road Index (IRI) in offline mode. Although this method provides easy installation and quantitative calculation results of road flatness for dry or wet states, it does not consider the snow or icy road states.

In the paper [9], a statistical model for estimating road surface state based on the values related to the slide friction coefficient is introduced. Based on the estimated the slide friction coefficient calculated from vehicle motion data and meteorological data is predicted for several hours in advance. However, this system does not consider the other factors such as road surface temperature and humidity.

In the paper [10], road surface temperature forecasting model based on heat balance, so called SAFF model is introduced to forecast the surface temperature distribution on dry road. Using the SAFF model, the calculation time is very short and its accuracy is higher than the conventional forecasting method. However, the cases of snow and icy road in winter are not considered.

In the paper [11], blizzard state in winter road is detected using on-board camera and AI technology. The consecutive ten images captured by commercial based camera with 1280×720 are averaged by pre-filtering function and then decided whether the averaged images are blizzard or not by Convolutional Neural Network. The accuracy of precision and F-score are high because the video images of objective road are captured only in the daytime and the contrast of those images are almost stable. It is required to test this method in all the time.

In the paper [12], road surface state analysis method is introduced for automobile tire sensing by using quasi-electric field technology. Using this quasi-electric field sensor, the changes of road state are precisely observed as the change of the electrostatic voltage between the tire and earth. In this experiment, dry and wet states can be identified.

In the paper [13], a road surface state decision system is introduced based on near infrared (NIR) sensor. In this system, three different wavelengths of NIR laser sensors is used to determine the qualitative paved road states such as dry, wet, icy, and snowy states as well as the quantitative friction coefficient. Although this system can provides realtime decision capability among those road states, decision between wet and icy states sometimes makes mistakes due to only use of NIR laser wavelength.

With all of the systems mentioned above, since only single sensor is used, the number of the road states are limited and cannot be shared with other vehicles in realtime. For those reasons, construction of communication infrastructure is essential to work out at challenged network environment in at inter-mountain areas. In the followings, a new road state information platform is proposed to overcome those problems.

3   Road State Sensing Information Platform

In order to resolve those problems in previous session, we introduce a new generation wide area road surface state information platform based on crowd sensing and V2X technologies as shown in Fig. 1. The wide area road surface state information platform mainly consists of multiple roadside wireless nodes, namely Smart Relay Shelters (SRS), Gateways, and mobile nodes, namely Smart Mobile Box (SMB). Each SRS or SMB is furthermore organized by a sensor information part and communication network part. The vehicle has sensor information part includes various sensor devices such as semi-electrostatic field sensor, an acceleration sensor, gyro sensor, temperature sensor, humidity sensor, infrared sensor and sensor server. Using those sensor devices, various road surface states such as dry, rough, wet, snowy and icy roads can be quantitatively decided as showed in Fig. 2. The humidity and frozen sensor data as well as GPS data and carries and exchanges to other smart node as message ferry while moving from one end to another along the roads. On the other hand, SRS not only collects and stores sensor data from its own sensors in its database server but exchanges the sensor data from SMB in vehicle nodes when it passes through the SRS in roadside wireless node by V2X communication protocol. Therefore, both sensor data at SRS and SMB are periodically uploaded to cloud system through the Gateway and synchronized. Thus, SMB performs as mobile communication means even through the communication infrastructure is challenged environment or not prepared.

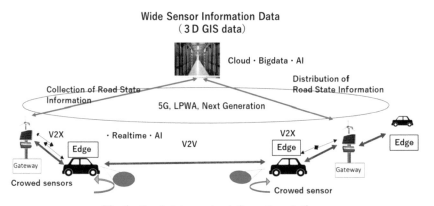

Fig. 1. Road state sensing information platform

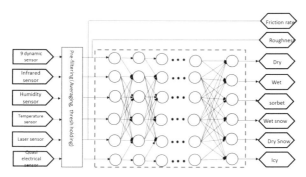

Fig. 2. Road state decision system

4   EV Autonomous Driving System

Figure 3 shows a typical EV driving vehicle and its basic control functions. The vehicle runs by own battery energy charged on Li-Ion battery on the electric magnet lines which installed under several ten cm depth from the ground. Thus, the EV car can keep driving on this road by detecting the magnetic force line by various guide sensors. The edge computer in EV control system analyzes the position of guide line and drive on the preset collect route. When the EV drives on the magnet sensor, voltage deviations are generated on magnet sensor, the edge computer can analyze this voltage and control the action of EV. It is also prepared override function by which the EV can automatically change from automatic operation to manual and manually switch back to automatic operation to keep safety of autonomous driving. Furthermore, the obstacle detection function is also existed. By memorizing the driving space state of the own vehicle beforehand using 3D stereo vision camera, the objects on the running road can be identified as obstacle, otherwise the objects are not regarded as obstacle such as the stone on the pavement.

Various sensors including dynamic accelerator, gyro sensor, infrared temperature sensor, humidity sensor, quasi electrical static sensor, camera and GPS measure the time series physical sensor data. Then, those sensor data are processed by the road surface decision unit (Machine Learning) and the current road state can be identified in realtime. Next, those road state data are input to the ECU to calculate the amount of breaking and steering, and sent to the braking and steering components to optimally control the speed and direction of the EV. This close loop of the measuring EV speed and direction, sensing road data, deciding road state, computing and controlling braking/steering processes is repeated within a several msec.

On the other hand, those road state data also transmitted to the road state server in cloud computing system through the edge computing by V2Xcommunication protocol and processed to organized wide the road state GIS platform. Those data are distributed to all of the running EVs to know the head road state of the current location. From the received the head state of the current location, the EV can look ahead road state and predict proper target set values of speed and direction of the EV. Thus, by combining the control of both the current and feature speed and direction of EV, more correct and safer automotive driving can be attained (Fig 4).

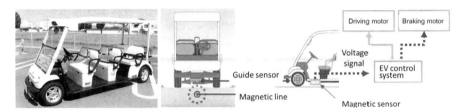

Fig. 3. EV autonomous driving system

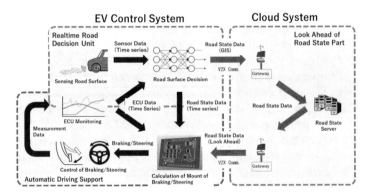

Fig. 4. Automatic EV control system

5   Prototype System

In order to verify the effects and usefulness of the proposed system, a prototype system which is based on the EV based is considered constructed and those functional and performance are evaluated. Figure 5 shows EV prototype system which is based on electro-magnetic induction line technology. The EV direction is inducted by the electro-magnetic induction line which is embedded in the ground and can be safely and reliably drive along the line. Only motor is needed to control the EV speed be considering the ahead road state data from SRS along the street. The EV of the prototype is made of YAMAHA, 7 limited persons and runs max. 12 km/h. The prototype also includes sensor server system and Communication server System, Smart Mobility Base station (SMB) for mobility and Smart Rely Shelter (SRS) for roadside station. We currently use GW-900D of Planex COMM. for Wi-Fi communication of 2.4 GHz as the prototype of two-wavelength communication, and OiNET-923 of Oi Electric Co., Ltd. for 920 MHz band communication respectively. GW-900D is a commercially available device (IEE802.11ac), and the rated bandwidth is 866 Mbps. On the other hand, the OiNET-923 has a communication distance of 2 km at maximum and a bandwidth of 50 kbps to 100 kbps.

On the other hand, in sensor server system, several sensor including BL-02 of Biglobe as 9 axis dynamic sensor and GPS, CS-TAC-40 of Optex as far-infrared temperature sensor, HTY7843 of azbil as humidity and temperature sensor and RoadEye of RIS system and quasi electrical static field sensor for road surface state are used. Those sensor data are synchronously sampled with every 10 ms. and averaged every 1 s to reduce sensor noise by another Raspberry Pi3 Model B+ as sensor server. Then those data are sent to Intel NUC Core i7 which is used for sensor data storage and data analysis by AI based road state decision. Both sensor and communication servers are connected to Ethernet switch.

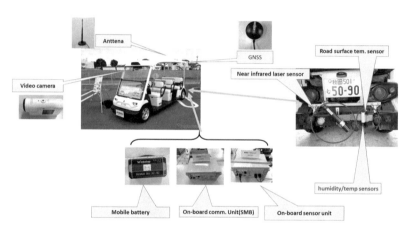

Fig. 5. Prototype system for road sensing and analysis

5.1 Results

In order to evaluate the sensing function and accuracy of sensing road surface condition, both sensor and communication servers and various sensors are set to the EV. The experiments were carried out in the small snow village in northern part of Japan, Kami-Koani village where the leased line based autonomous driving are actually implemented. We ran this vehicle about 4 h for 2 days on January 27 to 28, 2021 to evaluate decision accuracy in realtime on the winter road with various road conditions such as dry, wet, snowy, damp and icy condition along the leased line village. The experimental results are indicated in the Fig. 6 and Fig. 7. Currently, we are evaluating the road surface decision function using the video camera to compere the decision state and the actual road surface state.

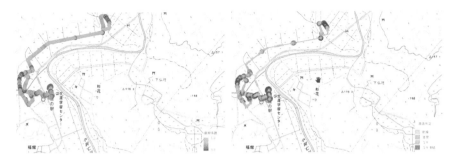

Fig. 6. Road state and friction rate on daytime (left) and evening (right)

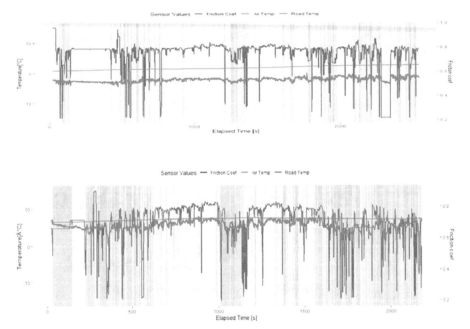

Fig. 7. Friction rate, air temperature and road surface temperature on daytime (upper) and evening (lower)

6   Conclusions and Future Works

In this paper, we introduce a wide area road state information platform decision system for not only both urban area but local snow and icy roads area where these states derive very serious traffic accidents. In our system in-vehicle sensor server system can detects and determine various road states such as dry, wet, damp, snowy, icy or even the roughness and friction rate on road surface by AI based decision making method. This road state information can be exchanged and shared with the many vehicles through V2X communication method. Furthermore, this road state information can be collected into the global cloud from smart road stations (SRS) though network gateway as big data to predict the future road state by combining with weather condition. Through the experimentation of our prototyped system, we could confirm that a wide area road state information GIS system can be realized.

As the future works of our research, more sensing data for wider areas, in local and urban areas, on different weather conditions should be corrected as training data for AI decision model to improve the decision accuracy. Social experiment for the actual driving operations such as bus, taxi, renter car enterprises is also required. Furthermore, wide area road state GIS by combining those road state data and open data with weather and geological data should be predicated and indicated as GIS system near future.

Acknowledgement. The research was supported by The research was supported by Japan Keiba Association Grant Numbers 2021M-198, JSPS KAKENHI Grant Numbers JP 20K11773, Strategic Information and Communications R&D Promotion Program Grant Number 181502003 by Ministry of Affairs and Communication and Communication and Communication and Strategic Research Project Grant by Iwate Prefectural University in 2021.

References

1. SAE International: Taxonomy and Definitions for Terms Related to Driving Automation Systems, for On-Road Motor Vehicles, J3016_201806, June 2018
2. Police department in Hokkaido: The Actual State of Winter Typed Traffic Accidents, November 2018. https://www.police.pref.hokkaido.lg.jp/info/koutuu/fuyumichi/blizzard.pdf
3. Takahashi, N., Tokunaga, R.A., Sato, T., Ishikawa, N.: Road surface temperature model accounting for the effects of surrounding environment. J. Jpn. Soc. Snow Ice **72**(6), 377–390 (2010)
4. Fujimoto, A., Nakajima, T., Sato, K., Tokunaga, R., Takahashi, N., Ishida, T.: Route-based forecasting of road surface temperature by using meshed air temperature data. In: JSSI&JSSE Joint Conference, pp. 1–34, September 2018
5. Saida, A., Sato, K., Nakajima, T., Tokunaga, R., Sato, G.: A study of route based forecast of road surface condition by melting and feezing mass estamation method using weather mesh data. In: JSSI&JSSE Joint Conference, pp. 2–57, September 2018
6. Hoshi, T., Saida, A., Nakajima, T., Tokunaga, R., Sato, M., Sato, K.: Basic Consideration of Wide-Scale Road Surface Snowy and Icy Conditions using Weather Mesh Data. Monthly report of Civil Engineering Research Institute for Cold Region, No. 800, pp. 28–34, January 2020
7. Fujita, S., Tomiyama, K., Abliz, N., Kawamura, A.: Development of a roughness data collection system for urban roads by use of a mobile profilometer and GIS. J. Jpn. Soc. Civ. Eng. **69**(2), I_90–I_97 (2013)
8. Yagi, K.: A measuring method of road surface longitudinal profile from sprung acceleration and verification with road profiler. J. Jpn. Soc. Civ. Eng. **69**(3), I_1–I_7 (2013)
9. Mizuno, H., Nakatsuji, T., Shirakawa, T., Kawamura, A.: A statistical model for estimating road surface conditions in winter. In: The Society of Civil Engineers, Proceedings of Infrastructure Planning (CD-ROM), December 2006
10. Saida, A., Fujimoto, A., Fukuhara, T.: Forecasting model of road surface temperature along a road network by heat balance method. J. Civ. Eng. Jpn. **69**(1), 1–11 (2013)
11. Okubo, K., Takahashi, J., Takechi, H., Sakurai, T., Kokubu, T.: Possibility of Blizzard Detection by On-board Camera and AI technology. Monthly report of Civil Engineering Research Institute for Cold Region, No. 798, pp. 32–37, November 2019

12. Takiguchi, K., et al.: Trial of quasi-electrical field technology to automobile tire sensing. In: Annual conference on Automobile Technology Association, pp. 417–20145406, May 2014
13. Casselgren, J., Rosendahl, S., Eliasson, J.: Road surface information system. In: Proceedings of the 16th SIRWEC Conference, Helsinki, 23–25 May 2012. http://sirwec.org/wp-content/uploads/Papers/2012-Helsinki/66.pdf

A Soldering Motion Analysis System for Danger Detection Considering Object Detection and Attitude Estimation

Tomoya Yasunaga[1], Tetsuya Oda[1(✉)], Nobuki Saito[2], Aoto Hirata[2], Chihiro Yukawa[1], Yuki Nagai[1], and Masaharu Hirota[3]

[1] Department of Information and Computer Engineering, Okayama University of Science (OUS), 1-1 Ridaicho, Kita-ku, Okayama 700–0005, Japan
{t18j091yt,t18j097yc,t18j057ny}@ous.jp, oda@ice.ous.ac.jp
[2] Graduate School of Engineering, Okayama University of Science (OUS), Okayama, 1-1 Ridaicho, Kita-ku, Okayama 700–0005, Japan
{t21jm01md,t21jm02zr}@ous.jp
[3] Department of Information Science, Okayama University of Science (OUS), 1-1 Ridaicho, Kita-ku, Okayama 700–0005, Japan
hirota@mis.ous.ac.jp

Abstract. It is important for the society to support the employment of people with disabilities. However, at the work sites, it is necessary to ensure safety while soldering for persons with disabilities and to support instructors. In this paper, in order to solve these problems, we propose a soldering motion analysis system for danger detection based on object detection and attitude estimation. Also, we show the experimental results for dangerous detection during soldering iron. The experimental results show that the proposed system has a good accuracy for detecting dangerous situations.

1 Introduction

In many countries around the world, measures to employ people with disabilities are being promoted in order to realize a society where it is normal for people with disabilities to work together. In Japan, the Law for Employment Promotion of Persons with disabilities has been enacted as part of the measures for employment of persons with disabilities. Companies are required to employ persons with disabilities at a specific ratio depending on the company scale. Also, the vocational rehabilitation such as vocational training, employment agency and workplace adaptation support is provided to persons with disabilities and consideration is given to provide a detailed support according to the characteristics of each disability.

Many persons with disabilities, especially persons with mental disabilities, have difficulty at complicated tasks such as those that require parallel tasks and those that need to respond flexibly. On the other hand, they are good at simple work such as factory work. In fact, some Japanese factories are instructing the soldering of electrical parts for persons with disabilities. However, it takes more time and efforts for people with disabilities to obtain soldering skills compared with general people. In order to get

L. Barolli et al. (Eds.): EIDWT 2022, LNDECT 118, pp. 301–307, 2022.
https://doi.org/10.1007/978-3-030-95903-6_32

the skills, it is necessary to remember and repeat the same work continuously which takes time. Instructors must also monitor the soldering work of persons with disabilities to prevent accidents such as the poor joining of parts and burns. At work sites, it is necessary to ensure safety during soldering by persons with disabilities and to support instructors [1–5].

In this paper, in order to solve these problems, we propose a soldering motion analysis system for danger detection based on object detection [6–10] and attitude estimation [12–16]. Also, we show the experimental results to detect the dangerous situation while holding a soldering iron and the dangerous pose during soldering based on image recognition.

The structure of the paper is as follows. In Sect. 2, we describe the proposed system. In Sect. 3, we present the experimental results. Finally, conclusions and future work are given in Sect. 4.

2   Proposed System

The structure of the proposed system is shown in Fig. 1. The proposed system can be used for dangerous situation detection during the soldering iron, hand tracking [17–20] and the pose estimation [21] of the upper body in real time. The stereo camera is used to decide the soldering iron holding, while the front camera is used to decide the pose during soldering. The stereo camera of the proposed system enables the derivation of distance between the camera and object using triangulation from the parallax of the images obtained by the two cameras.

The soldering iron holding is decided as a danger situation or not by collision detection considering the danger area and the three-dimensional coordinates of the fingertip. The danger situation indicates that the fingertip is in the danger area. The danger area is a cuboid by considering the soldering iron tip obtained by object detection based on YOLOv5 [22–24]. The center of the danger area is the center coordinate of the soldering iron tip and the height of the danger area is using an arbitrary distance.

The three-dimensional coordinates of the fingertip consist of the two-dimensional coordinates (x, y) and the z coordinates on the Z-axis. The (x, y) are the coordinates of the fingertip based on the hand tracking. The z coordinate indicates the distance from the camera to the fingertip derived by triangulation. The three-dimensional coordinates of the soldering iron tip consist of a, b and z coordinates. The a and b indicate the diagonal coordinates of the start point a (x_a, y_a) and end point b (x_b, y_b) of the rectangle considering the soldering iron tip.

The proposed system can detect whether or not the upper body pose during soldering is dangerous. In the proposed system, when soldering, the correct pose is one in which the body is parallel to the desk and the dangerous pose is one in which the body is leaned forward or tilted to the left or right beyond a certain angle. The pose is estimated based on key points (x, y, z) obtained by skeleton estimation using the MediaPipe [25]. When the key points of the shoulder are closer to the front camera than a certain distance, it is decided as a forward leaning pose. Also, when the difference between the Z coordinates in the three-dimensional coordinates of both shoulders is more than a certain value, it is decided as a pose that is tilted to the left or right.

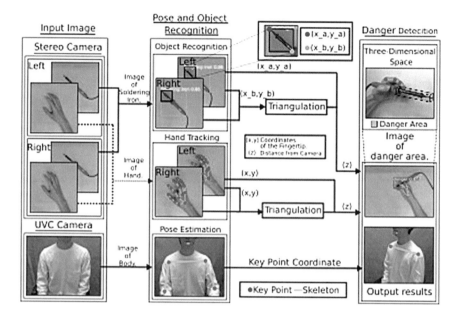

Fig. 1. Proposed system structure.

3 Experimental Results

In Fig. 2 is shown the experimental environment and the way to hold the soldering. The experimental environment consists of a tabletop with no objects and a tabletop with work tools. The patterns of the experimental environment and the way of holding are described below.

1. Figure 2(a) shows the state where there are no objects around and the correct way of holding.
2. Figure 2(b) shows the correct way to hold the object and there are tools around.
3. Figure 2(c) shows a dangerous holding position with no objects around.
4. Figure 2(d) shows a dangerous holding position with tools around.
5. Figure 2(e) shows a situation while touching the dangerous area every 5 s and there are no objects around.
6. Figure 2(f) shows a situation while touching the dangerous area every 5 s and there are tools around.

The dangerous area is defined as the area where the distance from the center of the soldering iron tip is 5.0 [cm]. In Fig. 2(a) and Fig. 2(b) are shown the percentage of correct answers, where each frame is the ratio of the judgment of danger detection with the total number of frames in 60 s. In Fig. 3 are shown the patterns of the upper body pose. The experimental results of Fig. 2(a) to Fig. 2(b) and Fig. 3 are shown in Fig. 4. The correct rate in Fig. 4 is the averages of the correct rate measured 10 times for every 70 frames, where one frame is about 0.015 [s]. Figure 4(a) shows the correct holding of soldering iron with and without other work tools. The correct holding answer rate is

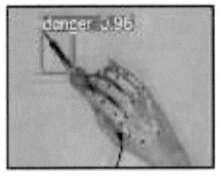

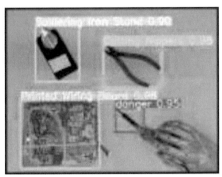

(a) The state where there are no objects around and the correct way of holding.

(b) The correct way to hold the object and there are tools around.

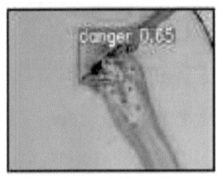

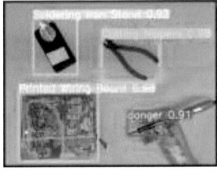

(c) Shows a dangerous holding position with no objects around.

(d) Shows a dangerous holding position with tools around.

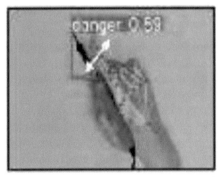

(e) Touching the dangerous area every 5 seconds and there are no objects around.

(f) Touching the dangerous area every 5 seconds and there are tools around.

Fig. 2. Patterns of soldering iron holding.

100.0 [%] and the danger holding rate of soldering iron with and without other work tools is about 83.2 [%] and 84.1 [%], respectively. From the experimental results, we can confirm that there is no difference in recognition accuracy due to differences in the environment.

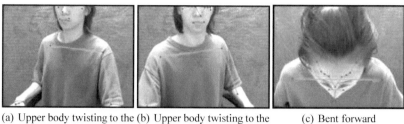

(a) Upper body twisting to the
right.

(b) Upper body twisting to the
left.

(c) Bent forward

Fig. 3. Patterns of upper body pose.

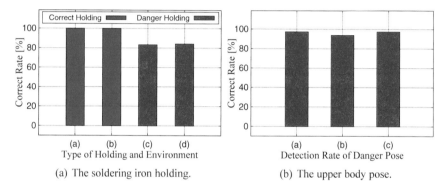

(a) The soldering iron holding.

(b) The upper body pose.

Fig. 4. Experimental results for soldering iron holding and upper body pose.

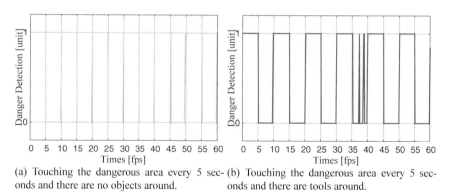

(a) Touching the dangerous area every 5 seconds and there are no objects around.

(b) Touching the dangerous area every 5 seconds and there are tools around.

Fig. 5. Experimental results for during touching dangerous area.

Figure 2(e) and Fig. 2(f) show the experimental scenario in which a finger is repeatedly moved into the danger zone of the soldering iron every 5 [s] for 60 [s]. Figure 5(a)
and Fig. 5(b) show the results of the danger detection for Fig. 2(e) and Fig. 2(f), respectively. Figure 5(a) shows that the results of danger detection are accurate with the movement of the finger. Figure 5(b) shows almost accurate results, but there are some errors
because the soldering iron is recognized as Cutting Nippers in 35 [s] to 40 [s].

For danger detection based on pose estimation, the front camera is used to determine whether a pose during soldering is dangerous or not. A dangerous pose is defined as the upper body being bent forward or twisted sideways by more than 35 [°]. The experiment is performed on the pattern of upper body poses shown in Fig. 3. Figure 4(b) shows the experimental results for each pattern of the upper body pose. It can be seen that all three patterns of the upper body have a high accuracy of about 95 [%] or more.

4   Conclusions

In this paper, we proposed a soldering motion analysis system for object detection and pose estimation. From experimental results, we found that by object detection and pose estimation can be decided the danger situation during soldering iron. In the future, we would like to improve the proposed system by adding new functions.

Acknowledgement. This work was supported by JSPS KAKENHI Grant Number JP20K19793 and Grant for Promotion of OUS Research Project (OUS-RP-20-3).

References

1. Yasunaga, T., et al.: Object detection and pose estimation approaches for soldering danger detection. In: Proceedings of the IEEE 10-th Global Conference on Consumer Electronics, pp. 776–777 (2021)
2. Toyoshima, K., et al.: Proposal of a haptics and LSTM based soldering motion analysis system. In: Proceedings of the IEEE 10-th Global Conference on Consumer Electronics, pp. 1–2 (2021)
3. Hirota, Y., Oda, T., Saito, N., Hirata, A., Hirota, M., Katatama, K.: Proposal and experimental results of an ambient intelligence for training on soldering iron holding. In: Barolli, L., Takizawa, M., Enokido, T., Chen, H.-C., Matsuo, K. (eds.) BWCCA 2020. LNNS, vol. 159, pp. 444–453. Springer, Cham (2021). https://doi.org/10.1007/978-3-030-61108-8_44
4. Oda, T., et al.: Design and implementation of an iot-based e-learning testbed. Int. J. Web Grid Serv. **13**(2), 228–241 (2017)
5. Liu, Y., et al.: Design and implementation of testbed using IoT and P2P technologies: improving reliability by a fuzzy-based approach. Int. J. Commun. Netw. Distrib. Syst. **19**(3), 312–337 (2017)
6. Papageorgiou, C., et al.: A general framework for object detection. In: Proceedings of the IEEE 6th International Conference on Computer Vision, pp. 555–562 (1998)
7. Felzenszwalb, P., et al.: Object detection with discriminatively trained part-based models. IEEE Trans. Pattern Anal. Mach. Intell. **32**(9), 1627–1645 (2009)
8. Obukata, R., et al.: Design and evaluation of an ambient intelligence testbed for improving quality of life. Int. J. Space-Based Situated Comput. **7**(1), 8–15 (2017)
9. Oda, T., Ueda, C., Ozaki, R., Katayama, K.: Design of a deep q-network based simulation system for actuation decision in ambient intelligence. In: Barolli, L., Takizawa, M., Xhafa, F., Enokido, T. (eds.) WAINA 2019. AISC, vol. 927, pp. 362–370. Springer, Cham (2019). https://doi.org/10.1007/978-3-030-15035-8_34
10. Obukata, R., Oda, T., Elmazi, D., Ikeda, M., Barolli, L.: Performance evaluation of an ami testbed for improving qol: evaluation using clustering approach considering parallel processing. In: BWCCA 2016. LNDECT, vol. 2, pp. 623–630. Springer, Cham (2017). https://doi.org/10.1007/978-3-319-49106-6_61

11. Yamada, M., et al.: Evaluation of an iot-based e-learning testbed: performance of olsr protocol in a nlos environment and mean-shift clustering approach considering electroencephalogram data. Int. J. Web Inf. Syst. **13**(1), 2–13 (2017)

12. Toshev, A., Szegedy, C.: DeepPose: human pose estimation via deep neural networks. In: Proceedings of the 27-th IEEE/CVF Conference on Computer Vision and Pattern Recognition (IEEE/CVF CVPR-2014), pp. 1653–1660 (2014)

13. Haralick, R., et al.: Pose estimation from corresponding point data. IEEE Trans. Syst. **19**(6), 1426–1446 (1989)

14. Fang, H., et al.: Rmpe: regional multi-person pose estimation. In: Proceedings of the IEEE International Conference on Computer Vision, pp. 2334–2343 (2017)

15. Xiao, B., et al.: Simple baselines for human pose estimation and tracking. In: Proceedings of the European Conference on Computer Vision (ECCV), pp. 466–481 (2018)

16. Martinez, J., et al.: A simple yet effective baseline for 3d human pose estimation. In: Proceedings of the IEEE International Conference on Computer Vision, pp. 2640–2649 (2017)

17. Zhang, F., et al.: MediaPipe Hands: On-device Real-time Hand Tracking, arXiv preprint arXiv:2006.10214 (2020)

18. Shin, J., et al.: American sign language alphabet recognition by extracting feature from hand pose estimation. Sensors (2021)

19. Hirota, Y., et al.: Proposal and experimental results of a DNN based real-time recognition method for ohsone style fingerspelling in static characters environment. In: Proceedings of the IEEE 9-th Global Conference on Consumer Electronics

20. Erol, A., et al.: Vision-based hand pose estimation: a review. Comput. Vis. Image Underst. 52–73 (2007)

21. Abolmaali, S., et al.: Pill ingestion action recognition using mediapipe holistic to monitor elderly patients. Int. Supply Chain Technol. J. **7**(11) (2021)

22. Redmon, J., et al.: You only look once: unified, real-time object detection. In: Proceedings of the 29-th IEEE/CVF Conference on Computer Vision and Pattern Recognition (IEEE/CVF CVPR-2016), pp. 779–788 (2016)

23. Zhou, F., et al.: Safety helmet detection based on YOLOv5. In: Proceedings of the IEEE International Conference on Power Electronics, Computer Applications (ICPECA), pp. 6–11 (2021)

24. Yu-Chuan, B., et al.: Using improved YOLOv5s for defect detection of thermistor wire solder joints based on infrared thermography. In: Proceedings of the 5th International Conference on Automation, Control and Robots (ICACR), pp. 29–32 (2021)

25. Lugaresi, C., et al.: MediaPipe: A Framework for Building Perception Pipelines, arXiv preprint arXiv:1906.08172 (2019)

Performance Evaluation of a Soldering Training System Based on Haptics

Kyohei Toyoshima[1], Tetsuya Oda[1(✉)], Chihiro Yukawa[1], Tomoya Yasunaga[1], Aoto Hirata[2], Nobuki Saito[2], and Leonard Barolli[3]

[1] Department of Information and Computer Engineering, Okayama University of Science (OUS), 1-1 Ridaicho, Kita-ku, Okayama 700–0005, Japan
{t18j056tk,t18j097yc,t18j091yt}@ous.jp, oda@ice.ous.ac.jp
[2] Graduate School of Engineering, Okayama University of Science (OUS), Okayama, 1-1 Ridaicho, Kita-ku, Okayama 700–0005, Japan
{t21jm02zr,t21jm01md}@ous.jp
[3] Department of Informantion and Communication Engineering, Fukuoka Insitute of Technology, 3-30-1 Wajiro-Higashi-ku, Fukuoka 811-0295, Japan
barolli@fit.ac.jp

Abstract. The soldering techniques are one of the industrial techniques required in electronic device manufacturing plants. However, a framework for quantifying soldering techniques has not been established, therefore it hard to evaluate the training of trainees. Also, safety is increasingly important in the workplace for persons with physical, intellectual, mental and other disabilities. The haptics is able to transmit the power generated in the virtual space to the manipulator. Therefore, we perform soldering virtual training based on haptics and analyze the soldering motion based on Long Short-Term Memory (LSTM) using training data in a virtual space. In this paper, we propose and evaluate a soldering training system based on haptics. The experimental results show that the proposed system is able to detect dangerous movements in the soldering motion.

1 Introduction

The soldering techniques are one of the industrial techniques required in electronic device manufacturing plants. They are very important techniques because they affect the product quality. For soldering training, it is hard for trainers to instruct a large number of trainees at the same time because it may cause accidents in production sites. Also, a framework for quantifying soldering techniques has not been established, therefore it hard to evaluate the training of trainees. The safety is increasingly important in the workplace for persons with physical, intellectual, mental and other disabilities. Furthermore, technical training for persons with disabilities requires the explanation of the detailed work procedures and the burden on the trainer is increased by monitoring to prevent accidents. On the other hand, the application of haptics to virtual space can

L. Barolli et al. (Eds.): EIDWT 2022, LNDECT 118, pp. 308–315, 2022.
https://doi.org/10.1007/978-3-030-95903-6_33

reduce the occurrence of accidents during training. The haptics is able to transmit the power generated in the virtual space to the manipulator. The haptics enables training similar to that in real space and is using a wide range of fields including medicine [1,2], welfare [3], sports [4] and so on [5–9]. Therefore, we implement a haptics based virtual space that can perform soldering and collect the soldering techniques training data. We analyze the soldering motion based on Long Short-Term Memory (LSTM) [10–14]. The LSTM is a method of Recurrent Neural Network (RNN) [15–17], which is used for natural language processing [18–21] and acoustic analysis [22] since it can predict time series. In addition, it can learn long-term dependencies.

In this paper, we propose and evaluate a haptics a soldering training system based on haptics. For evaluation of the proposed system, we perform the detection of anomalies for motions of the soldering technique.

The structure of the paper is as follows. In Sect. 2, we present the proposed system. In Sect. 3, we describe the experimental results. Finally, conclusions and future work are given in Sect. 4.

2 Proposed System

In this section, we present the proposed system. The structure of the proposed system is shown in Fig. 1. We implement the virtual space of the proposed system using Chai3d [23,24] which is a C++ framework for haptic, visualization and real-time interaction. The interaction with the real and virtual space is used the Novint Falcon which is a haptic device with 3 Degrees of Freedom (3-DoF). We design the touchable device of Novint Falcon [25,26] considering the soldering iron for soldering training. In addition, we can use arbitrary printed wired board objects as soldering targets in the virtual space. The soldering iron tip in the virtual space moves along with the manipulation of the soldering iron mounted to the Novint Falcon.

In the proposed system, we focus on the trajectory of the soldering iron tip and the moving distance in 0.1 [s] for analyzing the soldering technique. The trajectory of the soldering iron tip shows the difference in the way the trainee and the trainer use the soldering iron. From the moving distance in 0.1 [s], we can see the dangerous movements that may lead to injury or damage of the trainee while performing the soldering. Also, the moving distance shows the soldering procedure performed by the trainee. The trainee should perform the same procedure as the trainer. Therefore, the three-dimensional coordinates of Novint Falcon grip indicating the soldering iron tip are collected at 0.1 [s] intervals during the training. In order to detect dangerous behaviors that derived during the soldering training, the proposed system detects the anomaly based on LSTM and Mahalanobis distance [27].

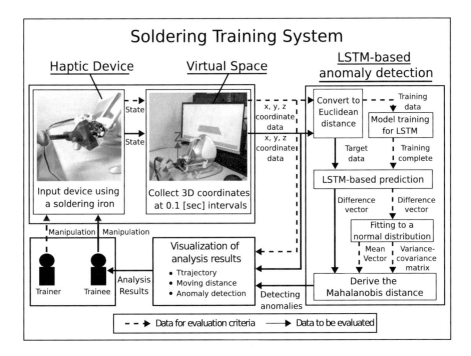

Fig. 1. Proposed system.

For LSTM training, the proposed system collects the training data of a reference trainer and is using the time series data converted to Euclidean distance of 0.1 [s] as relative location information. Then, the average of the movements data for the trainer collected in the same environment is used as the datasets for normal movements. The proposed system derives the difference vectors between the observed values and the predicted values by LSTM from the trainer and trainee data, respectively. Then, the difference vector of the trainer is matched to the normal distribution. The maximum likelihood estimator of the normal distribution is obtained by the mean vector and the variance-covariance matrix based on the maximum likelihood method. In addition, anomalies are detected by deriving the Mahalanobis distance from the mean vector and variance-covariance matrix and the difference vector of the trainee. In order to derive anomalies in soldering motion quantitatively, the proposed system labels anomalies and derives thresholds based on F-scores [28] which can be adapted to the labeled data.

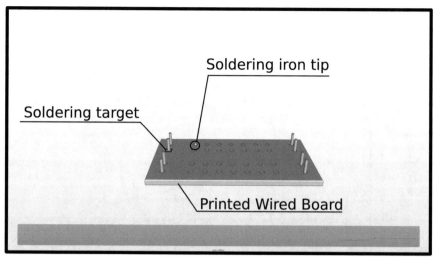

(a) Environment of the soldering training.

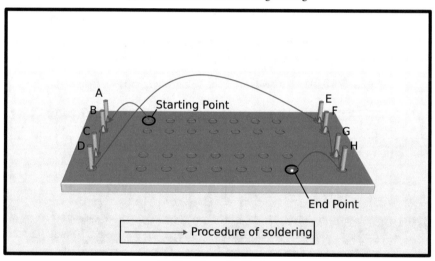

(b) Procedure of soldering.

Fig. 2. Experimental scenario.

3   Experimental Results

In this section, we describe the experimental results. The experimental scenario is shown in Fig. 2. In Fig. 2(a) shows the environment of the soldering training and in Fig. 2(b) the soldering procedure. The trainee performs the motion of touching the soldering iron tip to one of the soldering targets on the printed wired board for about 5 [s]. In the soldering training, the training starts when the soldering iron tip is at the starting point. The trainee performs soldering in

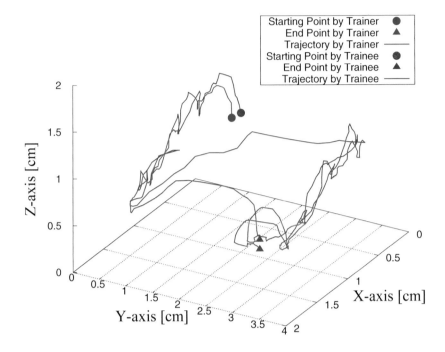

Fig. 3. Visualization results of the soldering iron tip trajectory.

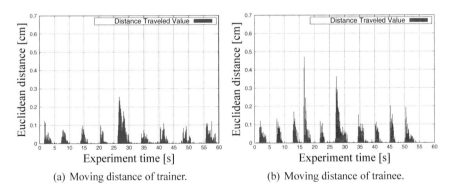

(a) Moving distance of trainer.   (b) Moving distance of trainee.

Fig. 4. Visualization results of moving distance.

the order from A to H. Then, the training ends when the soldering iron tip is at the end point. In adidition, the trainer represents the evaluation criteria and the trainee represents the evaluation target. The trainee manipulates the touchable device to move the soldering iron tip in the virtual space. The experiment time is 60 [s]. The proposed system predicts 600 points obtained in the experimental results using LSTM and detects the anomaly for the distance in 0.1 [s] interval. In order to detect anomalies based on LSTM, we consider the deliberately of large movements and different soldering procedures of trainee training data.

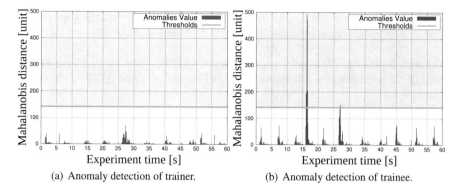

Fig. 5. Vsualization results of anomaly detection.

The experimental results are shown in Fig. 3, Fig. 4 and Fig. 5. In Fig. 3 shows the visualization results of the soldering iron tip trajectory by the trainer and trainee. It can be seen that the trainer moves the soldering iron tip less than the trainee and the soldering procedure of the trainee is different from the trainer. In Fig. 4 shows the visualization of the distance moved of the soldering iron tip in 0.1 [s] by the trainer and trainee. It can be seen that the moving distance of the soldering iron tip by the trainee compared to the trainer has a large movement between 15 [s] to 20 [s] and 25 [s] to 30 [s]. In Fig. 5(a) and Fig. 5(b) show the anomalies associated with the distance moved between 0.1 [s] of the training performed by the trainer and trainee, respectively. The results of the trainee show that anomalies exceeding the threshold are detected at about 16 [s] and 27 [s]. The movement at about 16 [s] is the large movement performed intentionally when the soldering iron tip is touching the soldering target. Also, the soldering procedure from D to H at about 27 [s] by the trainee is different from the procedure from D to E by the trainer. On the other hand, the results by the trainer show that no anomalies exceeding the threshold are detected. From the experimental results, we can see the effectiveness of the LSTM based anomaly detection for the large movements and soldering procedures of the trainees.

4   Conclusions

In this paper, we proposed a soldering training system based on haptics. Also, we presented the experimental results of detecting anomalies with trainee data considering large movements and different procedures. From the experimental results, we conclude the followings.

- The proposed system can perform virtual training based on haptics.
- The experimental results shows that the system can analyze soldering motions and detect dangerous movements.
- The proposed system is a good approach for soldering training and can support persons with disabilities.

In the future, we would like to improve the soldering training system for persons with disabilities by considering different scenarios.

Acknowledgement. This work was supported by JSPS KAKENHI Grant Number JP20K19793 and Grant for Promotion of OUS Research Project (OUS-RP-20-3).

References

1. Joseph, F.J., et al.: Neurosurgical simulator for training aneurysm microsurgery-a user suitability study involving neurosurgeons and residents. Acta Neurochir. **162**, 2313–2321 (2020)
2. Chen, X., et al.: A patient-specific haptic drilling simulator based on virtual reality for dental implant surgery. Int. J. Comput. Assist. Radiol. Surg. **13**(11), 1861–1870 (2018)
3. Bortone, I., et al.: Wearable haptics and immersive virtual reality rehabilitation training in children with neuromotor impairments. IEEE Trans. Neural Syst. Rehabil. Eng. **26**(7), 1469–1478 (2018)
4. Handa, T., Azuma, M., Shimizu, T., Kondo, S.: A ball-type haptic interface to enjoy sports games. In: Kajimoto, H., Lee, D., Kim, S.-Y., Konyo, M., Kyung, K.-U. (eds.) AsiaHaptics 2018. LNEE, vol. 535, pp. 284–286. Springer, Singapore (2019). https://doi.org/10.1007/978-981-13-3194-7_63
5. Khosravi, H., et al.: Simulating mass in virtual reality using physically-based hand-object interactions with vibration feedback. In: Proceedings of the Graphics Interface, pp. 241–248 (2021)
6. Liu, L., et al.: Haptic technology and its application in education and learning. In: Proceedings of the 10th International Conference on Ubi-Media Computing and Workshops (Ubi-Media), pp. 1–6 (2017)
7. Salazar, S., et al.: Altering the stiffness, friction, and shape perception of tangible objects in virtual reality using wearable haptics. IEEE Trans. Haptics **13**(1), 167–174 (2020)
8. Oda, T., et al.: Design and implementation of an iot-based e-learning testbed. Int. J. Web Grid Serv. **13**(2), 228–241 (2017)
9. Hirota, Y., et al.: Proposal and experimental results of a dnn based real-time recognition method for ohsone style fingerspelling in static characters environment. In: Proceedings of the IEEE GCCE-2020, pp. 476–477 (2020)
10. Ji, Y., et al.: A method for LSTM-based trajectory modeling and abnormal trajectory detection. IEEE Access **8**, 104063–104073 (2020)
11. Toyoshima, K., et al.: Proposal of a haptics and LSTM based soldering motion analysis system. In: Proceedings of the IEEE GCCE-2021, pp. 774–775 (2021)
12. Sherstinsky, A., et al.: Fundamentals of recurrent neural network (RNN) and long short-term memory (LSTM) network. Physica D **404**, 1–43 (2020)
13. Hochreiter, S., et al.: Long short-term memory. Neural Comput. **9**(8), 1735–1780 (1997)
14. Hirota, Y., Oda, T., Saito, N., Hirata, A., Hirota, M., Katatama, K.: Proposal and experimental results of an ambient intelligence for training on soldering iron holding. In: Barolli, L., Takizawa, M., Enokido, T., Chen, H.-C., Matsuo, K. (eds.) BWCCA 2020. LNNS, vol. 159, pp. 444–453. Springer, Cham (2021). https://doi.org/10.1007/978-3-030-61108-8_44

15. Ishitaki, T., et al.: Application of deep recurrent neural networks for prediction of user behavior in tor networks. In: Proceedings of the IEEE AINA-2017, pp. 238–243 (2017)
16. Ishitaki, T., et al.: A neural network based user identification for tor networks: data analysis using friedman test. In: Proceedings of the IEEE AINA-2016, pp. 7–13 (2016)
17. Oda, T., et al.: A neural network based user identification for tor networks: comparison analysis of activation function using friedman test. In: Proceedings of the CISIS-2016, pp. 477–483 (2016)
18. Yao, L., et al.: An improved LSTM structure for natural language processing. In: Proceedings of the IEEE International Conference of Safety Produce Informatization (IICSPI), pp. 565–569 (2018)
19. Nagai, Y., et al.: Approach of a Word2Vec based tourist spot collection method considering COVID-19. In: Proceedings of the BWCCA-2020, pp. 67–75 (2020)
20. Nagai, Y., et al.: Approach of an emotion words analysis method related COVID-19 for twitter. In: Proceedings of the IEEE GCCE-2021, pp. 1–2 (2021)
21. Nagai, Y., et al.: Approach of a Japanese co-occurrence words collection method for construction of linked open data for COVID-19. In: Proceedings of the IEEE GCCE-2020, pp. 478–479 (2020)
22. Wang, Q., et al.: Speaker diarization with LSTM. In: Proceedings of the IEEE International Conference on Acoustics, Speech and Signal Processing (ICASSP) (2018)
23. Rodríguez-Vila, B., et al.: A low-cost pedagogical environment for training on technologies for image-guided robotic surgery. In: Lhotska, L., Sukupova, L., Lacković, I., Ibbott, G.S. (eds.) World Congress on Medical Physics and Biomedical Engineering 2018. IP, vol. 68/2, pp. 821–824. Springer, Singapore (2019). https://doi.org/10.1007/978-981-10-9038-7_151
24. Battagli, E., et al.: TcHand: visualizing hands in CHAI3D. In: Proceedings of the IEEE World Haptics Conference (WHC), p. 354 (2021)
25. Jose, J., et al.: Design of a bi-manual haptic interface for skill acquisition in surface mount device soldering. Solder. Surf. Mount Technol. **31**(2), 133–142 (2019)
26. Ivanov, V., Strelkov, S., Klygach, A., Arseniev, D.: Medical training simulation in virtual reality. In: Voinov, N., Schreck, T., Khan, S. (eds.) Proceedings of International Scientific Conference on Telecommunications, Computing and Control. SIST, vol. 220, pp. 177–184. Springer, Singapore (2021). https://doi.org/10.1007/978-981-33-6632-9_15
27. McLachlan, G.: The mahalanobis distance. Chemom. Intell. Lab. Syst. **50**(1), 1–18 (2000)
28. Malhotra, P., et al.: Long short term memory networks for anomaly detection in time series. In: European Symposium on Artificial Neural Networks, vol. 23, no. 56, pp. 89–94 (2015)

Performance Evaluation of WMNs by WMN-PSOHC Hybrid Simulation System Considering Two Instances and Normal Distribution of Mesh Clients

Shinji Sakamoto[1(⊠)] and Leonard Barolli[2]

[1] Department of Information and Computer Science, Kanazawa Institute of Technology, 7-1 Ohgigaoka, Nonoichi, Ishikawa 921-8501, Japan
shinji.sakamoto@ieee.org
[2] Department of Information and Communication Engineering, Fukuoka Institute of Technology, 3-30-1 Wajiro-Higashi, Higashi-Ku, Fukuoka 811-0295, Japan
barolli@fit.ac.jp

Abstract. Wireless Mesh Networks (WMNs) have many good features and they are becoming an important networking infrastructure. However, WMNs have some problems such as node placement, security, transmission power and so on. To solve these problems, we have implemented a hybrid simulation system based on PSO and HC called WMN-PSOHC. In this paper, we evaluate the performance of WMNs by using WMN-PSOHC considering two instances: Instance 1 and Instance 2. Simulation results show that WMN-PSOHC performs better for Instance 1 compared with Instance 2.

1 Introduction

In this work, we deal with node placement problem in WMNs. We consider the version of the mesh router nodes placement problem in which we are given a grid area where to deploy a number of mesh router nodes and a number of mesh client nodes of fixed positions (of an arbitrary distribution) in the grid area. The objective is to find a location assignment for the mesh routers to the cells of the grid area that maximizes the network connectivity and client coverage. Network connectivity is measured by Size of Giant Component (SGC) of the resulting WMN graph, while the user coverage is simply the number of mesh client nodes that fall within the radio coverage of at least one mesh router node and is measured by Number of Covered Mesh Clients (NCMC). Node placement problems are known to be computationally hard to solve [13]. In some previous works, intelligent algorithms have been recently investigated [3,8,9]. We already implemented a Particle Swarm Optimization (PSO) based simulation system, called WMN-PSO [6]. Also, we implemented a simulation system based on Hill Climbing (HC) for solving node placement problem in WMNs, called WMN-HC [5].

L. Barolli et al. (Eds.): EIDWT 2022, LNDECT 118, pp. 316–323, 2022.
https://doi.org/10.1007/978-3-030-95903-6_34

In our previous work [6,7], we presented a hybrid intelligent simulation system based on PSO and HC. We called this system WMN-PSOHC. In this paper, we evaluate the performance of WMNs by using WMN-PSOHC considering two instances and Normal distribution of mesh clients.

The rest of the paper is organized as follows. We present our designed and implemented hybrid simulation system in Sect. 2. In Sect. 3, we introduce WMN-PSOHC Web GUI tool. The simulation results are given in Sect. 4. Finally, we give conclusions and future work in Sect. 5.

2   Proposed and Implemented Simulation System

2.1   Particle Swarm Optimization

In Particle Swarm Optimization (PSO) algorithm, a number of simple entities (the particles) are placed in the search space of some problem or function and each evaluates the objective function at its current location. The objective function is often minimized and the exploration of the search space is not through evolution [4]. However, following a widespread practice of borrowing from the evolutionary computation field, in this work, we consider the bi-objective function and fitness function interchangeably. Each particle then determines its movement through the search space by combining some aspect of the history of its own current and best (best-fitness) locations with those of one or more members of the swarm, with some random perturbations. The next iteration takes place after all particles have been moved. Eventually the swarm as a whole, like a flock of birds collectively foraging for food, is likely to move close to an optimum of the fitness function.

Each individual in the particle swarm is composed of three $\mathcal{D}$-dimensional vectors, where $\mathcal{D}$ is the dimensionality of the search space. These are the current position $\vec{x}_i$, the previous best position $\vec{p}_i$ and the velocity $\vec{v}_i$.

The particle swarm is more than just a collection of particles. A particle by itself has almost no power to solve any problem; progress occurs only when the particles interact. Problem solving is a population-wide phenomenon, emerging from the individual behaviors of the particles through their interactions. In any case, populations are organized according to some sort of communication structure or topology, often thought of as a social network. The topology typically consists of bidirectional edges connecting pairs of particles, so that if j is in i's neighborhood, i is also in j's. Each particle communicates with some other particles and is affected by the best point found by any member of its topological neighborhood. This is just the vector $\vec{p}_i$ for that best neighbor, which we will denote with $\vec{p}_g$. The potential kinds of population "social networks" are hugely varied, but in practice certain types have been used more frequently.

In the PSO process, the velocity of each particle is iteratively adjusted so that the particle stochastically oscillates around $\vec{p}_i$ and $\vec{p}_g$ locations.

2.2   Hill Climbing

Hill Climbing (HC) algorithm is a heuristic algorithm. The idea of HC is simple. In HC, the solution s' is accepted as the new current solution if $\delta \leq 0$ holds, where $\delta = f(s') - f(s)$. Here, the function f is called the fitness function. The fitness function gives points to a solution so that the system can evaluate the next solution s' and the current solution s.

The most important factor in HC is to define effectively the neighbor solution. The definition of the neighbor solution affects HC performance directly. In our WMN-PSOHC system, we use the next step of particle-pattern positions as the neighbor solutions for the HC part.

2.3   WMN-PSOHC System Description

In following, we present the initialization, particle-pattern, fitness function and router replacement methods.

Initialization
Our proposed system starts by generating an initial solution randomly, by *ad hoc* methods [14]. We decide the velocity of particles by a random process considering the area size. For instance, when the area size is $W \times H$, the velocity is decided randomly from $-\sqrt{W^2 + H^2}$ to $\sqrt{W^2 + H^2}$. Our system can generate many client distributions. In this paper, we consider Chi-square distribution of mesh clients as shown in Fig. 1.

Particle-pattern
A particle is a mesh router. A fitness value of a particle-pattern is computed by combination of mesh routers and mesh clients positions. In other words, each particle-pattern is a solution as shown is Fig. 2. Therefore, the number of particle-patterns is a number of solutions.

Fitness function
One of most important thing is to decide the determination of an appropriate objective function and its encoding. In our case, each particle-pattern has an own fitness value and compares other particle-patterns fitness value in order to share information of global solution. The fitness function follows a hierarchical approach in which the main objective is to maximize the SGC in WMN. Thus, we use α and β weight-coefficients for the fitness function and the fitness function of this scenario is defined as:

$$\text{Fitness} = \alpha \times \text{SGC}(\boldsymbol{x}_{ij}, \boldsymbol{y}_{ij}) + \beta \times \text{NCMC}(\boldsymbol{x}_{ij}, \boldsymbol{y}_{ij}).$$

Router replacement methods
A mesh router has x, y positions and velocity. Mesh routers are moved based on velocities. There are many router replacement methods in PSO field [2,10–12]. In this paper, we use Linearly Decreasing Inertia Weight Method (LDIWM). In LDIWM, C_1 and C_2 are set to 2.0, constantly. On the other hand, the ω parameter is changed linearly from unstable region ($\omega = 0.9$) to stable region ($\omega = 0.4$) with increasing of iterations of computations [1,12].

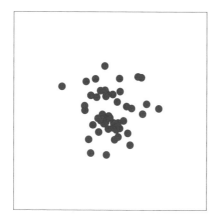

Fig. 1. Normal distribution of mesh clients.

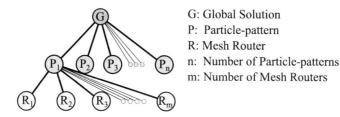

G: Global Solution
P: Particle-pattern
R: Mesh Router
n: Number of Particle-patterns
m: Number of Mesh Routers

Fig. 2. Relationship among global solution, particle-patterns and mesh routers.

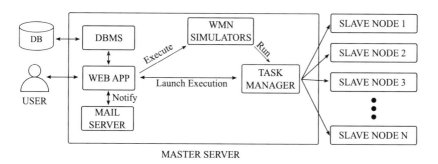

Fig. 3. System structure for web interface.

3 WMN-PSOHC Web GUI Tool

The Web application follows a standard Client-Server architecture and is implemented using LAMP (Linux + Apache + MySQL + PHP) technology (see Fig. 3). We show the WMN-PSOHC Web GUI tool in Fig. 4. Remote users

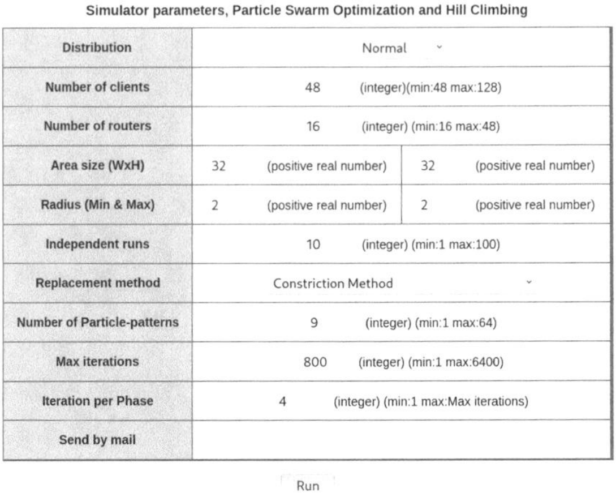

Simulator parameters, Particle Swarm Optimization and Hill Climbing

Distribution	Normal ⌄		
Number of clients	48 (integer)(min:48 max:128)		
Number of routers	16 (integer) (min:16 max:48)		
Area size (WxH)	32	(positive real number)	32 (positive real number)
Radius (Min & Max)	2	(positive real number)	2 (positive real number)
Independent runs	10 (integer) (min:1 max:100)		
Replacement method	Constriction Method ⌄		
Number of Particle-patterns	9 (integer) (min:1 max:64)		
Max iterations	800 (integer) (min:1 max:6400)		
Iteration per Phase	4 (integer) (min:1 max:Max iterations)		
Send by mail			

Run

Fig. 4. WMN-PSOHC Web GUI tool.

(clients) submit their requests by completing first the parameter setting. The parameter values to be provided by the user are classified into three groups, as follows.

- Parameters related to the problem instance: These include parameter values that determine a problem instance to be solved and consist of number of router nodes, number of mesh client nodes, client mesh distribution, radio coverage interval and size of the deployment area.
- Parameters of the resolution method: Each method has its own parameters.
- Execution parameters: These parameters are used for stopping condition of the resolution methods and include number of iterations and number of independent runs. The former is provided as a total number of iterations and depending on the method is also divided per phase (e.g., number of iterations in a exploration). The later is used to run the same configuration for the same problem instance and parameter configuration a certain number of times.

Table 1. Instances parameters.

Parameters	Instance 1	Instance 2
Area size	32×32	64×64
Number of mesh routers	16	32
Number of mesh clients	48	96

Table 2. Parameter settings.

Parameters	Values
Clients distribution	Normal distribution
Instances	Instance 1, Instance 2
Total iterations	800
Iteration per phase	4
Number of particle-patterns	9
Radius of a mesh router	From 2.0 to 3.0
Fitness function weight-coefficients (α, β)	0.7, 0.3
Replacement method	LDIWM

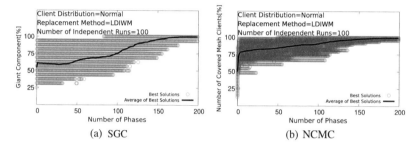

(a) SGC   (b) NCMC

Fig. 5. Simulation results of WMN-PSOHC for Instance 1.

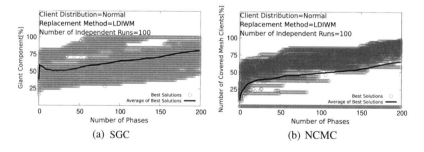

(a) SGC   (b) NCMC

Fig. 6. Simulation results of WMN-PSOHC for Instance 2.

4   Simulation Results

In this section, we show simulation results using WMN-PSOHC system. In this work, we consider Normal distribution of mesh clients. We consider the number of particle-patterns 9. We conducted simulations 100 times in order to avoid the effect of randomness and create a general view of results. The total number of iterations is considered 800 and the iterations per phase is considered 4. We consider two instances: Instance 1 and Instance 2 as shown in Table 1. We show the parameter setting for WMN-PSOHC in Table 2.

We show the simulation results of Instance 1 in Fig. 5 The value of SGC and NCMC for 200 phases has reached 100% for both parameters. We show the simulation results of Instance 2 in Fig. 6. Comparing with Instance 1, the performance of instance 2 decreases. The value of SGC and NCMC for 200 do not reach 100% for both parameters. This is because the search space of Instance 2 is larger than Instance 1. Also, we set the same number of total iterations and iteration per phase for both instances. However, for the instance 2, the WMN-PSOHC needs more time to get the same performance with Instance 1.

5   Conclusions

In this work, we evaluated the performance of WMNs by using a hybrid simulation system based on PSO and HC (called WMN-PSOHC) considering two instances and Normal distribution of mesh clients.

Simulation results show that WMN-PSOHC performs better for Instance 1 compared with Instance 2. This is because the WMN-PSOHC needs more time to get the same performance with Instance 1.

In our future work, we would like to evaluate the performance of the proposed system for different parameters and scenarios.

References

1. Barolli, A., Sakamoto, S., Ohara, S., Barolli, L., Takizawa, M.: Performance analysis of WMNs by WMN-PSOHC-DGA simulation system considering linearly decreasing inertia weight and linearly decreasing Vmax replacement methods. In: Barolli, L., Nishino, H., Miwa, H. (eds.) INCoS 2019. AISC, vol. 1035, pp. 14–23. Springer, Cham (2020). https://doi.org/10.1007/978-3-030-29035-1_2
2. Clerc, M., Kennedy, J.: The particle swarm-explosion, stability, and convergence in a multidimensional complex space. IEEE Trans. Evol. Comput. **6**(1), 58–73 (2002)
3. Ozera, K., Bylykbashi, K., Liu, Y., Barolli, L.: A Fuzzy-based approach for cluster management in VANETs: performance evaluation for two fuzzy-based systems. Internet of Things **3**, 120–133 (2018)
4. Poli, R., Kennedy, J., Blackwell, T.: Particle swarm optimization. Swarm Intell. **1**(1), 33–57 (2007)
5. Sakamoto, S., Lala, A., Oda, T., Kolici, V., Barolli, L., Xhafa, F.: Analysis of WMN-HC simulation system data using Friedman test. In: The Ninth International Conference on Complex, Intelligent, and Software Intensive Systems (CISIS-2015), pp. 254–259. IEEE (2015)
6. Sakamoto, S., Oda, T., Ikeda, M., Barolli, L., Xhafa, F.: Implementation and evaluation of a simulation system based on particle swarm optimisation for node placement problem in wireless mesh networks. Int. J. Commun. Netw. Distrib. Syst. **17**(1), 1–13 (2016)
7. Sakamoto, S., Ozera, K., Ikeda, M., Barolli, L.: Implementation of intelligent hybrid systems for node placement problem in WMNs considering particle swarm optimization, hill climbing and simulated annealing. Mob. Netw. Appl. **23**(1), 27–33 (2018). https://doi.org/10.1007/s11036-017-0897-7

8. Sakamoto, S., Barolli, A., Barolli, L., Okamoto, S.: Implementation of a web interface for hybrid intelligent systems. Int. J. Web Inf. Syst. **15**(4), 420–431 (2019)

9. Sakamoto, S., Barolli, L., Okamoto, S.: WMN-PSOSA: an intelligent hybrid simulation system for WMNs and Its performance evaluations. Int. J. Web Grid Serv. **15**(4), 353–366 (2019)

10. Schutte, J.F., Groenwold, A.A.: A study of global optimization using particle swarms. J. Global Optim. **31**(1), 93–108 (2005). https://doi.org/10.1007/s10898-003-6454-x

11. Shi, Y.: Particle swarm optimization. IEEE Connections **2**(1), 8–13 (2004)

12. Shi, Y., Eberhart, R.C.: Parameter selection in particle swarm optimization. In: Porto, V.W., Saravanan, N., Waagen, D., Eiben, A.E. (eds.) EP 1998. LNCS, vol. 1447, pp. 591–600. Springer, Heidelberg (1998). https://doi.org/10.1007/BFb0040810

13. Wang, J., Xie, B., Cai, K., Agrawal, D.P.: Efficient mesh router placement in wireless mesh networks. In: Proceedings of IEEE International Conference on Mobile Adhoc and Sensor Systems (MASS-2007), pp. 1–9 (2007)

14. Xhafa, F., Sanchez, C., Barolli, L.: Ad hoc and neighborhood search methods for placement of mesh routers in wireless mesh networks. In: Proceedings of 29th IEEE International Conference on Distributed Computing Systems Workshops (ICDCS-2009), pp. 400–405 (2009)

The Principal Dimensions Optimization of Large Ships Based on Improved Firefly Algorithm

Jianghao Yin and Na Deng[(✉)]

School of Computer Science, Hubei University of Technology, Wuhan, China

Abstract. Aiming at the principal dimensions optimization for large ships, we construct mathematics model, and propose an improved firefly algorithm. Firstly, the fitness is calculated, according to which, the firefly is divided into three populations; and then the elite strategy is used to realize the variable step mechanism to strengthen the elite individual's convergence speed; afterwards, Levy flight rules are introduced in to improve the firefly's position update formula to prevent the search from stagnating in the later stages of the iteration; later on, the negative reinforcement strategy is added. Finally, this paper compares the improved firefly algorithm, the standard firefly algorithm, and the particle swarm algorithm. The results prove that the improved firefly algorithm has better optimization performance in the principal dimensions optimization of large ships.

1 Introduction

When designing large ships, the selection of principal dimensions is one of the most important issues [1–4]. The choice of principal dimensions will directly affect the quality and performance of the ship. When designing ship dimensions, in order to optimize the design plan, various factors such as economy, technology, practicability, seakeeping quality, and work efficiency must be considered. Therefore, the principal dimensions optimization of ships is a multi-objective, multi-parameter nonlinear optimization problem.

Many scholars at home and abroad establish mathematical models and use intelligent optimization algorithms to solve the optimization problem of ship's principal dimensions. Literature [1] uses an improved artificial bee colony algorithm to optimize the principal dimensions of large ships, but the optimization efficiency needs to be improved; literature [2] uses particle swarm algorithm to optimize the ship's principal dimensions design plan, but particle swarm algorithm is easy to fall into the local optimal solution; literature [3] uses genetic algorithm to design the ship's principal dimensions, but the commonly used genetic algorithm programming is complicated. In literature [4], the firefly algorithm is applied to the principal dimensions optimization of deep sea mining vessel and achieved good results.

Firefly Algorithm (FA) is an intelligent algorithm that simulates the biological characteristics of fireflies in nature. It was first elaborated and proposed by Yang [5] of Cambridge University in the United Kingdom. This algorithm has the advantages of fewer parameters and easy implementation. It has good applications in ship principal dimensions optimization, PID parameter optimization, path planning problem, cluster analysis and other fields. However, the algorithm also has problems such as easy to fall into the local optimal solution in the later stage, and the solution accuracy is not high. In response to these problems, literature [6] introduces chaos into the Firefly algorithm to increase the mobility of its global search, thereby achieving robust global optimization; literature [7] proposes an improved Firefly algorithm based on gender differences to improve the algorithm. The performance of male fireflies in the male population is guided by two randomly selected female fireflies to conduct a global search, and the female fireflies move to the best male fireflies to perform a local search. This algorithm improves the accuracy of the search; literature [8] proposes a hybrid firefly algorithm based on the vector angle learning mechanism combines the advantages of the firefly algorithm and the differential evolution algorithm through the vector angle learning mechanism, which improves the search ability of the firefly algorithm.

The rest of this paper is organized as follows: Sect. 2 establishes a mathematical model for optimization of ship's principal dimensions from the perspectives of seakeeping quality, economy, safety, and practicability. Section 3 introduces the standard firefly algorithm. Section 4 proposes an improved firefly algorithm ELNFA (Elite Leavy Negative Firefly Algorithm) to solve the principal dimensions optimization problem of large ships. Section 5 uses the improved firefly algorithm to solve the calculation example, and compare the results with the particle swarm algorithm (PSO) and the standard firefly algorithm (FA). In Sect. 6 the research is summarized and the future work is given.

2 The Establishment of Principal Dimensions Optimization Model

2.1 The Establishment of Variables and Objective Function

In the principal dimensions design process of large ships, some factors need to be considered. In this paper, we mainly select the designed waterline length L_s, designed waterline width B_s, draft d, molded depth D, block coefficient C_b, standard displacement Δ as variables X:

$$X = (L_s, B_s, d, D, C_b, \Delta)^T \tag{1}$$

The schematic diagram is shown in Fig. 1.

Taking into account the economy, practicability, safety, and seakeeping quality of the ship, the following five indicators are selected as the objective function of the principal dimensions optimization:

1. Empty ship mass

The empty ship mass W will affect the cost of shipbuilding. In the optimization of ship's principal dimensions, minimimize empty ship mass is one of the optimization objectives. The empty ship mass of large ships mainly includes the hull steel mass W_h, the outfitting ship mass W_t, and the mechanical and electrical equipment mass W_m. This paper assumes that the outfitting ship mass W_t and the electromechanical equipment mass W_m are fixed values and there is no room for optimization. Therefore, finding the

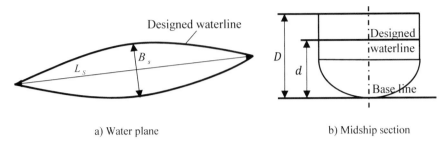

a) Water plane   b) Midship section

Fig. 1. Schematic diagram of ship structure.

minimum value of the empty ship mass W is converted to the optimization of the hull steel mass W_h. The calculation method of hull steel mass [2] is as follows:

$$W_h = 0.034 \times L_s^{1.7} \times B_s^{0.7} \times D^{0.7} \times C_b^{0.5} \tag{2}$$

where L_s is designed waterline length, B_s is designed waterline width, D is molded depth, C_b is block coefficient.

2. Effective power

In the optimization of ship's principal dimensions, ship speed can be used as an important criterion for measuring practicability. Ship speed can be converted into the effective power P of the engine. The effective power P should be maximized. The calculation formula [9] is as follows:

$$P = \frac{v^{\frac{2}{3}} \times \Delta^3}{v_0^{\frac{2}{3}} \times \Delta_0^3} \times P_0 \tag{3}$$

where P_0 is the effective power of the mother ship, v_0 is the speed of the mother ship, Δ_0 is the displacement of the mother ship, v is the speed of the ship, and Δ is the displacement of the ship.

3. Collision rate

When the mass of the ship is large, its own inertia is large, the motion state is difficult to change, and the probability of accidents caused by wind and waves and other weather factors during navigation is greater. In this paper, collision rate R is used to measure the safety of ships. The lower the collision rate, the higher the safety. Literature [10] proves that the relationship between ship's tonnage and collision rate is:

$$R = 0.014 \times \ln(W_h) + 0.009 \tag{4}$$

where W_h is the hull steel mass.

4. Rolling period and pitching period

We select the rolling period T_ϕ and the pitching period T_θ to measure the seakeeping quality of the ship. When the ship rolls from the initial equilibrium position through the maximum inclination position of the left and right sides and then back to the original initial equilibrium position, is deemed as a roll cycle. The time it takes for each roll cycle to complete is called the roll period. The calculation method of the rolling period T_ϕ [4] is as follows:

$$T_\phi = \frac{2 \times C \times B_S}{\sqrt{GM}} \tag{5}$$

where C is the roll coefficient of the ship, and GM is the initial metacentric height, the calculation method is as follows:

$$GM = A_1 \times d + \frac{A_2 \times B_w^2}{d} - 0.55 \times D \tag{6}$$

where $A_1 = \frac{C_w}{C_w + C_b}$; $A_2 = 0.0106 \times C_w + 0.072 \times C_w^2$,
where C_w is water surface coefficient, $C_w = 0.5375 \times C_b + 0.4474$.
The calculation method of pitching period T_θ [5] is as follows:

$$T_\theta = 2.01 \times \sqrt{(0.77 \times C_b + 0.26)(0.92 \times d + 0.44 \times B_s)} \tag{7}$$

2.2 Unified Objective Function

Combined with the above objective function, this paper establishes the ship's principal dimensions optimization model as:

$$F(x) = \begin{cases} min(W_h) \\ max(P) \\ min(R) \\ max(T_\phi) \\ max(T_\theta) \end{cases} \tag{8}$$

In this paper, the minimum deviation method is adopted, and the objective function obtained after processing is:

$$min(F) = \frac{W_h - W_{hmin}}{W_{h0} - W_{hmin}} + \frac{P_{max} - P}{P_{max} - P_0} + \frac{R - R_{min}}{R_0 - R_{min}} + \frac{T_{\phi max} - T_\phi}{T_{\phi max} - T_{\phi 0}} + \frac{T_{\theta max} - T_\theta}{T_{\theta max} - T_{\theta 0}} \tag{9}$$

where $W_{h0}, P_0, R_0, T_{\phi 0}, T_{\theta 0}$ are the ship's initial steel mass, initial effective power, initial collision rate, initial rolling period, and initialing pitching period, $W_{hmin}, P_{max}, R_{min}, T_{\phi max}, T_{\theta max}$ are the ship's minimum steel mass, maximum effective power, minimum collision rate, maximum pitching period, and maximum rolling period.

The size restrictions [9] are as follows:

$$t = \begin{cases} L_s \in [290, 340]m \\ B_s \in [65, 76]m \\ d \in [7, 13]m \\ D \in [27, 35]m \\ C_d \in [0.5, 0.65] \\ \Delta \in [5 \times 10^4, 10^5]t \end{cases} \tag{10}$$

where m is meters and t is tons.

In order to ensure the normal work of large ships, this article also makes the following constraints:

Length draft ratio: $d/L_S \in (0.35, +\infty)$,
Length breadth ratio: $L_S/B_S \in (7.1, 8.1)$,
Breadth draft ratio: $B_S/d \in (3.4, 4.1)$,
Draft depth ratio: $D/d \in (2, +\infty)$,
Rolling period and pitching period: $T_\phi \geq 11$ s, $T_\theta \geq 2$ s,
$L_S/\nabla^{1/3} \in (7.0, 7.9)$, where ∇ is the drainage volume.

3   Standard Firefly Algorithm

The firefly algorithm is a bionic algorithm which simulates that in the firefly population, each individual will move to a brighter individual. The idea of the standard firefly algorithm is to randomly allocate a certain number of fireflies in the feasible region. Each firefly represents a solution in the feasible region. The brightness of the firefly is related to the objective function of the problem to be solved, and is proportional to attractiveness. In the iterative process, each firefly moves to the attractive individual within the decision radius, and eventually all fireflies will gather near the one with maximum brightness. We describe the variables involved in the algorithm as follows:

The relative brightness of fireflies is:

$$I = I_0 \times e^{-\gamma r_{ij}} \tag{11}$$

where I_0 is the maximum brightness of the firefly, which is related to the fitness of the firefly itself, and γ is the light absorption coefficient. When light propagates in the medium, the light intensity will decrease as the distance increases. r_{ij} is the distance between two fireflies, usually calculated by Cartesian distance:

$$r_{ij} = ||x_i - x_j|| \tag{12}$$

where x_i, x_j are the coordinates of the firefly i and j.

The relative attractiveness of fireflies is:

$$\beta = \beta_0 \times e^{-\gamma r_{ij}^2} \tag{13}$$

where β_0 is the maximum attraction, usually set as a constant, r_{ij} is the distance between two fireflies.

The iterative formula for finding the optimal solution is:

$$x_i = x_i + \beta \times (x_j - x_i) + \alpha \times \left(rand - \frac{1}{2} \right) \tag{14}$$

It can be seen from the principle and evolution mechanism of the Firefly algorithm that the parameters of this algorithm are simple, the convergence speed is fast, and the evolution method is simple.

4 Improved Firefly Algorithm

4.1 Multi-group Elimination Mechanism

In the standard firefly algorithm, there is usually only one population. The evolution of a single population is not conducive to the communication between the populations. It often falls into a local optimal solution, and the search efficiency is not high. Literature [11] proposes a multi-group co-evolution model to improve the accuracy and evolution efficiency of the algorithm. This paper proposes a multi-group elimination mechanism based on this, and divides the original population into three groups into an elite group and two candidate populations. After initialization, the fitness of the fireflies is calculated and sorted. The top third enters the elite population, and the last two thirds enter the two candidate populations in turn. The iterative search within the population is conducive to finding the local maximum value, and the communication between the populations is conducive to the global search.

We denote the firefly at the i-th iteration as x_i, when x_i moves to x_j firefly individual, a variable step size mechanism is introduced in. When x_j belongs to the candidate populations, β does not change. When it belongs to the candidate group, $\beta = 1.25 \times \beta$. The variable step size mechanism can speed up the convergence speed.

After completing an iteration, the fitness of each firefly is recalculated. Individuals in the candidate population whose fitness is greater than the elite population enter the elite population, and the individuals in the elite population whose fitness is less than the candidate population are eliminated to the candidate population.

4.2 Levy Flight Mechanism

The firefly algorithm has the advantages of fewer parameters, easy implementation, and fast convergence speed. However, the firefly algorithm is easy to fall into the local optimal solution in the later stage of the search process, and the global search ability is not strong. The cuckoo algorithm mainly relies on Levy flight to generate new solutions. Levy flight has infinite mean and variance, and its step length is mainly characterized by multiple short-distance moves and accidental long-distance moves. Its characteristic is that there will be a large sudden change during the flight, thus it can be used to describe the situations that when an object is moving, the step size is suddenly increased, and the direction of movement is random.

Based on the above characteristics, the introduction of Levy flight in the search process can increase the global search capability. Inspired by the update mechanism of Levy flight in the cuckoo algorithm, this article introduces this update mechanism into the firefly algorithm to obtain an improved firefly algorithm based on Levy flight. In the position update formula, a factor θ that obeys Levy distribution is introduced. After the improvement, the formula is as follows:

$$x_i = x_i + \beta \times (x_j - x_i) \times Levy(\theta) + \alpha \times \left(rand - \frac{1}{2}\right) \tag{15}$$

where *Levy*(θ) is the search path that obeys the Levy distribution.

4.3 Negative Reinforcement Strategy

The idea of the firefly algorithm is that organisms tend to repeat past strategies. It includes favorable strategies used by the organism itself, as well as favorable strategies observed elsewhere. Literature [12] proposes a negative reinforcement particle swarm optimization, and verified the effectiveness of this algorithm in improving the optimization performance. In this paper, the negative reinforcement strategy is introduced into the firefly algorithm, so that each firefly can evade the strategy that has been proved to be harmful. Each firefly not only moves to an individual that is better than itself, but also moves far away from the worst individual. So the formula is changed to:

$$x_i = x_i + \beta \times (x_j - x_i) \times Levy(\theta) + \alpha \times \left(rand - \frac{1}{2}\right) - \phi \times (x_i - x_w) \tag{16}$$

where ϕ is the coefficient of the negative reinforcement term, which will determine how far each firefly will avoid the largest position in the population, x_w is the worst position in the current population.

The position of the firefly in the standard firefly algorithm is always updated to the position of the favorable solution, and in the negative reinforcement strategy, an item far away from the harmful position is added. Therefore, it is necessary to find a balance point between the positive reinforcement and the negative reinforcement to maximize the efficiency of optimization. This paper selects 4 test functions:

(1) The second Schaffer function

$$f(x) = 0.5 + \frac{\sin^2\left(x_1^2 - x_2^2\right) - 0.5}{\left[1 + 0.001\left(x_1^2 + x_2^2\right)\right]^2} \tag{17}$$

The function is usually evaluated on the square $x_i \in [-100, 100]$, for all i = 1, 2.
(2) Schewefel function

$$f(x) = 418.9829d - \sum_{i=1}^{d} x_i \sin\left(\sqrt{|x_i|}\right) \tag{18}$$

where d is the dimension, and it takes the value 20 in the test. The function is usually evaluated on the hypercube $x_i \in [-500, 500]$, for all i = 1, ..., d.

(3) The fourthSchaffer function

$$f(x) = 0.5 + \frac{\cos^2\left(\sin\left(\left|x_1^2 - x_2^2\right|\right)\right) - 0.5}{\left[1 + 0.001\left(x_1^2 + x_2^2\right)\right]^2} \tag{19}$$

The function is usually evaluated on the square $x_i \in [-100, 100]$, for all $i = 1, 2$.

(4) Rotated Hyper-Ellipsoid function

$$f(x) = \sum_{i=1}^{d} \sum_{j=1}^{i} x_j^2 \tag{20}$$

The function is usually evaluated on the hypercube $x_i \in [-65.536, 65.536]$, for all $i = 1, \ldots, d$.

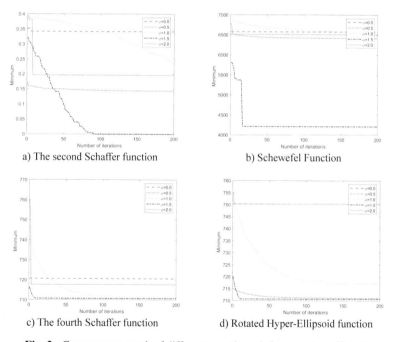

a) The second Schaffer function

b) Schewefel Function

c) The fourth Schaffer function

d) Rotated Hyper-Ellipsoid function

Fig. 2. Convergence graph of different negative reinforcement coefficients.

It can be seen from Fig. 2. that the optimization results are much better when using the reinforcement strategy ($\phi > 0$) than when not using it ($\phi = 0$). At the same time, in the four test functions, when $\phi = 1.5$, we can get the best optimization effect. So we set ϕ as 1.5 in this paper when considering the principal dimensions optimization problems of the ships.

The workflow of the improved Firefly algorithm is as follows (Fig. 3):

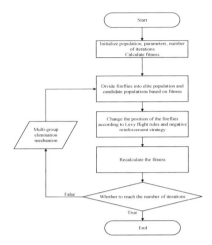

Fig. 3. Flow chart of ELNFA.

*Step*1: Initialize γ, α, β_0, n and other parameters.
*Step*2: Initialize firefly position.
*Step*3: Calculate the fitness of each firefly, and sort according to the fitness.
    Divide the fireflies into an elite population and two candidate populations.
*Step*4: Perform variable step movement between and within the population, as well as levy flight and negative reinforcement strategies.
*Step*5: Recalculate the fitness.
*Step*6: The worst two fireflies in the elite population are eliminated to the candidate populations, and the top one of the two candidate populations enters the elite population.
*Step*7: Judge whether the loop termination condition is met, if it is met, exit, else, go to *Step*3.

5   Experiment and Analysis

In this paper, the literature [9] is selected as a calculation example. The improved firefly algorithm (ELNFA), the standard firefly algorithm (FA), and the particle swarm algorithm (PSO) are used to solve the calculation example. Set the number of fireflies N as 30, and the maximum number of iterations t_{max} as 200. Since there are 6 variables, the dimension n is set as 6. The final optimized results are shown in Table 1 and Table 2.

Table 1 shows the comparison between the ELNFA, the particle swarm algorithm, the standard firefly algorithm and the main scale design scheme obtained in the literature [9]. Table 2 shows the comparison between the initial values of the indicators selected in this paper and the optimization results of the particle swarm algorithm, the standard firefly algorithm, and the ELNFA. It can be seen from the experimental results that the effective power is 225157.3 W, which is 0.6% higher than the original value; the rolling period is 12.6 s, which is 7% higher than the original value; the pitching period is 9.04 s, which is 1.9% higher than the original value. After using the improved Firefly algorithm, the effective power, rolling period, and pitching period have been significantly improved.

Table 1. Comparison of principal dimensions design schemes.

	L_s/m	B_s/m	d/m	D/m	C_b	Δ/t
Initial value	289.7	40.8	10.5	28.3	–	68615.00
PSO	325.3	42.6	11.8	32.6	0.63	65416.66
FA	323.7	44.0	11.3	31.7	0.51	83152.17
ELNFA	303.1	42.0	10.6	33.6	0.59	69287.29

Table 2. Index comparison.

	W_h/t	R	P/W	T_ϕ/s	T_θ/s
Initial value	22109.3	0.014905	223698.5	11.70	8.87
PSO	27450.3	0.015208	216691.9	11.93	9.45
FA	27534.2	0.015212	254272.5	12.31	8.86
ELNFA	24076.7	0.015025	225157.3	12.60	9.04

And compared with the standard firefly algorithm, the improved firefly algorithm has also improved the convergence speed. The improved firefly algorithm converges after about 10 iterations, while the standard firefly algorithm converges after about 150 iterations.

6   Conclusion and Future Work

In this paper, aiming at the optimization of the principal dimensions of large ships, we select empty ship mass, effective power, collision rate, rolling period and pitching period as evaluation indicators to construct a mathematical model. An improved firefly algorithm is proposed to solve the mathematical model. Firstly, we propose a multi-group elimination mechanism, which divides fireflies into elite population and candidate populations, and implements a variable step strategy between populations that can strengthen the influence of elite individuals on the convergence speed; then the Levy flight rule is introduced in. Inspired by the cuckoo algorithm, we improve the firefly's position update formula to prevent the search from stagnating in the later stages of the iteration; after that, the negative reinforcement strategy is added. Finally, we apply the improved firefly algorithm, standard firefly algorithm, and particle swarm algorithm to the calculation example. The results prove that the improved firefly algorithm is better than the standard firefly algorithm and particle swarm algorithm, and in the optimization of the principal dimensions of large ships, the effective power, rolling period, and pitching period have all been improved to a certain extent. The next step is to expand the application fields of the ELNFA algorithm, and continuously improve the convergence and efficiency in the application.

Acknowledgement. The research was supported by the National Natural Science Foundation of China under grant number 61902116.

References

1. Wang, W.Q., Huang, S., Hou, Y.H., Hu, Y.L.: Optimal principal parameters for large naval ship based on modified artificial bee colony algorithm. J. Wuhan Univ. Technol. **35**(6), 58–62 (2012)
2. Hou, L.: Application of multi-objective particle swarm optimization (MOPSO) in study of ship's principal parameters. J. Ship Mech. **15**(7), 784–790 (2011)
3. Chen, X.Q., Tan, J.H.: Principal dimensions optimization of 300000 DW T FPSO based genetic algorithm. J. Wuhan Univ. Technol. (Transp. Sci. Eng.) **33**(3), 426–429 (2009)
4. Zhuo, H.M., Chen, Q.Q.: Optimization of principal dimensions based on uniform orthogonal firefly algorithm for mining ships. Ship Ocean Eng. **49**(5), 82–86 (2020)
5. Yang, X.S.: Firefly algorithms for multimodal optimization. In: Watanabe, O., Zeugmann, T. (eds.) SAGA 2009. LNCS, vol. 5792, pp. 169–178. Springer, Heidelberg (2009). https://doi.org/10.1007/978-3-642-04944-6_14
6. Gandomi, A.H., Yang, X.-S., Talatahari, S., Alavi, A.H.: Firefly algorithm with chaos. Commun. Nonlinear Sci. Numer. Simul. **18**(1), 89–98 (2013)
7. Wang, C.F., Song, W.X.: A novel firefly algorithm based on gender difference and its convergence. Appl. Soft Comput. **80**, 107–124 (2019)
8. Xu, C., Meng, H., Wang, Y.: A novel hybrid firefly algorithm based on the vector angle learning mechanism. IEEE Access **8**, 205741–205754 (2020). https://doi.org/10.1109/ACCESS.2020.3037802
9. Wang, W.Q., Wang, C., Huang, S., Hou, Y.H., Hu, Y.L.: Scheme generation and optimal selection of principal dimensions for large naval ships. J. Shanghai Jiaotong Univ. (Chin. Ed.) **47**(6), 916–922 (2013)
10. Guan, Z.J.: Elementary analysis of marine traffic accidents. J. Dalian Marit. Univ. **23**(1), 48–53 (1997)
11. Liu, Z., Lu, H.J., Ren, J.C.: A diversity-enhanced hybrid firefly algorithm. J. Shanxi Univ. (Nat. Sci. Ed.). **44**(2), 249–256 (2021)
12. Selvakumar, A.I., Thanushkodi, K.: A new particle swarm optimization solution to nonconvex economic dispatch problems. IEEE Trans. Power Syst. **22**(1), 42–51 (2007). https://doi.org/10.1109/TPWRS.2006.889132

Improved Butterfly Optimization Algorithm Fused with Beetle Antennae Search

Jianghao Yin and Na Deng[⊠]

School of Computer Science, Hubei University of Technology, Wuhan, China

Abstract. Aiming at the problems that the butterfly optimization algorithm has low optimization accuracy and easy to fall into the local optimal solution, this paper proposes an improved butterfly algorithm fused with the beetle search (BAS-LBOA). Firstly, chaotic map initialization is used to replace random initialization, so that the butterfly individuals are more evenly distributed in the solution space. At the same time the chaotic mapping function is used to replace the random function to generate the parameters that control the global and local search. Then the Levy flight rule is added which can prevent the search from falling into the latter part of the iteration. Afterwards, inspired by the beetle search algorithm, the beetle antennae search is used to improve the random search of the butterfly optimization algorithm. Finally, the BAS-LBOA algorithm is compared with multiple intelligent algorithms, and the results prove that the BAS-LBOA algorithm has better optimization performance.

1 Introduction

The swarm intelligence optimization algorithm is a random search evolution algorithm based on probability, which mainly simulates the group behaviors of biological foraging, cooperation and competition in nature [1]. Because of its robustness, scalability, and ease of implementation, swarm intelligence optimization algorithms are widely used in PID parameter tuning, image threshold segmentation, path planning and other issues. In recent years, many swarm intelligence optimization algorithms have been proposed, such as particle swarm algorithm (PSO), differential evolution algorithm (DE), firefly algorithm (FA), artificial bee colony algorithm (ABC), gray wolf optimization algorithm (GWO) and so on.

The butterfly optimization algorithm (BOA) is a new meta-heuristic intelligent optimization algorithm proposed by Arora et al. [2] in 2018. This algorithm simulates the foraging behavior of butterflies to solve the global optimization problem, and has been used in feature selection and image threshold segmentation. But the butterfly optimization algorithm also has the problem that the optimization accuracy is not high and it is easy to fall into the local optimal solution. For this reason, many domestic and foreign scholars have proposed lots of improvement methods. Literature [3] proposes an MPBOA, which combine BOA with mutualism and parasitism phases of the symbiosis organisms search algorithm to enhance BOA's global and local search capabilities; literature [4] embed the bidirectional search into the butterfly optimization algorithm, and makes greedy selection when selecting the direction, which speeds up the convergence

L. Barolli et al. (Eds.): EIDWT 2022, LNDECT 118, pp. 335–345, 2022.
https://doi.org/10.1007/978-3-030-95903-6_36

speed of the butterfly optimization algorithm; literature [5] adds reverse guidance in the global search to broaden the search space, and at the same time introduces the neighborhood search weight factor, which increases the convergence speed of the algorithm, then introduces an information sharing mechanism for problems that are easy to fall into local optimal solutions. The algorithm has good performance in terms of convergence speed, convergence accuracy and robustness; literature [6] introduces the cross entropy method into the butterfly optimization algorithm, which avoids falling into the local optimal solution and enhances the global search ability; literature [7] uses cubic mapping to initialize the butterfly population, then uses a nonlinear control strategy to improve the power exponent, and finally mixes the particle swarm algorithm with the butterfly optimization algorithm to improve the algorithm's global search ability; literature [8] combines BOA with particle swarm algorithm, and introduces adaptive inertial weights and adversarial learning average elite strategy, which makes the algorithm converge faster and easily jump out of the local optimal solution.

The beetle antennae search algorithm (BAS) is a new intelligent algorithm proposed in 2017 [9]. It is an algorithm developed by the characteristics of beetle foraging. Specifically, the beetle can use their two antennae to perceive the strength of the smell when foraging. When the odor intensity perceived by the left antennae is greater than that of the right antennae, the longhorn beetle searches for food to the left, and vice versa. At present, the BAS has been applied to practical problems such as community detection, path planning, anchor node layout and so on. Because the BAS has only one individual, it increases the possibility of falling into a local optimal solution when searching for high-dimensional functions. Literature [10] adds inertia weight to BAS, which makes the algorithm perform a global search in the early stage and a local search in the later stage. This strategy improves the search accuracy of the algorithm; literature [11] combines the beetle antennae algorithm with the artificial bee colony algorithm, and proposes a BAS-ABC algorithm, which avoids randomness and improves the convergence speed of the ABC algorithm; literature [12] combines the beetle antennae search algorithm with the genetic algorithm, introduces the search characteristics of the beetle algorithm into the genetic algorithm, and adopts the strategy of dynamically adjusting the mutation probability, which improves the local search ability of the algorithm. It has good performance in multi-task assignment of multiple UAVs.

Aiming at the problems of the butterfly algorithm's low optimization accuracy and easy to fall into local optimal solutions, this paper proposes an improved butterfly optimization algorithm fused with beetle antennae search. The rest of this paper is organized as follows: Sect. 2 introduces the standard butterfly optimization algorithm. Section 3 introduces the improvement strategy of this paper and proposes a butterfly optimization algorithm that integrates the beetle search (BAS-LBOA). Section 4 compares the improved algorithm with the standard butterfly optimization algorithm, particle swarm algorithm, and differential evolution algorithm to prove the superiority of the improved algorithm by 5 standard functions. Section 5 summarizes the full paper and proposes future work.

2 Standard Butterfly Optimization Algorithm

Butterfly optimization algorithm is a kind of bionic intelligent algorithm that imitates the foraging behavior of butterflies in nature. In the butterfly optimization algorithm, each butterfly represents a solution in the feasible domain, and the fitness of the butterfly is related to the position of the butterfly in the solution space. The communication between butterflies relies on the fragrance emitted by each individual butterfly, and the fragrance in the air can attract the butterflies to each other. Each butterfly will move randomly or towards the butterfly that emits more fragrance. There are two types of butterfly movement in the search process. The first is that when a butterfly perceives the scent of other butterflies, it will move towards the butterfly with the largest scent. This process is called global search. The second is that when a butterfly cannot perceive the scent of other butterflies, it will move randomly. This process is called a local search. The switching of the two search modes is realized by the conversion probability p.

The formula for calculating the scent of butterflies is as follows:

$$f = cI^a \tag{1}$$

where f is the size of the fragrance, I is the stimulus intensity, α is the power exponent, c is the perception form, the calculation method is as follows:

$$c^{t+1} = c^t + \left(b/\left(c^t \cdot \text{MaxIter}\right)\right) \tag{2}$$

where c^t and c^{t+1} are the perception forms in the t-th and $t + 1$-th iterations, b is a constant with a value of 0.025, and MaxIter is the maximum number of iterations.

The location update formula is as follows:

$$X_i^{t+1} = \begin{cases} X_i^t + \left(r^2 \cdot g^* - X_i^t\right) \cdot f_i, r < p \\ X_i^t + \left(r^2 \cdot X_j^t - X_k^t\right) \cdot f_i, r \geq p \end{cases} \tag{3}$$

where X_i^t and X_i^{t+1} represent the solution vector of the i-th butterfly in the t and $t + 1$ iterations, r is a random number between [0, 1], g^* is the optimal solution in the current iteration, f_i is the fragrance of the i-th butterfly, X_j^t and X_k^t respectively represent the solution vector of the i-th and k-th butterfly in the t-th iteration, r is a random number between [0, 1], when $r < p$, a global search is performed, and when $r \geq p$, a local search is performed.

3 Improved Butterfly Optimization Algorithm

3.1 Chaos Mapping Mechanism

The standard BOA algorithm is completely random when initializing individuals, and cannot guarantee that the initial population is evenly distributed in the solution space. Chaotic mapping has the unique properties of randomness, ergodicity, and regularity, so it is often used as a global optimization mechanism to avoid falling into local extrema during the search process in the optimization design field. This paper selects Logistic chaotic map instead of random population initialization. The general form of Logistic chaotic map is as follows:

$$\alpha_{t+1} = \alpha_t \times \sigma(1 - \alpha_t) \tag{4}$$

where σ is the Logistic chaotic mapping function parameter, α_t is the function value of the t-th iteration of the Logistic mapping function. Research shows that when the value of σ is 4, the chaotic effect is better.

In the standard BOA algorithm, the value of r and p is used to determine whether to perform a global search or a local search in each iteration. The value of r is generated randomly and has a large randomness. Using chaotic mapping function instead of random function to generate r value can effectively avoid the local optimal risk caused by random function [13], so this paper uses Logistic mapping function to generate r value.

3.2 Levy Flight Mechanism

Levy flight is characterized by long-term short-distance movement and occasional long-distance movement to ensure that the movement does not stay in a certain local area. Levy flight has been widely used in intelligent algorithms to increase the diversity of populations and expand search capabilities. Therefore, this paper introduces Levy flight into the butterfly optimization algorithm to enhance the search ability of each butterfly.

The global search formula after introducing Levy flight is as follows:

$$X_i^{t+1} = X_i^t + \left(r^2 \cdot g^* - X_i^t\right) \cdot f_i \oplus Levy(\lambda) \tag{5}$$

where $\oplus$ is the point-to-multiplication, $Levy(\lambda)$ is the random search path, which satisfies the Levy distribution and in this paper λ takes 1.5.

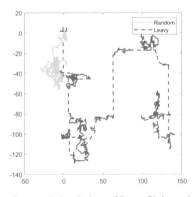

Fig. 1. Comparison and simulation of Levy flight and random walk.

Figure 1 simulates the Levy flight and the path of the random walk 1000 times. It can be seen that the Levy flight has a larger search range than the random walk.

3.3 Beetle Antennae Search Mechanism

The beetle antennae search algorithm has the characteristics of fast solving speed and high accuracy. The algorithm is inspired by the foraging characteristics of beetle. When the beetle is foraging for food, the beetle does not know the specific location of the food. It relies on two tentacles to detect the smell of the food and adjust its own direction. When the odor received by the right side is greater than the odor received by the left side, the beetle moves to the right, and vice versa.

The position of the long-horned beetle in the D-dimensional space is denoted as $X = (x_1, x_2, \ldots, x_n)$, and the calculation formula for the position of the beetle is as follows:

$$\begin{cases} X_r = X + l * \vec{d} \\ X_l = X - l * \vec{d} \end{cases} \tag{6}$$

where X_r is the position of the right antenna of the beetle, X_l is the position of the left antenna of the beetle, l is the distance between the centroid and the beetle, and $\vec{d}$ is a random unit vector, the calculation method is as follows:

$$\vec{d} = \frac{rands(D, 1)}{||rands(D, 1)||} \tag{7}$$

where D is the dimension of the solution space. According to the concentration of the two tentacles, the next position of beetle is:

$$X^t = X^{t-1} - \delta^t \times \vec{d} \times sign(f_l - f_r) \tag{8}$$

where δ^t is the step size of the beetle in the t-th iteration, f_l and f_r are the fitness of the left and right antennae, and $sign()$ is the sign function:

$$sign(x) = \begin{cases} 1, & \text{if } x > 0 \\ 0, & \text{if } x = 0 \\ -1, & \text{otherwise} \end{cases} \tag{9}$$

Different from the butterfly optimization algorithm and the particle swarm algorithm, there is only one individual in the beetle antennae search algorithm, which emphasizes the individual search strategy while ignoring the influence of the group. In the butterfly optimization algorithm, the two search modes are switched by the conversion probability p. In the global search process, the butterfly will move towards the individual with the largest fragrance, while in the local search, the butterfly moves randomly. The solution generated by the random movement often does not have better fitness. Therefore, this paper introduces the beetle search algorithm in the butterfly optimization algorithm to enhance the search ability of the algorithm. Specifically, beetle search is used instead of local search. The conversion probability p controls the switch between global search and beetle search. When r is less than p, the butterfly individual performs a global search, and when r is greater than p, the butterfly individual performs beetle search.

Based on the above improvement strategies, the improved butterfly optimization algorithm flow chart is shown in Fig. 2.

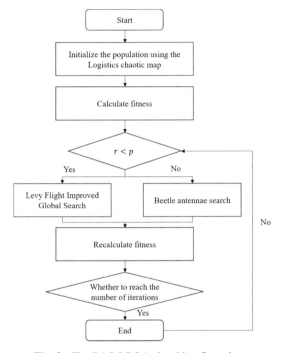

Fig. 2. The BAS-LBOA algorithm flow chart.

Step1: Logistic chaotic mapping initial population and parameter r;

Step2: Calculate population fitness;

Step3: When $r < p$, perform a global search that introduces Levy flight, otherwise perform the beetle search.

Step4: Recalculate fitness;

Step5: Judge whether it reaches the maximum number of iterations, if it reaches the maximum number of iterations, exit the iteration, otherwise skip to Step 3.

4   Experiment and Analysis

4.1   Standard Test Function

In order to test the optimization performance of the BAS-LBOA algorithm in this paper, select f_1: Sphere function, f_2: Schwefel 2.22 function, f_3: the second Schaffer function, f_4: Rastrigin function, f_5: Griewank function, where f_1 and f_2 are unimodal functions for testing the algorithm convergence speed and accuracy, f_3, f_4, f_5 are multimodal functions for testing the ability of the function to jump out of the local optimal solution. The theoretical optimal values of these 5 functions are all 0, and the specific details of the functions are shown in Table 1.

Table 1. Standard test function.

Function	Expression	Dim	Range	Min
f_1	$f_1(x) = \sum_{i=1}^{D} x_i^2$	30	$[-100, 100]$	0
f_2	$f_2(x) = \sum_{i=1}^{D} |x_i| + \prod_{i=1}^{n} |x_i|$	30	$[-10, 10]$	0
f_3	$f_3(x) = 0.5 + \dfrac{\sin^2\left(x_1^2 - x_2^2\right) - 0.5}{\left[1 + 0.001\left(x_1^2 + x_2^2\right)\right]^2}$	2	$[-100, 100]$	0
f_4	$f_4(x) = \sum_{i=1}^{D} \left[x_i^2 - 10\cos(2\pi x_i) + 10 \right]$	30	$[-5.12, 5.12]$	0
f_5	$f_{5(x)} = \sum_{i=1}^{d} \dfrac{x_i^2}{4000} - \prod_{i=1}^{d} \cos\left(\dfrac{x_i}{\sqrt{i}}\right) + 1$	30	$[-600, 600]$	0

4.2   Comparison of the BAS-LBOA and BOA Algorithm

First of all, this paper will compare the BAS-LBOA with the standard butterfly optimization algorithm to verify the effectiveness of the improved strategy. In the two algorithms, the number of iterations is set to 500, the number of butterflies is set to 30, the conversion probability p is 0.8, the power index α is 0.01, the perception form c is 0.1, and the operating environment is Matlab2018b under Windows 10, each test function uses the algorithm to run independently 20 times. The calculation results are shown in Table 2.

Table 2. The BAS-LBOA and BOA operation results.

Function	Algorithm	Max	Min	Mean	Variance
f_1	BOA	1.46E−11	1.19E−11	1.28E−11	5.56E−25
	BAS-LBOA	4.72E−23	2.84E−31	**3.76E−24**	**1.28E−46**
f_2	BOA	6.04E−09	9.98E−10	4.50E−09	1.96E−18
	BAS-LBOA	1.13E−13	2.36E−16	**3.33E−14**	**1.49E−27**
f_3	BOA	1.45E−09	1.80E−13	1.61E−10	1.22E−19
	BAS-LBOA	6.66E−16	0.00E+00	**4.44E−17**	**2.39E−32**
f_4	BOA	8.08E−05	0.00E+00	7.68E−06	4.43E−10
	BAS-LBOA	0.00E+00	0.00E+00	**0.00E+00**	**0.00E+00**
f_5	BOA	2.13E−10	1.16E−12	1.57E−11	2.16E−21
	BAS-LBOA	0.00E+00	0.00E+00	**0.00E+00**	**0.00E+00**

Among the five standard test functions, the optimization results and stability of the BAS-LBOA are better than those of the BOA. Compared with the original algorithm, the improved algorithm's optimization accuracy and the ability to jump out of the local optimal solution have been significantly improved. Select the best results in 20 experiments to make a convergence graph, as shown in Fig. 3.

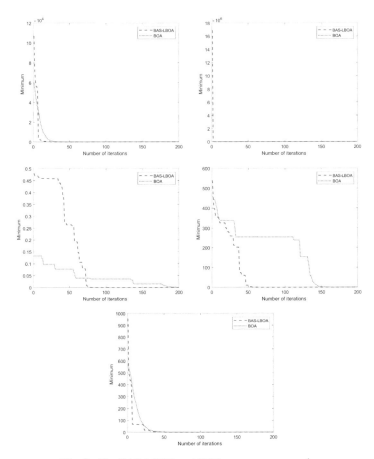

Fig. 3. The BAS-LBOA and BOA convergence graph.

The convergence speed of the BAS-LBOA algorithm in the five standard test functions is higher than that of the standard BOA algorithm. When the f_1, f_2, f_4, and f_5 functions are optimized, the improved algorithm reaches convergence in about 50 iterations.

4.3 Comparison of the BAS-LBOA with Other Intelligent Algorithms

In order to better evaluate the proposed the BAS-LBOA algorithm, this paper compares the improved algorithm with other intelligent algorithms. The algorithms involved in the comparison include the particle swarm algorithm and the differential evolution algorithm. The parameter settings in the BAS-LBOA algorithm are same as the Sect. 4.2 summary; in the particle swarm algorithm, the maximum number of iterations is set to 500, the number of particles is 30, and the individual learning factor c_1 is 2, the social learning factor c_2 is 2, the inertia weight w is 0.9, and the maximum velocity of the particle v_{max} is 6. In the differential evolution algorithm, the maximum number of iterations is set to 500, the number of individuals is set to 30, and the mutation operator F is 0.4, and the crossover operator CR is 0.1. Each test function uses each algorithm to run independently 20 times, and the calculation results are shown in Table 3.

Table 3. The BAS-LBOA and other intelligent algorithms operation results.

Function	Algorithm	Max	Min	Mean	Variance
f_1	PSO	7.68E+00	2.29E+00	4.47E+00	1.65E+00
	DE	1.95E−04	6.77E−05	1.15E−04	1.40E−09
	BAS-LBOA	4.72E−23	2.84E−31	**3.76E−24**	**1.28E−46**
f_2	PSO	2.84E+01	7.44E+00	1.33E+01	3.43E+01
	DE	2.37E−06	7.43E−07	1.28E−06	1.67E−13
	BAS-LBOA	1.13E−13	2.36E−16	**3.33E−14**	**1.49E−27**
f_3	PSO	1.06E−08	6.30E−12	3.05E−09	1.33E−17
	DE	2.26E−108	9.53E−114	**1.41E−109**	**2.55E−217**
	BAS-LBOA	6.66E−16	0.00E+00	4.44E−17	2.39E−32
f_4	PSO	1.82E+02	1.14E+01	9.73E+01	3.00E+03
	DE	4.88E−07	9.91E−08	2.94E−07	1.05E−14
	BAS-LBOA	0.00E+00	0.00E+00	**0.00E+00**	**0.00E+00**
f_5	PSO	3.94E−01	1.18E−01	2.29E−01	5.10E−03
	DE	7.50E−03	1.78E−07	3.70E−03	2.74E−06
	BAS-LBOA	0.00E+00	0.00E+00	**0.00E+00**	**0.00E+00**

It can be seen from the above table that in the tests of f_1, f_2, f_4, and f_5, the optimization ability and stability of the BAS-LBOA algorithm are better than those of PSO and DE algorithm. Although the average and variance of the optimization results of the DE algorithm are slightly better than those of the BAS-LBOA algorithm, in 20 independent experiments, 18 times of the BAS-LBOA algorithm's optimization results reached the optimal value of 0. The 20 optimization results of the differential evolution algorithm do not reach the optimal value.

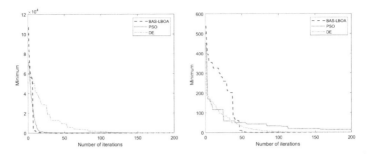

Fig. 4. The BAS-LBOA and other algorithms convergence graph.

This paper selects f_1 and f_4 to make the convergence images which are shown in Fig. 4. When the three algorithms optimize these two functions, the convergence speed of BAS-LBOA is the fastest, and the optimization accuracy is higher than that of PSO and DE.

5   Conclusion

This paper proposes an improved butterfly algorithm that integrates the search of beetles. First, the logistic chaotic map initialization is used to replace random initialization to make the butterfly individuals more uniform in the solution space. At the same time, the chaotic map function is used to replace the random function to generate control global and local search parameters, this strategy can avoid falling into the local optimal solution. Then the Levy flight rule is introduced in to improve the position update formula which can prevent the search from stagnation in the later iterations. Afterwards, inspired by the beetle antennae search algorithm, the beetle antennae search is used to improve the random search of the butterfly optimization algorithm and enhance the search ability of the algorithm. Finally, five functions, including Sphere function, Schwefel 2.22 function, the second Schaffer function, Rastrigin function, and Griewank function, are selected as the test functions. This paper selects the PSO algorithm and the DE algorithm to compare with the BAS-LBOA algorithm. Experimental results show that the BAS-LBOA algorithm has higher optimization accuracy and the ability to jump out of local optimal solutions than the BOA algorithm, PSO algorithm, and DE algorithm. In the future, we will continue to improve the optimization accuracy of the algorithm in this paper, and apply the algorithm to the optimization problem of industrial production.

Acknowledgement. The research was supported by the National Natural Science Foundation of China under grant number 61902116.

References

1. Li, Y.L., Wang, S.Q., Chen, Q.R., Wang, X.G.: Comparative study of several new swarm intelligence optimization algorithms. Comput. Eng. Appl. **56**(22), 1–12 (2020)

2. Arora, S., Singh, S.: Butterfly optimization algorithm: a novel approach for global optimiza-tion. Soft. Comput. **23**(3), 715–734 (2018). https://doi.org/10.1007/s00500-018-3102-4
3. Sharma, S., Saha, A.K., Majumder, A., Nama, S.: MPBOA - a novel hybrid butterfly opti-mization algorithm with symbiosis organisms search for global optimization and image seg-mentation. Multimedia Tools Appl. **80**(8), 12035–12076 (2021). https://doi.org/10.1007/s11 042-020-10053-x
4. Sharma, T.K.: Enhanced butterfly optimization algorithm for reliability optimization prob-lems. J. Ambient. Intell. Humaniz. Comput. **12**(7), 7595–7619 (2020). https://doi.org/10. 1007/s12652-020-02481-2
5. Luo, J., Tian, Q., Xu, M.: Reverse guidance butterfly optimization algorithm integrated with information cross-sharing. J. Intell. Fuzzy Syst. **41**(2), 3463–3484 (2021)
6. Li, G., Shuang, F., Zhao, P., et al.: An improved butterfly optimization algorithm for engineering design problems using the cross-entropy method. Symmetry **11**(8), 1049 (2019)
7. Zhang, M., Long, D., Qin, T., Yang, J.: A chaotic hybrid butterfly optimization algorithm with particle swarm optimization for high-dimensional optimization problems. Symmetry **12**(11), 1800 (2020)
8. Tong, L., Dong, M., Ai, B., et al.: A simple butterfly particle swarm optimization algorithm with the fitness-based adaptive inertia weight and the opposition-based learning average elite strategy. Fund. Inform. **163**(2), 205–223 (2018)
9. Liao, L.F., Yang, H.: Review of beetle antennae search. Comput. Eng. Appl. **57**(12), 54–64 (2021)
10. Lin, M., Li, Q., Wang, F., Chen, D.: An improved beetle antennae search algorithm and its application on economic load distribution of power system. IEEE Access **8**, 99624–99632 (2020)
11. Cheng, L., Yu, M., Yang, J., Wang, Y.: An improved artificial bee colony algorithm based on beetle antennae search. In: 2019 Chinese Control Conference (CCC), pp. 2312–2316 (2019)
12. Wang, Z., Wang, B., Wei, Y., Liu, P., Zhang, L.: Cooperative multi-task assignment of multiple UAVs with improved genetic algorithm based on beetle antennae search. In: 2020 39th Chinese Control Conference (CCC), pp. 1605–1610 (2009)
13. Li, T.L., Liu, F.A.: WSN butterfly localization optimization with chaotic mapping in wireless sensor networks. Comput. Eng. Des. **40**(06), 1729–1733 (2019)

A Delaunay Edge and CCM-Based SA Approach for Mesh Router Placement Optimization in WMN: a Case Study for Evacuation Area in Okayama City

Aoto Hirata[1], Tetsuya Oda[2(✉)], Nobuki Saito[1], Tomoya Yasunaga[2], Kengo Katayama[2], and Leonard Barolli[3]

[1] Graduate School of Engineering, Okayama University of Science (OUS), Okayama, 1-1 Ridaicho, Kita-ku, Okayama 700–0005, Japan
{t21jm02zr,t21jm01md}@ous.jp

[2] Department of Information and Computer Engineering, Okayama University of Science (OUS), 1-1 Ridaicho, Kita-ku, Okayama 700–0005, Japan
{oda,katayama}@ice.ous.ac.jp, t18j091yt@ous.jp

[3] Department of Information and Communication Engineering, Fukuoka Institute of Technology, 3-30-1 Wajiro-Higashi, Higashi-Ku, Fukuoka 811-0295, Japan
barolli@fit.ac.jp

Abstract. The Wireless Mesh Networks (WMNs) enables routers to communicate with each other wirelessly in order to create a stable network over a wide area at a low cost and it has attracted much attention in recent years. There are different methods for optimizing the placement of mesh routers. In our previous work, we proposed a Coverage Construction Method (CCM), CCM-based Hill Climbing (HC) and CCM-based Simulated Annealing (SA) system for mesh router placement problem considering normal and uniform distributions of mesh clients. In this paper, we propose a Delaunay edge and CCM-based Simulated Annealing (SA) approach for mesh router placement problem. The proposed method focuses on realistic mesh client placement rather than randomly generated mesh clients with normal or uniform distributions. In the case study, we chose evacuation areas in Okayama City, Japan, as the target to be covered by mesh routers. From the simulation results, we found that the proposed method was able to cover the evacuation area.

1 Introduction

The Wireless Mesh Networks (WMNs) [1–4] are wireless network technologies that enables routers to communicate with each other wirelessly to create a stable network over a wide area at a low cost and it has attracted much attention in recent years. The placement of the mesh routers has a significant impact on cost, communication range and operational complexity. Therefore, there are many research works to optimize the placement of these mesh routers. In our previous work [5–12], we proposed and evaluated different meta-heuristics such as Genetic Algorithms (GA) [13], Hill Climbing (HC) [14], Simulated Annealing (SA) [15], Tabu Search (TS) [16] and Particle Swarm Optimization (PSO) [17] for mesh router placement optimization. Also, we proposed a Coverage Construction Method (CCM) [18], CCM-based Hill Climbing (HC) [19] and CCM-based Simulated Annealing (SA) system. The CCM is able to rapidly create a group of mesh routers with the radio communication ranges of the mesh routers linked to each other. The CCM-based HC system covered many mesh clients generated by normal and uniform distributions. We also showed that in the two islands model, the CCM-based HC system was able to find two islands and covered many mesh clients [20]. The CCM-based HC system adapted to varying number of mesh clients, number of mesh routers and area size [21,22]. The CCM-based SA system was able to cover many mesh clients in normal distribution compared with CCM.

In this paper, we propose a Delaunay edge and CCM-based SA approach. As evaluation metrics, we consider the Size of Giant Component (SGC) and the Number of Covered Mesh Clients (NCMC).

The structure of the paper is as follows. In Sect. 2, we define the mesh router placement problem. In Sect. 3, we describe the proposed method. In Sect. 4, we present the case study. Finally, conclusions and future work are given in Sect. 5.

2   Mesh Router Placement Problem

We consider a two-dimensional continuous area to deploy a number of mesh routers and a number of mesh clients of fixed positions. The objective of the problem is to optimize a location assignment for the mesh routers to the two-dimensional continuous area that maximizes the network connectivity and mesh clients coverage. Network connectivity is measured by the SGC, while the NCMC is the number of mesh clients that is within the radio communication range of at least one mesh router. An instance of the problem consists as follows.

- An area $Width \times Height$ which is the considered area for mesh router placement. Positions of mesh routers are not pre-determined and are to be computed.
- The mesh router has radio communication range defining thus a vector of routers.
- The mesh clients are located in arbitrary points of the considered area defining a matrix of clients.

Algorithm 1. The method for randomly generating mesh routers.

Output: Placement list of mesh routers.
1: Set *Number of mesh routers*.
2: Generate mesh router [0] randomly in considered area.
3: $i \leftarrow 1$.
4: **while** $i < Number\ of\ mesh\ routers$ **do**
5:    Generate mesh router [i] randomly in considered area.
6:    **if** SGC is maximized **then**
7:       $i \leftarrow i + 1$.
8:    **else**
9:       Delete mesh router [i].
10:   **end if**
11: **end while**

Algorithm 2. Coverage construction method.

Input: Placement list of mesh clients.
Output: Placement list of best mesh routers.
1: Set *Number of loop for CCM*.
2: i, *Current NCMC*, *Best NCMC* $\leftarrow 0$.
3: *Current mesh routers* $\leftarrow$ Algorithm 1.
4: *Best mesh routers* $\leftarrow$ *Current mesh routers*.
5: **while** $i < Number\ of\ loop\ for\ CCM$ **do**
6:    *Current NCMC* $\leftarrow$ *NCMC of Current mesh routers*.
7:    **if** *Current NCMC* $>$ *Best NCMC* **then**
8:       *Best NCMC* $\leftarrow$ *Current NCMC*.
9:       *Best mesh routers* $\leftarrow$ *Current mesh routers*.
10:   **end if**
11:   $i \leftarrow i + 1$.
12:   *Current mesh routers* $\leftarrow$ Algorithm 1.
13:   *Current NCMC* $\leftarrow 0$.
14: **end while**

3 Proposed Method

In this section, we describe the proposed method. In Algorithm 1, Algorithm 2 and Algorithm 3 are shown pseudo codes of CCM and CCM-based SA.

3.1 CCM for Mesh Router Placement Optimization

In our previous work, we proposed a CCM [18] for mesh router placement optimization problem. The pseudo code of randomly generating mesh routers method used in the CCM is shown in Algorithm 1 and the pseudo code of the CCM is shown in Algorithm 2. The CCM searches the solution with maximized SGC. Among the solutions generated, the mesh router placement with the highest NCMC is the final solution.

We describe the operation of the CCM in follow. First, the mesh clients are generated in the considered area. Next, randomly is determined a single point coordinate to be mesh router 1. Once again, randomly determine a single point coordinate to be mesh router 2. Each mesh router has a radio communication range. If the radio communication ranges of the two routers do not overlap, delete router 2 and randomly determine a single point coordinate and make it as mesh router 2. This process is repeated until the radio communication ranges of two mesh routers overlaps. If the radio communication ranges of the two mesh routers overlap, generate next mesh routers. If there is no overlap in radio communication range with any mesh router, the mesh router is removed and generated randomly again. If any of the other mesh routers have overlapping radio communication ranges, generate next mesh routers. Continue this process until the setting number of mesh routers.

By this procedure is created a group of mesh routers connected together without the derivation of connected component using Depth First Search (DFS) [23]. However, this method only creates a population of mesh routers at a considered area, but does not take into account the location of mesh clients. So, the procedure should be repeated for a setting number of loops. Then, determine how many mesh clients are included in the radio communication range group of the mesh router. The placement of the mesh router with the highest number of mesh clients covered during the iterative process is the solution of the CCM.

3.2   CCM-Based SA for Mesh Router Placement Optimization

In this subsection, we describe the CCM-based SA. The pseudo code of the CCM-based SA for mesh router placement problem is shown in Algorithm 3. The SA is one of the local search algorithms, which is inspired by the cooling process of metals. In SA, local solutions are derived by transitioning states and repeating the search for neighboring solutions. SA also transitions states, according to the decided state transition probability if the current solution is worse than the previous one. SA requires solution evaluation and temperature values to decide state transition probabilities. The evaluation of placement ($Eval$), the temperature (T) and the state transition probability (STP) in the proposed method are shown in Eq. (1), Eq. (2) and Eq. (3).

$$Eval \leftarrow 10 \times$$
$$(NCMC \ with \ the \ best \ results \ so \ far - NCMC \ of \ current \ solution) \quad (1)$$

$$T \leftarrow Initial \ Temperature +$$
$$(Final \ Temperature - Initial \ Temperature) \times \frac{Current \ number \ of \ loops}{Number \ of \ loops \ for \ SA} \quad (2)$$

$$STP \leftarrow e^{-\frac{Eval}{T}} \quad (3)$$

Proposed method performs neighborhood search by changing the placement of one mesh router. The solution is basically updated when the SGC is maximized and the NCMC is larger than the previous one. In SA, the solution is also

---

Algorithm 3. CCM-based SA.

Input: Placement list of mesh clients.
Output: Placement list of best mesh routers.
1: Set $Number\ of\ loop\ for\ SA,\ Initial\ Temperature,\ Final\ Temperature.$
2: $Current\ number\ of\ loop \leftarrow 0.$
3: $Current\ mesh\ routers,\ Best\ mesh\ routers \leftarrow$ Algorithm 2 ($Placement\ list\ of\ mesh\ clients$).
4: $Current\ NCMC,\ Best\ NCMC \leftarrow NCMC\ of\ Current\ mesh\ routers.$
5: **while** $Current\ number\ of\ loop < Number\ of\ loop\ for\ SA$ **do**
6:     Randomly choose an index of $Current\ mesh\ routers.$
7:     Randomly change coordinate of $Current\ mesh\ router\ [choosed\ index].$
8:     $Current\ NCMC \leftarrow NCMC\ of\ Current\ mesh\ router.$
9:     **if** SGC is maximized **then**
10:         $r \leftarrow$ Randomly generate in (0.0, 100.0).
11:         $Eval \leftarrow 10 \times (Best\ NCMC - Current\ NCMC)$
12:         $T \leftarrow Initial\ Temperature + (Final\ Temperature - Initial\ Temperature) \times \frac{Current\ number\ of\ loops}{Number\ of\ loops\ for\ SA}$
13:         **if** $e^{-\frac{Eval}{T}} \geq 1.0$ **then**
14:             $Best\ NCMC \leftarrow Current\ NCMC.$
15:             $Best\ mesh\ routers \leftarrow Current\ mesh\ routers.$
16:         **else if** $e^{-\frac{Eval}{T}} > r$ **then**
17:             $Best\ NCMC \leftarrow Current\ NCMC.$
18:             $Best\ mesh\ routers \leftarrow Current\ mesh\ routers.$
19:         **else**
20:             Restore coordinate of $Current\ mesh\ routers\ [choosed\ index].$
21:         **end if**
22:     **else**
23:         Restore coordinate of $Current\ mesh\ routers\ [choosed\ index].$
24:     **end if**
25:     $Current\ number\ of\ loops \leftarrow Current\ number\ of\ loops + 1.$
26: **end while**

---

updated depending on the STP when the SGC is maximized but the NCMC is decreased. In the proposed method, the solution is updated with a probability of STP [%] in that case.

We describe the operation of the proposed method in following. First, we randomly select one of the mesh routers in the group of mesh routers as the initial solution obtained by the CCM and change the placement of the chosen mesh router randomly. Then, we decide the NCMC for all mesh routers and the SGC. The SGC is derived by creating an adjacency list of mesh routers and using DFS to confirm which mesh routers are connected to which mesh routers. If the SGC is maximized and the NCMC is greater than that of previous one, then the changed placement of mesh routers is the current solution. If the SGC is maximized the NCMC is less than that of previous NCMC, restore the placement of changed mesh router. But in this case, the changed placement of mesh routers is the current solution with a probability of STP [%]. This process is repeated until the setting number of loops.

3.3 Delaunay Edge and CCM-Based SA

In this subsection, we describe the proposed method, Delaunay edge and CCM-based SA. In the previous methods, the purpose of simulations was to cover mesh clients that were randomly generated based on normal or uniform distributions. The proposed method focuses on a more realistic scenario. In the proposed method, Voronoi decomposition is performed to divide the regions where mesh clients in the problem area are close to each other before performing the CCM. Each region obtained by Voronoi decomposition is called a Voronoi cell and the mesh clients of each Voronoi cell are connected by lines based on the adjacency of these Voronoi cells. This line is called a Delaunay edge. This Delaunay edge is used for the CCM and the raster coordinates of Delaunay edges are derived and listed. In the previous CCM, when mesh routers were randomly placed, the placement area was the entire problem area. In the proposed method, the placement area of the CCM is the area defind by the coordinates of the listed Delaunay edges. By continuing to place a group of mesh routers on Delaunay edges, the probability of finding a mesh client at a distant location is increased.

Table 1. Parameters and value for simulations.

Width of considered area	260
Hight of considered area	180
Number of mesh routers	256
Radius of radio communication range of mesh routers	4
Number of mesh clients	3089
Number of loop for CCM	3000
Number of loop for SA	100000
Initial tempreature	100
Final tempreature	1

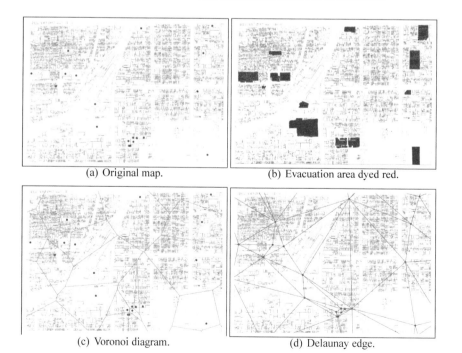

(a) Original map.   (b) Evacuation area dyed red.

(c) Voronoi diagram.   (d) Delaunay edge.

Fig. 1. Visualizing the problem area.

Table 2. Simulation results of Delaunay edge based CCM and Delaunay edge and CCM-based SA.

Method	Best SGC	Average SGC	Best NCMC	Average NCMC [%]
Delaunay Edge based CCM	256	256	2118	64.788
Delaunay Edge and CCM-based SA	256	256	3089	95.161

4   Case Study

4.1   Simulation Setting

The parameters used for simulation are shown in Table 1. We deployed mesh clients based on geographic information. The placement area is around Okayama Station in Okayama City, Okayama Prefecture, Japan. The mesh clients are the buildings used as evacuation area. We used the GIS application, QGIS, to display the geographic information. We also used the shapefiles of buildings from Open Street Map and the information of evacuation areas from the open data released by Okayama City [24]. The original map image is shown in Fig. 1(a). The red points in Fig. 1(a) indicate the points where evacuation areas are located. Figure 1(b) shows buildings designated as evacuation areas painted in red, and these red buildings are considered mesh clients. Figure 1(c) and Fig. 1(d) show the Voronoi diagram and the derived Delaunay edge. The Delaunay edges used in the proposed system are converted into information that can be used in the proposed system by extracting color pixels from images. Figure 2(b) shows the converted evacuation area and Fig. 2(b) shows the converted Delaunay edge. The Delaunay edge in Fig. 2(b) becomes placement area of the CCM in this simulation. We performed the simulations 100 times.

4.2   Simulation Results

The simulation results are shown in Table 2. We also show the simulation results of the best SGC, avg. SGC, the best NCMC and avg. NCMC. In the simulation, the SGC is maximized and the proposed method is able to cover most mesh clients. The visualization results of proposed method are shown in Fig. 3. The Delaunay edge and CCM-based SA can cover all mesh clients. We can see that the Delaunay edge based CCM was able to find distant mesh clients and Delaunay edge and CCM-based SA was able to find the remaining mesh clients.

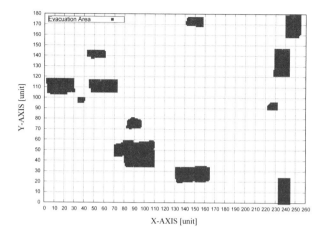

(a) Converted evacuation area.

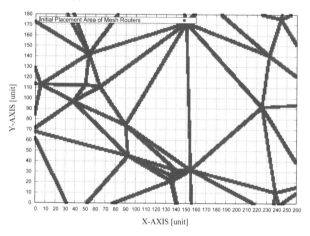

(b) Converted Delaunay edge.

Fig. 2. Converted information for the proposed system.

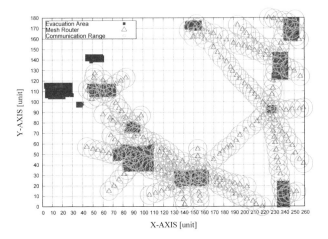

(a) Result of Delaunay edge based CCM.

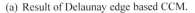

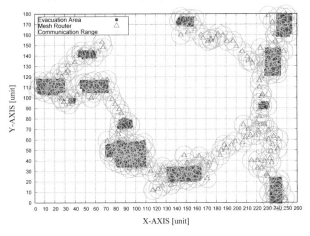

(b) Result of Delaunay edge and CCM-based SA.

Fig. 3. Visualization results.

5   Conclusions

In this paper, we proposed a Delaunay edge and CCM-based SA approach for mesh router placement problem in WMNs. From the simulation results, we found that the proposed method covered more mesh clients compared with the CCM. In the future, we would like to consider other local search algorithms and Genetic Algorithms.

Acknowledgement. This work was supported by JSPS KAKENHI Grant Number JP20K19793 and Grant for Promotion of OUS Research Project (OUS-RP-20-3).

References

1. Akyildiz, I.F., et al.: Wireless mesh networks: a survey. Comput. Netw. **47**(4), 445–487 (2005)
2. Oda, T., et. al.: Implementation and experimental results of a WMN testbed in indoor environment considering LoS scenario. In: Proceedings of The IEEE 29th International Conference on Advanced Information Networking and Applications (IEEE AINA-2015), pp.37–42. IEEE (2015)
3. Jun, J., et al.: The nominal capacity of wireless mesh networks. IEEE Wirel. Commun. **10**(5), 8–15 (2003)
4. Oyman, O., et al.: Multihop relaying for broadband wireless mesh networks: from theory to practice. IEEE Commun. Mag. **45**(11), 116–122 (2007)
5. Oda, T., et al.: Evaluation of WMN-GA for different mutation operators. Int. J. Space-Based Situated Comput. **2**(3), 149–157 (2012)
6. Oda, T., et al.: WMN-GA: a simulation system for WMNs and its evaluation considering selection operators. J. Ambient. Intell. Humaniz. Comput. **4**(3), 323–330 (2013). https://doi.org/10.1007/s12652-011-0099-2
7. Ikeda, M., et. al.: Analysis of WMN-GA simulation results: WMN performance considering stationary and mobile scenarios. In: Proceedings of The 28th IEEE International Conference on Advanced Information Networking and Applications (IEEE AINA-2014), pp. 337–342. IEEE (2014)
8. Oda, T., et. al.: Analysis of mesh router placement in wireless mesh networks using Friedman test. In: Proceedings of The IEEE 28th International Conference on Advanced Information Networking and Applications (IEEE AINA-2014), pp. 289–296. IEEE (2014)
9. Oda, T., et al.: Effect of different grid shapes in wireless mesh network-genetic algorithm system. Int. J. Web Grid Serv. **10**(4), 371–395 (2014)
10. Oda, T., et al.: Analysis of mesh router placement in wireless mesh networks using Friedman test considering different meta-heuristics. Int. J. Commun. Netw. Distrib. Syst. **15**(1), 84–106 (2015)
11. Oda, T., et al.: A genetic algorithm-based system for wireless mesh networks: analysis of system data considering different routing protocols and architectures. Soft. Comput. **20**(7), 2627–2640 (2016). https://doi.org/10.1007/s00500-015-1663-z
12. Sakamoto, S., Ozera, K., Oda, T., Ikeda, M., Barolli, L.: Performance evaluation of intelligent hybrid systems for node placement in wireless mesh networks: a comparison study of WMN-PSOHC and WMN-PSOSA. In: Barolli, L., Enokido, T. (eds.) IMIS 2017. AISC, vol. 612, pp. 16–26. Springer, Cham (2018). https://doi.org/10.1007/978-3-319-61542-4_2
13. Holland, J.H.: Genetic algorithms. Sci. Am. **267**(1), 66–73 (1992)
14. Skalak, D.B.: Prototype and feature selection by sampling and random mutation hill climbing algorithms. In: Proceedings of The 11th International Conference on Machine Learning (ICML-1994), pp. 293–301 (1994)
15. Kirkpatrick, S., et al.: Optimization by simulated annealing. Science **220**(4598), 671–680 (1983)
16. Glover, F.: Tabu search: a tutorial. Interfaces **20**(4), 74–94 (1990)

17. Kennedy, J., Eberhart, R.: Particle swarm optimization. In: Proceedings of The IEEE International Conference on Neural Networks (ICNN-1995), pp. 1942–1948 (1995)

18. Hirata, A., et. al.: Approach of a solution construction method for mesh router placement optimization problem. In: Proceedings of The IEEE 9th Global Conference on Consumer Electronics (IEEE GCCE-2020), pp. 467–468. IEEE (2020)

19. Hirata, A., Oda, T., Saito, N., Hirota, M., Katayama, K.: A coverage construction method based hill climbing approach for mesh router placement optimization. In: Barolli, L., Takizawa, M., Enokido, T., Chen, H.-C., Matsuo, K. (eds.) BWCCA 2020. LNNS, vol. 159, pp. 355–364. Springer, Cham (2021). https://doi.org/10.1007/978-3-030-61108-8_35

20. Hirata, A., et. al.: Simulation results of CCM based HC for mesh router placement optimization considering two Islands model of mesh clients distributions. In: Proceedings of The 9th International Conference on Emerging Internet, Data and Web Technologies (EIDWT-2021), pp. 180–188 (2021)

21. Hirata, A., et al.: A coverage construction and hill climbing approach for mesh router placement optimization: simulation results for different number of mesh routers and instances considering normal distribution of mesh clients. In: Barolli, L., Yim, K., Enokido, T. (eds.) CISIS 2021. LNNS, vol. 278, pp. 161–171. Springer, Cham (2021). https://doi.org/10.1007/978-3-030-79725-6_16

22. Hirata, A., et al.: A CCM-based HC system for mesh router placement optimization: a comparison study for different instances considering normal and uniform distributions of mesh clients. In: Barolli, L., Chen, H.-C., Enokido, T. (eds.) NBiS 2021. LNNS, vol. 313, pp. 329–340. Springer, Cham (2022). https://doi.org/10.1007/978-3-030-84913-9_33

23. Tarjan, R.: Depth-first search and linear graph algorithms. SIAM J. Comput. $\mathbf{1}(2)$, 146–160 (1972)

24. Integrated GIS for all of Okayama Prefecture. http://www.gis.pref.okayama.jp/pref-okayama/OpenData. 16 Nov 2021

FPGA Implementation of a Interval Type-2 Fuzzy Inference for Quadrotor Attitude Control

Tomoaki Matusi[1], Tetsuya Oda[1(✉)], Chihiro Yukawa[1], Tomoya Yasunaga[1], Nobuki Saito[2], Aoto Hirata[2], and Leonard Barolli[3]

[1] Department of Information and Computer Engineering, Okayama University of Science (OUS), 1-1 Ridaicho, Kita-ku, Okayama 700–0005, Japan
{t19j077mt,t18j097yc,t18j091yt}@ous.jp, oda@ice.ous.ac.jp
[2] Graduate School of Engineering, Okayama University of Science (OUS), Okayama, 1-1 Ridaicho, Kita-ku, Okayama 700–0005, Japan
{t21jm01md,t21jm02zr}@ous.jp
[3] Department of Information and Communication Engineering, Fukuoka Insitute of Technology, 3-30-1 Wajiro-Higashi-ku, Fukuoka 811-0295, Japan
barolli@fit.ac.jp

Abstract. In this paper, we propose an FPGA implementation of Interval Type-2 Fuzzy Inference (IT2FI) for attitude control of quadrotors. Since IT2FI is suitable for parallel processing, the processing time can be significantly reduced even on low-power FPGA devices. In the proposed system, the IT2FI module is applied to each of the roll, pitch and yaw axis, and the operation amount of the quadrotor motor is obtained from each output value.

1 Introduction

The Quadrotor is expected to be used in different fields, such as surveillance, surveying, military, disaster relief and agriculture. A Quadrotor attitude control requires stabilization of the tilt and rotation speed of the roll, pitch and yaw axis [1–6]. However, quadrotors have nonlinear dynamic behavior, so the design of attitude control using an rigorous mathematical model is complicated. Therefore, the attitude control of quadrotors is often based on linear control such as Proportional-Integral-Differential (PID) control [7] and Linear-Quadratic-Regulator (LQR) control [8]. The linear control is relatively simple but limitations in handling the uncertainty of nonlinear systems. On the other hand, the fuzzy inference based control is also used for attitude control of quadrotors [9–11]. A fuzzy inference based controller can control a time-varying nonlinear system empirically and intuitively. In particular, a Interval Type-2 Fuzzy Inference (IT2FI) based control is suitable for handling uncertainties such as disturbances and noise. In fact, IT2FI has better control results than conventional Type-1 Fuzzy Inference (T1FI) in different fields [12–16]. However, IT2FI has a disadvantage because requires more processing time than T1FI and other conventional control methods. Therefore, we implement the control fuzzy inference to adjust the attitude angle of each axis in Field-Programmable Gate Array (FPGA) [17–20].

The Advantage of FPGA implementation is reduction of processing time and power consumption. For quadrotors, where power consumption is directly related to

L. Barolli et al. (Eds.): EIDWT 2022, LNDECT 118, pp. 357–365, 2022.
https://doi.org/10.1007/978-3-030-95903-6_38

flight time, various systems using small, low-power FPGA devices have been implemented [21–25]. In addition, there are applications that implement fuzzy inference based control in FPGA [26,27]. A fuzzy inference is suitable for parallel processing, so it can significantly reduce the computation time even on low-power FPGA devices. In addition, the use of memory blocks in the FPGA allows easy adjustment of the fuzzy membership functions. In our previous work, we implemented a T1FI system for quadrotor attitude control [28]. IT2FI requires more processing time than T1FI, but Upper Membership Function (UMF) and Lower Membership Function (LMF) processing and type reduction algorithms can be processed in parallel on FPGA [29]. The proposed system implements a fuzzy inference based attitude angle adjuster for the roll, pitch and yaw axis on a FPGA device.

In this paper, we propose the IT2FI based quadrotor attitude control system and its FPGA implementation. In addition, we simulate the output values from the input values defined in the fuzzy inference.

The structure of the paper is as follows. In Sect. 2, we describe the proposed fuzzy interface system. In Sect. 3, we introduce our implementation method in FPGA. In Sect. 4, we present the simulation results. Finally, conclusions and future work are given in Sect. 5.

2   Proposed System

The gyro sensor in the proposed system is assumed to be a BNO055 which can output both angle and angular velocity. In the proposed system, we consider a IT2FI for each roll, pitch and yaw axis. Three input values of the fuzzy inference for each axis are shown in following.

1. The difference between the angle value read from the gyro sensor and the target angle value, assuming that the target angle value is an external input ($\Delta\theta$).
2. The angular velocity (θ/s).
3. The difference between the current angular rate and the previous angular rate, which is calculated every time the angular rate output of the gyro sensor is updated ($\Delta\theta/s$).

Let y be the output value of each axis, and the output values f_1, f_2, f_3 and f_4 of four motors to Electoric Speed Controller (ESC) are defined as follows.

- $f_1 = y_{roll} + y_{pitch} - y_{yaw}$
- $f_2 = -y_{roll} + y_{pitch} + y_{yaw}$
- $f_3 = y_{roll} - y_{pitch} + y_{yaw}$
- $f_4 = -y_{roll} - y_{pitch} - y_{yaw}$

For attitude angle control, we define the membership function and fuzzy rule-base considering that the three defined input values result in zero. In addition, we define the fuzzy rule-base considering that the effects on the output are in the order $\Delta\theta$, θ/s and $\Delta\theta/s$. As an example, we set up a fuzzy inference module for the roll axis using the fuzzy membership function shown in Fig. 1 and the fuzzy rule-base shown in Table 1. The number of each rule is assigned in the order of decreasing value in the result part.

We consider Positive (*P*), Zero (*Z*) and Negative (*N*) membership functions for the slope and speed of the angle from each axis, with the right direction being positive. The proposed system uses Takagi-Sugeno-Kang fuzzy inference, where the consequent part of the fuzzy rule-base is constant for both UMF and LMF. Therefore, the conseqent part of each rule is a constant crisp interval. The rules for the consequent part are shown in Table 2. The $\underline{y}$ is LMF, $\bar{y}$ is UMF, and the larger absolute value is UMF.

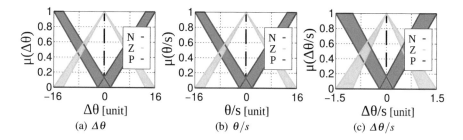

(a) $\Delta\theta$ (b) θ/s (c) $\Delta\theta/s$

Fig. 1. Membership functions.

Table 1. Fuzzy rule-base.

No.	$\Delta\theta$	θ/s	$\Delta\theta/s$	Out	NO.	$\Delta\theta$	θ/s	$\Delta\theta/s$	Out	No.	$\Delta\theta$	θ/s	$\Delta\theta/s$	Out
1	N	N	N	VVVL	10	Z	N	N	VL	19	P	N	N	VH
2	N	N	Z	VVVL	11	Z	N	Z	VL	20	P	N	Z	VVH
3	N	N	P	VVVL	12	Z	N	P	L	21	P	N	P	VVH
4	N	Z	N	VVVL	13	Z	Z	N	L	22	P	Z	N	VVH
5	N	Z	Z	VVVL	14	Z	Z	Z	M	23	P	Z	Z	VVVH
6	N	Z	P	VVL	15	Z	Z	P	H	24	P	Z	P	VVVH
7	N	P	N	VVL	16	Z	P	N	H	25	P	P	N	VVVH
8	N	P	Z	VVL	17	Z	P	Z	VH	26	P	P	Z	VVVH
9	N	P	P	VL	18	Z	P	P	VH	27	P	P	P	VVVH

Table 2. Consequent rule.

Rule	$[\underline{y},\bar{y}]$	Rule	$[\underline{y},\bar{y}]$	Rule	$[\underline{y},\bar{y}]$
VVVL	$[-5.0,-7.0]$	L	$[0.0,-2.0]$	VH	$[2.0,4.0]$
VVL	$[-4.0,-6.0]$	M	$[0.0,0.0]$	VVH	$[4.0,6.0]$
VL	$[-2.0,-4.0]$	H	$[0.0,2.0]$	VVVH	$[5.0,7.0]$

3   FPGA Implementation

For FPGA device implementation, we considered the 5CEBA4F23C7N and the hardware design language is VHDL. In Fig. 2, we show a FPGA implementation of the fuzzy inference module for one axis. First, the truth value of each rule is obtained simultaneously from the UMF and LMF of all rules using an address decoder and Read Only Memory (ROM) for membership functions. The truth value is obtained by the look up table scheme using the memory array of the FPGA device. The lower 11 bits are taken from the signed 16 bit binary input value and this is used as the ROM address to obtain the truth value of each membership function. In the case when a value falls outside the range where the lower 11 bits can be used as a value, a conditional expression is used to output a specific address. Then, the selector module outputs the truthfulness μ of each rule in one clock unit, and pipeline processing is performed thereafter. The t-norm uses the min operation and the type reduction algorithm uses the Enhanced Iterative Algorithm with Stop Condition (EIASC) algorithm [30]. The algorithm for obtaining the output values in each axis is shown in Algorithm 1. The rule number is n, and for each rule, the consequent part is $\overline{y_n}$, y_n, UMF truthfulness is $\overline{\mu_n}$, LMF truthfulness is $\underline{\mu_n}$, and so on. In addition, eight variables, a, b, c, d, yl, yr, L and R are defined to obtain the output values. The controller module sends signals such as ROM and Random Access Memory (RAM) addresses, I/O permission, and reset to each module according to the value of the internal counter.

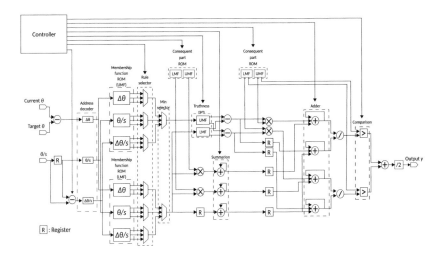

Fig. 2. FPGA implementation of fuzzy inference module for quadrotor attitude control.

Algorithm 1. A Type reduction Algorithm for proposed system using EIASC.

1: *Initializing variables.*
$$a \leftarrow 0, \quad b \leftarrow 0, \quad c \leftarrow 0, \quad d \leftarrow 0,$$
$$yl \leftarrow 0, \quad yr \leftarrow 0, \quad L \leftarrow 0, \quad R \leftarrow 27 \ (Total \ number \ of \ rules).$$

2: *Set values for a, b, c and d.*
 for $i = 1 \ to \ 27 \ (All \ rules)$ **do**
  $a \leftarrow a + y_i \, \underline{\mu_i}.$
  $b \leftarrow b + \underline{\mu_i}.$
  $c \leftarrow c + \overline{y_i} \, \underline{\mu_i}.$
  $d \leftarrow d + \underline{\mu_i}.$
 end for

3: *EIASC for computing yl.*
 while $yl \leq y_{L+1}$ **or** $L > 27$ **do**
  $L \leftarrow L + 1.$
  $a \leftarrow a + \underline{y_L}(\overline{\mu_L} - \underline{\mu_L}).$
  $b \leftarrow b + \overline{\mu_L} - \underline{\mu_L}.$
  $yl \leftarrow a \, / \, b.$
 end while

4: *EIASC for computing yr.*
 while $yr \geq y_R$ **or** $R < 0$ **do**
  $c \leftarrow c + \overline{y_R}(\overline{\mu_R} - \underline{\mu_R}).$
  $d \leftarrow d + \overline{\mu_R} - \underline{\mu_R}.$
  $yr \leftarrow c \, / \, d.$
  $R \leftarrow R - 1.$
 end while

Output: $\dfrac{yl + yr}{2}.$

4 Simulation Results

In Fig. 3, Fig. 4 and Fig. 5 are shown the simulation results of the output values in the roll axis for the cases $\Delta \theta = 0$, $\theta/s = 0$ and $\Delta \theta/s = 0$. The effects increase in older $\Delta \theta$, θ/s and $\Delta \theta/s$, and the simulation results are in line with the objectives of rule-based design. In particular, the case $\Delta \theta = 0$, and the output are close to 0. Table 3 shows the resources used to implement the proposed system on the three axis and derive the output to the four motors. Block Memory Bits (BMB) is used for setting membership functions, setting parameters for fuzzy rule-based postcases, and temporarily storing the truthfulness of each rule. These are used heavily to save on Adaptive Logic Modules (ALMs) usage, especially for setting up membership functions. Since 5CEBA4F23C7N is equipped with Digital Signal Processing Blocks (DSP Blocks), we used them for the

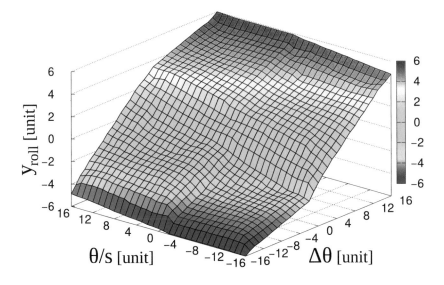

Fig. 3. Simulation results for the case $\Delta\theta/s = 0$.

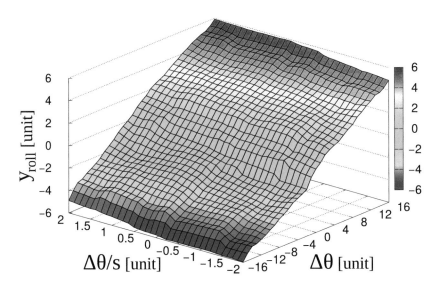

Fig. 4. Simulation results for the case $\theta/s = 0$.

multiplication process. A total of 12 DSP Blocks are used, four for each axis. The proposed system has a processing time of about 1080 [ns] and the system clock is 50 [Mhz].

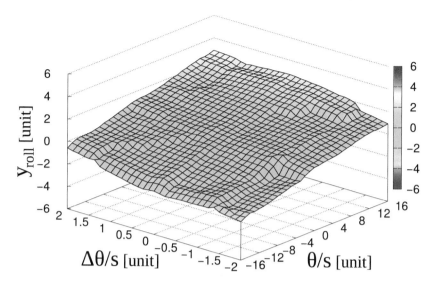

Fig. 5. Simulation results for the case $\Delta\theta = 0$.

Table 3. FPGA resource usage of the proposed system.

Resource	Utilization/Available (%)
Adaptive Logic Modules (ALMs)	$2,643/18,480$ (14%)
Block Memory Bits (BMB)	$670,976/3,153,920$ (21%)
Digital Signal Processing Blocks (DSP Blocks)	$12/66$ (18%)

5 Conclusions

In this paper, we proposed a IT2FI module for a quadrotor attitude control implemented in an FPGA. From the simulation results, we found that the proposed system is capable of producing the outputs in line with the objective of fuzzy rule-base and membership functions to attitude control of quadrotor. In addition, we found that the proposed system can compute the output value from the input value In the future, we will consider another fuzzy membership function and fuzzy rule-base and implement it on a quadrotor.

Acknowledgement. This work was supported by JSPS KAKENHI Grant Number JP20K19793 and Grant for Promotion of OUS Research Project (OUS-RP-20-3).

References

1. Oda, T., et al.: Design and implementation of a simulation system based on deep Q-network for mobile actor node control in wireless sensor and actor networks. In: Proceedings of The IEEE 31st International Conference on Advanced Information Networking and Applications Workshops, pp. 195–200 (2017)

2. Saito, N., Oda, T., Hirata, A., Hirota, Y., Hirota, M., Katayama, K.: Design and implementation of a DQN based AAV. In: Barolli, L., Takizawa, M., Enokido, T., Chen, H.-C., Matsuo, K. (eds.) BWCCA 2020. LNNS, vol. 159, pp. 321–329. Springer, Cham (2021). https://doi.org/10.1007/978-3-030-61108-8_32

3. Saito, N., Oda, T., Hirata, A., Toyoshima, K., Hirota, M., Barolli, L.: Simulation results of a DQN based AAV testbed in corner environment: a comparison study for normal DQN and TLS-DQN. In: Barolli, L., Yim, K., Chen, H.-C. (eds.) IMIS 2021. LNNS, vol. 279, pp. 156–167. Springer, Cham (2022). https://doi.org/10.1007/978-3-030-79728-7_16

4. Saito, N., et al.: A tabu list strategy based DQN for AAV mobility in indoor single-path environment: implementation and performance evaluation. Internet Things **14**, 100394 (2021)

5. Saito, N., Oda, T., Hirata, A., Yukawa, C., Kulla, E., Barolli, L.: A LiDAR based mobile area decision method for TLS-DQN: improving control for AAV mobility. In: Barolli, L. (ed.) 3PGCIC 2021. LNNS, vol. 343, pp. 30–42. Springer, Cham (2022). https://doi.org/10.1007/978-3-030-89899-1_4

6. Saito, N., Oda, T., Hirata, A., Toyoshima, K., Hirota, M., Barolli, L.: A movement adjustment method for DQN-based autonomous aerial vehicle. In: Barolli, L., Chen, H.-C., Miwa, H. (eds.) INCoS 2021. LNNS, vol. 312, pp. 136–148. Springer, Cham (2022). https://doi.org/10.1007/978-3-030-84910-8_15

7. Salih, A.L., et al.: Flight PID controller design for a UAV quadrotor. Sci. Res. Essays **5**(23), 3660–3667 (2010)

8. Bouabdallah, S., et al.: PID vs LQ Control Tech-Niques Applied to an Indoor Micro Quadrotor. In: Proceedings of The IEEE International Conference on Intelligent Robots and Systems, pp. 2451–2456 (2004)

9. Raza, S., Gueaieb, W.: Fuzzy logic based quadrotor flight controller. In: Proceedings of The 6th International Conference on Informatics in Control, Automation and Robotics, pp. 105–112 (2009)

10. Iswanto, et al.: Trajectory and altitude controls for autonomous hover of a quadrotor based on fuzzy algorithm. In: Proceedings of The 8th International Conference on Information Technology and Electrical Engineering, pp. 1–6 (2016)

11. Sarabakha, A., et al.: Type-2 fuzzy logic controllers made even simpler: from design to deployment for UAV. IEEE Trans. Industr. Electron. **65**(6), 5069–5077 (2018)

12. Naik, K.A., Gupta, C.P.: Performance comparison of Type-1 and Type-2 fuzzy logic systems. In: Proceedings of The 4th International Conference on Signal Processing, Computing and Control, pp. 72–76 (2017)

13. Baklouti, N., Alimi, A.M.: Interval type-2 fuzzy logic control of mobile robots. J. Intell. Learn. Syst. Appl. **4**(4), 291–302 (2012)

14. Linda, O., Manic, M.: Interval type-2 fuzzy voter design for fault tolerant systems. Inf. Sci. **181**, 2933–2950 (2011)

15. Huang, J., et al.: Interval type-2 fuzzy logic modeling and control of a mobile two-wheeled inverted pendulum. IEEE Trans. Fuzzy Syst. **26**(4), 2030–2038 (2018)

16. Imam, R., Kharisma, B.: Design of interval type-2 fuzzy logic based power system stabilize. Int. J. Electr. Electron. Eng. **3**(10), 593–600 (2009)

17. Takano, K., Oda, T., Kohata, M.: Design of a DSL for converting rust programming language into RTL. In: Barolli, L., Okada, Y., Amato, F. (eds.) EIDWT 2020. LNDECT, vol. 47, pp. 342–350. Springer, Cham (2020). https://doi.org/10.1007/978-3-030-39746-3_36

18. Takano, K., et. al.: Approach of a coding conventions for warning and suggestion in transpiler for rust convert to RTL. In: Proceedings of The IEEE 9th Global Conference on Consumer Electronics, pp. 789–790 (2020)

19. Takano, K., et al.: PC process migration using FPGAs in ring networks. IEICE Commun. Express **9**(5), 141–145 (2020)

20. Takano, K., Oda, T., Ozaki, R., Uejima, A., Kohata, M.: Implementation of process migration method for PC-FPGA hybrid system. In: Barolli, L., Takizawa, M., Enokido, T., Chen, H.-C., Matsuo, K. (eds.) BWCCA 2020. LNNS, vol. 159, pp. 204–210. Springer, Cham (2021). https://doi.org/10.1007/978-3-030-61108-8_20

21. Oleynikova, H., et al.: Reactive avoidance using embedded stereo vision for MAV flight. In: Proceedings of The IEEE International Conference on Robotics and Automation, pp. 50–56 (2015)

22. Honegger, D., et al.: Real-time and low latency embedded computer vision hardware based on a combination of FPGA and mobile CPU. In: International Conference on Intelligent Robots and Systems, pp. 4930–4935 (2014)

23. Jiang, Q., et al.: Attitude and heading reference system for quadrotor based on MEMS sensors. In: Proceedings of The 2nd International Conference on Instrumentation and Measurement, Computer, Communication and Control, pp. 1090–1093 (2012)

24. Lechekhab, T.E., et al.: Robust error-based active disturbance rejection control of a quadrotor. Aircr. Eng. Aerosp. Technol. 93(1), 89–104 (2020)

25. Krajník, T., et al.: A simple visual navigation system for an UAV. In: Proceedings of The IEEE 9th International Multi-Conference on Systems, Signals and Devices, pp. 1–6 (2012)

26. Hamed, B.M., El-Moghany, M.S.: Fuzzy controller design using FPGA for sun tracking in solar array system. Int. J. Intell. Syst. Technol. Appl. 4(1), 46–52 (2012)

27. Alvarez, J., et al.: FPGA Implementation of a Fuzzy Controller for Automobile DC-DC Converters. In: Proceedings of The IEEE International Conference on Field Programmable Technology, pp. 237–240 (2006)

28. Matsui, T., et al.: FPGA implementation of a fuzzy inference based quadrotor attitude control system. In: Proceedings of The IEEE 10th Global Conference on Consumer Electronics, pp. 770–771 (2021)

29. Melgarejo, M.A., Pena-Reyes, C.A.: Hardware architecture and FPGA implementation of a type-2 fuzzy system. In: Proceedings of The 14th ACM Great Lakes Symposium on VLSI, pp. 458–461 (2004)

30. Wu, D., Nie, M.: Comparison and practical implementation of type-reduction algorithms for type-2 fuzzy sets and system. In: Proceedings of International Conference on Fuzzy Systems, pp. 2131–2138 (2011)

Design of a Robot Vision System for Microconvex Recognition

Chihiro Yukawa[1], Tetsuya Oda[1(✉)], Nobuki Saito[2], Aoto Hirata[2],
Tomoya Yasunaga[1], Kyohei Toyoshima[1], and Kengo Katayama[1]

[1] Department of Information and Computer Engineering, Okayama University
of Science (OUS), 1-1 Ridaicho, Kita-ku, Okayama-shi 700-0005, Japan
{t18j097yc,t18j091yt,t18j056tk}@ous.jp,
{oda,katayama}@ice.ous.ac.jp
[2] Graduate School of Engineering, Okayama University of Science (OUS),
1-1 Ridaicho, Kita-ku, Okayama 700–0005, Japan
t21jm01md,t21jm02zr@ous.jp

Abstract. The goals of Industry 4.0 are to achieve a higher level of operational efficiency and productivity, as well as a higher level of automatization. Also, the measurement at the nano-level on the surface of the target object by a machine has been considered for automation, but there are problems such as the need for high cost and a large amount of time for measurement. In this paper, we propose a robot vision system based on an intelligent algorithm for recognizing microconvexes on arbitrary surfaces. The proposed system is inexpensive, make quick measurement and is capable of autonomously recognizing microconvexes to improve the efficiency of production processes.

1 Introduction

The automation is being promoted by Industry 4.0 [1] in the manufacturing industry to improve the efficiency of production processes. The measurement of nano-level microconvexes on the target surface by machines is being considered for automation [2–8]. However, there are problems such as the high cost and a lot of time required for measurement. In addition, most of the processing skills related to microconvexes and convexities have been performed manually by craftsmen who have acquired them through many years of experience. Furthermore, it is essential to hand down the skills to the next generation, and since it takes a great deal of time and experience to acquire the skills as artisans, continuous employment is required from the young generation. Therefore, there are few people who want to become artisans, and it is difficult to hire young people, resulting in a chronic labor shortage. On the other hand, the skills of artisans can be technically reproduced by applying current machine and control technologies. In addition, it is thought that robotization will become possible by imitating the skills of artisans by intelligent algorithms. Currently, research on robot vision thrive for the automation of inspection [9–12].

L. Barolli et al. (Eds.): EIDWT 2022, LNDECT 118, pp. 366–374, 2022.
https://doi.org/10.1007/978-3-030-95903-6_39

In this paper, we propose a robot vision system based on an intelligent algorithm for recognizing microconvexes on arbitrary target surfaces. The proposed system is inexpensive, make quick measurement and is capable of autonomously recognizing microconvexes to improve the efficiency of production processes.

The structure of the paper is as follows. In Sect. 2, we present the proposed system. In Sect. 3, we discuss the experimental results. Finally, conclusions and future work are given in Sect. 4.

2   Proposed System

The robotic vision system for recognizing microconvexes on the arbitrary target surface is shown in Fig. 1. The proposed system considers images from various angles with the 6 degrees of freedom. For the robot arm, we propose a vibration suppression method based on fuzzy inference [13–17] to improve the recognition rate of microconvexes. In addition, an electron microscope is mounted to the edge of the robot arm. The images received by the electron microscope are reconstructed by stitching. Then, the stitched images are sent to a Deep Learning Network (Conventional Neural Netork: CNN) for object recognition.

2.1   Servo Motors Vibration Reduction Method

The robot arm has a very low accuracy of movement [18]. But, by using the training datasets for image recognition, it is possible to improve the recognition rate by suppressing the vibration of the robot arm [19]. Therefore, we propose a method for reduction of vibration of the robot arm considering sensing, fuzzy inference, and servomotor control to improve the recognition rate of microconvexes. The sensing module sends to the Jetson the acceleration values X, Y and Z-$axis$ from the accelerometer GY-521 mounted to the edge of the robot arm. Also, it sends the sensing data to controller via serial transmission. In the fuzzy inference are inserted the values of error and angle. We use Interval Type-2 Fuzzy Sets (IT2FS) [20,21], and the output is determined by the EIASC method [22,23]. The error is the average squared error of the X, Y and Z-$axis$, where the true value is the average value of the acceleration received from the accelerometer. The IT2FS is outputs gain. The gain affects the responsiveness of the servo mechanism. The servo motor control decides the angle based on the gain of fuzzy inference output. In the vibration suppression method, the gain is gradually decreased from the starting position to the specified angle to suppress the vibration. The gain of the robot arm is gradually reduced from the start position (Fig. 2 and Table 1).

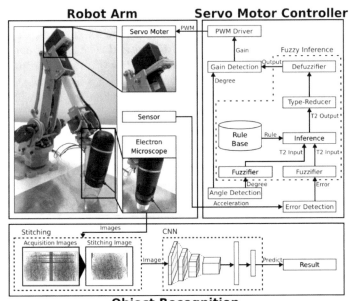

Fig. 1. Proposed system.

Table 1. Fuzzy rule-base.

Angle of robot arm	Error of acceleration	Gain of servo motor
High	High	High
High	Middle	High
High	Low	Low
Middle	High	High
Middle	Middle	Middle
Middle	Low	Low
Low	High	Middle
Low	Middle	Middle
Low	Low	Low

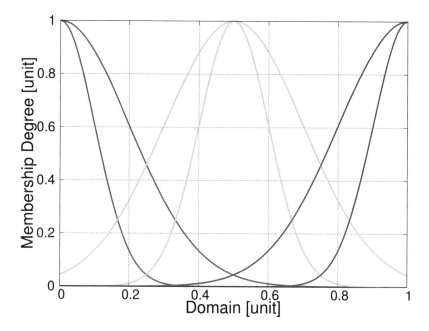

Fig. 2. Membership function of IT2FS.

2.2 Image Stitching

In image stitching, the feature points are detected in multiple images by the electron microscope, and matching is performed based on the feature points. It is possible to obtain a single high-resolution image of the entire acquired object by stitching each image. Since the lighting environment is expected to change depending on the imaging angle, the Scale-Invariant Feature Transform [24] is effective for scaling and lighting changes to detect feature points. It is possible to obtain the distribution and to prevent duplicate recognition of the microconvexs on the target surface by recognizing a single image generated from stitching results.

2.3 Object Detection

The proposed system recognizes microconvexes on the surface by YOLOv5. We use a deep learning model that can perform object detection and class classification. A large number of images containing microconvexes are required to train the model. However, since there are only a few factories and other production sites to process microconvexes, it is predicted that there will be few images that

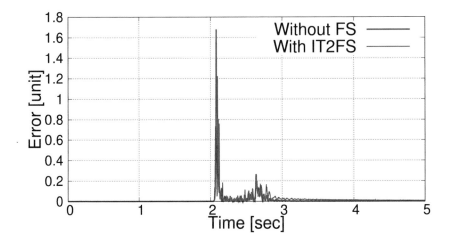

Fig. 3. Fuzzy inference based vibration reduction.

can be used for model training. Therefore, we create a dataset of images containing microconvexes for training YOLOv5. In addition, we consider transfer learning [25, 26], which is a method of transferring a model that has already been trained for another problem to improve the recognition rate of microconvexes and reduce the training time.

3   Exprimental Results

3.1   Case of Fuzzy Inference

In this section, we compare the vibration reduction of the robot arm without fuzzy inference and with IT2FS. In the vibration reduction experiment, the servo motor is moved 90 [*deg.*] after about 2 [*s*] of starting the measurement. The results of vibration reductions are shown in Fig. 3. The error value at around 2 [*s*] is the vibration at the beginning of the movement. It can be seen that the acceleration value at about 2.5 [*s*] in Fig. 3 is reduced by the gain using fuzzy inference. The IT2FS has smaller error compared with the case without fuzzy inference.

3.2   Case of Object Detection

In the object recognition experiment, we recognize microconvexes on the metal surface as the target. We use 90 training data and the 10 test data, and also performed transfer learning for the training of YOLOv5. The learning results are shown in Fig. 4. The maximum recognition rate of the microconvexes is 93.86 [%]. The mAP reached its maximum value at around 500 iterations and then

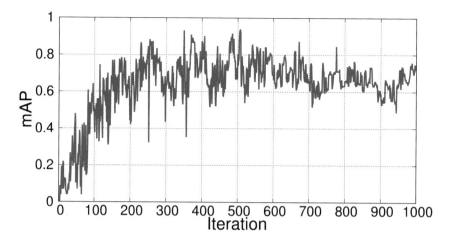

Fig. 4. Recognition rate results.

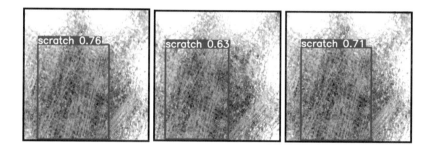

Fig. 5. Object detection.

decreased. This may be due to overfitting caused by insufficient training data. The recognition results of microconvexes in the image taken by the electron microscope are shown in Fig. 5. In addition, the recognition results of microconvexes in the stitching image are shown in Fig. 6. By stitching, the proposed system can recognize the edges, which cannot be recognized by each image taken with an electron microscope.

Fig. 6. Object detection of stitching image.

4   Conclusions

In this paper, we proposed a robot vision system based on an intelligent algorithm for recognizing microconvexes on arbitrary surfaces. The experimental results have shown that the proposed system can reduce the servo Motors vibration by using IT2FS. In addition, the microconvexes can recognize the image taken by the electron microscope on the surface of the target object by YOLOv5. In order to improve the recognition rate, we will implement the system by Vision Transformer [27], which has good recognition rate in transfer learning.

Acknowledgement. This work was supported by JSPS KAKENHI Grant Number 20K19793 and Grant for Promotion of OUS Research Project (OUS-RP-20-3).

References

1. Dalenogare, L., et al.: The expected contribution of industry 4.0 technologies for industrial performance. Int. J. Prod. Econ. (IJPE-2018) **204**, 383–394 (2018)
2. Shang, L., et al.: Detection of rail surface defects based on CNN image recognition and classification. In: The IEEE 20th International Conference on Advanced Communication Technology (ICACT), pp. 45–51 (2018)
3. Li, J., et al.: Real-time detection of steel strip surface defects based on improved yolo detection network. IFAC-PapersOnLine **51**(21), 76–81 (2018)

4. Oda, T., et al.: Design and implementation of a simulation system based on deep Q-Network for mobile actor node control in wireless sensor and actor networks. In: Proceedings of The IEEE 31st International Conference on Advanced Information Networking and Applications Workshops, pp. 195-200 (2017)

5. Saito, N., Oda, T., Hirata, A., Hirota, Y., Hirota, M., Katayama, K.: Design and implementation of a DQN based AAV. In: Barolli, L., Takizawa, M., Enokido, T., Chen, H.-C., Matsuo, K. (eds.) BWCCA 2020. LNNS, vol. 159, pp. 321–329. Springer, Cham (2021). https://doi.org/10.1007/978-3-030-61108-8_32

6. Saito, N., Oda, T., Hirata, A., Toyoshima, K., Hirota, M., Barolli, L.: Simulation results of a DQN based AAV testbed in corner environment: a comparison study for normal DQN and TLS-DQN. In: Barolli, L., Yim, K., Chen, H.-C. (eds.) IMIS 2021. LNNS, vol. 279, pp. 156–167. Springer, Cham (2022). https://doi.org/10.1007/978-3-030-79728-7_16

7. Saito, N., et al.: A tabu list strategy based DQN for AAV mobility in indoor single-path environment: implementation and performance evaluation. Internet Things 14, 100394 (2021)

8. Saito, N., Oda, T., Hirata, A., Yukawa, C., Kulla, E., Barolli, L.: A LiDAR based mobile area decision method for TLS-DQN: improving control for AAV mobility. In: Barolli, L. (ed.) 3PGCIC 2021. LNNS, vol. 343, pp. 30–42. Springer, Cham (2022). https://doi.org/10.1007/978-3-030-89899-1_4

9. Wang, H., et al.: Automatic illumination planning for robot vision inspection system. Neurocomputing 275, 19–28 (2018)

10. Zuxiang, W., et al.: Design of safety capacitors quality inspection robot based on machine vision. In: 2017 First International Conference on Electronics Instrumentation and Information Systems (EIIS), pp. 1–4 (2017)

11. Li, J., et al.: Cognitive visual anomaly detection with constrained latent representations for industrial inspection robot. Appl. Soft Comput. 95, 106539 (2020)

12. Ruiz-del-Solar, J., et al.: A survey on deep learning methods for robot vision. arXiv preprint arXiv:1803.10862 (2018)

13. Matsui, T., et al.: FPGA implementation of a fuzzy inference based quadrotor attitude control system. In: Proceedings of IEEE GCCE-2021, pp. 691–692 (2021)

14. Saito, N., et al.: Approach of fuzzy theory and hill climbing based recommender for schedule of life. In: Procedings of LifeTech-2020, pp. 368–369 (2020)

15. Ozera, K., et al.: A fuzzy approach for secure clustering in MANETs: effects of distance parameter on system performance. In: Proceedings of IEEE WAINA-2017, pp. 251–258 (2017)

16. Elmazi, D., et al.: Selection of Secure Actors in Wireless Sensor and Actor Networks Using Fuzzy Logic. In: Proceedings of BWCCA-2015, pp. 125–131 (2015)

17. Elmazi, D., et al.: Selection of rendezvous point in content centric networks using fuzzy logic. In: Proceedings of NBiS-2015, pp. 345–350 (2015)

18. Zaeh, M.F., et al.: Improvement of the machining accuracy of milling robots. Prod. Eng. Res. Devel. 8(6), 737–744 (2014). https://doi.org/10.1007/s11740-014-0558-7

19. Yukawa, C., et al.: Design of a fuzzy inference based robot vision for CNN training image acquisition. In: Proceedings of IEEE GCCE-2020, pp. 871–872 (2021)

20. Liang, Q., et al.: Interval type-2 fuzzy logic systems: theory and design. IEEE Trans. Fuzzy Syst. 8(5), 535–550 (2000)

21. Mendel, J.M.: Interval type-2 fuzzy logic systems made simple. IEEE Trans. Fuzzy Syst. 14(6), 808–821 (2006)

22. Dongrui, W., et al.: Comparison and practical implementation of type-reduction algorithms for type-2 fuzzy sets and systems. In: 2011 IEEE International Conference on Fuzzy Systems (FUZZ-IEEE 2011), pp. 2131–2138 (2011)

23. Mendel, J.M.: On KM algorithms for solving type-2 fuzzy set problems. IEEE Trans. Fuzzy Syst. **21**(3), 426–446 (2012)

24. Wang, Y., et al.: Automatic image stitching using SIFT. In: Proceedings of International Conference on Audio, Language and Image Processing (IEEE-2008), pp. 568-571 (2008)

25. Yosinski, J., et al.: How transferable are features in deep neural networks? *arXiv preprint* arXiv:1411.1792 (2014)

26. Zhuang, F., et al.: A comprehensive survey on transfer learning. Proc. IEEE **109**(1), 43–76 (2020)

27. Dosovitskiy, A., et al.: An image is worth 16×16 words: transformers for image recognition at scale. *arXiv preprint* arXiv:2010.11929 (2020)

Path Control Algorithm for Weeding AI Robot

Misato Shiba and Hiroyoshi Miwa[(✉)]

Graduate School of Science and Technology, Kwansei Gakuin University,
2-1 Gakuen, Sanda-shi, Hyogo, Japan
{shiba.m,miwa}@kwansei.ac.jp

Abstract. This paper addresses a path control algorithm for an AI robot that performs automatic weeding work. An AI robot can grasp the situation (locations of weeds to be weeded, crops not to be weeded, and obstacles, etc.) of the limited range of the circumference of the AI robot by installed cameras and sensors. The AI robot carries out the weeding work simultaneously, while the path which removes all weeds in a short time is determined and renewed by collecting the situation of the whole object region by moving. Since the object area is wide, it is generally necessary to carry out the weeding work continuously, because the weed may grow again or it may grow from the new place, even if it is the plot which weeded once. In this way, the AI robot basically carries out the weeding work of the whole object region continuously without manual intervention. In this paper, some weeding path control algorithms suitable for such an AI robot are designed, and several algorithms are evaluated by simulation.

1 Introduction

The development of AI (Artificial Intelligence) technology is remarkable in recent years. Especially, an AI robot is an effective tool for improving the efficiency of our daily work. AI robots are not only used in factories like industrial robots, but also active in various fields such as agriculture, manufacturing, and they are coming into our daily life as daily electric appliances like cleaning robots.

We focus on AI robots in the field of agriculture. Although agriculture is a very important industry for the large role of stable supply of food to the nation, in present Japan, the aging rapidly advances due to the shortage of successors, and the drastic decrease of the production can not be avoided by the decrease of producers. Therefore, it is necessary to efficiently produce even with a smaller number of people. For that purpose, various solutions such as plant factory are considered. In this paper, we address an AI robot that performs automatic weeding work.

Weeding work is indispensable, since the use of agricultural chemicals must be controlled by the recent increase of interest for the safety of food. Frequent weeding work is necessary, since the growth of weeds is rapid. This workload and burden on the body is large, and this reduction is required. The algorithm

L. Barolli et al. (Eds.): EIDWT 2022, LNDECT 118, pp. 375–385, 2022.
https://doi.org/10.1007/978-3-030-95903-6_40

of weeding work is considerably different by crops. The development of AI robot for general weeding work is not yet easy, because the kinds of weeds, the way of growing, the density, etc. are greatly different depending on the object crops such as the boundary between paddy fields and causeways, leaf vegetables, root vegetables, orchards, etc.

This paper deals with a weeding path control algorithm for an AI robot that performs automatic weeding work in a relatively flat but wide area, especially in orchards. The AI robot can grasp the situation (locations of weeds to be weeded, crops not to be weeded, and obstacles) of the limited range of the circumference of the AI robot by installed cameras and sensors. The AI robot carries out the weeding work simultaneously, while it collects the situation of the whole object region by moving, and determines and updates the path to weed all weeds in a short time. Since the object area is wide, it is generally necessary to carry out the weeding work continuously, because the weed may grow again or it may grow from the new place, even if it is the plot which weeded once. In this way, the AI robot basically carries out the weeding work of the whole object region continuously without manual intervention. In this paper, we design some weeding path control algorithms suitable for such AI robots, and we evaluate the algorithm are by simulation.

2   Related Works

AI robots for weeding work are widely used from theoretical research to development of actual machines. Many weeding robots have been developed based on the same theory as cleaning robots. We describe theoretical research and actual developed weeding and cleaning AI robots.

In recent years, there are two types of cleaning robots: a reflection action type in which a cleaning robot [1] runs based on information obtained from sensors while running without having information on the environment, and a planning action type in which a cleaning robot runs after teaching information such as the shape of a sweep area to a robot in advance [2]. Many of the cleaning robots on the market do not have the knowledge of the cleaning route and adopt the reflection action type. In the reflex action type, the efficiency is not always good, because it may clean the same place many times. There is the cleaning robots [3] using SLAM (Simultaneous Localization and Mapping) [4]. SLAM is a technology which simultaneously carries out self-position estimation and map construction. By using SLAM, it is possible to learn the situation in the room while moving the whole room, and to distinguish the cleaned area from the other area. The efficiency is improved by using SLAM.

Some AI robots for weeding have already been used practically. The AI robot [5] disperses herbicides between weeds and crops by analyzing photographed images. However, since organic agriculture is promoted in Japan and the use of herbicides is not desirable, the actual machine developed in Japan carries out weeding by cutters, etc. The Miimo [6] developed by Honda Motor Co., Ltd. can recognize a work area, automatically charge a battery, and avoid obstacles, and carries out weeding by moving in such a way as to cover the whole area by three running patterns. Though the actual machine can weed all weeds because it moves the whole area, it does not weed through the shortest route.

In the development of a weeding AI robot and a cleaning AI robot, the control to determine the moving route for efficient weeding and cleaning is important. Therefore, the research on an algorithm which determines a moving route has also been carried out. The research of [7] aims to find a short moving route of a lawnmower for a given area of turf, such that all the grass is mowed. The problem of finding such route is NP-hard in general, because it includes the Traveling Salesman Problem. An algorithm using an approximation algorithm for the Traveling Salesman Problem has been proposed [8].

3   Path Control Algorithm for Weeding AI Robot

In this section, we describe path control algorithms for weeding AI robot.

An AI robot collects the situation of the limited range of the circumference, and the weeding work is also carried out simultaneously, while it determines and updates the path as needed, while it moves. In addition, it is necessary to charge at a charging station within a fixed time. In such situation, the purpose of the path control of an AI robot is to weed all weeds in a short time by collecting the situation of the whole object region. In this paper, for the sake of simplicity, we assume that the path which weeds the whole region only once is determined.

We consider the following model (Fig. 1). First, a region on a two-dimensional plane are divided by squares (cells). The locations of weeds, obstacles, crops and charging stations are specified by a set of cells, respectively. The size of an AI robot is assumed to coincide with one cell. Assume that an AI robot can obtain all the information of the area consisting of 5×5 cells centered on the AI robot.

An AI robot has four movement directions of up, down, left and right of the present position. When the AI robot enters a cell of weed, the weed is removed, and the AI robot cannot enter the cell of obstacles and crops. Once the AI robot is charged, it must return to the cell with the charging station within a limited time again. In the initial state, only the shape of the whole region and the position of the charging station is given to the AI robot, and the AI robot can discriminate whether it is crop not to be removed or a weed to be removed. The present position of the AI robot can be grasped by GPS. In such a situation, it is a purpose to determine the moving path which minimizes the time from the initial position of the AI robot to weeding all weeds in the whole region. Since the cell information is only partially obtained, it is necessary to determine the path while collecting the information.

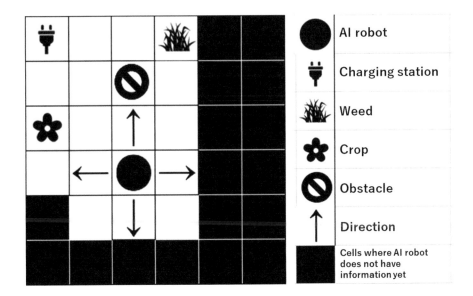

Fig. 1. Model

To realize the above purpose, we design the following algorithms; the algorithm which combines an information collection algorithm and an approximation algorithm of TSP (Traveling Salesman Problem); the algorithm which combines an information collection algorithm and the nearest neighbor algorithm of TSP; the algorithm using the depth first search algorithm; the algorithm which carries out information collection and weeding simultaneously.

3.1   Algorithm Using Approximation Algorithm of TSP

We describe the algorithm which combines an information collection algorithm and an approximation algorithm of TSP.

First, we describe the information collection algorithm.

Repeat the following procedure.

1. Record "visit" to the current cell of the AI robot.
2. Record "obstacle" in the cell when an obstacle cell is found.
3. If the current cell of the AI robot is a weed cell, the weed is removed.
4. To move from left to right, repeat the following until the AI robot reaches the right edge of the area:
   a. Finds unvisited cells adjacent to the current cell in right, bottom, top, and left order of precedence.
   b. If an unvisited cell is found, put the current cell in the stack and make the unvisited cell the current cell.
   c. Otherwise, fetch the previous cell in the stack and set the previous cell as the current cell.

5. To move from right to left, repeat the following until the AI robot reaches the left edge of the area:
   a. Finds unvisited cells adjacent to the current cell in left, bottom, top, or right order of precedence.
   b. If an unvisited cell is found, put the current cell in the stack and make the unvisited cell the current cell.
   c. Otherwise, fetch the previous cell in the stack and set the previous cell as the current cell.

Next, we describe the Christofides algorithm [9], an approximation algorithm of TSP (Traveling Salesman Problem). The path output by this algorithm is known to have a weight less than 1.5 of the optimal solution. That is, the Christofides algorithm is a 1.5 approximation algorithm.

1. Find the minimum spanning tree of a given complete graph.
2. Find the minimum weight matching between the vertices whose degree is odd number.
3. Make the graph that consists of the minimum spanning tree and the minimum weight matching.
4. Find an Euler cycle in the graph.
5. Make a cycle by skipping already-visited-vertices while proceeding along the Euler cycle.

We describe the algorithm. First, the information collection algorithm is carried out and then the path is determined so that the found all cells of weed which have not been visited yet are visited along the path determined by the Christofides algorithm. We show an example in Fig. 2.

3.2   Algorithm Using Nearest Neighbor Algorithm of TSP

We describe the algorithm which combines an information collection algorithm and the nearest neighbor algorithm for TSP. The information collection algorithm is the same algorithm in the previous section. First, the information collection algorithm is carried out and then the path is determined so that the found all cells of weed which have not been visited yet are visited along the order of the weed cell nearest from the current cell. We show an example in Fig. 3. The distance between a found cell of weed and the current cell is checked for all found all cells of weed ((a), (b), (c), and (d) in Fig. 3), and then the nearest cell of weed ((d) in Fig. 3) is chosen.

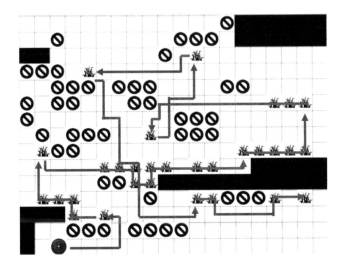

Fig. 2. Example of path by algorithm using approximation algorithm of TSP

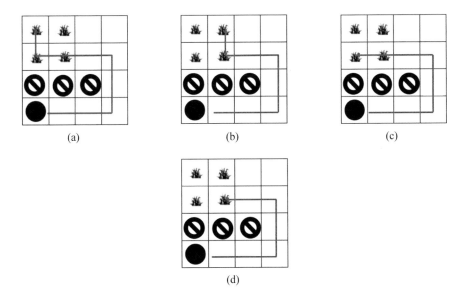

Fig. 3. Algorithm using nearest neighbor algorithm of TSP

3.3 Algorithm Using Depth First Search

We describe the algorithm using the depth first search algorithm. The AI robot moves so as to visit all cells except obstacles.

Repeat the following procedure.

1. Record "visit" to the current cell of the AI robot.

2. Record "obstacle" in the cell when an obstacle cell is found.
3. If the current cell of the AI robot is a weed cell, the weed is removed.
4. Finds unvisited cells adjacent to the current cell in top, right, left, and bottom order of precedence.
5. If an unvisited cell is found, put the current cell in the stack and make the unvisited cell the current cell.
6. Otherwise, fetch the previous cell in the stack and set the previous cell as the current cell.

3.4 Algorithm Collecting and Weeding Simultaneously

We describe the algorithm which carries out information collection and weeding simultaneously.

Repeat the following procedure.

1. Record "weed" in the cell when a weed cell is found.
2. Record "obstacle" in the cell when an obstacle cell is found.
3. If the current cell of the AI robot is a weed cell, the weed is removed.
4. Choose the nearest weed cell among the found all weed cells and move to the cell along the shortest path.

4   Performance Evaluation

In this section, we evaluate the algorithms by simulation.

We use the weeding ratio, the search ratio, and the work time as the evaluation measures. The weeding ratio is defined as the ratio of the number of the weed cells which has been weeded to the number of the initial weed cells. The search ratio is defined as the ratio of the number of the searched cells to the number of the total cells, where a searched cell is the cell whose information is retrieved. The work time is defined as the number of cells that the AI robot visited until the algorithm halts.

We compare the weeding ratio, the search ratio, and the work time of the following algorithms; the algorithm which combines an information collection algorithm and an approximation algorithm of TSP (AppTSP); the algorithm which combines an information collection algorithm and the nearest neighbor algorithm of TSP (NNTSP); the algorithm using the depth first search algorithm (DFS); the algorithm which carries out information collection and weeding simultaneously (Simul).

The size of the whole region to be the weeding object was set at 20×15, and we prepare 100 regions with different positions with weed cells of 30% of the whole region, obstacle cells which cannot pass through 20% of the whole region. The simulation experiments are carried out 100 times for each algorithm.

4.1   Evaluation of Weeding Ratio

We show the result of the weeding ratios of the algorithms in Fig. 4. All the weeding ratios of 100 experiments are shown for each algorithm.

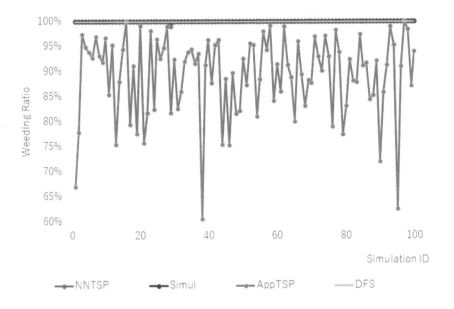

Fig. 4. Weeding ratios of algorithms

More than 60% of weeds existing in the whole region have been weeded by any algorithm. Since the weeding ratio of AppTSP varies largely and low in comparison with the other algorithms, the performance of AppTSP is worse than the other algorithms. The weeding ratio of NNTSP is mostly over 95%, and those of DFS and Simul are mostly 100%. From the viewpoint of the weeding ratio, the performance of DFS and Simul are good.

4.2   Evaluation of Search Ratio

We show the result of the search ratios of the algorithms in Fig. 5. All the search ratios of 100 experiments are shown for each algorithm.

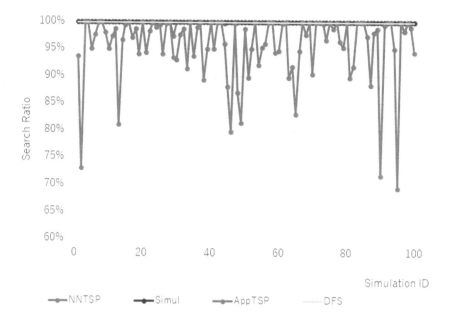

Fig. 5. Search ratios of algorithms

More than 65% of the total cells in the whole region have been searched by any algorithm. Since the search ratio of AppTSP varies largely and low in comparison with the other algorithms, the performance of AppTSP is worse than the other algorithms. The search ratio of NNTSP is over 95%, and those of DFS and Simul are mostly 100%. From the viewpoint of the search ratio, the performance of DFS and Simul are also good.

4.3 Evaluation of Work Time

We show the result of the work times of the algorithms in Fig. 6. All the work times of 100 experiments are shown for each algorithm.

The work time of AppTSP varies largely but is short in comparison with the other algorithms. The work time of Simul is almost same as AppTSP but is stable. The second best is NNTSP, and DFS needs much time. From the viewpoint of the work time, the performance of Simul is good.

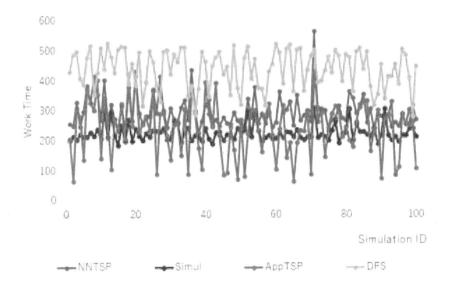

Fig. 6. Work time of algorithms

5   Conclusions

This paper dealt with a weeding path control algorithm for AI robots that perform automatic weeding work in a relatively flat but wide area, especially in orchards. We designed some path control algorithms suitable for weeding AI robots, and we evaluated the algorithm are by simulation.

We compared the weeding ratio, the search ratio, and the work time of the following algorithms; the algorithm which combines an information collection algorithm and an approximation algorithm of TSP (AppTSP); the algorithm which combines an information collection algorithm and the nearest neighbor algorithm of TSP (NNTSP); the algorithm using the depth first search algorithm (DFS); the algorithm which carries out information collection and weeding simultaneously (Simul).

As for the weeding ratio and the search ratio, the algorithms Simul, DFS, and NNTSP achieved high ratios; as for the work time, the algorithms Simul and AppTSP achieved short work time but the work time of AppTSP is not stable. As a result, the performance of the algorithm Simul is efficient for weeding path control algorithm, since it can weed much within short time.

Acknowledgements. This work was partially supported by the Japan Society for the Promotion of Science through Grants-in-Aid for Scientific Research (B) (17H01742).

References

1. https://www.irobot-jp.com/product/e5/

2. Okada, Y., et al.: Motion Control Algorithm of Home Cleaner Robot. In: Proceedings of the 48th Japan Automatic Control Conference (2005). (in Japanese)
3. https://www.irobot-jp.com/product/900series/
4. Cadena, C., et al.: Past, present, and future of simultaneous localization and mapping: towards the robust-perception age. IEEE Trans. Rob. **32**(6), 1309–1332 (2016)
5. https://www.ecorobotix.com/en/
6. Kakawami, T., Kobayashi, H., Yamamura, M., Maruyama, S.: Development of robotic lawnmower miimo. Honda R&D Tech. Revi., **25**(2) (2013)
7. Arkin, E.M., Fekete, S.P., Mitchell, J.S.B.: Approximation algorithms for lawn mowing and milling. Comput. Geom. **17**(1–2), 25–50 (2000)
8. Arkin, E.M., Hassin, R.: Approximation algorithms for the geometric covering salesman problem. Discret. Appl. Math. **55**(3), 197–218 (1994)
9. Goodrich, M.T., Tamassia, R.: The Christofides Approximation Algorithm, Algorithm Design and Applications. Wiley, pp. 513–514 (2015)

Performance Analysis of RIWM and RDVM Router Replacement Methods for WMNs by WMN-PSOSA-DGA Hybrid Simulation System Considering Stadium Distribution of Mesh Clients

Admir Barolli[1], Shinji Sakamoto[2], and Leonard Barolli[3(✉)]

[1] Department of Information Technology, Aleksander Moisiu University of Durres, L.1, Rruga e Currilave, Durres, Albania
[2] Department of Information and Computer Science, Kanazawa Institute of Technology, 7-1 Ohgigaoka, Nonoichi, Ishikawa 921-8501, Japan
[3] Department of Information and Communication Engineering, Fukuoka Institute of Technology, 3-30-1 Wajiro-Higashi, Higashi-Ku, Fukuoka 811-0295, Japan
barolli@fit.ac.jp

Abstract. Wireless Mesh Networks (WMNs) have many advantages, for example easy maintenance, low upfront cost and high robustness. However, WMNs have some problems to be solved such as node placement problem, hidden terminal problem and so on. In our previous work, we implemented a simulation system to solve the node placement problem in WMNs considering Particle Swarm Optimization (PSO), Simulated Annealing (SA) and Distributed Genetic Algorithm (DGA), called WMN-PSOSA-DGA. In this paper, we compare the performance of Random Inertia Weight Method (RIWM) and Rational Decrement of Vmax Method (RDVM) for WMNs by using WMN-PSOSA-DGA hybrid simulation system considering Stadium distribution of mesh clients. Simulation results show that RDVM has better performance than RIWM.

1 Introduction

Wireless Mesh Networks (WMNs) are gaining a lot of attention because of their low cost nature that makes them attractive for providing wireless Internet connectivity. A WMN is dynamically self-organized and self-configured, with the nodes in the network automatically establishing and maintaining mesh connectivity among themselves (creating, in effect, an ad hoc network). This feature brings many advantages to WMNs such as low up-front cost, easy network maintenance, robustness and reliable service coverage [1, 6, 13–15].

Mesh node placement in WMN can be seen as a family of problems, which are shown to be computationally hard to solve for most of the formulations [9, 25]. We consider the version of the mesh router nodes placement problem in which we are

L. Barolli et al. (Eds.): EIDWT 2022, LNDECT 118, pp. 386–394, 2022.
https://doi.org/10.1007/978-3-030-95903-6_41

given a grid area where to deploy a number of mesh router nodes and a number of mesh client nodes of fixed positions (of an arbitrary distribution) in the grid area. The objective is to find a location assignment for the mesh routers to the cells of the grid area that maximizes the network connectivity and client coverage. Network connectivity is measured by Size of Giant Component (SGC) of the resulting WMN graph, while the user coverage is simply the number of mesh client nodes that fall within the radio coverage of at least one mesh router node and is measured by Number of Covered Mesh Clients (NCMC). Node placement problems are known to be computationally hard to solve [10, 26]. In previous works, some intelligent algorithms have been investigated for node placement problem [8, 12, 27].

In [19], we implemented a Particle Swarm Optimization (PSO) and Simulated Annealing (SA) based simulation system, called WMN-PSOSA. Also, we implemented another simulation system based on Genetic Algorithm (GA), called WMN-GA [2, 11], for solving node placement problem in WMNs. Then, we designed a hybrid intelligent system based on PSO, SA and DGA, called WMN-PSOSA-DGA [18].

In this paper, we compare the performance of Random Inertia Weight Method (RIWM) and Rational Decrement of Vmax Method (RDVM) for WMNs by using WMN-PSOSA-DGA hybrid simulation system considering Stadium distribution of mesh clients.

The rest of the paper is organized as follows. We present our designed and implemented hybrid simulation system in Sect. 2. The simulation results are given in Sect. 3. Finally, we give conclusions and future work in Sect. 4.

2   Proposed and Implemented Simulation System

Distributed Genetic Algorithms (DGAs) are capable of producing solutions with higher efficiency (in terms of time) and efficacy (in terms of better quality solutions). They have shown their usefulness for the resolution of many computationally hard combinatorial optimization problems. Also, Particle Swarm Optimization (PSO) and Simulated Annealing (SA) are suitable for solving NP-hard problems.

2.1   Velocities and Positions of Particles

WMN-PSOSA-DGA decides the velocity of particles by a random process considering the area size. For instance, when the area size is $W \times H$, the velocity is decided randomly from $-\sqrt{W^2 + H^2}$ to $\sqrt{W^2 + H^2}$. Each particle's velocities are updated by simple rule.

For SA mechanism, next positions of each particle are used for neighbor solution s'. The fitness function f gives points to the current solution s. If $f(s')$ is larger than $f(s)$, the s' is better than s so the s is updated to s'. However, if $f(s')$ is not larger than $f(s)$, the s may be updated by using the probability of $\exp\left[\frac{f(s')-f(s)}{T}\right]$. Where T is called the "Temperature value" which is decreased with the computation so that the probability to update will be decreased. This mechanism of SA is called a cooling schedule and the next Temperature value of computation is calculated as $T_{n+1} = \alpha \times T_n$. In this paper, we set the starting temperature, ending temperature and number of iterations. We calculate α as:

$$\alpha = \left(\frac{\text{SA ending temperature}}{\text{SA starting temperature}} \right)^{1.0/\text{number of iterations}}.$$

It should be noted that the positions are not updated but the velocities are updated in the case when the solusion s is not updated.

2.2   Routers Replacement Methods

A mesh router has x, y positions and velocity. Mesh routers are moved based on velocities. There are many router replacement methods. In this paper, we use RIWM and RDVM.

Constriction Method (CM)

CM is a method which PSO parameters are set to a week stable region ($\omega = 0.729, C_1 = C2 = 1.4955$) based on analysis of PSO by M. Clerc et al. [3,7,21].

Random Inertia Weight Method (RIWM)

In RIWM, the ω parameter is changing randomly from 0.5 to 1.0. The C_1 and C_2 are kept 2.0. The ω can be estimated by the week stable region. The average of ω is 0.75 [5,23].

Linearly Decreasing Inertia Weight Method (LDIWM)

In LDIWM, C_1 and C_2 are set to 2.0, constantly. On the other hand, the ω parameter is changed linearly from unstable region ($\omega = 0.9$) to stable region ($\omega = 0.4$) with increasing of iterations of computations [4,24].

Linearly Decreasing Vmax Method (LDVM)

In LDVM, PSO parameters are set to unstable region ($\omega = 0.9, C_1 = C_2 = 2.0$). A value of V_{max} which is maximum velocity of particles is considered. With increasing of iteration of computations, the V_{max} is kept decreasing linearly [5,20,22].

Rational Decrement of Vmax Method (RDVM)

In RDVM, PSO parameters are set to unstable region ($\omega = 0.9, C_1 = C_2 = 2.0$). The V_{max} is kept decreasing with the increasing of iterations as:

$$V_{max}(x) = \sqrt{W^2 + H^2} \times \frac{T - x}{x}.$$

Where, W and H are the width and the height of the considered area, respectively. Also, T and x are the total number of iterations and a current number of iteration, respectively [17].

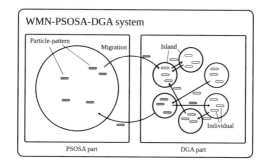

Fig. 1. Model of WMN-PSOSA-DGA migration.

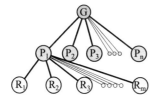

G: Global Solution
P: Particle-pattern
R: Mesh Router
n: Number of Particle-patterns
m: Number of Mesh Routers

Fig. 2. Relationship among global solution, particle-patterns and mesh routers in PSOSA part.

2.3 DGA Operations

Population of individuals: Unlike local search techniques that construct a path in the solution space jumping from one solution to another one through local perturbations, DGA use a population of individuals giving thus the search a larger scope and chances to find better solutions. This feature is also known as "exploration" process in difference to "exploitation" process of local search methods.

Selection: The selection of individuals to be crossed is another important aspect in DGA as it impacts on the convergence of the algorithm. Several selection schemes have been proposed in the literature for selection operators trying to cope with premature convergence of DGA. There are many selection methods in GA. In our system, we implement 2 selection methods: Random method and Roulette wheel method.

Crossover operators: Use of crossover operators is one of the most important characteristics. Crossover operator is the means of DGA to transmit best genetic features of parents to offsprings during generations of the evolution process. Many methods for crossover operators have been proposed such as Blend Crossover (BLX-α), Unimodal Normal Distribution Crossover (UNDX), Simplex Crossover (SPX).

Mutation operators: These operators intend to improve the individuals of a population by small local perturbations. They aim to provide a component of randomness in the neighborhood of the individuals of the population. In our system, we implemented two mutation methods: uniformly random mutation and boundary mutation.

Escaping from local optimal: GA itself has the ability to avoid falling prematurely into local optimal and can eventually escape from them during the search process. DGA has one more mechanism to escape from local optimal by considering some islands. Each island computes GA for optimizing and they migrate its gene to provide the ability to avoid from local optimal.

Convergence: The convergence of the algorithm is the mechanism of DGA to reach to good solutions. A premature convergence of the algorithm would cause that all individuals of the population be similar in their genetic features and thus the search would result ineffective and the algorithm getting stuck into local optimal. Maintaining the diversity of the population is therefore very important to this family of evolutionary algorithms.

In following, we present fitness function, migration function, particle pattern and gene coding.

2.4 Fitness and Migration Functions

The determination of an appropriate fitness function, together with the chromosome encoding are crucial to the performance. Therefore, one of most important thing is to decide the determination of an appropriate objective function and its encoding. In our case, each particle-pattern and gene has an own fitness value which is comparable and compares it with other fitness value in order to share information of global solution. The fitness function follows a hierarchical approach in which the main objective is to maximize the SGC in WMN. Thus, the fitness function of this scenario is defined as:

$$\text{Fitness} = 0.7 \times \text{SGC}(\boldsymbol{x}_{ij}, \boldsymbol{y}_{ij}) + 0.3 \times \text{NCMC}(\boldsymbol{x}_{ij}, \boldsymbol{y}_{ij}).$$

Our implemented simulation system uses Migration function as shown in Fig. 1. The Migration function swaps solutions between PSOSA part and DGA part.

2.5 Particle-Pattern and Gene Coding

In order to swap solutions, we design particle-patterns and gene coding carefully. A particle is a mesh router. Each particle has position in the considered area and velocities. A fitness value of a particle-pattern is computed by combination of mesh routers and mesh clients positions. In other words, each particle-pattern is a solution as shown is Fig. 2.

A gene describes a WMN. Each individual has its own combination of mesh nodes. In other words, each individual has a fitness value. Therefore, the combination of mesh nodes is a solution.

Fig. 3. Stadium distribution.

Table 1. WMN-PSOSA-DGA parameters.

Parameters	Values
Clients distribution	Stadium distribution
Area size	32.0×32.0
Number of mesh routers	24
Number of mesh clients	48
Number of GA islands	16
Number of Particle-patterns	32
Number of migrations	200
Evolution steps	320
Radius of a mesh router	$2.0 - 3.5$
Selection method	Roulette wheel method
Crossover method	SPX
Mutation method	Boundary mutation
Crossover rate	0.8
Mutation rate	0.2
SA Starting value	10.0
SA Ending value	0.01
Total number of iterations	64000
Replacement method	RIWM, RDVM

3   Simulation Results

In this section, we show simulation results. In this work, we analyze the performance of WMNs by using the WMN-PSOSA-DGA hybrid intelligent simulation system considering Stadium distribution [16] as shown in Fig. 3.

We carried out the simulations 10 times in order to avoid the effect of randomness and create a general view of results. We show the parameter setting for WMN-PSOSA-DGA in Table 1.

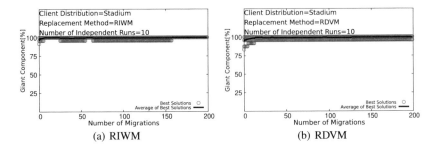

Fig. 4. Simulation results of WMN-PSOSA-DGA for SGC.

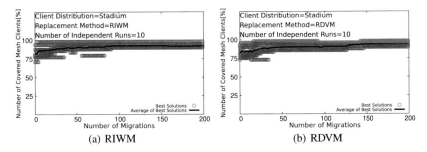

Fig. 5. Simulation results of WMN-PSOSA-DGA for NCMC.

We show simulation results in Fig. 4 and Fig. 5. We consider number of mesh routers 24. We see that the SGC can be maximized for both replacement methods. However, for NCMC, the performance of RDVM is better compared with the performance of RIWM.

4 Conclusions

In this work, we evaluated the performance of WMNs by using a hybrid simulation system based on PSO, SA and DGA (called WMN-PSOSA-DGA) considering Stadium distribution of mesh clients. Simulation results show that RDVM achieved the better performance compared with the case of RIWM.

In our future work, we would like to evaluate the performance of the proposed system for different parameters and patterns.

References

1. Akyildiz, I.F., Wang, X., Wang, W.: Wireless mesh networks: a survey. Comput. Netw. **47**(4), 445–487 (2005)
2. Barolli, A., Sakamoto, S., Ozera, K., Barolli, L., Kulla, E., Takizawa, M.: Design and implementation of a hybrid intelligent system based on particle swarm optimization and distributed genetic algorithm. In: International Conference on Emerging Internetworking, pp. 79–93. Springer, Data & Web Technologies (2018). https://doi.org/10.1007/978-3-319-75928-9_7

3. Barolli, A., Sakamoto, S., Durresi, H., Ohara, S., Barolli, L., Takizawa, M.: A comparison study of constriction and linearly decreasing vmax replacement methods for wireless mesh networks by WMN-PSOHC-DGA simulation system. In: International Conference on P2P, pp. 26–34. Parallel, Grid, Cloud and Internet Computing, Springer (2019)

4. Barolli, A., Sakamoto, S., Ohara, S., Barolli, L., Takizawa, M.: Performance analysis of WMNs by WMN-PSOHC-DGA simulation system considering linearly decreasing inertia weight and linearly decreasing vmax replacement methods. In: Barolli, L., Nishino, H., Miwa, H. (eds.) INCoS 2019. AISC, vol. 1035, pp. 14–23. Springer, Cham (2020). https://doi.org/10.1007/978-3-030-29035-1_2

5. Barolli, A., Sakamoto, S., Ohara, S., Barolli, L., Takizawa, M.: Performance Analysis of WMNs by WMN-PSOHC-DGA Simulation System Considering Random Inertia Weight and Linearly Decreasing Vmax Router Replacement Methods. In: Barolli, L., Hussain, F.K., Ikeda, M. (eds.) CISIS 2019. AISC, vol. 993, pp. 13–21. Springer, Cham (2020). https://doi.org/10.1007/978-3-030-22354-0_2

6. Barolli, A., Sakamoto, S., Ohara, S., Barolli, L., Takizawa, M.: Performance evaluation of WMNs using WMN-PSOHC-DGA considering evolution steps and computation time. In: Barolli, L., Okada, Y., Amato, F. (eds.) EIDWT 2020. LNDECT, vol. 47, pp. 127–137. Springer, Cham (2020). https://doi.org/10.1007/978-3-030-39746-3_14

7. Clerc, M., Kennedy, J.: The particle swarm-explosion, stability, and convergence in a multidimensional complex space. IEEE Trans. Evol. Comput. **6**(1), 58–73 (2002)

8. Girgis, M.R., Mahmoud, T.M., Abdullatif, B.A., Rabie, A.M.: Solving the wireless mesh network design problem using genetic algorithm and simulated annealing optimization methods. Int. J. Comput. Appl. **96**(11), 1–10 (2014)

9. Hirata, A., Oda, T., Saito, N., Hirota, M., Katayama, K.: A coverage construction method based hill climbing approach for mesh router placement optimization. In: International Conference on Broadband and Wireless Computing, pp. 355–364. Springer, Communication and Applications (2020). https://doi.org/10.1007/978-3-030-61108-8_35

10. Maolin, T., et al.: Gateways placement in backbone wireless mesh networks. Int. J. Commun. Netw. Syst. Sci. **2**(1), 44 (2009)

11. Matsuo, K., Sakamoto, S., Oda, T., Barolli, A., Ikeda, M., Barolli, L.: Performance analysis of WMNs by WMN-GA simulation system for two WMN architectures and different TCP congestion-avoidance algorithms and client distributions. Int. J. Commun. Netw. Distrib. Syst. **20**(3), 335–351 (2018)

12. Naka, S., Genji, T., Yura, T., Fukuyama, Y.: A hybrid particle swarm optimization for distribution state estimation. IEEE Trans. Power Syst. **18**(1), 60–68 (2003)

13. Ohara, S., Barolli, A., Sakamoto, S., Barolli, L.: Performance analysis of WMNs by WMN-PSODGA simulation system considering load balancing and client uniform distribution. In: Barolli, L., Xhafa, F., Hussain, O.K. (eds.) IMIS 2019. AISC, vol. 994, pp. 25–38. Springer, Cham (2020). https://doi.org/10.1007/978-3-030-22263-5_3

14. Ohara, S., Durresi, H., Barolli, A., Sakamoto, S., Barolli, L.: A hybrid intelligent simulation system for node placement in WMNs considering load balancing: a comparison study for exponential and normal distribution of mesh clients. In: International Conference on Broadband and Wireless Computing, pp. 555–569. Springer, Communication and Applications (2019). https://doi.org/10.1007/978-3-030-33506-9_50

15. Ohara, S., Qafzezi, E., Barolli, A., Sakamoto, S., Liu, Y., Barolli, L.: WMN-PSODGA-An intelligent hybrid simulation system for WMNs considering load balancing: a comparison for different client distributions. Int. J. Distrib. Syst. Technol. (IJDST) **11**(4), 39–52 (2020)
16. Sakamoto, S., Oda, T., Bravo, A., Barolli, L., Ikeda, M., Xhafa, F.: WMN-SA system for node placement in WMNS: evaluation for different realistic distributions of mesh clients. In: The IEEE 28th International Conference on Advanced Information Networking and Applications (AINA-2014), IEEE, pp. 282–288 (2014)
17. Sakamoto, S., Oda, T., Ikeda, M., Barolli, L., Xhafa, F.: Implementation of a new replacement method in WMN-PSO simulation system and its performance evaluation. In: The 30th IEEE International Conference on Advanced Information Networking and Applications (AINA-2016), pp. 206–211 (2016). https://doi.org/10.1109/AINA.2016.42
18. Sakamoto, S., Barolli, A., Barolli, L., Takizawa, M.: Design and implementation of a hybrid intelligent system based on particle swarm optimization, hill climbing and distributed genetic algorithm for node placement problem in WMNs: a comparison study. In: The 32nd IEEE International Conference on Advanced Information Networking and Applications (AINA-2018), pp. 678–685. IEEE (2018)
19. Sakamoto, S., Ozera, K., Ikeda, M., Barolli, L.: Implementation of intelligent hybrid systems for node placement problem in WMNs considering particle swarm optimization, hill climbing and simulated annealing. Mob. Netw. Appl. **23**(1), 27–33 (2018). https://doi.org/10.1007/s11036-017-0897-7
20. Sakamoto, S., Ohara, S., Barolli, L., Okamoto, S.: Performance evaluation of WMNs by WMN-PSOHC system considering random inertia weight and linearly decreasing Vmax replacement methods. In: Barolli, L., Nishino, H., Enokido, T., Takizawa, M. (eds.) NBiS - 2019 2019. AISC, vol. 1036, pp. 27–36. Springer, Cham (2020). https://doi.org/10.1007/978-3-030-29029-0_3
21. Sakamoto, S., Ohara, S., Barolli, L., Okamoto, S.: Performance evaluation of WMNs WMN-PSOHC system considering constriction and linearly decreasing inertia weight replacement methods. In: International Conference on Broadband and Wireless Computing, pp. 22–31. Springer, Communication and Applications (2019). https://doi.org/10.1007/978-3-030-33506-9_3
22. Schutte, J.F., Groenwold, A.A.: A study of global optimization using particle swarms. J. Global Optim. **31**(1), 93–108 (2005). https://doi.org/10.1007/s10898-003-6454-x
23. Shi, Y.: Particle swarm optimization. IEEE Connections **2**(1), 8–13 (2004)
24. Shi, Y., Eberhart, R.C.: Parameter selection in particle swarm optimization. In: Porto, V.W., Saravanan, N., Waagen, D., Eiben, A.E. (eds.) EP 1998. LNCS, vol. 1447, pp. 591–600. Springer, Heidelberg (1998). https://doi.org/10.1007/BFb0040810
25. Vanhatupa, T., Hannikainen, M., Hamalainen, T.: Genetic algorithm to optimize node placement and configuration for WLAN planning. In: The 4th IEEE International Symposium on Wireless Communication Systems, pp. 612–616 (2007)
26. Wang, J., Xie, B., Cai, K., Agrawal, D.P.: Efficient mesh router placement in wireless mesh networks. In: Proceedings of IEEE International Conference on Mobile Ad hoc and Sensor Systems (MASS-2007), pp. 1–9 (2007)
27. Yaghoobirafi, K., Nazemi, E.: An autonomic mechanism based on ant colony pattern for detecting the source of incidents in complex enterprise systems. Int. J. Grid Util. Comput. **10**(5), 497–511 (2019)

An Energy-Efficient Process Replication to Reduce the Execution of Meaningless Replicas

Tomoya Enokido[1]([✉]), Dilawaer Duolikun[2], and Makoto Takizawa[3]

[1] Faculty of Business Administration, Rissho University, 4-2-16, Osaki, Shinagawa-ku, Tokyo 141-8602, Japan
eno@ris.ac.jp
[2] Department of Advanced Sciences, Faculty of Science and Engineering, Hosei University, 3-7-2, Kajino-cho, Koganei-shi, Tokyo 184-8584, Japan
[3] Research Center for Computing and Multimedia Studies, Hosei University, 3-7-2, Kajino-cho, Koganei-shi, Tokyo 184-8584, Japan
makoto.takizawa@computer.org

Abstract. In server cluster systems like cloud computing systems, process replication approaches are widely used to provide available and reliable distributed application services. However, process replication approaches imply the large amount of electric energy consumption than non-replication approaches. In this paper, an *RATB-DSTFMRT (Redundant Active Time-Based algorithm with Differentiating Starting Time and Forcing Meaningless Replicas to Terminate)* algorithm is newly proposed to reduce the total electric energy consumption of a server cluster by not only making each pair of process replicas not simultaneously start but also forcing meaningless replicas to terminate. The evaluation results show the total electric energy consumption of a server cluster can be reduced in the RATB-DSTFMRT algorithm.

Keywords: The RATB-DSTFMRT algorithm · Process replication · Energy consumption · Green computing · Server cluster

1 Introduction

In current information systems, a huge amount of data is gathered from various types of devices like sensors [1] and manipulated to provide various types of distributed applications like vehicle network services [2]. Server cluster systems [3–6] equipped with virtual machines [7] like a cloud computing system [8] are widely used to implement these distributed applications since a server cluster system can provide scalable, available, and reliable computing systems. A server cluster system is composed of a large number of physical servers and multiple virtual machines are supported by each physical server. Application processes are performed on virtual machines supported by each physical server to more efficiently utilize computation resources of each physical server in a server cluster

© The Author(s), under exclusive license to Springer Nature Switzerland AG 2022
L. Barolli et al. (Eds.): EIDWT 2022, LNDECT 118, pp. 395–405, 2022.
https://doi.org/10.1007/978-3-030-95903-6_42

system. In a server cluster system, some physical servers might stop by fault [9]. Then, virtual machines supported by the physical servers also stop and application processes performed on the virtual machines cannot successfully terminate. As a result, application services stop due to the fault of the physical servers. In order to prevent an application service from stopping by physical server faults, process replication approaches [10–13] are widely used. In the process replication approaches, multiple replicas of each application process are performed on multiple virtual machines which are performed on different physical servers in a server cluster system. Then, an application service can be continuously provided even if some physical servers stop by fault. However, a large amount of electric energy is consumed in a server cluster system than non-replication approaches since multiple replicas of each application process are performed on multiple virtual machines which are performed on different physical servers.

In our previous studies, the *RATB-DSTPR (Redundant Active Time-Based algorithm with Differentiating Starting Time of Process Replicas)* [14] algorithm was proposed to reduce the total electric energy consumption of a server cluster and the average response time of each request process by making each pair of process replicas not simultaneously start. Suppose a replica of a request process successfully terminates on a virtual machine and a load balancer receives a reply of the replica from the virtual machine. Here, the other replicas of the request process still being performed on other virtual machines are *meaningless* since the request process can commit without performing meaningless replicas. In the RATB-DSTPR algorithm, a load balancer serially forwards a request process to each selected virtual machine every δ time unit [msec]. The load balancer stops forwarding the request process to selected virtual machines if the load balancer has received a reply of the request process. As a result, the number of replicas for a request process can be fewer than the required redundancy of the request process. Hence, in the RATB-DSTPR algorithm, the total electric energy consumption of a server cluster can be reduced.

In this paper, an *RATB-DSTFMRT (Redundant Active Time-Based algorithm with Differentiating Starting Time and Forcing Meaningless Replicas to Terminate)* algorithm is newly proposed to furthermore reduce the total electric energy consumption of a server cluster and the average response time of each process by not only making each pair of replicas not simultaneously start but also forcing meaningless replicas to terminate. In the RATB-DSTFMRT algorithm, each time a replica r^i of a request process p^i successfully terminates on a virtual machine VM_{kt} performed on a server s_t, the virtual machine VM_{kt} sends a termination notification message of the replica r^i to every other selected virtual machine VM_{ku} for performing replicas of the same request process p^i. On receipt of the termination notification message of the replica r^i from the virtual machine VM_{kt}, the meaningless replica is forced to terminate on each virtual machine VM_{ku}. As a result, the total electric energy consumption of a server cluster can be more reduced in the RATB-DSTFMRT algorithm than the RATB-DSTPR algorithm. In addition, the response time of each replica can be more reduced in the RATB-DSTFMRT algorithm than the RATB-DSTPR algorithm since the computation resources of each virtual machine to perform meaningless replicas can be used to perform other replicas. Evaluation results show the average total processing electric energy consumption of a server cluster and the average response time of each process can be more reduced in the RATB-DSTFMRT algorithm than the RATB-DSTPR algorithm.

In Sect. 2, a system model of this paper is discussed. In Sect. 3, the RATB-DSTFMRT algorithm is newly proposed. In Sect. 4, the RATB-DSTFMRT algorithm is evaluated compared with the RATB-DSTPR algorithm.

2   System Model

2.1   A Server Cluster System

A server cluster S is composed of multiple physical servers s_1, ..., s_n ($n \geq 1$). Each server s_t is equipped with a multi-core CPU. Let nc_t ($nc_t \geq 1$) be the total number of cores in a server s_t and C_t be a set of homogeneous cores c_{1t}, ..., $c_{nc_t t}$ in a server s_t. Let ct_t ($ct_t \geq 1$) be the total number of threads on each core c_{ht} in a server s_t and nt_t ($nt_t \geq 1$) be the total number of threads in a server s_t, i.e. $nt_t = nc_t \cdot ct_t$. Let TH_t be a set of threads th_{1t}, ..., $th_{nt_t t}$ in a server s_t. A set V_t of virtual machines VM_{1t}, ..., $VM_{nt_t t}$ are supported by a server s_t. Each virtual machine VM_{kt} is exclusively performed on one thread th_{kt} in a server s_t. In this paper, a term *process* stands for a *computation type application processes* (*computation processes*) which mainly consume CPU resources of a virtual machine VM_{kt}. Let NF be the maximum number of physical servers which concurrently stop by fault in the server cluster S. A notation rd^i shows the *redundancy* of a process p^i. Here, $NF + 1 \leq rd^i \leq n$. Let VMS^i be a subset of virtual machines in the server cluster S to redundantly perform the request process p^i, i.e. $rd^i = |VMS^i|$. Each time a load balancer K receives a request process p^i from a client cl^i, the load balancer K selects a subset VMS^i of virtual machines for the request process p^i and forwards the process p^i to every virtual machine in the subset VMS^i. On receipt of a request process p^i, a virtual machine VM_{kt} creates and performs a replica p^i_{kt} of a process p^i. Then, the virtual machine VM_{kt} sends a reply r^i_{kt} to the load balancer K. The load balancer K takes only the first reply r^i_{kt} and ignores every other reply.

2.2   Computation Model of a Virtual Machine

A virtual machine VM_{kt} is *active* iff (if and only if) at least one replica is performed on the virtual machine VM_{kt}. Otherwise, a virtual machine VM_{kt} is *idle*. A core c_{ht} is *active* iff at least one virtual machine VM_{kt} is active on a thread th_{kt} in the core c_{ht}. Otherwise, a core c_{ht} is *idle*. Replicas which are being performed on a virtual machine VM_{kt} at time τ are *current*. Let $CP_{kt}(\tau)$ be a set of current replicas being performed on a virtual machine VM_{kt} at time τ and $NC_{kt}(\tau)$ be the number of current replicas performed on the virtual machine VM_{kt} at time τ. Let $minT^i_{kt}$ be the minimum computation time of a replica p^i_{kt}

performed on a virtual machine VM_{kt}. We assume $minT_{1t}^i = \cdots = minT_{nt_t t}^i$ in a server s_t and $minT^i = minT_{kt}^i$ on the fastest server s_t. We assume one virtual computation step [vs] is performed for one time unit [tu] on a virtual machine VM_{kt} in the fastest server s_t. The maximum computation rate $Maxf_{kt}$ of a virtual machine VM_{kt} in the fastest server s_t is 1 [vs/msec]. $Maxf = max(Maxf_{k1}, ..., Maxf_{kn})$. A replica p_{kt}^i is considered to be composed of VS_{kt}^i virtual computation steps, i.e. $VS_{kt}^i = minT_{kt}^i \cdot Maxf = minT_{kt}^i$ [vs].

The computation rate $f_{kt}^i(\tau)$ of a replica p_{kt}^i performed on a virtual machine VM_{kt} at time τ is given as Eq. (1) based on our experiment [15]:

$$f_{kt}^i(\tau) = \alpha_{kt}(\tau) \cdot VS^i / (minT_{kt}^i \cdot NC_{kt}(\tau)) \cdot \beta_{kt}(nv_{kt}(\tau)). \tag{1}$$

A notation $\alpha_{kt}(\tau)$ $(0 \leq \alpha_{kt}(\tau) \leq 1)$ shows the *computation degradation ratio* of a virtual machine VM_{kt} at time τ. The computation degradation ratio $\alpha_{kt}(\tau)$ is assumed to be $\varepsilon_{kt}^{NC_{kt}(\tau)-1}$ where $0 \leq \varepsilon_{kt} \leq 1$. $\alpha_{kt}(\tau) = 1$ if $NC_{kt}(\tau) = 1$. $\alpha_{kt}(\tau_1) \leq \alpha_{kt}(\tau_2) \leq 1$ if $NC_{kt}(\tau_1) \geq NC_{kt}(\tau_2)$. Let $nv_{kt}(\tau)$ be the total number of active virtual machines on a core which performs a virtual machine VM_{kt} at time τ. A notation $\beta_{kt}(nv_{kt}(\tau))$ shows the *performance degradation ratio* of a virtual machine VM_{kt} at time τ $(0 \leq \beta_{kt}(nv_{kt}(\tau)) \leq 1)$ where multiple virtual machines are active on the same core. $\beta_{kt}(nv_{kt}(\tau)) = 1$ if $nv_{kt}(\tau) = 1$. $\beta_{kt}(nv_{kt}(\tau_1)) \leq \beta_{kt}(nv_{kt}(\tau_2))$ if $nv_{kt}(\tau_1) \geq nv_{kt}(\tau_2)$.

Suppose a replica p_{kt}^i is started on a virtual machine VM_{kt} at time st_{kt}^i and terminates at time et_{kt}^i, respectively. The total computation time T_{kt}^i [msec] of a replica p_{kt}^i is et_{kt}^i - st_{kt}^i and $VS^i = \sum_{\tau=st_{kt}^i}^{et_{kt}^i} f_{kt}^i(\tau)$ [vs]. The computation laxity $lc_{kt}^i(\tau)$ [vs] of a replica p_{kt}^i at time τ is VS^i - $\sum_{x=st_{kt}^i}^{\tau} f_{kt}^i(x)$.

2.3 Power Consumption Model of a Physical Server

The *PCSV (Power Consumption Model of a Server with Virtual machines)* model [16] is proposed in our previous studies. A notation $E_t(\tau)$ shows the electric power [W] of a server s_t at time τ. Let $maxE_t$ and $minE_t$ be the maximum and minimum electric power [W] of a server s_t, respectively. A notation $ac_t(\tau)$ shows the total number of active cores in a server s_t at time τ. Let $minC_t$ be the electric power [W] where at least one core c_{ht} is active on a server s_t and cE_t be the electric power [W] consumed by a server s_t to make one core active. In the PCSV model, the electric power $E_t(\tau)$ [W] of a server s_t at time τ is given as Eq. (2) [16]:

$$E_t(\tau) = minE_t + \sigma_t(\tau) \cdot (minC_t + ac_t(\tau) \cdot cE_t). \tag{2}$$

Here, $\sigma_t(\tau) = 1$ if at least one core is active in a server s_t at time τ. Otherwise, $\sigma_t(\tau) = 0$.

The *processing power* $PE_t(\tau)$ [W] of a server s_t at time τ to perform replicas of computation processes on virtual machines is $E_t(\tau)$ - $minE_t$. The total processing electric energy consumption $TPE_t(\tau_1, \tau_2)$ [J] of a server s_t between time τ_1 and τ_2 is $\sum_{\tau=\tau_1}^{\tau_2} PE_t(\tau)$.

3   Energy-Efficient Process Replication

3.1   The RATB-DSTPR Algorithm

In our previous studies, the *RATB-DSTPR* (*Redundant Active Time-Based algorithm with Differentiating Starting Time of Process Replicas*) algorithm [14] is proposed to reduce the total electric energy consumption of a server cluster by differentiating the starting time of each replica. A notation $iACT_{ht}(\tau)$ shows the *increased active time* [11] of a core c_{ht} in a server s_t at time τ. The increased active time $iACT_{ht}(\tau)$ shows a period of time when a core c_{ht} in a server s_t is active to perform every current replica on every active virtual machine performed on threads allocated to the core c_{ht} at time τ. In the RATB-DSTPR algorithm, the increased active time $iACT_{ht}(\tau)$ of each core c_{ht} in a server s_t at time τ is estimated based on the response time of each replica p_{kt}^i performed on each virtual machine VM_{kt} in the server s_t.

The total processing electric energy laxity $tpel_t(\tau)$ [J] [11] shows how much electric energy a server s_t consumes to perform every current replica on every active virtual machine in the server s_t at time τ. The total processing electric energy laxity $tpel_t(\tau)$ of a server s_t at time τ is given as Eq. (3):

$$tpel_t(\tau) \;=\; minC_t \;+\; \sum_{h=1}^{nc_t}(iATC_{ht}(\tau) \cdot cE_t). \qquad (3)$$

Let $TPEL^i(\tau)$ be the total processing electric energy laxity of a server cluster S at time τ where rd^i replicas of a new request process p^i is allocated to a set VMS^i of virtual machines in a server cluster S. In the RATB-DSTPR algorithm, each time a load balancer K receives a new request process p^i, the load balancer K selects a set VMS^i of virtual machines where the total processing electric energy laxity $TPEL^i(\tau)$ is the minimum for the new request process p^i.

In the RATB-DSTPR algorithm, the number of replicas for a process p^i is fewer than the redundancy rd^i by making each pair of process replicas p_{kt}^i not simultaneously start while rd^i replicas are performed for every request process p^i in an active replication [11]. As a result, in the RATB-DSTPR algorithm, the total electric energy consumption of a server cluster can be reduced. In the RATB-DSTPR algorithm, a replica is modeled to be a sequence of operations. **[Definition]** A subsequence of a replica p_{ku}^i to be performed after another replica p_{kt}^i successfully terminates is a *meaningless*.

A notation δ^i shows the inter-request time [msec] to forward a request process p^i to each virtual machine VM_{kt} in a subset VMS^i. In the RATB-DSTPR algorithm, a load balancer K serially forwards a request process p^i to each virtual machine VM_{kt} in the subset VMS^i every δ^i time unit. Suppose a load balancer K receives a request process p^i and the redundancy rd^i of the process p^i is three $(rd^i = 3)$ as shown in Fig. 1. Suppose the load balancer K selects a subset VMS^i $(= \{VM_{kt}, VM_{ku}, VM_{kv}\})$ of virtual machines. At time τ_1, the load balancer K forwards the request process p^i to the virtual machine VM_{kt}. At time τ_2 $(= \tau_1 + \delta^i)$, the load balancer K forwards the request process p^i to the second virtual machine VM_{ku}. Suppose the load balancer K receives a first reply r^i_{kt} from the virtual machine VM_{kt} while forwarding the request process p^i to the virtual machine VM_{ku}. Here, a replica p^i_{ku} performed on the virtual machine VM_{ku} is meaningless since the request process p^i can commit without performing the replica p^i_{ku}. On the other hand, the load balancer K stops forwarding the third request to the virtual machine VM_{kv} since the load balancer K can commit the request process p^i without performing the replica p^i_{kv}. In the RATB-DSTPR algorithm, the total processing electric energy consumption of a server cluster can be reduced by making each pair of process replicas not simultaneously start since the execution of meaningless replicas can be reduced.

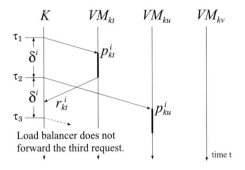

Fig. 1. The stop forwarding.

3.2   The RATB-DSTFMRT Algorithm

In this paper, an *RATB-DSTFMRT (Redundant Active Time-Based algorithm with Differentiating Starting Time and Forcing Meaningless Replicas to Terminate)* algorithm is newly proposed to reduce the total electric energy consumption of a server cluster and the average response time of each replica by not only making each pair of process replicas not simultaneously start but also forcing meaningless replicas to terminate. The RATB-DSTFMRT algorithm can furthermore reduce the total electric energy consumption of a server cluster and the average response time of each process than the RATB-DSTPR algorithm by forcing meaningless replica to terminate. In the RATB-DSTFMRT algorithm, each time a load balancer K receives a new request process p^i, the load balancer K selects a subset VMS^i of virtual machines and serially forwards the request process p^i to the virtual machines in the subset VMS^i by using the same way as the RATB-DSTPR algorithm. Each time a replica p^i_{kt} successfully terminates on a virtual machine VM_{kt}, the virtual machine VM_{kt} sends a termination notification $TN(p^i_{kt})$ message of the replica p^i_{kt} to both the load balancer K and every other virtual machine in the subset VMS^i. The termination notification $TN(p^i_{kt})$ includes a reply r^i_{kt}. On receipt of $TN(p^i_{kt})$ from a virtual machine

VM_{kt}, a meaningless replica p^i_{ku} is forced to terminate on each virtual machine VM_{ku} by the following **Termination** procedure:

Termination($TN(p^i_{kt})$) {
    if a replica p^i_{ku} is performing, p^i_{ku} is forced to terminate;
    else $TN(p^i_{kt})$ is neglected;
}

    Suppose a replica p^i_{kt} successfully terminates on a virtual machine VM_{kt} at time τ_3 while another replica p^i_{ku} is still being performed on another virtual machine VM_{ku} as shown in Fig. 2. Here, the virtual machine VM_{kt} sends a termination notification $TN(p^i_{kt})$ message of the replica p^i_{kt} to both the load balancer K and the virtual machine VM_{ku}. On receipt of a $TN(p^i_{kt})$ message from the virtual machine VM_{kt}, the replica p^i_{ku} is forced to terminate on the virtual machine VM_{ku} since the replica p^i_{ku} is meaningless and the request process p^i can commit without performing the replica p^i_{ku}.

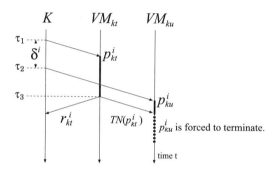

Fig. 2. Forcing meaningless replica to terminate.

    The total processing electric energy consumption of a server cluster can be more reduced in the RATB-DSTFMRT algorithm than the RATB-DSTPR algorithm by forcing meaningless replicas to terminate. In addition, the computation resources of each virtual machine to perform meaningless replicas can be used to perform other replicas. As a result, the response time of each replica can be more reduced in the RATB-DSTFMRT algorithm than the RATB-DSTPR algorithm.

4   Evaluation

We evaluate the RATB-DSTFMRT algorithm in terms of the total processing electric energy consumption [KJ] of a homogeneous server cluster S and the average response time [msec] of each process p^i compared with the RATB-DSTPR algorithm [11]. A homogeneous server cluster S is composed of three physical servers s_1, s_2, and s_3 ($n = 3$). Parameters of every physical server s_t and virtual machine VM_{kt} are obtained from the experiment [16] as shown in Tables 1 and 2. Every physical server s_t is equipped with a dual-core CPU ($nc_t = 2$). Two threads are bounded for each core in each server s_t ($ct_t = 2$) and the total number nt_t of threads in each server s_t is four. Hence, there are twelve virtual machines in the server cluster S. We assume the fault probability fr_t for every server s_t is the same $fr = 0.1$.

Table 1. Homogeneous cluster S ($t = 1, 2, 3$).

Server	nc_t	ct_t	nt_t	$minE_t$	$minC_t$	cE_t	$maxE_t$
s_t	2	2	4	14.8 [W]	6.3 [W]	3.9 [W]	33.8 [W]

Table 2. Parameters of virtual machines ($k = 1, ..., 4$ and $t = 1, 2, 3$).

Virtual machine	$Max f_{kt}$	ε_{kt}	$\beta_{kt}(1)$	$\beta_{kt}(2)$
VM_{kt}	1 [vs/msec]	1	1	0.6

The number m of request processes $p^1, ..., p^m$ ($0 \leq m \leq 18{,}000$) are issued to the server cluster S. The starting time st^i of each process p^i is randomly selected in a unit of one millisecond [msec] between 1 and 10,000 [ms]. The minimum computation time $minT_{kt}^i$ of every replica p_{kt}^i is 1 [ms]. The delay time $d_{K,kt}$ of every pair of a load balancer K and every virtual machine VM_{kt} is 1 [ms] in the server cluster S. The minimum response time $minRT_{kt}^i$ of every replica p_{kt}^i is $2d_{K,kt} + minT_{kt}^i = 2 \cdot 1 + 1 = 3$ [ms]. We assume the redundancy rd^i for each process p^i is the same $rd = 3$ and $NF = rd - 1 = 2$.

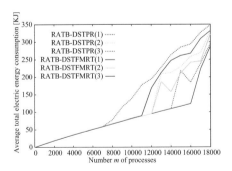

Fig. 3. Total processing electric energy consumption [KJ].

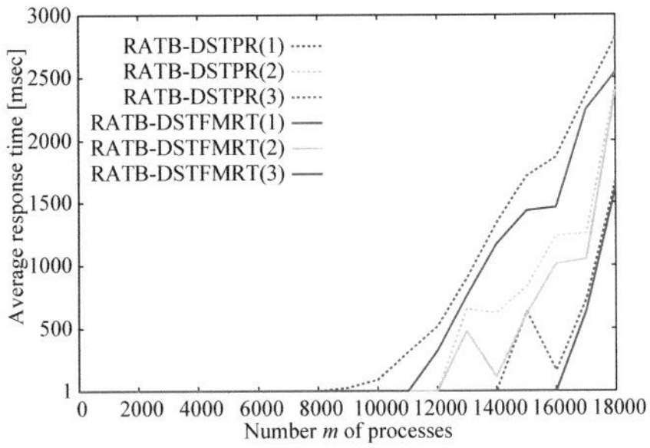

Fig. 4. Average response time [msec].

Figure 3 shows the average total processing electric energy consumption [KJ] of the server cluster S to perform the total number m of processes in the RATB-DSTPR and RATB-DSTFMRT algorithms. In Fig. 3, RATB-DSTPR(δ) and

RATB-DSTFMRT(δ) stand for the total processing electric energy consumption of the server cluster S in the RATB-DSTPR and RATB-DSTFMRT algorithms with inter-request time δ ($= \{1, 2, 3\}$) [ms], respectively. The longer inter-request time δ, the smaller total processing electric energy is consumed in the server cluster S in the RATB-DSTPR(δ) and RATB-DSTFMRT(δ) algorithms. For $1 \leq \delta \leq 3$, the average total processing electric energy consumption can be more reduced in the RATB-DSTFMRT(δ) algorithm than the RATB-DSTPR algorithm since the execution of meaningless replicas can be more reduced by forcing meaningless replica to terminate in the RATB-DSTFMRT(δ) algorithm.

Figure 4 shows the average response time [msec] of each process in the RATB-DSTPR(δ) and RATB-DSTFMRT(δ) algorithms. In Fig. 4, RATB-DSTPR(δ) and RATB-DSTFMRT(δ) stand for the average response time of each process in the RATB-DSTPR and RATB-DSTFMRT algorithms with inter-request time δ ($= \{1, 2, 3\}$) [ms], respectively. The longer inter-request time δ, the shorter average response time each process takes in the RATB-DSTPR(δ) and RATB-DSTFMRT(δ) algorithms. In the RATB-DSTFMRT(δ) algorithm, the execution of meaningless replicas can be more reduced than the RATB-DSTPR(δ) algorithm by forcing meaningless replica to terminate. As a result, the computation resources of each virtual machine to perform meaningless replicas can be used to perform other replicas. Therefore, the average response time of each process can be more reduced in the RATB-DSTFMRT(δ) algorithm than the RATB-DSTPR(δ) algorithm.

From the evaluation, the average total processing electric energy consumption of a homogeneous server cluster S and the average response time of each process are shown to be more reduced in the RATB-DSTFMRT algorithm than the RATB-DSTPR algorithm for $rd = 3$ and $1 \leq \delta \leq 3$. Following the evaluation, the RATB-DSTFMRT algorithm is more useful in a homogeneous server cluster than the RATB-DSTPR algorithm.

5   Concluding Remarks

In this paper, the RATB-DSTFMRT algorithm was proposed to reduce the total electric energy consumption of a server cluster and the average response time of each process by not only making each pair of process replicas not simultaneously start but also forcing meaningless replicas to terminate. In the RATB-DSTFMRT algorithm, the average total processing electric energy consumption of a server cluster and the average response time of each process can be more reduced than the RATB-DSTPR algorithm since the execution of meaningless replicas can be more reduced by forcing meaningless replicas to terminate. Following the evaluation results, we showed the RATB-DSTFMRT algorithm is more useful than the RATB-DSTPR algorithm in a homogeneous server cluster.

References

1. Oma, R., Nakamura, S., Enokido, T., Takizawa, M.: Fault-tolerant strategies in the tree-based fog computing model. Int. J. Distrib. Syst. Technol. **11**(4), 72–91 (2020)
2. Bylykbashi, T., Qafzezi, E., Ikeda, M., Matsuo, K., Barolli, L.: Fuzzy-based driver monitoring system (FDMS): implementation of two intelligent FDMSs and a testbed for safe driving in VANETs. Futur. Gener. Comput. Syst. **105**, 665–674 (2020)
3. Enokido, T., Aikebaier, A., Takizawa, M.: Process allocation algorithms for saving power consumption in peer-to-peer systems. IEEE Trans. Ind. Electron. **58**(6), 2097–2105 (2011)
4. Enokido, T., Aikebaier, A., Takizawa, M.: A model for reducing power consumption in peer-to-peer systems. IEEE Syst. J. **4**(2), 221–229 (2010)
5. Enokido, T., Aikebaier, A., Takizawa, M.: An extended simple power consumption model for selecting a server to perform computation type processes in digital ecosystems. IEEE Trans. Ind. Inform. **10**(2), 1627–1636 (2014)
6. Enokido, T., Takizawa, M.: Integrated power consumption model for distributed systems. IEEE Trans. Ind. Electron. **60**(2), 824–836 (2013)
7. KVM: Main Page - KVM (Kernel Based Virtual Machine) (2015). http://www.linux-kvm.org/page/Mainx_Page
8. Natural Resources Defense Council (NRDS): Data center efficiency assessment - scaling up energy efficiency across the data center industry: evaluating key drivers and barriers (2014). http://www.nrdc.org/energy/files/data-center-efficiency-assessment-IP.pdf
9. Lamport, R., Shostak, R., Pease, M.: The byzantine generals problems. ACM Trans. Program. Lang. Syst. **4**(3), 382–401 (1982)
10. Schneider, F.B. : Replication Management Using the State-Machine Approach. Distributed Systems, 2nd edn., pp. pp. 169–197. ACM Press (1993)
11. Enokido, T., Duolikun, D., Takizawa, M.: An energy-efficient process replication algorithm based on the active time of cores. In: Proceedings of the 32nd IEEE International Conference on Advanced Information Networking and Applications (AINA-2018), pp. 165–172 (2018)
12. Enokido, T., Duolikun, D., Takizawa, M.: The improved redundant active time-based (IRATB) algorithm for process replication an energy-efficient redundant execution algorithm by terminating meaningless redundant processes. In: Proceedings of the 35th IEEE International Conference on Advanced Information Networking and Applications (AINA-2021), pp. 172–180 (2021)
13. Enokido, T., Duolikun, D., Takizawa, M.: The redundant active time-based algorithm with forcing meaningless replica to terminate. In: Barolli, L., Yim, K., Enokido, T. (eds.) CISIS 2021. LNNS, vol. 278, pp. 206–213. Springer, Cham (2021). https://doi.org/10.1007/978-3-030-79725-6_20
14. Enokido, T., Duolikun, D., Takizawa, M.: An energy-efficient process replication by differentiating starting time of process replicas in virtual machine environments. In: Barolli, L. (ed.) BWCCA 2021. LNNS, vol. 346, pp. 57–66. Springer, Cham (2022). https://doi.org/10.1007/978-3-030-90072-4_6

15. Enokido, T., Takizawa, M.: An energy-efficient process replication algorithm in virtual machine environments. In: BWCCA 2016. LNDECT, vol. 2, pp. 105–114. Springer, Cham (2017). https://doi.org/10.1007/978-3-319-49106-6_10
16. Enokido, T., Takizawa, M.: Power consumption and computation models of virtual machines to perform computation type application processes. In: Proceedings of the 9th International Conference on Complex, Intelligent and Software Intensive Systems (CISIS-2015), pp. 126–133 (2015)

A Byzantine Fault Tolerant Protocol for Realizing the Blockchain

Akihito Asakura[1]($^{\boxtimes}$), Shigenari Nakamura[2], Dilawaer Duolikun[4],
Tomoya Enokido[3], Kuninao Nashimoto[1], and Makoto Takizawa[4]

[1] Department of Advanced Sciences, Hosei University, Tokyo, Japan
`akihito.asakura.7c@stu.hosei.ac.jp, nao@hosei.ac.jp`
[2] Tokyo Metropolitan Industrial Technology Research Institute, Tokyo, Japan
`nakamura.shigenari@iri-tokyo.jp`
[3] Faculty of Business Administration, Rissho University, Tokyo, Japan
`eno@ris.ac.jp`
[4] Research Center for Computing and Multimedia Studies, Hosei University,
Tokyo, Japan
`makoto.takizawa@computer.org`

Abstract. The Blockchain is now used in various applications like Bitcoin. Here, a database named *ledger* is fully replicated on every node, which is a chain of blocks. Each block is composed of transactions and is appended to the ledger at each node. Nodes may tamper with transactions, i.e. nodes suffer from Byzantine faults. Consensus protocols tolerant to Byzantine faults are required to make every replica of the ledger mutually consistent. In one approach, consensus protocols like PoW (Proof of Work) are proposed. However, nodes consume large energy to make a consensus. In another approach, the PBFT (Practical Byzantine Fault Tolerant) protocols are proposed in order to reduce the computation and communication overheads in scalable systems. Here, a leader node plays a role of centralized coordinator. However, the leader node consumes larger computation resources to communicate with member nodes. In this paper, we newly propose a DBFT (Decentralized BFT) protocol. Here, a leader node divides a proposal on a new block to smaller segments, each of which is sent to a subleader. The DBFT protocol is composed of three phases, prepare, precommit, and commit ones. In the evaluation, we show the DBFT protocol supports better scalability and performance.

Keywords: Blockchain · Byzantine fault tolerant (BFT) protocol · PBFT protocol · DBFT protocol

1 Introduction

Information systems are getting scalable and consume electric energy [14–19]. In order to increase the reliability and performance, databases and processes are distributed and replicated to multiple nodes [20,21]. A blockchain [1] is a typical

L. Barolli et al. (Eds.): EIDWT 2022, LNDECT 118, pp. 406–416, 2022.
https://doi.org/10.1007/978-3-030-95903-6_43

example of scalable system and is a distributed, fully replicated, append-only database named *ledger*. There is no centralized coordinator. A ledger is realized in a chain of blocks. Each block is composed of transactions. A transaction is a record of data, e.g. who purchases what item with what price from whom at what time. A block B_{i-1} preceding a block B_i means B_{i-1} is created before B_i and every transaction in the block B_{i-1} commits before the succeeding block B_i. In the block B_i, the transactions and a nonce N_i are encrypted by using the nonce N_{i-1} of the preceding block B_{i-1}. In traditional database systems, once data is updated on a node, a replica of the data has to be updated on every node in order to keep the mutual consistency among the replicas under assumption each node only suffers from stop-fault. The replicas are synchronized by using commitment protocols like 2PC (two-phase commitment) [5] and 3PC (three-phase commitment) [6] protocols. In scalable systems, some nodes might tamper with blocks, i.e. each node may suffer from Byzantine faults [7]. In scalable systems including millions of nodes, it is difficult, maybe impossible to adopt the commitment protocols due to communication and processing overheads.

In the Blockchain, consensus protocols like PoW (Proof of Work) [1] and PoS (Proof of Stake) [4] are used to make a consensus among nodes by using cryptography technologies [2] and taking advantage of the scalability. Here, only one node which can earliest obtain the nonce N_{i-1} of a last block B_{i-1} can create a block B_i following the block B_{i-1}. It takes time to do the calculation and nodes consume electric energy.

In another approach, all the proper nodes make a consensus on a new block by taking advantage of the Byzantine fault tolerant (BFT) protocol [7]. The number of messages exchanged among n nodes in the BFT protocol is $O(n^f)$ and it takes $O(f)$ phases to make a consensus assuming at most f nodes get faulty. In the BFT protocol [7], $n \geq 3f + 1$ has to be held to make a consensus among all the proper nodes. However, it is difficult to adopt the BFT protocol to scalable systems due to large communication and computation overhead. In order to adopt the BFT protocol to a scalable system, the PBFT (Practical BFT) protocol [8–10, 22, 23] is proposed. Here, messages are encrypted [24] and a leader node sends a proposal, i.e. a new block of transactions to all the member nodes. Each member node broadcasts a vote for the proposal and every node collects votes and makes a consensus decision. Here, $O(n^2)$ messages are transmitted and it takes two phases to make a consensus. In the PBFT protocols, a leader node is required to consume large computation resources.

In this paper, we newly propose a DBFT (Decentralized Byzantine Fault Tolerant) protocol which is composed of three phases, *prepare*, *precommit*, and *commit* phases. In order to reduce the computation load of a leader node, at the first prepare phase, the leader node receives a block B from a client and divides the block B to segments, and sends each segment to a subleader node. Each subleader node sends the received segment to all the nodes. Thus, each node can obtain the block B by receiving the segments from the subleader nodes. In the evaluation, we show the DBFT protocol provides the better scalability and performance than the HotStuff protocol [9].

In Sect. 2, we present the system model. In Sect. 3, we overview the BFT protocol. In Sect. 4, we propose the DBFT protocol. In Sect. 5, we evaluate the DBFT protocol.

2   System Model

A system is composed of nodes $p_1, \ldots, p_n$ ($n \geq 1$) which are interconnected in networks. Each node p_i holds a full replica of an append-only database named *ledger* L. The ledger L is a chain of blocks $B_1, \ldots, B_{i-1}, B_i, \ldots$. Here, a block B_{i-1} precedes a block B_i, i.e. B_i is appended after the block B_{i-1}. The newest block B_i is referred to as *last* block in the ledger L. Each block B_i is composed of transactions $Tx_{i1}, \ldots, Tx_{ik_i}$ ($k_i \geq 1$). Every transaction Tx_{ih} in a block B_i commits after every transaction in a block B_{i-1}. An application on each node issues a transaction which is distributed to every node. A node collects uncommitted transactions and creates a block B to be appended to the ledger L. If every node makes a consensus on the block B_i, the block B_i is appended to a ledger replica in every node.

It is critical for every proper node to make a consensus on appending the block B_i to the ledger L, i.e. the block B_i is stored as the last block. In order to append a block in the ledger L, all the proper nodes have to make a consensus by using consensus protocols like the PoW (Proof of Work) [1] under assumption more than half of the total mining power is controlled by proper nodes. The PoW is the first consensus protocol which works on a permissionless network. Here, each block B_i is generated by a node named *miner*. Each miner node calculates cryptographic hash values made of transactions $Tx_{i-1,1}, \ldots, Tx_{i-1,k_{i-1}}$ ($k_{i-1} \geq 1$) in the previous block B_{i-1} and the number N_{i-1} used at once, i.e. nonce Fig. 1. This process is called *mining*. The hash value with 0 of the specified number of digits from the left, e.g. 00000000000000000000020b7a7f86... has to be calculated. Each miner node repeats the hash calculation until the condition holds and then generates a new block with a different nonce. The first miner node who finishes the calculation is rewarded and a block created by the miner node is appended to the ledger L. Another miner node may finish the calculation and append another block to the ledger L. Here, the longest chain is taken as a main chain, i.e. ledger if there are multiple chains of blocks. The other chains of blocks are discarded in every node. Each proper miner node is assumed to perform the mining process on the proper chain. Hence, a malicious miner node needs at least 51% of the total mining power on the network to tamper with a chain. It is probabilistically guaranteed that the block is not tampered with in scalable systems. However, the huge energy is consumed in the mining process and the throughput decreases and the latency increases as the number of transactions increases [3]. Secondly, the PoW does not support the deterministic finality, i.e. there is possibility blocks are tampered with by malicious nodes if they can control more than 51% of the total mining power on the network. In order to improve the performance, the PoS (Proof of Stake) [4] is proposed.

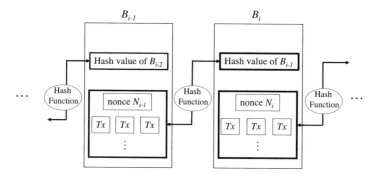

Fig. 1. Blockchain.

In another approach, all the proper nodes make a consensus on appending a new block by taking advantage of the BFT (Byzantine Fault Tolerant) protocol [7]. The communication and computation complexity of the BFT protocol is $O(n^f)$ where at most f nodes may get faulty out of n nodes. In order to improve the performance of the BFT protocol, types of the PBFT (Practical BFT) protocols [8–10,22,23] are proposed. There is a leader node which distributes a proposal of a new block to every member node, collects votes from the member nodes, and then makes decision on a consensus, i.e. every proper node appends the block the replica of the ledger L to reduce the communication overhead. The PBFT protocol is based on the asynchronous model [25,26].

3   The Byzantine Fault Tolerant (BFT) Protocols

In a group of multiple nodes, all the proper nodes are required to make a consensus on appending a new block to the ledger in presence of faulty nodes. Here, nodes may suffer from not only stop-fault but also commission and omission fault, i.e. Byzantine fault [7]. In the BFT (Byzantine Fault Tolerant) protocol [7] to make a consensus among multiple nodes, there are one leader node p_1 and $(n-1)$ member nodes $p_2, \ldots, p_n$ ($n \geq 4$). At most f nodes are assumed to get faulty in n nodes of the group.

First, the leader node p_1 sends a proposal value v_1 to all the nodes in the BFT protocol. On receipt of the value v_1, each member node p_i keeps in record the value v_1 and forwards the value $v_i(= v_1)$ to the other member nodes $p_2, \ldots, p_{i-1}, p_{i+1}, \ldots, p_n$. Then, each member node p_i receives a value v_j from another member node p_j as shown in Fig. 2. If the member node p_i does not receive any value from a node p_j, v_j is $\perp$. A proper node p_j forwards a value v_i to the other member nodes on receipt of the value v_1. However, a faulty node p_i might send no value or send a different value $v_i(\neq v_1)$ to a member node. Then, for a value v_j from each node p_j, a node p_i forwards the value v_j to every node except the nodes p_1 and p_j as presented for the node p_1. The node p_i receives values as a value sent by the node p_j from the nodes and takes a majority value of the values as v_i. The procedure is recursively applied. Thus, each proper node p_i takes the same value at the fth phases. Therefore, the total number of phases for a consensus is $f + 1$. The total number of messages is $(n - 1)(n - 2) \ldots (n - (f + 1))$, i.e. $O(n^f)$. The number n of nodes should be larger than $3f$ for all the proper nodes to make a consensus, i.e. $n \geq 3f + 1$.

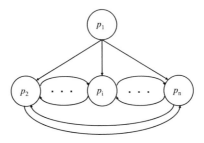

Fig. 2. BFT protocol.

The PBFT (Practical BFT) protocol [8] is proposed to improve the performance of the BFT protocol. Here, there are a leader node p_1 and member nodes $p_2, \ldots, p_n$ ($n \geq 4$). Differently from the BFT protocol, each member node p_i receives a value v_1 from the leader node p_1 and then sends the value v_i ($= v_1$) to the leader node p_1 and all the member nodes. Each node p_i takes a majority value of values which p_i receives. In addition, each message is hashed by the MAC (Message Authentication Code) [24]. Each pair of nodes p_i and p_j share their secret keys and each of the nodes p_i and p_j verifies whether or not the hash value of a received message matches the hash value of its own message by using the secret key. In the acknowledge message, the digital signatures used in the BFT protocol is replaced with a single MAC in the PBFT protocol. In the PBFT protocol, the message complexity is improved to $O(n^2)$ and it takes three phases, i.e. pre-prepare, prepare, commit phases while $O(n^f)$ messages and $f + 1$ phases in the BFT. In the preprepare phase, a leader node p_1 sends a proposal value v_1 after receiving requests from its clients. In the prepare phase, each member node p_i broadcasts a prepare-vote to the other members after receiving the proposal v_1. In the commit phase, each member node verifies the received prepare-votes. Once each node receives $2f + 1$ valid prepare-votes, the node broadcasts a commit-vote. Finally, each node commits the value v_1 and stores v_1 into the log once the node receives $2f + 1$ valid commit-votes. If a node could not receive $2f + 1$ valid votes, each node broadcasts the view-change message to make a new view which includes no faulty node. In the PBFT protocol, the communication complexity is $O(n^2)$ since votes are broadcast by every member node in the prepare and commit phases. If some node is detected to be faulty, a membership view of each node has to be changed. In the view-change case, the communication complexity is $O(n^3)$.

The HotStuff protocol [9] is the first type of the PBFT protocol which takes both of the optimistic responsiveness [27] and the linear view-change mechanism. The HotStuff protocol supports the liner communication complexity $O(n)$ both in the normal case and the view-change case. The threshold signature is used. The Chained HotStuff protocol [9] is a pipelined version of the HotStuff protocol. Here, the a four-phase commitment is adopted to realize the optimistic responsiveness. The HotStuff and Chained HotStuff protocols are composed of four phases, *prepare*, *precommit*, *commit*, and *decide* phases. In the first prepare phase of the Chained HotStuff, the leader node p_i sends a proposal pr to all the member nodes Fig. 3. Every member node sends a vote message to the leader node p_i. The leader node p_i makes the first QC (Quorum Certificate) composed of the votes. The QC guarantees the uniqueness of the proposal from the leader node p_i. A member node p_j takes over the leader node and receives the QC from the leader node p_i. In the second precommit phase, the leader node p_j sends the first QC to all the member nodes and each member node sends a vote to the leader node p_j. The leader node p_j makes the second QC which guarantees that each member node receives the first QC from the leader node p_j, and sends it to the new leader node p_k. In the third commit phase, the new leader node p_k receives the second QC from the leader node p_j and sends it to all the member nodes. Here, every member node is locked on the second QC. The third QC is made in a same way as the second phase and sent to the new leader node p_l. In the fourth decide phase, the new leader node p_l sends the third QC to all the member nodes. After receiving the third QC, every member node commits and executes the commands in the proposal and replies the result of the executions to the client. Then, every member node sends the new leader node p_m a new view message to start the next consensus round. The new view message is included the first QC. The third QC guarantees that each member node receives the second QC of the leader node p_k. Thus, the proposal pr is held by every member node if every other member node receives the valid third QC from the leader node p_l. Therefore, a leader node is only required to choice the latest first QCs from the new view messages from member nodes to start the next consensus phase. The new view message is also sent by each member node which does not receive a valid message from the leader node p_l in every phase. Moreover, the HotStuff and Chained HotStuff protocols adopt the tree data structure like the data structure of the Blockchain. Therefore, the completion of the commit phase is proven in the decide phase and all the nodes are locked to the second QC which is a set of the $(n - f)$ first votes. In the Chained HotStuff protocol, requests from clients are sent to leaders in the pipeline manner as shown in Fig. 3.

4   A Decentralized BFT (DBFT) Protocol

A consensus group G is composed of n nodes $p_1, \ldots, p_n$ $(n \geq 4)$ interconnected in a network N. All the proper nodes are required to make a consensus in presence of at most f $(< n/3)$ faulty nodes. In this paper, we newly propose a DBFT (Decentralized BFT) protocol to realize the Blockchain. The DBFT protocol is composed of three phases, *prepare*, *precommit*, and *commit* phases Fig. 4. The group G is composed of a *leader* node $(p_l \in G)$, g $(< n)$ subleader nodes

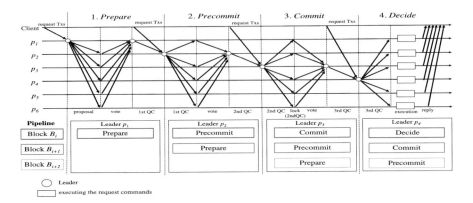

Fig. 3. Chained HotStuff protocol.

$p_{b_1}, \ldots, p_{b_g}$ ($\in G$), and $(n - 1 - g)$ member nodes. We assume the network N is reliable.

First, a leader node p_l receives a block B of transactions from a client c and divides the block B into segments $B_1, \ldots, B_g$ and sends each segment B_i to a subleader p_{b_i} ($i = 1, \ldots, g$). Here, each segments B_i is signed by the leader node p_l. After receiving a segment B_i from the leader node p_l, each subleader node p_{b_i} signs the segment B_i and sends it to all the $(n - 1)$ nodes except the leader node p_l. Thus, each member node p_m receives the segments $B_1, \ldots, B_g$ from the g subleader nodes $p_{b_1}, \ldots, p_{b_g}$, respectively, and obtains the block B by merging the segments. A member node p_m and subleader nodes $p_{b_1}, \ldots, p_{b_{i-1}}, p_{b_{i+1}}, \ldots, p_{b_g}$ check if each signed segment B_i is valid by using the public key of a subleader node p_{b_i}. If all the segments are valid, each member node p_m obtains a full copy of the block B. Each of the member node p_m and all subleader nodes $p_{b_1}, \ldots, p_{b_g}$ sends a prepare *vote* (PPV) message to the leader node p_l. If some segment B_i is not valid, each node except the leader node p_l does not send a PPV message to the leader node p_l and sends a *new view message* to the next leader node after time-out. The leader node p_l collects PPV messages from the member nodes. This is the *prepare* phase.

From here, in the group G, there are a leader node p_l and $(n - 1)$ member nodes. There is no subleader node. Each subleader node p_{b_i} becomes a member of the member node. Once the leader node p_l receives PPV messages from more than two third of the n nodes including the leader node, the leader node p_l sends a prepare $QC(Quorum\ Certificate)$ (PQC) message to all the member nodes. On receipt of the PQC message from the leader node p_l, each member node p_k $(\in(G - \{p_l\}))$ sends a *precommit vote* (PCV) message to the leader node p_l. The leader node p_l receives PCV messages from the member nodes. This is the *precommit* phase.

The leader node p_l collects PCV messages from more than two third of the n nodes including the leader node. Once the leader node p_l receives PCV messages from more than two third of the n nodes including the leader node, the leader node p_l sends a *precommit QC* (CQC) message to all the member nodes. On receipt of the CQC message from the leader node p_l, each member node p_k stores the block B to the replica of the ledger L. That is, every transaction in the block B is committed. Then, the node p_k sends a *committed* (C) message to the client c. This is the *commit* phase.

As discussed in the Chained HotStuff, a membership view of the group G is changed if some node does not receive valid messages.

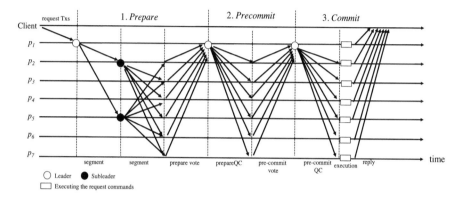

Fig. 4. DBFT protocol.

5   Evaluation

We evaluate the DBFT protocol compared with the Chained HotStuff protocol [9] in terms of scalability, total execution time, and number of messages. We consider a group G of n nodes $p_1, \ldots, p_n$ in the evaluation. In the group G, one node p_l is a leader node. In the DBFT protocol, there are g subleader nodes and the other $(n - g - 1)$ member nodes at the prepare phase. In the Chained HotStuff protocol and the precommit and commit phases of the DBFT protocol, a group G is composed of one leader node p_l and $(n - 1)$ member nodes.

In scalable systems like the Blockchain, it is critical to make clear the scalability of the system. The *scaling factor* (SF) is proposed as the scalability metric to evaluate the PBFT protocols [23]. Let th be the throughput [req/sec] of the protocol, i.e. the number req of requests received/sent by nodes for one second and br be the number [bit] of bits of each request. Thus, each node receives/sends the number B $(= th \times br)$ of bits. The workload $W_{type}^{rec/snd}$ [bit/sec] shows the total number of bits received (rec)/sent (snd) for one second by the node of *type* $(\in \{l(leader), sl(subleader), m(member)\})$ divided by B. The workload $W_{type}^{rec/snd}$ is given as (the number of bits received/sent by a node p_i) × (the number of nodes which send to/receive from the node p_i)/B. For example, in the DBFT protocol, if a leader node p_l sends g segments to g subleader nodes $p_{b_1}, \ldots, p_{b_g}$, the workload W_l^{snd} of the leader node p_l is $((B/g) \times g)/B = 1$. The scaling factor SF

is the heaviest workload $\max\left\{W_{type}^{rec} + W_{type}^{snd} \mid type \in \{l, sl, m\}\right\}$ of nodes, i.e.
$SF = \max(W_l^{rec} + W_l^{snd}, W_{sl}^{rec} + W_{sl}^{snd}, W_m^{rec} + W_m^{snd})$. In the DBFT protocol,
each member node receives all g segments $B_1, \ldots, B_g$ of a block B from the subleader nodes $p_{b_1}, \ldots, p_{b_g}$, respectively, and sends only a vote message to a leader
node p_l. Here, a vote message is not considered in the scaling factor because the
size of vote message is so small that it can be neglected compared with the segment. Therefore, the workload W_m^{rec} and W_m^{snd} of a member node to receive and
send messages are $W_m^{rec} = ((B/g) \times g)/B = 1$ and $W_m^{snd} = 0$, respectively.

In the DBFT protocol, each member node receives a full replica of a block B
from the subleader nodes. Therefore, the workload W_m^{rec} and W_m^{snd} of a member
node to receive and send messages are $W_m^{rec} = 1$ and $W_m^{snd} = 0$, respectively.

Table 1 shows the scaling factor and performance of the DBFT protocol compared with the Chained HotStuff [9] protocol. The scaling factor of the Chained
HotStuff protocol is n. In the paper [23], the SF of HotStuff is $n-1$. In this paper,
we calculated the SF of the Chained HotStuff including the workload of receiving
requests from a client on a leader node. Therefore, the SF is n. The scaling factor of
the DBFT protocol is 2 for $n-2 \leq g$ and $(n-2)/g+1$ for $n-2 > g$. Figure 5 shows
the scaling factor of the DBFT and Chained HotStuff protocols for number n of
nodes. As shown in the figure, the scaling factor of the DBFT protocol is smaller
than the Chained HotStuff protocol. This means, the DBFT protocol can be more
efficiently used for each consensus in scalable systems than the Chained HotStuff
protocol. In addition, the larger the number g of subleader nodes, the smaller the
scaling factor in the DBFT protocol.

Table 1. Scaling factor.

Protocols	Scaling factor		Number of messages	Number of rounds
	$n - 2 \leq g$	$n - 2 > g$		
DBFT	2	$\frac{n-2}{g} + 1$	$(g+5)n - g - 3$	8
Chained HotStuff	n		$8n$	12

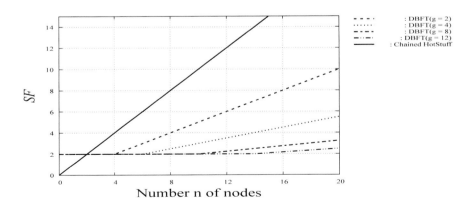

Fig. 5. Scaling factor for number n of nodes.

The number of messages transmitted is $(g + 5)n - g - 3$ in the DBFT protocol while $8n$ in the Chained HotStuff protocol. In the DBFT protocol, more umber of messages are transmitted than the Chained HotStuff protocol. The number of rounds to reach a consensus is 8 in the DBFT protocol while 12 in the Chained HotStuff protocol. It takes shorter time for every proper node to make a consensus in the DBFT protocol than the Chained HotStuff protocol.

6   Concluding Remarks

In this paper, we newly proposed the DBFT protocol to efficiently realize the Blockchain. Here, a leader node divides a new block to smaller segments and sends each segment to a subleader. Then, a subleader sends a segment to all the nodes. Each subleader holds a full replica of the block by receiving segments from other subleaders. In the evaluation, we showed the scaling factor of the DBFT protocol is better than the Chained HotStuff protocol. It takes shorter time to make a consensus in the DBFT protocol than the Chained HotStuff protocol while more number of messages are transmitted in the DBFT protocol.

Faulty nodes are detected by using messages received from the nodes. Some malicious node can send correct messages by stealing the messages sent by proper nodes. These nodes are *implicitly faulty* [11–13]. We are now considering a new type of PBFT protocol which is tolerant of implicitly faulty nodes.

References

1. Nakamoto, S.: Bitcoin: a peer-to-peer electronic cash system (2008)
2. Jeong, K., Lee, Y., Sung, J., Hong, S.: Security analysis of HMAC/NMAC by using fault injection. J. Appl. Math. **2013** (2013). https://doi.org/10.1155/2013/101907
3. Best, R.: Bitcoin average energy consumption per transaction compared to that of VISA as of October 21, 2021 (2021). https://www.statista.com/statistics/881541/bitcoin-energy-consumption-transaction-comparison-visa/
4. Sunny, K., Scott, N.: PPCoin: peer-to-peer crypto-currency with proof-of-stake. Self-published paper (2012). https://bitcoin.peryaudo.org/vendor/peercoin-paper.pdf
5. Gray, J.N.: Notes on database operating systems. Lect. Notes Comput. Sci. **60**, 466–471 (1978)
6. Skeen, D., Stonebraker, M.: A formal model of crash recovery in a distributed system. IEEE Trans. Softw. Eng. **SE9**(3), 219–228 (1983)
7. Lamport, L., Shostak, R., Pease, M.: The Byzantine generals problem. ACM Trans. Program. Lang. Syst. **4**(3), 382–401 (1982)
8. Castro, M., Liskov, B.: Practical Byzantine fault tolerance. In: Proceedings of the Third Symposium on Operating Systems Design and Implementation, pp. 1–14 (1999)
9. Yin, M., Malkih, D., Reiter, M., Gueta, G., Abraham, I.: HotStuff: BFT consensus with linearity and responsiveness. In: Proceedings of the 2019 ACM Symposium on Principles of Distributed Computing, pp. 347–356 (2019)
10. The ZILLIQA team: The ZILLIQA Technical Whitepaper (2017). https://docs.zilliqa.com/whitepaper.pdf

11. Ishii, H., Oma, R., Nakamura, S., Enokido, T., Takizawa, M.: Fault detection of process replicas on reliable servers. In: Barolli, L., Nishino, H., Enokido, T., Takizawa, M. (eds.) NBiS - 2019 2019. AISC, vol. 1036, pp. 423–433. Springer, Cham (2020). https://doi.org/10.1007/978-3-030-29029-0_40

12. Ishii, H., Nakamura, S., Saito, T., Enokido, T., Takizawa, M.: An energy-based algorithm for detecting implicitly faulty replicas of a process. In: Barolli, L., Poniszewska-Maranda, A., Enokido, T. (eds.) CISIS 2020. AISC, vol. 1194, pp. 288–299. Springer, Cham (2021). https://doi.org/10.1007/978-3-030-50454-0_27

13. Ishii, H., Oma, R., Nakamura, S., Enokido, T., Takizawa, M.: Algorithm for detecting implicitly faulty replicas based on the power consumption model. In: Barolli, L., Hellinckx, P., Enokido, T. (eds.) BWCCA 2019. LNNS, vol. 97, pp. 483–493. Springer, Cham (2020). https://doi.org/10.1007/978-3-030-33506-9_43

14. Duolikun, D., Enokido, T., Takizawa, M.: Static and dynamic group migration algorithms of virtual machines to reduce energy consumption of a server cluster. Trans. Comput. Collect. Intell. **XXXIII**, 144–166 (2019)

15. Enokido, T., Aikebaier, A., Takizawa, M.: Process allocation algorithms for saving power consumption in peer-to-peer systems. IEEE Trans. Ind. Electron. **58**(6), 2097–2105 (2011)

16. Enokido, T., Aikebaier, A., Takizawa, M.: A model for reducing power consumption in peer-to-peer systems. IEEE Syst. J. **4**(2), 221–229 (2010)

17. Enokido, T., Aikebaier, A., Takizawa, M.: An extended simple power consumption model for selecting a server to perform computation type processes in digital ecosystems. IEEE Trans. Ind. Inf. **10**(2), 1627–1636 (2014)

18. Enokido, T., Takizawa, M.: Integrated power consumption model for distributed systems. IEEE Trans. Ind. Electron. **60**(2), 824–836 (2013)

19. Enokido, T., Takizawa, M.: Power consumption and computation models of virtual machines to perform computation type application processes. In: Proceedings of the 9th International Conference on Complex, Intelligent, and Software Intensive Systems (CISiS-2015), pp. 126–133 (2015)

20. Enokido, T., Takizawa, M.: The redundant energy consumption laxity based algorithm to perform computation processes for IoT services. Internet Things **9**, 100165 (2020)

21. Enokido, T., Duolikun, D., Takizawa, M.: The improved redundant active time-based (IRATB) algorithm for process replication. In: Proceedings of the the 35th International Conference on Advanced Information Networking and Applications (AINA-2021), pp. 172–180 (2021)

22. Jalalzai, M., Niu, J., Feng, C., Gai, F.: Fast-HotStuff: a fast and robust BFT protocol for blockchains (2021). https://arxiv.org/abs/2010.11454

23. Hu, K., Guo, K., Tang, Q., Zhang, Z., Cheng, H., Zhao, Z.: Leopard: scaling BFT without sacrificing efficiency (2021). https://arxiv.org/abs/2106.08114

24. Tsudik, G.: Message authentication with one-way hash functions. Assoc. Comput. Mach. **22**(5), 29–38 (1992)

25. Fischer, M., Lynch, N., Paterson, M.: Impossibility of distributed consensus with one faulty process. J. ACM **32**(2), 374–382 (1985)

26. Dwork, C., Lynch, N., Stockmeyer, L.: Consensus in the presence of partial synchrony. J. ACM **35**(2), 288–323 (1988)

27. Pass, R., Shi, E.: Thunderella: blockchains with optimistic instant confirmation. In: Nielsen, J.B., Rijmen, V. (eds.) EUROCRYPT 2018, Part II. LNCS, vol. 10821, pp. 3–33. Springer, Cham (2018). https://doi.org/10.1007/978-3-319-78375-8_1

Performance Evaluation of a DQN-Based Autonomous Aerial Vehicle Mobility Control Method in an Indoor Single-Path Environment with a Staircase

Nobuki Saito[1], Tetsuya Oda[2(✉)], Aoto Hirata[1], Chihiro Yukawa[2], Masaharu Hirota[3], and Leonard Barolli[4]

[1] Graduate School of Engineering, Okayama University of Science (OUS), 1-1 Ridaicho, Kita-ku, Okayama 700–0005, Japan
{t21jm01md,t21jm02zr}@ous.jp

[2] Department of Information and Computer Engineering, Okayama University of Science (OUS), 1-1 Ridaicho, Kita-ku, Okayama 700–0005, Japan
oda@ice.ous.ac.jp, t18j097cy@ous.jp

[3] Department of Information Science, Okayama University of Science (OUS), 1-1 Ridaicho, Kita-ku, Okayama 700–0005, Japan
hirota@mis.ous.ac.jp

[4] Department of Information and Communication Engineering, Fukuoka Institute of Technology, 3-30-1 Wajiro-higashi, Higashi-ku, Fukuoka 811-0295, Japan
barolli@fit.ac.jp

Abstract. The Deep Q-Network (DQN) is one of the deep reinforcement learning algorithms, which uses deep neural network structure to estimate the Q-value in Q-learning. In the previous work, we designed and implemented a DQN-based Autonomous Aerial Vehicle (AAV) testbed and proposed a Tabu List Strategy based DQN (TLS-DQN). In this paper, we propose a DQN-based AAV mobility control method. The performance evaluation results show that the proposed method can decide and reach the destination in an indoor single-path environment with a staircase.

1 Introduction

The Unmanned Aerial Vehicle (UAV) is expected to be used in different fields such as aerial photography, transportation, search and rescue of humans, inspection, land surveying, observation and agriculture. Autonomous Aerial Vehicle (AAV) [1] has the ability to operate autonomously without human control and is expected to be used in a variety of fields, similar to UAV. So far many AAVs [2–4] are proposed and used practically. However, existing autonomous flight systems are designed for outdoor use and rely on location information by the Global Navigation Satellite System (GNSS) or others. On the other hand, in an environment where it is difficult to obtain position information from GNSS, it is necessary to determine a path without using position information. Therefore, autonomous

L. Barolli et al. (Eds.): EIDWT 2022, LNDECT 118, pp. 417–429, 2022.
https://doi.org/10.1007/978-3-030-95903-6_44

movement control is essential to achieve operations that are independent of the external environment, including non-GNSS environments such as indoor, tunnel and underground.

In [5–7] the authors consider Wireless Sensor and Actuator Networks (WSANs), which can act autonomously for disaster monitoring. A WSAN consists of wireless network nodes, all of which have the ability to sense events (sensors) and perform actuation (actuators) based on the sensing data collected by the sensors. WSAN nodes in these applications are nodes with integrated sensors and actuators that have the high processing power, high communication capability, high battery capacity and may include other functions such as mobility. The application areas of WSAN include AAV [8], Autonomous Surface Vehicle (ASV) [9], Heating, Ventilation, Air Conditioning (HVAC) [10], Internet of Things (IoT) [11], Ambient Intelligence (AmI) [12], ubiquitous robotics [13], and so on.

Deep reinforcement learning [14] is an intelligent algorithm that is effective in controlling autonomous robots such as AAV. Deep reinforcement learning is an approximation method using deep neural network for value function and policy function in reinforcement learning. Deep Q-Network (DQN) is a method of deep reinforcement learning using Convolution Neural Network (CNN) as a function approximation of Q-values in the Q-learning algorithm [14,15]. DQN combines the neural fitting Q-iteration [16,17] and experience replay [18], shares the hidden layer of the action value function for each action pattern and can stabilize learning even with nonlinear functions such as CNN [19,20]. However, there are some points where learning is difficult to progress for problems with complex operations and rewards, or problems where it takes a long time to obtain a reward.

In this paper, we propose a DQN-based AAV mobility control method. Also, we present the simulation results for AAV control using TLS-DQN [21] considering an indoor single-path environment with a staircase.

The structure of the paper is as follows. In Sect. 2, we show the DQN based AAV testbed. In Sect. 3, we describe the proposed method. In Sect. 4, we discuss the performance evaluation. Finally, conclusions and future work are given in Sect. 5.

2   DQN Based AAV Testbed

In this section, we discuss quadrotor for AAV and DQN for AAV mobility.

2.1   Quadrotor for AAV

For the design of AAV, we consider a quadrotor, which is a type of multicopter. Multicopter is high maneuverable and can operate in places that are difficult for people to enter, such as disaster areas and dangerous places. It also has the advantage of not requiring space for takeoffs and landings and being able to stop at mid-air during the flight, therefore enabling activities at fixed points.

Fig. 1. Snapshot of AAV.

Table 1. Components of quadrotor.

Component	Model
Propeller	15x5.8
Motor	MN3508 700kv
Electric speed controller	F45A 32bitV2
Flight controller	Pixhawk 2.4.8
Power distribution board	MES-PDB-KIT
Li-Po battery	22.2v 12000mAh XT90
Mobile battery	Pilot Pro 2 23000 mAh
ToF ranging sensor	VL53L0X
Raspberry Pi	3 Model B Plus
PVC pipe	VP20
Acrylic plate	5 mm

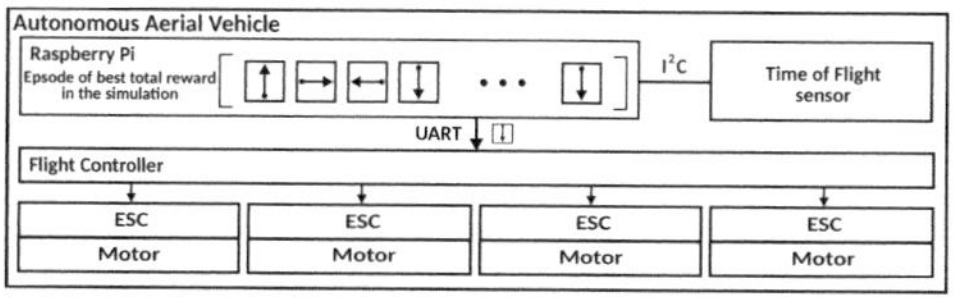

Fig. 2. AAV control system.

In Fig. 1 is shown a snapshot of the quadrotor used for designing and implementing an AAV testbed. The quadrotor frame is mainly composed of polyvinyl chloride (PVC) pipe and acrylic plate. The components for connecting the battery, motor, sensor, etc. to the frame are created using an optical 3D printer. Table 1 shows the components in the quadrotor. The size specifications of the quadrotor (including the propeller) are length 87 [cm], width 87 [cm], height 30 [cm] and weight 4259 [g].

In Fig. 2 is shown the AAV control system. The raspberry pi reads saved data of the best episode when carrying out the simulations by DQN and uses telemetry communication to send commands such as up, down, forward, back, left, right and stop to the flight controller. Also, multiple Time-of-Flight (ToF) range sensors using Inter-Integrated Circuit (I^2C) communication and General-Purpose Input Output (GPIO) are used to acquire and save flight data. The Flight Controller (FC) is a component that calculates the optimum motor rotation speed for flight based on the information sent from the built-in acceleration sensor and gyro sensor. The Electronic Speed Controller (ESC) is a part that controls the rotation speed of the motor in response to commands from FC. Through these sequences, AAV behaves and reproduces movement in simulation.

2.2   DQN for AAV Mobility

The DQN for moving control of AAV structure is shown in Fig. 3. In this work, we use the Deep Belief Network (DBN), because the computational complexity is smaller than CNN for DNN part in DQN. The environment is set as v_i. At each step, the agent selects an action a_t from the action sets of the mobile actuator nodes and observes a position v_t from the current state. The change of the mobile actuator node score r_t was regarded as the reward for the action. For the reinforcement learning, we can complete all of these mobile actuator nodes sequences m_t as Markov decision process directly, where sequences of observations and actions are $m_t = v_1, a_1, v_2, \ldots, a_{t-1}, v_t$. A method known as experience replay is used to store the experiences of the agent at each timestep, $e_t = (m_t, a_t, r_t, m_{t+1})$ in a dataset $D = e_1, \ldots, e_N$, cached over many episodes into a Experience Memory. By defining the discounted reward for the future by a factor γ, the sum of the future reward until the end would be $R_t = \sum_{t'=t}^{T} \gamma^{t'-t} r_{t'}$. T means the termination time-step of the mobile actuator nodes. After running experience replay, the agent selects and executes an action according to an ϵ-greedy strategy. Since using histories of arbitrary length as inputs to a neural network can be difficult, Q-function instead works on fixed length format of histories produced by a function ϕ. The target was to maximize the action value function $Q^*(m, a) = \max_\pi E[R_t | m_t = m, a_t = a, \pi]$, where π is the strategy for selecting of best action. From the Bellman equation (see Eq. (1)), it is possible to maximize the expected value of $r + \gamma Q^*(m', a')$, if the optimal value $Q^*(m', a')$ of the sequence at the next time step is known.

$$Q^*(m', a') = E_{m' \sim \xi}[r + \gamma_{a'} \max Q^*(m', a') | m, a]. \tag{1}$$

By not using iterative updating method to optimize the equation, it is common to estimate the equation by using a function approximator. Q-network in DQN is a neural network function approximator with weights θ and $Q(s, a; \theta) \approx Q^*(m, a)$. The loss function to train the Q-network is shown in Eq. (2):

$$L_i(\theta_i) = E_{s, a \sim \rho(.)}[(y_i - Q(s, a; \theta_i))^2]. \tag{2}$$

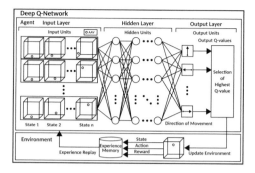

Fig. 3. DQN for AAV mobility control.

The y_i is the target, which is calculated by the previous iteration result θ_{i-1}. The $\rho(m, a)$ is the probability distribution of sequences m and a. The gradient of the loss function is shown in Eq. (3):

$$\nabla_{\theta_i} L_i(\theta_i) = E_{m,a \sim \rho(.); s' \sim \xi}[(y_i - Q(m, a; \theta_i))\nabla_{\theta_i} Q(m, a; \theta_i)]. \tag{3}$$

We consider tasks in which an agent interacts with an environment. In this case, the AAV moves step by step in a sequence of observations, actions and rewards. We took in consideration AAV mobility and consider 7 mobile patterns (up, down, forward, back, left, right, stop). In order to decide the reward function, we considered Distance between AAV and Obstacle (DAO) parameter.

The initial weights values are assigned as Normal Initialization [22]. The input layer is using AAV and the position of destination, total reward values in Experience Memory and AAV movements patterns. The hidden layer is connected with 256 rectifier units in Rectified Linear Units (ReLU) [23]. The output Q-values are the AAV movement patterns.

3 Proposed Method

In this section, we discuss the AAV mobility control method for DQN-based AAV.

3.1 LiDAR Based Mobile Area Decision Method

The proposed method decides the destination in TLS-DQN within the considered area based on the point cloud obtained by LiDAR and reduces the setting operation for the destination set manually by humans in TLS-DQN. In addition, the proposed method can decide the destination with less computation than using Simultaneous Localization and Mapping (SLAM) or other methods. In Algorithm 1, we consider as inputs the coordinates list of obstacles (*distance*, *angle*) obtained by LiDAR and the coordinates of LiDAR placement. The output is the *Destination* (X, Y), which may be local destination or global

---

Algorithm 1. LiDAR Based Mobile Area Decision Method.

Input: *Point Cloud List* ← The coordinates list of obstacles (*distance, angle*) obtained by LiDAR
(x_{LiDAR}, y_{LiDAR}) ← The coordinates of LiDAR Placement.

Output: *Destination* (X, Y).

1: **for** $i = 0$ to 360 **do**
2:    $x_{Point\ Cloud\ List}[i]$ ← *Point Cloud List*$[i][0] \times cos($*Point Cloud List*$[i][1])$.
3:    $y_{Point\ Cloud\ List}[i]$ ← *Point Cloud List*$[i][0] \times sin($*Point Cloud List*$[i][1])$.
4:    **if** *Point Cloud List*$[i][0]$ > *Any Distance* **then**
5:       *Distant Point Cloud*$[i]$ ← $(x_{Point\ Cloud\ List}[i],\ y_{Point\ Cloud\ List}[i])$.
6: $(x_{\min},\ x_{\max})$ ← Min. and Max. value for X-axis in the *Distant Point Cloud*.
7: $(y_{\min},\ y_{\max})$ ← Min. and Max. value for Y-axis in the *Distant Point Cloud*.
8: $(x_{center},\ y_{center})$ ← $\left(\frac{x_{\min}+\ x_{\max}}{2},\ \frac{y_{\min}+\ y_{\max}}{2}\right)$.
9: $flag \leftarrow 0$.
10: **for** $x = x_{LiDAR}$ **to** x_{center} **do**
11:    $y \leftarrow \left(\frac{y_{center}-\ y_{LiDAR}}{x_{center}-\ x_{LiDAR}}\right) \times (x - x_{LiDAR}) + y_{LiDAR}$.
12:    **for** $i = 0$ to 360 **do**
13:       **if** $\sqrt{(x - x_{Point\ Cloud\ List}[i])^2 + (y - y_{Point\ Cloud\ List}[i])^2}$ > *Any Distance*
          then
14:          *Destination* ← (x, y).
15:       **else**
16:          $flag \leftarrow 1$.
17:          **break**
18: **if** $flag = 0$ **then**
19:    *Destination is* `local destination`.
20: **else**
21:    *Destination is* `global destination`.

---

destination. The `global destination` indicates the destination in the considered area, and the `local destination` indicates the target passage points until the `global destination`. The Z-coordinate of destination is the median of the movable range in the Z-axis for the destination. In the proposed method, the destination is continuously decided by letting the LiDAR placement be the coordinate of reached destination when the AAV reached the destination.

3.2 TLS-DQN

The idea of the Tabu List Strategy (TLS) is motivated from Tabu Search (TS) proposed by F. Glover [24] to achieve an efficient search for various optimization problems by prohibiting movements to previously visited search area in order to prevent getting stuck in local optima.

Algorithm 2. Tabu List for TLS-DQN.

Require: The coordinate with the highest evaluated value in the section is (x, y, z).

1: **if** $(x_{before} \leq x_{current}) \wedge (x_{current} \leq x)$ **then**
2:     $tabu\ list \Leftarrow ((x_{min} \leq x_{before}) \wedge (y_{min} \leq y_{max}) \wedge (z_{min} \leq z_{max}))$
3: **else if** $(x_{before} \geq x_{current}) \wedge (x_{current} \geq x)$ **then**
4:     $tabu\ list \Leftarrow ((x_{before} \leq x_{max}) \wedge (y_{min} \leq y_{max}) \wedge (z_{min} \leq z_{max}))$
5: **else if** $(y_{before} \leq y_{current}) \wedge (y_{current} \leq y)$ **then**
6:     $tabu\ list \Leftarrow ((x_{min} \leq x_{max}) \wedge (y_{min} \leq y_{before}) \wedge (z_{min} \leq z_{max}))$
7: **else if** $(y_{before} \geq y_{current}) \wedge (y_{current} \geq y)$ **then**
8:     $tabu\ list \Leftarrow ((x_{min} \leq x_{max}) \wedge (y_{before} \leq y_{max}) \wedge (z_{min} \leq z_{max}))$
9: **else if** $(z_{before} \leq z_{current}) \wedge (z_{current} \leq z)$ **then**
10:     $tabu\ list \Leftarrow ((x_{min} \leq x_{max}) \wedge (y_{min} \leq y_{max}) \wedge (z_{min} \leq z_{before}))$
11: **else if** $(z_{before} \geq z_{current}) \wedge (z_{current} \geq z)$ **then**
12:     $tabu\ list \Leftarrow ((x_{min} \leq x_{max}) \wedge (y_{min} \leq y_{max}) \wedge (z_{before} \leq z_{max}))$

$$r = \begin{cases} 3 & (if\ (x_{current} = x_{global\ destinations}) \wedge \\ & (y_{current} = y_{global\ destinations}) \wedge \\ & (z_{current} = z_{global\ destinations})) \vee \\ & (((x_{before} < x_{current}) \wedge (x_{current} \leq x_{local\ destinations})) \vee \\ & ((x_{before} > x_{current}) \wedge (x_{current} \geq x_{local\ destinations})) \vee \\ & ((y_{before} < y_{current}) \wedge (y_{current} \leq y_{local\ destinations})) \vee \\ & ((y_{before} > y_{current}) \wedge (y_{current} \geq y_{local\ destinations})) \vee \\ & ((z_{before} < z_{current}) \wedge (z_{current} \leq z_{local\ destinations})) \vee \\ & ((z_{before} > z_{current}) \wedge (z_{current} \geq z_{local\ destinations}))). \\ -1 & (else). \end{cases} \quad (4)$$

In this paper, the reward value for DQN is decided by Eq. (4), where "x", "y" and "z" means X-axis, Y-axis and Z-axis, respectively. The `current` means the current coordinates of the AAV in the DQN, and the `before` means the coordinates before moving the action. The considered area is partitioned based on the `local destination` or `global destination` and a destination is set in each area. The tabu list in TLS is used when a DQN selects an action randomly or determined a reward for the action. If the tabu list includes the area in the direction of movement, the DQN will reselect the action. Also, if the reward is "3", the prohibited area is added to the tabu list based on the rule shown in Algorithm 2. The search by TLS-DQN is done in a wider range and is better than the search by random direction of move

3.3   Movement Adjustment Method

The movement adjustment method is used for reducing movement fluctuations caused by TLS-DQN. The Algorithm 3 inputs the movement of coordinates (X, Y, Z) in the episode of `Best` derived by TLS-DQN and generates the *Adjustment Point Coordinates List*. In Algorithm 3, the

---

Algorithm 3. Movement Adjustment Decision.

Input: *Movement Coordinates* ← The movement of coordinates (X, Y, Z) by TLS-DQN

Output: *Adjustment Point Coordinates List.*

1: *Number of divided list* ← *Any number.*
2: *Number of coordinates* ← $\frac{Number\ of\ Iterations\ in\ TLS-DQN}{Number\ of\ divided\ list}$.
3: $i \leftarrow 0, j \leftarrow 0$
4: **for** $k = 0$ **to** Number of coordinates in *Movement Coordinates* **do**
5:     *Divided List*$[j]$ ← *Movement Coordinates*$[k]$.
6:     $j \leftarrow j + 1$.
7:     **if** $j \geq$ *Number of coordinates* **then**
8:         $(x_{\min}, x_{\max})$ ← Min. and Max. values for X-axis in the *Divided List*.
9:         $(y_{\min}, y_{\max})$ ← Min. and Max. values for Y-axis in the *Divided List*.
10:         $(z_{\min}, z_{\max})$ ← Min. and Max. values for Z-axis in the *Divided List*.
11:         $(x_{center}, y_{center}, z_{center})$ ← $\left(\frac{x_{\min} + x_{\max}}{2}, \frac{y_{\min} + y_{\max}}{2}, \frac{z_{\min} + z_{\max}}{2}\right)$
12:         *Adjustment Point Coordinates List*$[i]$ ← $(x_{center}, y_{center}, z_{center})$.
13:         $i \leftarrow i + 1, j \leftarrow 0$

Number of divided lists indicates the number of divisions to the coordinate movements; the *Number of coordinates* indicates how many coordinates are included in the *Divided List*; and the $(x_{cneter}, y_{cneter}, z_{cneter})$ indicates the center coordinates derived from the maximum and minimum values of coordinates in X-axis, Y-axis and Z-axis included in the *Divided List*.

4   Performance Evaluation

In this section, we discuss the experimental results of LiDAR based decision method, the simulation results of TLS-DQN and the movement adjustment method.

4.1   Results of LiDAR Based Decision Method

The target environment is an indoor single-path environment with a staircase. Figure 4 shows snapshots of the area used in the simulation scenario and was taken on the ground floor of Building C5 at Okayama University of Science, Japan. In Fig. 4, Fig. 5 and Fig. 6, the blue filled area indicates the floor surface, the red filled area indicates the corner space and the green filled area indicates the staircase space. Figure 5 visualize the experimental results of the LiDAR based mobile area decision method. Figure 5 shown the visualization results of LiDAR based decision method, the obstacle obtained by LiDAR and the decided destination. Figure 6 shows the considered area based on the actual measurements of Fig. 5. The experimental results show that the path and the `global destination` is decided at the end of the path through the staircase space.

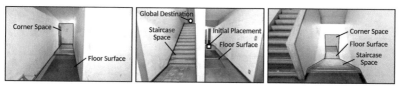

(a) From the initial place-ment to the corner space.

(b) From the corner space to the global destination.

(c) From the global destina-tion to the corner space.

Fig. 4. Snapshot of the considered area.

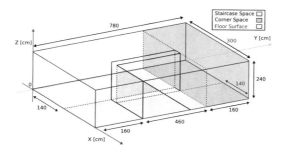

Fig. 5. Visualization results of LiDAR based decision method.

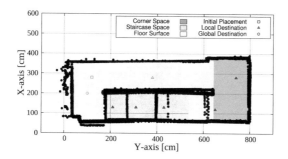

Fig. 6. Considered area for simulation.

4.2   Simulation Results of TLS-DQN

We consider for simulations the operations such as takeoffs, flights and landings between the initial position and the destination decided by the LiDAR based decision method. The target environment is an indoor single-path environment including a staircase. Table 2 shows the parameters used in the simulations. Figure 7 shows the change in reward value of the action in each iteration for Worst, Median, and Best episodes in TLS-DQN. For Best episodes, the reward value is increased much more than Median episodes. While, for Worst episodes, the reward value is decreased.

Table 2. Simulation parameters of TLS-DQN.

Parameters	Values
Number of episode	50000
Number of iteration	2000
Number of hidden layers	3
Number of hidden units	15
Initial weight value	Normal initialization
Activation function	ReLU
Action selection probability (ϵ)	$0.999 - (t/\text{Number of episode})$ $(t = 0, 1, 2, \ldots, \text{Number of episode})$
Learning rate (α)	0.04
Discount rate (γ)	0.9
Experience memory size	300×100
Batch size	32
Number of AAV	1

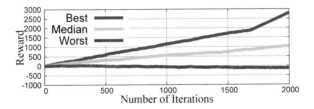

Fig. 7. Simulation results of rewards.

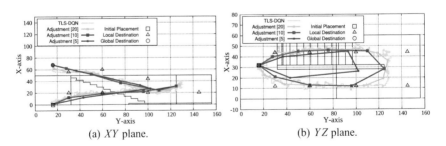

(a) XY plane.   (b) YZ plane.

Fig. 8. Visualization results of TLS-DQN and movement adjustment method.

Table 3. The distance of movement in the XY and YZ plane.

Plane	Minimum	TLS-DQN	Adjustment 20	Adjustment 10	Adjustment 5
XY	370.00	850.00	397.61	390.94	370.65
YZ	406.19	890.00	395.06	390.39	387.91

4.3   Results of Movement Adjustment Method

Figure 8 shows the visualization results on the XY and YZ planes for the movement of the Best episodes in TLS-DQN and the results of the movement adjustment method when the *Number of divided lists* is 20, 10 and 5, respectively. From the visualization results, the TLS-DQN has reached the destination. On the other hand, the movement adjustment method could not consider the obstacles when the *Number of divided lists* is 5 and 10. Table 3 shows the distance of movement derived from the total Euclidean distances between each coordinate. The performance evaluation shows that the movement adjustment method can decrease the distance of movement and the movement fluctuations for both XY and YZ planes.

5   Conclusions

In this paper, we proposed a DQN-based AAV mobility control method and presented simulation results for AAV control by TLS-DQN considering an indoor single-path environment with a staircase. From performance evaluation results, we conclude as follows.

- The proposed method can decide the mobile area and destination based on LiDAR for TLS-DQN.
- The visualization results of the movement show that the TLS-DQN can reach the destination.
- The proposed method is a good approach for indoor single-path environments.

In the future, we would like to improve the TLS-DQN for AAV mobility by considering different scenarios.

Acknowledgement. This work was supported by JSPS KAKENHI Grant Number JP20K19793 and Grant for Promotion of OUS Research Project (OUS-RP-20-3).

References

1. Stöcker, C., et al.: Review of the current state of UAV regulations. Remote Sens. **9**(5), 1–26 (2017)
2. Artemenko, O., et al.: Energy-aware trajectory planning for the localization of mobile devices using an unmanned aerial vehicle. In: Proceedings of the 25-th International Conference on Computer Communication and Networks (ICCCN-2016), pp. 1–9 (2016)
3. Popović, M., et al.: An informative path planning framework for UAV-based terrain monitoring. Auton. Robot. **44**, 889–911 (2020)
4. Nguyen, H., et al.: LAVAPilot: lightweight UAV trajectory planner with situational awareness for embedded autonomy to track and locate radio-tags. arXiv:2007.15860, pp. 1–8 (2020)

5. Oda, T., et al.: Design and implementation of a simulation system based on deep Q-network for mobile actor node control in wireless sensor and actor networks. In: Proceedings of the 31-th IEEE International Conference on Advanced Information Networking and Applications Workshops (IEEE AINA-2017), pp. 195–200 (2017)

6. Oda, T., Elmazi, D., Cuka, M., Kulla, E., Ikeda, M., Barolli, L.: Performance evaluation of a deep Q-network based simulation system for actor node mobility control in wireless sensor and actor networks considering three-dimensional environment. In: Barolli, L., Woungang, I., Hussain, O.K. (eds.) INCoS 2017. LNDECT, vol. 8, pp. 41–52. Springer, Cham (2018). https://doi.org/10.1007/978-3-319-65636-6_4

7. Oda, T., Kulla, E., Katayama, K., Ikeda, M., Barolli, L.: A deep Q-network based simulation system for actor node mobility control in WSANs considering three-dimensional environment: a comparison study for normal and uniform distributions. In: Barolli, L., Javaid, N., Ikeda, M., Takizawa, M. (eds.) CISIS 2018. AISC, vol. 772, pp. 842–852. Springer, Cham (2019). https://doi.org/10.1007/978-3-319-93659-8_77

8. Sandino, J., et al.: UAV framework for autonomous onboard navigation and people/object detection in cluttered indoor environments. Remote Sens. **12**(20), 1–31 (2020)

9. Moulton, J., et al.: An autonomous surface vehicle for long term operations. In: Proceedings of MTS/IEEE OCEANS, pp. 1–10 (2018)

10. Oda, T., Ueda, C., Ozaki, R., Katayama, K.: Design of a deep Q-network based simulation system for actuation decision in ambient intelligence. In: Barolli, L., Takizawa, M., Xhafa, F., Enokido, T. (eds.) WAINA 2019. AISC, vol. 927, pp. 362–370. Springer, Cham (2019). https://doi.org/10.1007/978-3-030-15035-8_34

11. Oda, T., et al.: Design and implementation of an IoT-based E-learning testbed. Int. J. Web Grid Serv. **13**(2), 228–241 (2017)

12. Hirota, Y., Oda, T., Saito, N., Hirata, A., Hirota, M., Katatama, K.: Proposal and experimental results of an ambient intelligence for training on soldering iron holding. In: Barolli, L., Takizawa, M., Enokido, T., Chen, H.-C., Matsuo, K. (eds.) BWCCA 2020. LNNS, vol. 159, pp. 444–453. Springer, Cham (2021). https://doi.org/10.1007/978-3-030-61108-8_44

13. Hayosh, D., et al.: Woody: low-cost, open-source humanoid torso robot. In: Proceedings of the 17-th International Conference on Ubiquitous Robots (ICUR-2020), pp. 247–252 (2020)

14. Mnih, V., et al.: Human-level control through deep reinforcement learning. Nature **518**, 529–533 (2015)

15. Mnih, V., et al.: Playing Atari with deep reinforcement learning. arXiv:1312.5602, pp. 1–9 (2013)

16. Lei, T., Ming, L.: A robot exploration strategy based on Q-learning network. In: IEEE International Conference on Real-time Computing and Robotics (IEEE RCAR-2016), pp. 57–62 (2016)

17. Riedmiller, M.: Neural fitted Q iteration – first experiences with a data efficient neural reinforcement learning method. In: Gama, J., Camacho, R., Brazdil, P.B., Jorge, A.M., Torgo, L. (eds.) ECML 2005. LNCS (LNAI), vol. 3720, pp. 317–328. Springer, Heidelberg (2005). https://doi.org/10.1007/11564096_32

18. Lin, L.J.: Reinforcement learning for robots using neural networks. In: Proceedings of Technical Report, DTIC Document (1993)

19. Lange, S., Riedmiller, M.: Deep auto-encoder neural networks in reinforcement learning. In: Proceedings of the International Joint Conference on Neural Networks (IJCNN-2010), pp. 1–8 (2010)

20. Kaelbling, L.P., et al.: Planning and acting in partially observable stochastic domains. Artif. Intell. **101**(1–2), 99–134 (1998)
21. Saito, N., et al.: A Tabu list strategy based DQN for AAV mobility in indoor single-path environment: implementation and performance evaluation. Internet Things **14**, 100394 (2021)
22. Glorot, X., Bengio, Y.: Understanding the difficulty of training deep feedforward neural networks. In: Proceedings of the 13-th International Conference on Artificial Intelligence and Statistics (AISTATS-2010), pp. 249–256 (2010)
23. Glorot, X., et al.: Deep sparse rectifier neural networks. In: Proceedings of the 14-th International Conference on Artificial Intelligence and Statistics (AISTATS-2011), pp. 315–323 (2011)
24. Glover, F.: Tabu search - part I. ORSA J. Comput. **1**(3), 190–206 (1989)

Practical Survey on MapReduce Subgraph Enumeration Algorithms

Xiaozhou Liu$^{(\boxtimes)}$, Yudi Santoso, Venkatesh Srinivasan, and Alex Thomo

University of Victoria, Victoria, BC, Canada
{xiaozhou,santoso,srinivas,thomo}@uvic.ca

Abstract. Subgraph enumeration is a basic task in many graph analyses. Therefore, it is necessary to get this task done within a reasonable amount of time. However, this objective is challenging when the input graph is very large, with millions of nodes and edges. Known solutions are limited in terms of scalability. Distributed computing is often proposed as a solution to improve scalability. However, it has to be done carefully to reduce the overhead cost and to really benefit from the distributed solution. In this work we provide a comprehensive overview of several Map-Reduce subgraph enumeration algorithms which currently represent the state of the art. We identify and describe the main conceptual approaches, giving insight on their advantages and limitations, and provide a summary of their similarities and differences.

1 Introduction

Finding occurrences of particular subgraphs in a network/graph is a tool for analyzing and understanding such network. This analytical tool has many applications in various fields, such as in biology [10,14] chemistry [8,21], social study [4,9], network classification [25], and more. Moreover, some applications require the enumeration of all subgraphs up to a certain degree. For example, Milenkovic and Przulj [14] used 2, 3, 4, and 5 node induced connected subgraphs (or graphlets) to analyse Protein-Protein-Interaction (PPI) networks.

Subgraph enumeration problem is challenging when the size of the input graph is large. This can be understood by analysing the complexity. The computational complexity grows exponentially by the order of the subgraph to enumerate. Suppose we want to find subgraphs of k nodes in a graph of n nodes, then there are $\binom{n}{k} \propto n(n-1)\dots(n-k+1)$ possible combinations to examine. If $n \gg k$, this is approximately n^k. On the other hand, an order of magnitude increase in n would increase the complexity by 10^k. In practice, not all combinations need to be checked. Nonetheless, it is generally true that the number of subgraphs grows rapidly with the size of the graph, and for enumeration the algorithm needs to touch each subgraph.

© The Author(s), under exclusive license to Springer Nature Switzerland AG 2022
L. Barolli et al. (Eds.): EIDWT 2022, LNDECT 118, pp. 430–444, 2022.
https://doi.org/10.1007/978-3-030-95903-6_45

Using a distributed platform is an obvious option to increase the computing power, where we can add more compute nodes to get more done within a fixed time budget. Known distributed solutions [5,16,18,24] had limited scalability, and were only designed to enumerate *triangle* subgraphs. The work of Park et al. improved upon these previous solutions and was able to enumerate triangles from graphs with billions edges [17]. The same authors, furthermore, generalized their work to support arbitrary subgraph query [19]. As argued in [19], their proposed method is superior to join-based methods of [1] and [11], because those can generate a large amount of intermediate data during the shuffle step. Liu et al. [13] identified the redundant computational work in Park et al.'s proposal [19], and devised a duplication-free distributed algorithm, enumerating quadrillions of 4-node graphlets.

We consider in this survey only works that solve the problem of induced subgraphs. For example, if we discover a clique of 4 nodes, we only enumerate the clique, but not any other 4-node subgraphs inside it. Small induced connected subgraphs, such as 4-cliques, are known as graphlets.

1.1   Contributions

We provide a practical review of several distributed algorithms for subgraph enumeration, focusing on the state-of-the-art works of [13,16–19]. We give a structured retrospective reflection on the algorithms for computing exact subgraph occurrences using the *MapReduce* paradigm[1]. We also identify and describe the main conceptual ideas, giving insight on their main advantages and possible limitations. We provide a complete overview table that classifies the key characteristics of the algorithms, highlighting their main similarities and differences.

2   Preliminaries

In the following we give some important notions that we use in this paper.

Subgraph. A graph $H(V_H, E_H)$ is called a *subgraph* of $G(V_G, E_G)$ if $V_H \subseteq V_G$ and $E_H \subseteq E_G$.

Induced subgraph. A subgraph $H(V_H, E_H) \subseteq G(V_G, E_G)$ is an induced subgraph if for every pair of nodes $u, v \in V_H$, edge $(u, v) \in E_H$ if and only if $(u, v) \in E_G$.

Graphlet. A small induced connected subgraph. There are two types of 3-node graphlets: wedge and triangle as shown in Fig. 1, and six types of 4-node graphlets, shown in Fig. 2: 3-path, 3-star, rectangle, tailed-triangle, diamond and 4-clique.

---

[1] Here we use the term MapReduce in a broad sense. The algorithms in this work can be implemented in newer frameworks such as Apache Spark as well.

Fig. 1. 3-node graphlets: a wedge and a triangle.

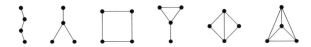

Fig. 2. 4-node graphlets: a 3-path, a 3-star, a rectangle or 4-cycle, a tailed-triangle, a diamond, and a 4-clique.

Edge-orientation. Assigning directions to edges in an undirected graph $G(V, E)$. Given a function η which determines a total ordering of the nodes in V, an undirected edge $(u, v) = (v, u)$ is orientated by η, such that if $\eta(u) < \eta(v)$, only (u, v) is listed but not (v, u). Edge-orientation transforms the undirected graph into a Directed Acyclic Graph (DAG).

Coloring. Applying a color function to each edge such that the entire graph can be partitioned into many subgraphs. First, define a function θ which maps vertex $u \in V$ to $\theta(u)$. Vertex u is said to have *color* $\theta(u)$. Edge $(u, v) \in E$ is said to have color $(i = \theta(u), j = \theta(v))$.

Edgeset. After coloring is applied to $G(V, E)$, edges of the same color are grouped together to form a subgraph, denoted by E_{ij}. E_{ij} contains all edges $(u, v) \in E$ with color (i, j).

Directed Edgeset. Edgeset of a directed graph, denoted by E_{ij}^*, representing all the edges of $\overrightarrow{E}$ of $\overrightarrow{G}(V, \overrightarrow{E})$, that originates from vertices of color i and points to vertices of color j. For $i \neq j$, $E_{ij}^* \cup E_{ji}^* = E_{ij}$; for $i = j$, $E_{ij}^* = E_{ij}$.

Sub-problem. Refers to the union of (directed) edgesets of particular colors where the distributed workers can discover all the subgraphs/graphlets in that union independently. It is the fundamental enumeration task assigned to each distributed worker.

Symmetrization. Refers to deriving graph $G^{\text{sym}}(V, E^{\text{sym}})$ from the original graph $\overrightarrow{G}(V, \overrightarrow{E})$, where $E^{\text{sym}} = \overrightarrow{E} \cup \{\overrightarrow{(v, u)} | \forall \overrightarrow{(u, v)} \in \overrightarrow{E}\}$. In other words, G^{sym} possesses the original edge as well as the reversed edge in $\overrightarrow{G}$.

MapReduce. A *de-facto* programming scheme for distributed processing [7]. The paradigm orchestrates the processing in three operations: Map, Shuffle and Reduce. Each operation facilitates its own duty, drastically reducing the complexity of implementation.

3   Serial Graphlet Enumerations

The distributed algorithms of interests in this work are composed using MapReduce paradigm, meaning all of them consist of a localized serial enumeration algorithm with a partition scheme.

All of the distributed algorithms, except one, are employing some well-known serialized triangle/subgraph enumeration such as CompactForward [12] and VF2 [6], the introduction of which are omitted here. In this section we introduce a single machine 4-node graphlet enumeration algorithm that was proposed recently, called *Simultaneous 4-node Graphlets Enumeration* (S4GE) [23].

3.1 4-Node Graphlet Enumeration

Algorithm 1. S4GE

Require: An oriented-by-degree, symmetrized graph $\overrightarrow{G}(V, \overrightarrow{E})$

1: **for all** $(\overline{u,v}) \in \overrightarrow{E}$ **do**
2:     **for all** $u' \in N(u)$ and $v' \in N(v)$ **do**
3:         **if** $(u' > u) \wedge (v' \geq u)$ **then**
4:             **if** $u' = v' > v$ **then**
5:                 EXPLORETRIANGLE (u, v, u')
6:             **if** $((u' < v') \vee (v' = u)) \wedge (u' > v)$ **then**
7:                 EXPLOREWEDGETYPE1 (v, u, u')
8:             **if** $(u' > v') \wedge (v' \neq u)$ **then**
9:                 EXPLOREWEDGETYPE2 (u, v, v')

Algorithm 2. EXPLORE TRIANGLE

Require: Triangle $(u, v, w)_3$ with $u < v < w$; symmetrized neighbourhood $N(u)$, $N(v)$ and $N(w)$.
1: $N^{>u}(u) \equiv \{z | z \in N(u), \eta(z) > \eta(u)\}$; $N^{>u}(v) \equiv \{z | z \in N(v), \eta(z) > \eta(u)\}$; $N^{>u}(w) \equiv \{z | z \in N(w), \eta(z) > \eta(u)\}$
2: **for all** $z \in N^{>u}(u) \cap N^{>u}(v) \cap N^{>u}(w)$ with $z > w$ **do**
3:     ENUMERATE4CLIQUE (u, v, w, z)
4: **for all** z in two sets and $z > $ opposite vertex **do**
5:     ENUMERATEDIAMOND (u, v, w, z)
6: **for all** z in one set only **do**
7:     ENUMERATETAILEDTRIANGLE (u, v, w, z)

The algorithm first discovers triangles and wedges in a similar fashion as CompactForward [12], then it proceeds to search for 4-node graphlets after the discovery of triangles and wedges. That is, it first discovers triangles and wedges, and for each triangle that it locates, it checks if this triangle is a part of any tailed triangles, diamonds and 4-cliques. Similarly, through wedges it checks for 3-paths, 3-stars and rectangles. The checks are done by intersecting the neighbourhoods of the three vertices of the triangles and wedges. The details can be found in [23]. S4GE has a running time $O(T_{3g} + (|\Delta| + |\angle|)d_{max}^{sym})$, where T_{3g} is the time to enumerate wedges and triangles, and d_{max}^{sym} is the maximum degree of the symmetrized input graph. The pseudo code of S4GE for discovering all the triangles and wedges is listed as Algorithm 1; the pseudo code of S4GE for discovering all the six types of 4-node graphlets are listed as Algorithms 2, 3, and 4.

Algorithm 3. EXPLORE WEDGE TYPE-1

Require: Wedge $(v, u, w)_1$ with $u < v < w$; symmetrized neighbourhood $N(u)$, $N(v)$ and $N(w)$.
1: $N^{>u}(u) \equiv \{z | z \in N(u), \eta(z) > \eta(u)\}$; $N^{>u}(v) \equiv \{z | z \in N(v), \eta(z) > \eta(u)\}$; $N^{>u}(w) \equiv \{z | z \in N(w), \eta(z) > \eta(u)\}$
2: **for all** $z \in N^{>u}(v) \cap N^{>u}(w)$ with $z \notin N^{>u}(u)$ **do**
3:     ENUMERATERECTANGLE (u, v, z, w)
4: **for all** $z \in N^{>u}(u)$ only **do**
5:     **if** $z > w$ **then**
6:         ENUMERATE3STAR (u, v, w, z)
7: **for all** $z \in N^{>u}(v)$ only **do**
8:     ENUMERATE3PATH (w, u, v, z)
9: **for all** $z \in N^{>u}(w)$ only **do**
10:     ENUMERATE3PATH (v, u, w, z)

Algorithm 4. EXPLORE WEDGE TYPE-2

Require: Wedge $(u, v, w)_2$ with $u < v$ and $u < w$; symmetrized neighbourhood $N(v)$ and $N(w)$.
1: $N^{>u}(v) \equiv \{z | z \in N(v), \eta(z) > \eta(u)\}$; $N^{>u}(w) \equiv \{z | z \in N(w), \eta(z) > \eta(u)\}$
2: **for all** $z \in N^{>u}(v)$ only **do**
3:     **if** $z > w$ **then**
4:         ENUMERATE3STAR (v, u, w, z)
5: **for all** $z \in N^{>u}(w)$ only **do**
6:     **if** $z \neq v$ **then**
7:         ENUMERATE3PATH (u, v, w, z)

---

4   Distributed Subgraph Enumeration

Park et al. [17] proposed three variants of distributed algorithms for triangle enumeration, and PTE_{Base} is the base version. PTE_{Base} partitions the undirected input graph into edgesets which are stored on a distributed file system. It defines a set of overlapping sub-problems such that each sub-problem can be solved independently. Each sub-problem is assigned to a distributed worker, and the worker proceeds to read the edgesets that are necessary to solve the sub-problem. PTE_{CD} improves on PTE_{Base} by using *color-direction* technique, which enable it to minimize the redundant computations. Finally, PTE_{SC} improves on PTE_{CD} by minimizing network traffic through a careful scheduling. However, with modern network technology and internal network connections networking cost has been of a less concern. Later on, Park et al. generalized PTE to support the enumeration of subgraphs of any order, and called their solution PSE [19].

4.1 PTE_{Base}

PTE_{Base} defines three types of sub-problems: type-1 sub-problems, S_i, that emits triangles with all three vertices of the same colors; type-2 sub-problems, S_{ij}, that emitts triangles of two different colors; and type-3 sub-problems, S_{ijk}, that emitts triangles with three different colors. The triangles emitted from type-1 sub-problems are called type-1 triangles, and similarly for type-2 and type 3 triangles.

PTE$_{Base}$ partitions the input graph into edgesets, E_{ij} for $(i, j) \in \{0, 1, \ldots, \rho - 1\}^2$. For $i = j$ there are $\binom{\rho}{1}$ of such edgesets E_{ii}, and $\binom{\rho}{2}$ E_{ij} for $i \neq j$. The Modulo operation of ρ is used to color an edge. Each edge (u, v) belongs to the edgeset $E_{(u\%\rho)(v\%\rho)}$.

PTE$_{Base}$ computes $\binom{\rho}{2}$ type-2 sub-problems S_{ij}, and $\binom{\rho}{3}$ type-3 sub-problems, S_{ijk}. Type-1 sub-problems S_i can be embedded into type-2 sub-problem S_{ij}, hence do not need to be computed separately. For example, with $\rho = 4$, the sub-problems are: S_0, S_1, S_2, S_3, S_{01}, S_{02}, S_{03}, S_{12}, S_{13}, S_{23}, S_{012}, S_{013}, S_{023} and S_{123}. However, since solving S_{01} requires $E_{00} \cup E_{01} \cup E_{11}$, S_0 can also be solved along the process, as S_0 only requires E_{00}. Notice that S_0 can be embedded not only in S_{01}, but also in S_{02} and S_{03}. To avoid duplicated embedding, type-1 triangles of color i is only emitted when $i + 1 = j\%\rho$ for a type-2 sub-problem S_{ij}.

PTE$_{Base}$ employs CompactForward [12] algorithm as the local serial algorithm. Because PTE$_{Base}$ partitions the input graph into $O(\rho^2)$ colored edgesets, each sub-problem is expected to process a subgraph of size $O(m/\rho^2)$, where m is the number of edges in the graph, and each distributed worker is expected to do $O((m/\rho^2)^{1.5}) = O(m^{1.5}/\rho^3)$ amount of work. Summing over all $O(\rho^3)$ sub-problems, PTE$_{Base}$ does $O(m^{1.5})$ amount of work overall, which is the same amount of work as applying CompactForward [12] on a single machine. The network read for PTE$_{Base}$ is $O(\rho m)$, since each of the $O(\rho^3)$ sub-problems needs to read $O(m/\rho^2)$ amount of data.

The pseudo code of PTE$_{Base}$ is given as Algorithm 5:

Algorithm 5. PTE$_{Base}$

Require: An undirected graph $G(V, E)$; the number of colors ρ
1: Construct $\vec{G}(V, \vec{E})$ by applying edge-orientation to $G(V, E)$
2: Partition $\vec{E}$ into edge sets E_{ij} using ρ colors
3: Generate all sub-problems $\{S\}$ for the given ρ
4: **for all** sub-problem $S_{Cs} \in \{S\}$ **do**   ▷ Distributed-for
5:     $E' \leftarrow E_{ij}$ from distributed storage.
6:     CompactForward (E')

4.2 PTE$_{CD}$

Park et al. [17] improve on PTE$_{Base}$ by introducing PTE$_{CD}$, *color-direction*. Without loss of generality, consider a type-3 sub-problem S_{ijk}. If we only care about enumerating all the triangles (u, v, w) with color (i, j, k), PTE$_{Base}$ does extra work when intersecting $\vec{N}(u)$ and $\vec{N}(v)$. CompactForward algorithm used in PTE$_{Base}$ considers all possible neighbouring vertices of u, from both E_{ij}^* and E_{ik}^*, and all possible neighbouring vertices of v, from both E_{ji}^* and E_{jk}^*. This is unnecessary since for any triangle (u, v, w) with color (i, j, k), given vertices u and v with color i and j respectively, we know that w must have color k, hence only need to intersect E_{ik}^* and E_{jk}^*.

PTE_{CD} exploits this fact, and lists all six possible color-direction cases for any type-3 sub-problem: for a type-3 sub-problem S_{ijk}, an arbitrary edge $\overrightarrow{(u,v)}$ of CompactForward algorithm can have colors of $\{(i,j),(j,i),(i,k),(k,i),(j,k),(k,j)\}$, and PTE_{CD} only needs to intersect $\overrightarrow{N}(u)$ and $\overrightarrow{N}(v)$ from edge sets $\{(E_{ik}^*, E_{jk}^*), (E_{jk}^*, E_{ik}^*), (E_{ij}^*, E_{kj}^*), (E_{kj}^*, E_{ij}^*), (E_{ji}^*, E_{ki}^*), (E_{ki}^*, E_{ji}^*)\}$ respectively. Similarly, PTE_{CD} lists all six cases for any type-2 sub-problem as well.

Park et al. prove that by adapting PTE_{CD}, the total number of operations is reduced by a factor of $2 - \frac{2}{\rho}$ compared to PTE_{Base}.

The pseudo code of PTE_{CD} is given as Algorithm 6 and Algorithm 7:

Algorithm 6. PTE_{CD}

Require: An undirected graph $G(V, E)$; the number of colors ρ
1: Construct $\overrightarrow{G}(V, \overrightarrow{E})$ by applying edge-orientation to $G(V, E)$
2: Partition $\overrightarrow{E}$ into directed edge sets E_{ij}^* using ρ
3: Generate all sub-problems $\{S\}$ given ρ
4: **for all** sub-problem $\in \{S\}$ **do**   ▷ Distributed-for
5:     **if** sub-problem of type S_{ij} **then**
6:         Read $E_{ii}^*, E_{ij}^*, E_{ji}^*$ and E_{jj}^*
7:         **if** $i + 1 == j\%\rho$ **then**
8:             COMPACTFORWARD$_{CD}$ $(E_{ii}^*, E_{ii}^*, E_{ii}^*)$ ▷ Type-1
9:         **else**
10:             **for all** (p,q,r) $\in \{(i,i,j),(i,j,i),(j,i,i),(j,j,i),(j,i,j),(i,j,j)\}$ **do**
11:                 COMPACTFORWARD$_{CD}$ $(E_{pq}^*, E_{pr}^*, E_{qr}^*)$ ▷ Type-2
12:     **if** sub-problem of type S_{ijk} **then**
13:         Read $E_{ij}^*, E_{ji}^*, E_{ik}^*, E_{ki}^*, E_{jk}^*$ and E_{kj}^*
14:         **for all** (p,q,r) $\in \{(i,j,k),(i,k,j),(j,i,k),(j,k,i),(k,i,j),(k,j,i)\}$ **do**
15:             COMPACTFORWARD$_{CD}$ $(E_{pq}^*, E_{pr}^*, E_{qr}^*)$ ▷ Type-3

Algorithm 7. COMPACTFORWARD$_{CD}$

Require: An edge-oriented edge set E_{ij}^*, E_{ik}^* and E_{jk}^*
1: **for all** $\overrightarrow{(u,v)} \in E_{ij}^*$ **do**
2:     **for all** $w \in \{\overrightarrow{N}(u)|\overrightarrow{N}(u) \in E_{ik}^*\} \cap \{\overrightarrow{N}(v)|\overrightarrow{N}(v) \in E_{jk}^*\}$ **do**
3:         ENUMERATE (u, v, w)

4.3   PSE

Park et al. generalise PTE_{Base} to support non-induced subgraph query of an arbitrary order, called PSE (Pre-partitioned subgraph Enumeration) [19]. PSE takes a query subgraph $G_q(V_q, E_q)$ of order k as input where $k = |V_q|$, and enumerates all the matching subgraphs to G_q. PSE employs VF2 algorithm [6] as the local serial algorithm for query graph matching.

PSE starts by defining $\sum_{l=1}^{k} \binom{\rho}{l}$ sub-problems. For example, with $\rho = 4$ and $k = 4$, PSE first defines the following sub-problems: S_0, S_1, S_2, S_3, S_{01}, S_{02}, S_{03}, S_{12}, S_{13}, S_{23}, S_{012}, S_{013}, S_{023}, S_{123} and S_{0123}.

PSE then makes the observations that solving all the sub-problems independently introduces duplicated emissions. Some sub-problems can be grouped together to reduce duplications. This can be shown by the following: continuing with the above example, $S_{012} = E_{00} \cup E_{11} \cup E_{22} \cup E_{01} \cup E_{02} \cup E_{12}$, $S_{01} = E_{00} \cup E_{01} \cup E_{11}$ and $S_0 = E_{00}$. It is clear that $S_0 \subset S_{01} \subset S_{012}$. In other words, enumerating the subgraphs from S_{012} also enumerates all the subgraphs from S_0 and S_{01}.

To avoid duplicated emissions, PSE introduces *sub-problem groups* and the *dominant sub-problem* of the sub-problem group. In the above example, S_0, S_{01} and S_{012} forms a sub-problem group; and since enumerating S_{012} is sufficient to enumerate all sub-problems of the group, S_{012} is the dominant sub-problem of the sub-problem group. In PSE, each sub-problem group becomes the fundamental computing task on each distributed worker; solving the dominant sub-problem of each sub-problem group is equivalent to solving all the sub-problems under this group.

PSE carefully groups the sub-problems to ensure a balanced workload distribution, such that all the sub-problem groups have similar number of sub-problems assigned. The process of generating such grouping can be described as:

1. Assign sub-problems $S_{c_0,c_1,\ldots,c_l}$ where $|\{c_0, c_1, \ldots, c_l\}| = k$ and $|\{c_0, c_1, \ldots, c_l\}| = k - 1$ to different groups. It is clear that these sub-problems can never dominate each other, hence belong to their own group. These sub-problems are the prospective dominant sub-problems of their groups.
2. Assign the rest of the sub-problems to the groups created in Step 1. The assignment is prioritized towards the group with the least number of sub-problems in that group, while ensuring that the subproblem is covered by the dominat sub-problem within that group. Ties are broken randomly.
3. Repeat Step 2 until all sub-problems are assigned.

The pseudo code of PSE is given in Algorithm 8:

Algorithm 8. PSE

Require: An undirected graph $G(V, E)$; a query graph $G_q(V_q, E_q)$; the number of colors ρ

1: Construct $\overrightarrow{G}(V, \overrightarrow{E})$ by applying edge-orientation to $G(V, E)$
2: Partition $\overrightarrow{E}$ into edgesets E_{ij} using ρ
3: Generate sub-problem groups $\{SGs\}$
4: **for all** sub-problem group $SG \in \{SGs\}$ **do**   ▷ Distributed-for
5:     Select dominant sub-problem S_d from SG
6:     $E' = $READEDGESETS$(d, |V_q|)$
7:     $R = $ VF2$(E', G_q(V_q, E_q))$
8:     $R' = $ DE-DUPLICATE(R) ▷ PSE emits duplicates
9:     Enumerate(R')

Park et al. showed that PSE requires at most $\binom{\rho-1}{k-2}|E|$ amount of network read, for k-order subgraph query with input graph E.

PSE by Park et al., when published, was widely considered as the SotA of the subgraph enumeration. However the scheme discovers duplicated subgraphs. To correct the enumerations, PSE post-filters the duplicated subgraphs from the final emission. This motivates Liu et al. [13] to derive a partitioning scheme that guarantees duplication-free from the onset.

4.4   Distributed 4-Node Graphlet Enumeration

Liu et al. introduced a partitioning scheme for S4GE algorithm, enabling it to be deployed in a distributed setting [13]. The partitioning scheme is named as D4GE, short for Distributed 4-node Graphlet Enumeration.

First, it is obvious that each enumerated subgraph can be mapped to only one colored tuple - for example, when enumerating 4-node graphlets, any discovered graphlet $(u, v, w, z)_4$ has a particular 4-tuple (i, j, k, l) representation, where $i = \theta(u)$, $j = \theta(v)$, $k = \theta(w)$, $l = \theta(z)$ and θ is the coloring function. The colored tuples are called the **color-assignments** of the corresponding graphlet, denoted by K_{ijkl}: the graphlet $(u, v, w, z)_4$ with color (i, j, k, l) has color-assignment K_{ijkl}.

Next, linearity is asserted over the 4-node graphlets query, such that $(u, v, w, z)_4$ is emitted if and only if $\eta(u) < \eta(v) < \eta(w) < \eta(z)$, where η is the edge-orientation function. The linearity on the emitted 4-node graphlets also implies ordering on the color-assignment of the graphlets: for an emitted graphlet $(u, v, w, z)_4$ with $\eta(u) < \eta(v) < \eta(w) < \eta(z)$, its color-assignment K_{ijkl} must satisfy $i \prec j \prec k \prec l$, where $i \prec j$ means i precedes j. On the other hand, the ordering of the color-assignments also imposes the linearity of the underlying graphlets: K_{ijkl} can only represent the graphlets $(u, v, w, z)_4$ with $\eta(u) < \eta(v) < \eta(w) < \eta(z)$. Both the graphlets and the color-assignments follow the same ordering in colors. Given the number of color ρ for 4-node graphlet enumeration, there are ρ^4 color-assignments.

While previous work employed the idea of color-direction to reduce the amount of work (PTE_{CD}), D4GE differentiates itself by the fact that D4GE exploits both the linearity of the DAG and the color-assignment, along with the fact that the color-assignment problem is essentially a combination problem. The unique relationship between any subgraph and its color-assignment guarantees the duplication-free nature of D4GE. In addition, previous works explicitly list all the ordered color-tuples in the algorithm, while D4GE utilizes combinations to generalize the color-assignments. This works not only for $k = 4$, but also to any order k (with ρ^k color-assignments).

The pseudo code of D4GE is given as Algorithms 9:

Algorithm 9. D4GE

Require: An undirected graph $G(V, E)$; the number of colors ρ
1: Construct $\vec{G}(V, \vec{E})$ by applying edge-orientation to $G(V, E)$
2: Symmetrise $\vec{G}(V, \vec{E})$ into $G^{\mathrm{sym}}(V, E^{\mathrm{sym}})$
3: Partition E^{sym} into directed edgeset E^*_{ij}.
4: Generate sub-problems $S_{ij} \cup S_{ijk} \cup S_{ijkl}$.
5: **for all** $S_{\mathrm{Cs}} \in S_{ij} \cup S_{ijk} \cup S_{ijkl}$ **do**   ▷ Distributed-for
6:     $K_s \leftarrow$ Grouped color-assignments under S_{Cs}
7:     $E_{\mathrm{map}} \leftarrow$ Read directed edgesets from distributed storage.
8:     **for all** $K_{ijkl} \in K_s$ **do**
9:         S4GE$_{\mathrm{CD}}$ $(E_{\mathrm{map}}, ijkl)$

D4GE groups the color-assignments into sub-problems (line 6 of Algorithm 9). Consider color-assignments K_{0001} and K_{0002}. By definition K_{0001} requires knowledge of $E^*_{00} \cup E^*_{01}$ and K_{0002} requires knowledge of $E^*_{00} \cup E^*_{02}$. If these two color-assignments are computed on two different workers, the partitioned edgeset E^*_{00} is then loaded twice. To address this, color-assignments K_{pqrs} are grouped into sub-problems. Inherently S_{ijkl} is used to denote a sub-problem. The color-assignments are grouped by the following rule: K_{pqrs} belongs to sub-problem S_{ijkl} if the *sorted* and *reduced* form of $\{p, q, r, s\}$ is $\{i, j, k, l\}$, where *sorted* means sorting $\{p, q, r, s\}$ in ascending order, and *reduced* means removing the duplicated colors from the sequence $\{p, q, r, s\}$. In the example of K_{2010}, the sorted and reduced form of $\{2, 0, 1, 0\}$ is $\{0, 1, 2\}$. Therefore K_{2010} belongs to S_{012}.

To fully cover all the color-assignments, D4GE generates $\binom{\rho}{2}$ number of S_{ij}, $\binom{\rho}{3}$ number of S_{ijk} and $\binom{\rho}{4}$ number of S_{ijkl}. Sub-problem S_{ijkl} contains all the ordered color-assignments K_{pqrs} where $p, q, r, s \in \{i, j, k, l\}$; sub-problem S_{ijk} contains all the ordered color-assignments K_{pqrs} where $p, q, r, s \in \{i, j, k\}$; sub-problem S_{ij} contains all the ordered color-assignments K_{pqrs} where $p, q, r, s \in \{i, j\}$. In the special case of ordered color-assignments K_{iiii} (omitted S_i) where all four colors are the same, we attach K_{iiii} to sub-problem S_{ij} where $i+1 = j\%\rho$. Each sub-problem is computed independently on a distributed worker.

Liu et al. in [13] proved that with the same serialized local algorithm S4GE, D4GE requires no more than $2\,m^{\mathrm{sym}}$ amount network read than PSE partitioning scheme. Under the same condition, Liu et al. showed that PSE does $3\left(\binom{\rho-1}{2} - \rho\right) d^{\mathrm{sym}}_{\max}(|\Delta_I| + |\angle_I|) - 6\,d^{\mathrm{sym}}_{\max}(|\Delta_{II}| + |\angle_{II}| + |\Delta_{III}| + |\angle_{III}|)$ amount of extra work when enumerating all six types of 4-node graphlets. Liu et al. argued that for real-world graphs, the number of wedges plus triangles is often a magnitude greater than the number of the edges, and for a reasonable-sized cluster, ρ is often set to a large value. Thus D4GE with S4GE$_{\mathrm{CD}}$ can often achieve greater performance improvement. In the Experiment chapter, Liu et al. showed that D4GE/S4GE$_{\mathrm{CD}}$ combo achieved a convincing 10-fold speedup compared against PSE/S4GE with over eight different datasets, with almost perfect scalability up-to 256 distributed workers. Lastly, Liu et al. proved D4GE/S4GE$_{\mathrm{CD}}$ combo's capability by enumerating the *indochina* dataset ([2,3]), emitting more than ten quadrillion (10×10^{15}) 4-node graphlets, which is the first of its kind.

4.4.1   S4GE$_{\mathrm{CD}}$

S4GE is modified accordingly so that it is able to enumerate all 4-node graphlets for an ordered color-assignment K_{ijkl}. This modified version is called S4GE$_{\mathrm{CD}}$.

Instead of enumerating on a complete graph, S4GE$_{\mathrm{CD}}$ now enumerates on a subgraph denoted by the color-assignment K_{ijkl}. The subgraph consists of a mapping between the ordered color 2-tuples (i, j) and the corresponding directed edgesets E_{ij}^*. For an ordered color-assignment, there are $\binom{4}{2} = 6$ such 2-tuples: (i, j), (i, k), (i, l), (j, k), (j, l) and (k, l). (i, j), (i, k), (j, k) and the corresponding edgesets are used to discover the wedge or triangle, and (i, l), (j, l), (k, l) and the corresponding edgesets are used to discover the graphlet after the base wedge or triangle have been discovered. S4GE$_{\mathrm{CD}}$ inherits the correctness from S4GE since the actual intersection logic is untouched, whereas S4GE$_{\mathrm{CD}}$ solely focuses on a particular edge-induced sub-set of the input graph, with all the edges pointing from color i to j, k, l, from j to k, l and from k to l.

The pseudocode for S4GE$_{\mathrm{CD}}$ is given in Algorithm 10, with the details of the explore-functions are given in Algorithms 11, 12 and 13 respectively.

Algorithm 10. S4GE$_{\mathrm{CD}}$

Require: A mapping from the colors of directed edge set to the edge set $E_{\mathrm{map}} \equiv \{(i, j) \mapsto E_{ij}^*\}$; ordered color-assignment $ijkl$

1: $E_{ij}^* \equiv E_{\mathrm{map}}[ij]$, $E_{ik}^* \equiv E_{\mathrm{map}}[ik]$, $E_{jk}^* \equiv E_{\mathrm{map}}[jk]$
2: **for all** $(u, v) \in E_{ij}^*$ **do**
3:      **if** $\eta(u) < \eta(v)$ **then**
4:          **for** $u' \in N(u) \subset E_{ik}^*$ and $v' \in N(v) \subset E_{jk}^*$ **do**
5:              **if** $(u' > u) \wedge (v' > u)$ **then**
6:                  **if** $u' = v' > v$ **then**
7:                      EXPLORETRIANGLE$_{\mathrm{CD}}$ $(u, v, u', E_{\mathrm{map}}, ijkl)$
8:                  **if** $(u' < v') \wedge (u' > v)$ **then**
9:                      EXPLOREWEDGE-1$_{\mathrm{CD}}$ $(v, u, u', E_{\mathrm{map}}, ijkl)$
10:                 **if** $u' > v'$ **then**
11:                     EXPLOREWEDGE-2$_{\mathrm{CD}}$ $(u, v, v', E_{\mathrm{map}}, ijkl)$

Algorithm 11. EXPLORETRIANGLE$_{\mathrm{CD}}$

Require: Given triangle (v, u, w); $E_{\mathrm{map}} \equiv \{(i, j) \mapsto E_{ij}^*\}$; color-assignment $ijkl$.

1: $E_{il}^* \equiv E_{\mathrm{map}}[il]$, $E_{jl}^* \equiv E_{\mathrm{map}}[jl]$, $E_{kl}^* \equiv E_{\mathrm{map}}[kl]$
2: $N^{>u}(u) \equiv \{z | z \in N(u)|_{E_{il}^*}, \eta(z) > \eta(u)\}$
3: $N^{>u}(v) \equiv \{z | z \in N(v)|_{E_{jl}^*}, \eta(z) > \eta(u)\}$
4: $N^{>u}(w) \equiv \{z | z \in N(w)|_{E_{kl}^*}, \eta(z) > \eta(u)\}$
5: Rest follows Algorithm 2 line 2.

Algorithm 12. EXPLOREWEDGE-1$_{\text{CD}}$

Require: Given wedge (v, u, w); $E_{\text{map}} \equiv \{(i, j) \mapsto E_{ij}^*\}$; color-assignment $ijkl$.
1: $E_{il}^* \equiv E_{\text{map}}[il]$, $E_{jl}^* \equiv E_{\text{map}}[jl]$, $E_{kl}^* \equiv E_{\text{map}}[kl]$
2: $N^{>u}(u) \equiv \{z | z \in N(u)|_{E_{il}^*}, \eta(z) > \eta(u)\}$
3: $N^{>u}(v) \equiv \{z | z \in N(v)|_{E_{jl}^*}, \eta(z) > \eta(u)\}$
4: $N^{>u}(w) \equiv \{z | z \in N(w)|_{E_{kl}^*}, \eta(z) > \eta(u)\}$
5: Rest follows Algorithm 3 line 2.

Algorithm 13. EXPLOREWEDGE-2$_{\text{CD}}$

Require: Given wedge (v, u, w); $E_{\text{map}} \equiv \{(i, j) \mapsto E_{ij}^*\}$; ordered color-assignment $ijkl$.
1: $E_{jl}^* \equiv E_{\text{map}}[jl]$, $E_{kl}^* \equiv E_{\text{map}}[kl]$
2: $N^{>u}(v) \equiv \{z | z \in N(v)|_{E_{jl}^*}, \eta(z) > \eta(u)\}$
3: $N^{>u}(w) \equiv \{z | z \in N(w)|_{E_{kl}^*}, \eta(z) > \eta(u)\}$
4: Rest follows Algorithm 4 line 2.

---

5 Summary

In this section we present to the audience a tabular overview of the characteristics of the distributed algorithms studied in this paper. Some of the key characteristics are listed in Table 1.

Table 1. Summary of the four mentioned subgraph enumeration algorithms. CW12 (Clueweb12), SD (SubDomain) and IC (Indochina) are referring to the largest graphs that have ever been processed using the respective algorithms.

Characteristics	PTE$_{\text{Base}}$	PTE$_{\text{CD}}$	PSE	D4GE
Runtime	$O(m^{3/2})$	$O(m^{3/2})$	Query-dependent	$O((|\Delta| + |\angle|)d_{max}^{\text{sym}})$
NetworkRead	$O(\rho m)$	$O(\rho m)$	$\binom{\rho-1}{k-2} * m$	$[\binom{\rho}{2} - \rho + 3 - \frac{2}{\rho}]m^{\text{sym}}$
Capability	3×10^{12}(CW12)	3×10^{12}(CW12)	0.27×10^{15}(SD)	2.8×10^{15}(IC)
Application	Triangle	Triangle	QueryAnswering	4-node Graphlet
SotA	N	Y	Y	Y

All four distributed algorithms are within the same runtime class of their serialized counterpart, namingly, the runtime of CompactForward for PTE$_{\text{Base}}$ and PTE$_{\text{CD}}$, VF2 for PSE and S4GE for D4GE. In the case of PSE, because the runtime complexity is dependent on the user-query, no exact expression is given by the authors. While not proved directly, the authors of D4GE presented a high correlation between the enumeration time against and the augmented runtime complexity. Overall the results are welcomed as none of the candidates incurs additional complexity to the enumeration tasks that are already challenging.

The network read of all four algorithms increases with respect to the number of colors ρ. This can be understood by the fact that all of the algorithms share the same definition of sub-problems, and sub-problems have overlaps, and the amount of overlaps can be parameterized by ρ. Without loss of generality, consider tri-color sub-problems S_{ijk}. Fixing the colors of i and j, there are precisely ρ number of variations in third color k, meaning that if ρ increases, the number of tri-color sub-problems also rises.

Next we summarize the maximum number of emitted subgraphs/graphlets and the respective input graphs of the candidate algorithms. This is used to indicate the capability of the candidate algorithms in their unique application. Both PTE_{Base} and PTE_{CD} are able to enumerate all triangles of the largest web-crawl graph $Clueweb12^2$, emitting more than three trillions of triangles within 10^3 min with 120 workers. PSE, with the 4-clique query, is able to discover 0.27 quadrillions of such subgraphs within 10^4 min with 120 workers, from the $SubDomain^3$ dataset. D4GE enumerates an astronomical 3-quadrillions 4-node graphlets from the $Indochina^4$ dataset, requiring 672 workers and 7×10^3 min of runtime.

We can conclude that all candidate algorithms in this work are extremely capable and performant in their respective applications. PTE_{Base}, although outperformed by its successor PTE_{CD}, inspires the invention of PSE. D4GE, while requires more computing power than the other candidates, is addressing an unprecedentedly challenging enumeration problem - that it enumerates all six-types of 4-node graphlets, and the graphlets are induced subgraphs, which are harder to enumerate [13].

6   Conclusion

Over the past decade, subgraph enumeration has been exposed under increased focuses, especially since the introduction of networks motifs [15] as an important tool for network analysis, as well as graphlets [20] which are now established measures for network alignment [22].

In this survey we explored several existing MapReduce-based methods to solve the subgraph enumeration problem of three different focuses: triangles, k-order subgraphs and 4-node graphlets. We presented the audiences with their main conceptual approaches, gave insight on their advantages and limitations, and provided a summary of their similarities and differences.

The aim of this work was to describe some of the state-of-the-art MapReduce algorithms, offering a thorough under-the-hood review. We restricted our focus in the MapReduce community because of the tried-and-true nature of MapReduce; the algorithms under this framework are often the most intuitive, practical and easy to implement.

---

[2] http://www.lemurproject.org/clueweb12/webgraph.php.
[3] http://webdatacommons.org/hyperlinkgraph/.
[4] http://law.di.unimi.it/webdata/indochina-2004/.

Last but not least, we provided more than two dozens of references allowing further exploration of any aspects that might be of particular interest to the audience.

References

1. Afrati, F.N., Fotakis, D., Ullman, J.D.: Enumerating subgraph instances using map-reduce. In: 2013 IEEE 29th International Conference on Data Engineering (ICDE), pp. 62–73. IEEE (2013)
2. Boldi, P., Rosa, M., Santini, M., Vigna, S.: Layered label propagation: a multiresolution coordinate-free ordering for compressing social networks. In: Srinivasan, S., Ramamritham, K., Kumar, A., Ravindra, M. P., Bertino, E., Kumar, R. (eds.) Proceedings of the 20th International Conference on World Wide Web, pp. 587–596. ACM Press (2011)
3. Boldi, P., Vigna, P.: The WebGraph framework I: compression techniques. In: Proceedings of the Thirteenth International World Wide Web Conference (WWW 2004), Manhattan, USA, pp. 595–601. ACM Press (2004)
4. Bröcheler, M., Pugliese, A., Subrahmanian, V.S.: Cosi: cloud oriented subgraph identification in massive social networks. In: 2010 International Conference on Advances in Social Networks Analysis and Mining, pp. 248–255. IEEE (2010)
5. Cohen, J.: Graph twiddling in a MapReduce world. Comput. Sci. Eng. $11(4)$, 29 (2009)
6. Cordella, L.P., Foggia, P., Sansone, C., Vento, M.: A (sub) graph isomorphism algorithm for matching large graphs. IEEE Trans. Pattern Anal. Mach. Intell. $26(10)$, 1367–1372 (2004)
7. Dean, J., Ghemawat, S.: MapReduce: simplified data processing on large clusters (2008)
8. Deshpande, M., Kuramochi, M., Wale, N., Karypis, G.: Frequent substructure-based approaches for classifying chemical compounds. IEEE Trans. Knowl. Data Eng. $17(8)$, 1036–1050 (2005)
9. Faust, K.: A puzzle concerning triads in social networks: graph constraints and the triad census. Soc. Netw. $32(3)$, 221–233 (2010)
10. Hu, H., Yan, X., Huang, Y., Han, J., Zhou, X.J.: Mining coherent dense subgraphs across massive biological networks for functional discovery. Bioinformatics, $21(\text{suppl\_1})$, i213–i221 (2005)
11. Lai, L., Qin, L., Lin, X., Chang, L.: Scalable subgraph enumeration in MapReduce. Proc. VLDB Endow. $8(10)$, 974–985 (2015)
12. Latapy, M.: Main-memory triangle computations for very large (sparse (power-law)) graphs. Theor. Comput. Sci. $407(1–3)$, 458–473 (2008)
13. Liu, X., Santoso, Y., Thomo, A., Srinivasan, V.: Distributed enumeration of four node graphlets at quadrillion-scale. In: SSDBM 2021: 33rd International Conference on Scientific and Statistical Database Management, pp. 85–96 (2021)
14. Milenković, T., Pržulj, N.: Uncovering biological network function via graphlet degree signatures. Cancer Inform. 6 (2008)
15. Milo, R., Shen-Orr, S., Itzkovitz, S., Kashtan, N., Chklovskii, D., Alon, U.: Network motifs: simple building blocks of complex networks. Science $298(5594)$, 824–827 (2002)

16. Park, H.-M., Chung, C.-W.: An efficient MapReduce algorithm for counting triangles in a very large graph. In: 22nd ACM International Conference on Information and Knowledge Management, CIKM'13, San Francisco, CA, USA, 27 October–1 November, pp. 539–548 (2013)

17. Park, H.-M., Myaeng, S.-H., Kang, U.: PTE: enumerating trillion triangles on distributed systems. In: Proceedings of the 22nd ACM SIGKDD International Conference on Knowledge Discovery and Data Mining, pp. 1115–1124. ACM (2016)

18. Park, H.M., Silvestri, F., Kang, U., Pagh, R.: MapReduce triangle enumeration with guarantees. In: Proceedings of the 23rd ACM International Conference on Conference on Information and Knowledge Management, CIKM 2014, Shanghai, China, 3–7 November 2014, pp. 1739–1748 (2014)

19. Park, H.M., Silvestri, F., Pagh, R., Chung, C.W., Myaeng, S.H., Kang, U.: Enumerating trillion subgraphs on distributed systems. ACM Trans. Knowl. Discov. Data (TKDD) **12**(6), 1–30 (2018)

20. Pržulj, N.: Biological network comparison using graphlet degree distribution. Bioinformatics **23**(2), e177–e183 (2007)

21. Ralaivola, L., Swamidass, S.J., Saigo, H., Baldi, P.: Graph kernels for chemical informatics. Neural Netw. **18**(8), 1093–1110 (2005)

22. Ribeiro, P., Paredes, P., Silva, M.E., Aparicio, D., Silva, F.: A survey on subgraph counting: concepts, algorithms, and applications to network motifs and graphlets. ACM Comput. Surv. **54**(2), 1–36 (2021)

23. Santoso, Y., Srinivasan, V., Thomo, A.: Efficient enumeration of four node graphlets at trillion-scale. In: 23rd EDBT, pp. 439–442 (2020)

24. Suri, S., Vassilvitskii, S.: Counting triangles and the curse of the last reducer. In: Proceedings of the 20th International Conference on World Wide Web, WWW '11. ACM, New York (2011)

25. Wong, S.W.H., Cercone, N., Jurisica, I.: Comparative network analysis via differential graphlet communities. Proteomics **15**(2–3), 608–617 (2015)

Identifying Vehicle Exterior Color by Image Processing and Deep Learning

Somayeh Abniki[1]([✉]), Kin Fun Li[1], and Tom Avant[2]

[1] Electrical and Computer Engineering, University of Victoria, Victoria, BC, Canada
{somayehabniki,kinli}@uvic.ca
[2] VINN Automotive, 844 Courtney Street, Victoria, BC, Canada
tom@vinnauto.com

Abstract. The vehicle's color is one of the factors considered in car purchasing. Hence, color extraction and identification from online vehicle images play an important role in the vehicle e-commerce marketplace. In this paper, we present a vehicle color identification methodology. Image processing techniques are employed to construct feature vectors, which are then used as input to deep neural networks to classify a vehicle's color into 14 classes. Local relative entropy is utilized as a measure of image segmentation to select the region of interest. Experiments are performed on an image dataset provided by an automobile e-commerce operator. Our implementation results are evaluated and discussed.

1 Introduction to Color Recognition

Color recognition plays a crucial role in many applications. Vehicle color recognition is an important research and application component in Intelligent Transport Systems [1–4]. With the proliferation of the Web, there is a great potential in using this technology for electronic commerce. VINN is an automobile e-commerce platform that advances the vehicle buying industry by utilizing artificial intelligence and machine learning techniques to empower both dealerships and customers to advertise, sell, search, find and purchase their ideal vehicle online. The company's main goal is to provide buyers with a service that meets their lifestyle, hence the visual aspects of vehicles are of prime importance. VINN has been collecting vehicle images for the past years. The data consists of two categories: an image set containing different types of vehicle images (exterior, interior, engine, wheel, etc.) from various perspectives (front, rear, back, etc.), and a database consisting of vehicle information, such as make, model, trim, exterior color, interior color, etc., as entered by the sellers. However, the entered information is often missing or inconsistent with the vehicle's images. Amongst the many vehicle attributes, color is one of the primary factors considered by car buyers. Therefore, there are many advantages of being able to recognize a vehicle's color, including:

- Dataset cleaning: to resolve incomplete and dirty data issues.
- Ground truthing the image sets: to tackle the inconsistency between the images and the description of the vehicle as entered by the sellers or dealers.

© The Author(s), under exclusive license to Springer Nature Switzerland AG 2022
L. Barolli et al. (Eds.): EIDWT 2022, LNDECT 118, pp. 445–457, 2022.
https://doi.org/10.1007/978-3-030-95903-6_46

- Auto filling the color field in the database: to fill the color data automatically when sellers or dealers upload the images of their vehicle, for a more streamlined C2C experience.

Image processing techniques and machine learning algorithms have become the most useful tools to achieve automation in color recognition [5, 6]. In this work, a hybrid approach is used to classify a vehicle's color from its images into 14 colors: Red, Black, Blue, Silver, White, Grey, Purple, Brown, Yellow, Orange, Green, Pink, Beige and Gold. Our method utilizes pre-trained models to localize the main vehicle in an image, then extracts the dominant color of the selected region of interest (ROI) by detecting the most frequent Hue, Saturation and Value component (HSV), and constructing feature vectors as input to various models to classify the color. We also experimented with relative local entropy as a measure of image segmentation. The segment with local relative entropy lower than a threshold is identified as the ROI, and its most frequent HSV component is extracted as dominant color.

However, due to the similarity between some of the colors (e.g., silver and white), the variations of vehicle colors, and the impact of environmental factors including lighting and reflections, it is difficult to identify the color precisely. It is a challenge to develop a robust and effective system for vision-based color recognition [7].

The rest of this paper is organized as follows: related works of vehicle color recognition techniques and image segmentation from an entropy-based perspective are presented in Sect. 2. Section 3 describes VINN's dataset and image set. The proposed methodology is detailed in Sect. 4. Section 5 presents the experimental results of the models introduced in Sect. 4. Lastly, in Sect. 6, we conclude and propose future works.

2 Related Works

In this section, we discuss prior work related to vehicle color recognition and image segmentation techniques with a focus on entropy as a measure of thresholding for ROI selection. A study conducted by Dule et al. [8] analyzes the performance of three classification methods (K-Nearest Neighbors, Artificial Neural Networks, and Support Vector Machines) using all possible combinations of sixteen color space components as feature sets on two ROIs (smooth hood piece and semi front vehicle). Feature are selected in the determined ROIs by three methods, histogram-based feature selection, pixel-based majority selection and pixel-based median selection. A vehicle is classified into one of the seven colors: black, grey, white, red, green, blue, and yellow.

Chen et al. [2] provide a BoW-based method to select the ROI for color recognition while focusing on vehicle localization. The BoW representation as introduced in [2] utilizes a large codebook quantized from color features to map these features into a higher dimensional subspace. Pre-processing using the haze removal method [9] and color contrast normalization method [10] is carried out to overcome image quality degradation. The ROI selection is performed by partitioning the vehicle image into subregions carrying different weights. A classifier is trained to learn the assigned weights. To deal with the multiclass issue, they train the classifier by a linear SVM to increase the efficiency and precision. Nonlinear SVM, though outperforms the linear one, is not used, due to its longer training time [2, 11]. The color of a vehicle is classified into eight classes: black, white, blue, yellow, green, red, grey, and cyan. They apply their models to two datasets which they built, a vehicle image dataset (15,601 vehicle images with half for training and half for testing) and a vehicle video dataset.

Rachmadi and Purnama [4] convert an input image to different color spaces: RGB, CIE Lab, CIE XYZ, and HSV, and input their dataset into a convolutional neural network (CNN) architecture to classify the vehicle color as one of the 8 classes: black, blue, cyan, grey, green, red, white, and yellow. Their CNN architecture consists of 2 base networks of 8 layers each giving a total of 16 layers including two convolutional layers with ReLU as the activation function. The base networks were followed by normalization and max pooling, two convolutional layers without pooling and normalization process, and one more convolutional layer with only pooling and no normalization. They follow the procedure introduced by Krizhevsky et al. [12] where after a number of iterations, the learning rate is decreased by a factor of 10. They test their approach on the image dataset provided by Chen in [2].

Shaded colors, such as navy blue as black, are often misclassified. A concept proposed in [13] has vehicles with different chromatic attributes to be trained separately. Their method is comprised of multiple steps. First, each foreground vehicle is extracted from its background. Then, color correction is performed to reduce lighting effects. Next, window parts of the car image is removed and the vehicle color is classified using the lower parts of the image like bumper and doors, making vehicle pixels of the same color more apparent. Finally, a tree-based classifier labels the vehicle into grey and non-grey subgroups, which is then followed by a detailed grey classifier identifying black, silver or white. In addition, a detailed color classifier for red, green, blue, and yellow, is designed for classifying vehicles into chromatic and nonchromatic classes. They have experimented with the NTOU Vehicle Dataset which contains 3,373 vehicle images as the training set and 16,648 vehicle images as the test set.

In [14], a deep-learning-based algorithm (a CNN architecture proposed by Krizhevsky et al. [12]) is adopted as the feature extractor, that computes a feature vector for each image. A linear SVM instead of a fully connected artificial neural network is employed as a color classifier. The spatial pyramid strategy is combined with the original convolutional neural network architecture to improve recognition accuracy [14]. The features are learnt from training data automatically, instead of adopting manually designed features. To assess the proposed process, Hu et al. use the vehicle color dataset provided by Chen [2] with nine-group splits of training and test. Their results show that the ratio of training data does not influence the final recognition accuracy for their proposed feature.

Kim et al. in [15] use the HSI (Hue Saturation Intensity) space to represent the color feature of the image by a histogram (from the normalized H, S and I component). Using a distance function to compute the similarity between a feature vector and templates, the color of a vehicle is classified into one of seven colors: black, silver, white, red, yellow, green, and blue. They use 700 images for their experiment, where there are 100 images for each color. Half of the images are used for templates, and the remaining 350 images are used for the test.

Chowdhury et al. [16] advocate Image segmentation by splitting the image into groups of homogeneous pixels based on ROI, as a universal step for image recognition. Many researchers favor the use of image segmentation to extract the most significant regions that contain the desired characteristics of the image, such as the dominant color. The threshold and seed-point selections are two parameters that have a major impact on the effectiveness of image segmentation [17].

There are many threshold selection methods such as basic global thresholding, clustering, region growing and entropy based. Entropy-based image segmentation uses entropy to measure how uniform the pixel characteristic is in order to separate the regions of the image into objects [18].

In [17], different entropy measures have been applied on both greyscale and color images. The entropy of an image is computed with a four-step algorithm in a plot of entropy versus grey level. This plot gives the minima points and the lowest minima can be employed as a threshold value. To reflect the contextual information between pixels, a thresholding method is proposed that constructs a two-dimensional histogram of a pixel's brightness and local relative entropy of its neighboring region [19]. The local relative entropy (LRE) measures the brightness difference between a pixel and its neighboring pixels. In [16], a multilevel thresholding image segmentation method is proposed based on the minimization of bi-entropy function. Then both the Shannon's entropy and the proposed method are employed in image segmentation. Their method considers both the spatial information and the grey information to reduce computation complexity. Pauzi in [20] introduces a real-time instrumentation system for visually impaired people to help them recognize color. Their color classification system uses an entropy algorithm based on the HSV and RGB color spaces, and labels a color into one of the ten colors: black, brown, cyan, red, orange, yellow, green, blue, magenta, grey and white. The results show that the HSV model classifies more accurately than the RGB one.

3   VINN's Dataset

VINN's dataset has more than 111,250 vehicle images. There are 35 features associated with each vehicle. For each vehicle there are between 1 to 30 unlabeled images of various parts such as interior, exterior, engine, wheels, lights, etc., and in different perspectives including front, back, rear, left-rear, and right-rear. As the goal of this project is exterior color recognition, there is a need to label the images as exterior for the training, validation, and test sets. Labelling has been done manually using a GUI (python application with Flask hosted on AWS) which displays an image and labels it based on user's key press to exterior, interior, not a car, engine, open trunk, etc. Finally, 2,530 images labelled as exterior are used as samples. To ensure a valid dataset, all these images have been checked manually for consistency between their color in the image and the corresponding 'exterior color' field in the dataset. The 14 color labels are Red, Black, Blue, Silver, White, Grey, Purple, Brown, Yellow, Orange, Green, Pink, Beige and Gold.

There exist other challenges in the dataset. The majority of the images are captured by smartphones in various resolutions at a close range mostly in a parking lot area. Environmental conditions often affect the quality of the images. Some images are impacted by sun light, shadows, and reflections of other objects, such as trees and people, in the area. Furthermore, in most of the images there are many other parked vehicles which necessitates a pre-processing step to differentiate and identify the target vehicle in the image as Fig. 1 illustrates.

(a) Sample Image

(b) Detected vehicles with their bounding boxes

Fig. 1. A sample image without (a) and with (b) vehicles detected

(a) Rear-left view

(b) Side view

(c) Rear view

Fig. 2. Sample images from different views

Another big challenge is that for each vehicle, there are several exterior shots from different views as shown in Fig. 2. Hence, a post-processing step is required to conclude the exact color of each vehicle based on the recognized colors, which may be different from all its images. Moreover, there are seven body types in the dataset including SUVs, trucks, sedans, hatchbacks, convertibles, coupes and vans, which may complicate the color recognition process. Obviously, similar to most other data-related projects, the lack of labelled data instances, in our case, color labelled images, is a major challenge in developing effective and efficient models. We overcome this lack of sufficient samples by increasing the number of images, by applying image augmentation technique to our image dataset.

4 Proposed Color Recognition Method

This section contains the details of the process carried out and the models applied to our image set for color recognition. The results of alternative solutions used in each step are compared. A flowchart of the proposed method is illustrated in Fig. 3.

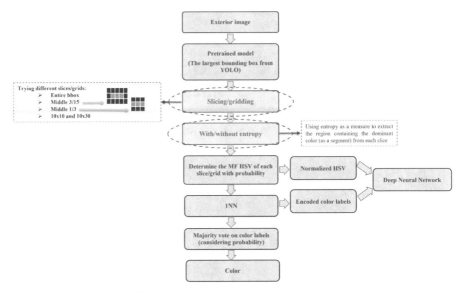

Fig. 3. Flowchart of the proposed method

4.1 Pre-processing

The majority of the images includes more than one vehicle, therefore we need to identify the main vehicle of interest in each image. We used YOLO v4 [21] as a pre-trained model [22] to localize all the vehicles in an image. After obtaining their coordinates, we computed the area of each detected vehicle in the image. We then selected the vehicle with the largest area as the main vehicle as shown in Fig. 1(b). Also, to ensure that the pre-trained model was working effectively in finding and localizing the main vehicles, and more importantly to set its appropriate confidence level, we checked all the output images visually. Since the RGB color space is not robust enough when images are captured in complex natural scenes, as artifacts of low illumination, strong lighting, and camera color bias, etc., a further pre-processing step is needed to convert an image to HSV color space for color recognition.

4.2 ROI Selection

We tested our method on different regions of an image. We first removed the top and bottom third of the bounding box of the vehicle as these regions contain windows and wheels, respectively. Then we sliced the remaining region into 5 horizontal subregions and selected the middle three slices for dominant color extraction. Another experiment worked with the entire middle third of the bounding box without any slicing. Subsequent experiments used the entire bounding box without removing the top and bottom subregions, while using different configurations to grid this region. These include 10×10 and 10×30 grids. We also tested the 10×10 and 10×30 grids on the middle-third of the bounding box to see how the model performed. Furthermore, some of these experiments have been applied on the regions with and without entropy-based image segmentation. Thus, for each region, the grey level local relative entropy (LRE) method proposed in [19] is utilized to calculate the entropy of each pixel by using the brightness value of all the pixels and the mean grey level value of the pixels in its $N \times N$ neighborhood. As shown in Eq. (1), the LRE of each pixel in its neighborhood is computed as:

$$LRE = \sum_{i=-(n-1)/2}^{(n-1)/2} \sum_{j=-(n-1)/2}^{(n-1)/2} I(x+i, y+j) \times \left| \log \frac{I(x+i,y+j)}{\tilde{I}(x,y)} \right| \tag{1}$$

where $\tilde{I}(x, y)$, the mean gray level value [19], is calculated from Eq. (2):

$$\tilde{I}(x,y) = \frac{1}{n^2} \sum_{i=-(n-1)/2}^{(n-1)/2} \sum_{j=-(n-1)/2}^{(n-1)/2} I(x+i, y+j) \tag{2}$$

The LRE is then normalized to a number between 0 and $L - 1$ (L is the highest brightness value in that neighborhood).

We tested both $N = 10$ and $N = 20$ as the number of adjacent pixels. Next, a thresholding step was performed on the pixel-wise entropy matrix (E) to generate a binary mask. We considered the mean entropy as a threshold value [23], to distinguish the color of different parts of the vehicle. A sample image with its output based on this approach is shown in Fig. 4.

Any output region from the previous step has its most frequent HSV component extracted with probability calculated as (3):

$$\text{Probability} = \frac{\text{number of pixels of MF HSV}}{\text{number of pixels in region}} \tag{3}$$

This probability was used as a measure in the majority voting on slice colors and then on vehicle images, for the cases when there are equal votes on slice or image color labels. We employed three models in this step to recognize color. The first model used normalized HSV vectors of all the slices with probabilities as a feature vector, to feed into a deep neural network. As an example, for a 10×30 grid image, we have 300 normalized H, S and V numbers and 300 probability numbers, resulting in a total of 1200 features. The second and third models used the assigned color label from each slice's MF HSV. This color label assignment is described below.

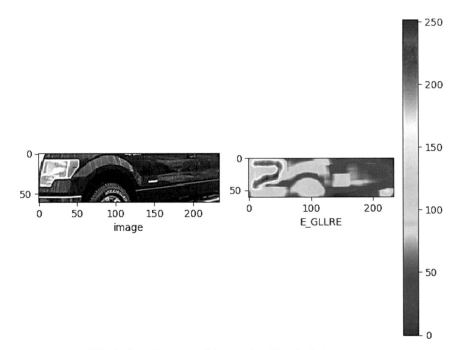

Fig. 4. Sample output of the grey level local relative entropy

4.3   Color Label Assignment

For color label assignment, we conducted a multi-stage approach in mapping the MF HSV to the color classes. First, we did a thresholding on the value component to differentiate black, and then another thresholding on saturation to distinguish white, grey and silver from other colors. To differentiate these three colors, we checked the hue to determine what color it was. At the next stage, we used a 1NN classifier with centroids referring to the other eight colors. To define the centroids and ranges of H, S and V components for each color, we utilized an HSV color thresholder script to determine the lower and upper bounds of each color using sliders. We also used a color wheel named Martian color [24] to fine tune the ranges and centroids. We combined the red and brown as a single color (red-brown) in 1NN as they have some overlap in hue. Similarly, we combined orange and brown as (orange-brown) in 1NN. If these two new colors were selected from the 1NN, then we checked their value to decide if it was brown, red or orange. As the hue is circular, any two pixels with hue 175 and 5 should have the same distance as from a pixel with hue 0^1. As a result, we had to change the distance function in 1NN to (4).

$$Hue\_Distance = 90 - ||Hue - Centre| - 90| \tag{4}$$

---

[1] Hue is scaled in the range of 0 to 179 for all the images. Even if the hue is in the range of 0 to 359, two pixels with hue 350 and 10 should have the same distance from a pixel with hue 0.

The output of this step was used as the input to the two other models. The first model performed a majority vote on the slice colors of each image and used the probability in case of equal votes.

The second model encoded the color labels to 14 bits using a one-hot encoder, while slice probabilities were concatenated as well to construct the feature vector. For example, for a 10×10 grid image, a feature vector of size 1500 is constructed. Then, this feature vector was used as the input to a deep neural network for color classification. After recognizing the color of each image, a majority vote using the image color labels and their probabilities from the images of the same vehicle, was carried out to determine the vehicle's color.

5   Experimental Results

The performance of the methods was evaluated by the precision of all color categories, with the micro average precision of all categories as the average precision (AP). In the feature learning stage, we constructed different feature vectors based on the number of grids selected from the images. To train the deep neural network, we used five dense layers with ReLU as an activation function and one last fully connected layer with a SoftMax activation function. The number of neurons in the last layer was equal to the number of colors, 14 in our case. The prediction output was a 14-length vector which we used to decode and mapped to the 14 color labels. To avoid overfitting, we applied dropout and l2 regularization as suggested in [12] since this is the most commonly used method to reduce the effect of overfitting in training deep learning networks.

We used 80% of our images for training, 10% for validation and 10% for testing. Each class consisted of different number of instances from 12 to 435. To tackle the imbalanced dataset issue, we applied the Synthetic Minority Oversampling Technique after our feature training step. The proposed techniques were implemented in Python on a Windows 10×64-based system. The experiments were carried out on a desktop computer with an Intel(R) Xeon(R) E-2176M CPU 2.71 GHz with 16-GB memory.

We summarized the results of our approaches in Table 1[2], showing each method with the configuration of the slice, the number of grids, and how they performed. As can be seen in Table 1, performance increases with the slicing of the vehicle image and the gridding of more subregions. Results using entropy show an improvement in comparison to non-entropy based experiments on the same region and configuration in the voting model. It may be helpful to experiment with additional configurations of the models, while considering entropy before the MF HSV extraction and color label assignment step. The only issue with entropy is the relatively long computation time which is a critical factor in many real-time applications. This issue could be resolved if the models were running on appropriate hardware configuration as confirmed in [4]. The precisions of each color category from the best results obtained and the corresponding confusion matrix are presented in Table 2 and Fig. 5, respectively.

We also investigated the misclassified images to identify the deficiency of the models. More than 15% of the grey images which are misclassified as black, have a very

---

[2] For the last two rows of experiments, the entire vehicle bounding box was gridded into 10×10 and 10×30.

Table 1. Average recognition performance of various models

		Method			
		Voting on color labels		DNN on normalized HSV	DNN on encoded color labels
		Without entropy	With entropy	With entropy	With entropy
Configuration	Entire vehicle bounding box	31%	39%	62%	64%
	Middle 3/15	53%	62%	66%	67%
	Middle 1/3	46%	54%	58%	65%
	10×10 grid	55%	61%	65%	72%
	10×30 grid	42%	63%	68%	71%

dark shade close to black. Therefore, the color assignment labeler may have incorrectly classified their slices as black. The same case appears in some other class samples which are misclassified as blue. Reviewing the samples show that these images have more bluish tones in their regions. This could be due to the reflection of the sky on the vehicle bodies. Another important issue is that we have some misclassified images from various colors to silver. This may have happened because of the very bright sunlight reflection on metallic paint or the color is too light so it becomes very close to another color. As previously mentioned, one challenge is having different images of the same vehicle from different viewpoints, which affects the model performance; however, performance seems to be better with a more flexible ROI selection process.

Table 2. Precision metrics of each color category from the best results.

	Precision	Recall	F1-score	Support
Red	0.73	0.83	0.78	42
Black	0.82	0.80	0.81	79
Yellow	0.63	0.50	0.56	24
Grey	0.67	0.79	0.72	58
White	0.83	0.90	0.86	78
Silver	0.70	0.66	0.68	29
Blue	0.72	0.84	0.77	49
Gold	0.50	0.58	0.54	19

(*continued*)

Table 2. (*continued*)

	Precision	Recall	F1-score	Support
Brown	0.59	0.48	0.53	21
Orange	0.67	0.78	0.72	23
Purple	0.60	0.33	0.43	27
Green	0.79	0.60	0.68	25
Beige	0.56	0.38	0.45	13
Pink	0.75	0.63	0.69	19
Accuracy			0.72	506

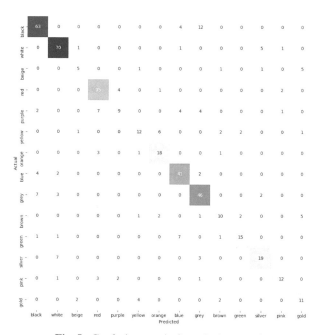

Fig. 5. Confusion matrix from the best results

6 Conclusion and Future Work

In this paper, we present a color recognition technique for vehicle exterior images. We classify the vehicle color into 14 classes. Our proposal is based on the HSV version of the images. After some pre-processing to detect the target vehicle in the image and its RoI, the slicing and gridding of the RoI are carried out, followed by extracting the most frequent HSV component. Three models are tested with various feature vectors as input. We assign a color label using 1NN to each slice of the MF HSV component in the first model, and determine the color based on a majority vote. The second and third models use the encoded color label and normalized HSV component, respectively, concatenated to the color probability of each grid as feature vectors to a deep neural network. Although experimental results show 72% of successful classification on average, the precision is low for some color classes. The causes of misclassification are discussed, with enhancements as future work presented.

References

1. Aarathi, K.S., Abraham, A.: Vehicle color recognition using deep learning for hazy images. In: International Conference on Inventive Communication and Computational Technologies (ICICCT), pp. 335–339 (2017). https://doi.org/10.1109/ICICCT.2017.7975215
2. Chen, P., Bai, X., Liu, W.: Vehicle color recognition on urban road by feature context. IEEE Trans. Intell. Transp. Sys. **15**(5), 2340–2346 (2014). https://doi.org/10.1109/TITS.2014.230 8897
3. Li, X., et al.: Vehicle color recognition using vector matching of template. In: 3rd International Symposium on Electronic Commerce and Security, pp. 189–193 (2010). https://doi.org/10. 1109/ISECS.2010.50
4. Rachmadi, R.F., Purnama, I.: Vehicle color recognition using convolutional neural network. http://arxiv.org/abs/1510.07391
5. Shobha, R.B.R.: Classification of vehicles using image processing techniques. Int. J. Recent Trends Eng. Res. **4**(3) 393–401 (2018). https://doi.org/10.23883/ijrter.2018.4144.qcmco
6. Tilakaratna, D.S.B., Watchareeruetai, U., Siddhichai, S., Natcharapinchai, N.: Image analysis algorithms for vehicle color recognition. In: International Electrical Engineering Congress, pp. 4–7 (2017). https://doi.org/10.1109/IEECON.2017.807588
7. Tsai, L.W., Hsieh, J.W., Fan, K.C.: Vehicle detection using normalized color and edge map. IEEE Tran. Image Process. **16**(3), 850–864 (2007). https://doi.org/10.1109/TIP.2007.891147
8. Dule, E., Gökmen, M., Beratoğlu, M.S.: A convenient feature vector construction for vehicle color recognition. In: 11th WSEAS International Conference on Neural Networks, pp. 250–255 (2010)
9. He, K., Sun, J., Tang, X.: Single image haze removal using dark channel prior. IEEE Trans. Pattern Anal. Mach. Intell. **33**(12) (2011)
10. Gonzalez, R.C., Woods, R.E., Eddis, S.L.: Digital Image Processing Using MATLAB, 2nd edn. Tata McGraw Hill, New York (2010)
11. Fan, R.-E., et al.: LIBLINEAR: a library for large linear classification. J. Mach. Learn. Res. **9**, 1871–1874 (2008). https://doi.org/10.1145/1390681.1442794
12. Krizhevsky, A., Sutskever, I., Hinton, G.E.: ImageNet classification with deep convolutional neural networks. Commun. ACM **60**, 84–90 (2012)
13. Hsieh, J.W., et al.: Vehicle color classification under different lighting conditions through color correction. IEEE Sens. J. **15**(2), 971–983 (2015). https://doi.org/10.1109/JSEN.2014. 2358079
14. Hu, C., et al.: Vehicle color recognition with spatial pyramid deep learning. IEEE Trans. Intell. Trans. Sys. **16**(5), 2925–2934 (2015). https://doi.org/10.1109/TITS.2015.2430892
15. Kim, K.J., Park, S.M., Choi, Y.J.: Deciding the number of color histogram bins for vehicle color recognition. In: 3rd IEEE Asia-Pacific Services Computing Conference, pp. 134–138 (2008). https://doi.org/10.1109/APSCC.2008.207
16. Chowdhury, K., Chaudhuri, D., Pal, A.K.: A new image segmentation technique using bi-entropy function minimization. Multimedia Tools Appl. **77**(16), 20889–20915 (2017). https:// doi.org/10.1007/s11042-017-5429-8

17. Sen, H., Agarwal, A.: A comparative analysis of entropy based segmentation with Otsu method for gray and color images. In: International conference of Electronics, Communication and Aerospace Technology, vol. 2017, pp. 113–118, January 2017. https://doi.org/10.1109/ICECA.2017.8203655

18. Zhang, H., Fritts, J.E., Goldman, S.A.: An entropy-based objective evaluation method for image segmentation. Int. Soc. Opt. Eng. 38–49 (2004)

19. Yang, W., Cai, L., Wu, F.: Image segmentation based on gray level and local relative entropy two dimensional histogram. PLoS ONE 15(3), 1–9 (2020). https://doi.org/10.1371/journal.pone.0229651

20. Pauzi, G.A.: Colour classification using entropy algorithm in real time colour recognition system for blindness people. KnE Eng. 1(1), 0–5 (2016). https://doi.org/10.18502/keg.v0i0.485

21. Bochkovskiy, A., Wang, C.-Y., Liao, H.-Y.M.: YOLOv4: optimal speed and accuracy of object detection (2020). https://arxiv.org/abs/2004.10934

22. Ponnusamy, A.: cvlib-high level computer vision library for Python (2018). https://www.cvlib.net/

23. AlSaeed, D.H., El-Zaart, A., Bouridane, A.: Minimum cross entropy thresholding using entropy-Li based on log-normal distribution for skin cancer images. In: 7th International Conference on Signal Image Technology & Internet-Based Systems, pp. 426–430 (2011). https://doi.org/10.1109/SITIS.2011.86

24. Mars, W.: Martian Colour Wheel. https://warrenmars.com

Author Index

A
Abniki, Somayeh, 445
Adhiatma, Ardian, 123
Alimuddin, Ais Prayogi, 163
Ampririt, Phudit, 153, 223, 236, 272
Arai, Yoshikazu, 291
Asakura, Akihito, 406
Assyilah, Fannisa, 114
Autarrom, Suphatchaya, 1
Avant, Tom, 445

B
Barolli, Admir, 223, 386
Barolli, Leonard, 130, 153, 183, 223, 236, 254,
    263, 272, 283, 308, 316, 346, 357, 386,
    417
Bylykbashi, Kevin, 153, 236, 272

C
Chantaranimi, Kittayaporn, 1
Charina, 163
Chen, Fadong, 21
Chen, Lei, 21, 68
Chompupoung, Anchan, 1

D
Deng, Jie, 31
Deng, Na, 324, 335
Duolikun, Dilawaer, 130, 395, 406

E
Enokido, Tomoya, 130, 142, 395, 406

F
Fachrunnisa, Olivia, 114

G
Gotoh, Yusuke, 212
Guo, Lian, 68

H
Hino, Takanori, 190
Hirata, Aoto, 283, 301, 308, 346, 357, 366, 417
Hirota, Masaharu, 102, 301, 417

I
Iiyama, Shota, 102
Ikeda, Makoto, 153, 183, 236, 254, 272
Ilham, Amil Ahmad, 163
Inoue, Yu, 212
Ishida, Tomoyuki, 11

J
Jiao, Lili, 31
Jinapook, Pichan, 1

K
Kagawa, Tomomichi, 190
Kanahara, Kazuho, 201
Katayama, Kengo, 201, 346, 366
Kato, Shigeru, 190
Kojima, Mutsuki, 11
Koroveshi, Andrea, 175
Kulla, Elis, 201, 223, 245, 254, 263
Kume, Shunsaku, 190
Kumeno, Hironori, 190

L
Li, Kin Fun, 445
Li, Qing, 68

L. Barolli et al. (Eds.): EIDWT 2022, LNDECT 118, pp. 459–460, 2022.
https://doi.org/10.1007/978-3-030-95903-6

Li, Ying, 52
Liang, Guangsheng, 68
Liu, Bin, 21
Liu, Jiasen, 52
Liu, Xiaozhou, 430
Liu, Zhaohua, 21, 68
Lumpoon, Pathathai Na, 1

M

Mahanan, Waranya, 1
Matsuo, Kazuma, 183
Matsuo, Keita, 153, 236, 245, 263, 272
Matusi, Tomoaki, 357
Miwa, Hiroyoshi, 42, 375

N

Na, Deng, 81, 91
Nagai, Yuki, 283, 301
Nakamura, Shigenari, 142, 406
Nashimoto, Kuninao, 406
Natwichai, Juggapong, 1
Nobuhara, Hajime, 190
Nurhidayati, 123

O

Oda, Tetsuya, 102, 201, 283, 301, 308, 346, 357,
　　366, 417
Ogiela, Marek R., 38
Ogiela, Urszula, 38

P

Pan, Xiaozhong, 63

Q

Qafzezi, Ermioni, 153, 236, 272

R

Ren, Yongjin, 31

S

Saito, Nobuki, 283, 301, 308, 346, 357, 366, 417
Saito, Yoshiya, 291
Sakamoto, Shinji, 223, 316, 386
Sakuraba, Akira, 291

Sangamuang, Sumalee, 1
Santoso, Yudi, 430
Shiba, Misato, 375
Shibata, Yositaka, 291
Shintani, Kuya, 245, 254
Spaho, Evjola, 175, 254
Srinivasan, Venkatesh, 430
Su, Yunxuan, 52
Sugunsil, Prompong, 1
Sukhvibul, Titipat, 1
Suyuti, Muhammad Zulfadly A., 163

T

Tahir, Zulkifli, 163
Takizawa, Makoto, 38, 130, 142, 395, 406
Thiengburanathum, Pree, 1
Thomo, Alex, 430
Tiansi, Du, 81, 91
Toyoshima, Kyohei, 308, 366
Tu, Zheng, 52

U

Uchida, Noriki, 291
Uejima, Akira, 201

W

Wang, Xu An, 52, 63
Wei, Lixian, 63
Weijie, Chen, 81, 91

Y

Yano, Hiroki, 42
Yasunaga, Tomoya, 283, 301, 308, 346, 357, 366
Yin, Jianghao, 324, 335
Yin, Wengang, 31
Yoneyama, Sumihiro, 42
Yukawa, Chihiro, 283, 301, 308, 357, 366, 417

Z

Zhang, Lili, 31
Zhao, Yize, 63
Zheng, Heding, 68
Zheng, Mengyao, 21
Zhou, Kui, 21

Printed in the United States
by Baker & Taylor Publisher Services